T0310163

ADAPTIVE AEROSERVOELASTIC CONTROL

Aerospace Series List

Adaptive Aeroservoelastic Control	Tewari	March 2016
Theory and Practice of Aircraft Performance	Kundu, Price and Riordan	November 2015
The Global Airline Industry, Second Edition	Belobaba, Odoni and Barnhart	July 2015
Modeling the Effect of Damage in Composite Structures: Simplified Approaches	Kassapoglou	March 2015
Introduction to Aircraft Aeroelasticity and Loads, 2nd Edition	Wright and Cooper	December 2014
Aircraft Aerodynamic Design: Geometry and Optimization	Sóbester and Forrester	October 2014
Theoretical and Computational Aerodynamics	Sengupta	September 2014
Aerospace Propulsion	Lee	October 2013
Aircraft Flight Dynamics and Control	Durham	August 2013
Civil Avionics Systems, 2nd Edition	Moir, Seabridge and Jukes	August 2013
Modelling and Managing Airport Performance	Zografos, Andreatta and Odoni	July 2013
Advanced Aircraft Design: Conceptual Design, Analysis and Optimization of Subsonic Civil Airplanes	Torenbeek	June 2013
Design and Analysis of Composite Structures: With Applications to Aerospace Structures, 2nd Edition	Kassapoglou	April 2013
Aircraft Systems Integration of Air-Launched Weapons	Rigby	April 2013
Design and Development of Aircraft Systems, 2nd Edition	Moir and Seabridge	November 2012
Understanding Aerodynamics: Arguing from the Real Physics	McLean	November 2012
Aircraft Design: A Systems Engineering Approach	Sadraey	October 2012
Introduction to UAV Systems 4e	Fahlstrom and Gleason	August 2012
Theory of Lift: Introductory Computational Aerodynamics with MATLAB and Octave	McBain	August 2012
Sense and Avoid in UAS: Research and Applications	Angelov	April 2012
Morphing Aerospace Vehicles and Structures	Valasek	April 2012
Gas Turbine Propulsion Systems	MacIsaac and Langton	July 2011
Basic Helicopter Aerodynamics, 3rd Edition	Seddon and Newman	July 2011
Advanced Control of Aircraft, Spacecraft and Rockets	Tewari	July 2011
Cooperative Path Planning of Unmanned Aerial Vehicles	Tsourdos *et al*	November 2010
Principles of Flight for Pilots	Swatton	October 2010
Air Travel and Health: A Systems Perspective	Seabridge *et al*	September 2010
Design and Analysis of Composite Structures: With applications to aerospace Structures	Kassapoglou	September 2010
Unmanned Aircraft Systems: UAVS Design, Development and Deployment Austin		April 2010
Introduction to Antenna Placement & Installations	Macnamara	April 2010
Principles of Flight Simulation	Allerton	October 2009
Aircraft Fuel Systems	Langton *et al*	May 2009
The Global Airline Industry	Belobaba	April 2009
Computational Modelling and Simulation of Aircraft and the Environment: Volume 1 - Platform Kinematics and Synthetic Environment	Diston	April 2009
Handbook of Space Technology	Ley, Wittmann Hallmann	April 2009
Aircraft Performance Theory and Practice for Pilots	Swatton	August 2008
Aircraft Systems, 3rd Edition	Moir & Seabridge	March 2008

Introduction to Aircraft Aeroelasticity And Loads	Wright & Cooper	December 2007
Stability and Control of Aircraft Systems	Langton	September 2006
Military Avionics Systems	Moir & Seabridge	February 2006
Design and Development of Aircraft Systems	Moir & Seabridge	June 2004
Aircraft Loading and Structural Layout	Howe	May 2004
Aircraft Display Systems	Jukes	December 2003
Civil Avionics Systems	Moir & Seabridge	December 2002

ADAPTIVE AEROSERVOELASTIC CONTROL

Ashish Tewari
Indian Institute of Technology,
Kanpur, India

This edition first published 2016

Registered office
John Wiley & Sons Ltd, The Atrium, Southern Gate, Chichester, West Sussex, PO19 8SQ, United Kingdom

For details of our global editorial offices, for customer services and for information about how to apply for permission to reuse the copyright material in this book please see our website at www.wiley.com.

Library of Congress Cataloging-in-Publication Data applied for.

ISBN: 9781118457634

A catalogue record for this book is available from the British Library.

Typeset in 10/12pt TimesLTStd by SPi Global, Chennai, India

Printed and bound in Singapore by Markono Print Media Pte Ltd

1 2016

For example, that the certain is worth more than the uncertain, that illusion is less valuable than 'truth', such valuations, in spite of their regulative importance for us, might notwithstanding be only superficial valuations, special kinds of maiserie, such as may be necessary for the maintenance of beings such as ourselves. Supposing, in effect, that man is not just the 'measure of things'.

—Friedrich Nietzsche in *Beyond Good and Evil.*

Contents

About the Author

Ashish Tewari is a Professor of Aerospace Engineering at the Indian Institute of Technology, Kanpur. He specializes in Flight Mechanics and Control, and is the single author of five previous books, including *Aeroservoelasticity – Modeling and Control* (Birkhäuser, Boston, 2015) and *Advanced Control of Aircraft, Spacecraft, and Rockets* (Wiley, Chichester, 2011). He is also the author of several research papers in aircraft and spacecraft dynamics and control systems. He is an Associate Fellow of the American Institute of Aeronautics and Astronautics (AIAA), and a Senior Member of the Institution of Electrical and Electronics Engineers (IEEE). Prof. Tewari holds PhD. and MS degrees in Aerospace Engineering from the University of Missouri-Rolla, and a B.Tech. degree in Aeronautical Engineering from the Indian Institute of Technology, Kanpur.

About the Authors

Series Editor's Preface

The field of aerospace is multidisciplinary and wide ranging, covering a large variety of products, disciplines and domains, not merely in engineering but also in many related supporting activities. These combine to enable the aerospace industry to produce exciting and technologically advanced vehicles. The wealth of knowledge and experience that has been gained by expert practitioners in the various aerospace fields needs to be passed onto others working in the industry, including those just entering from University.

The *Aerospace Series* aims to be a practical, topical and relevant series of books intended for people working in the aerospace industry, including engineering professionals and operators, allied professions such as commercial and legal executives, and also engineers in academia. The range of topics is intended to be wide ranging, covering design and development, manufacture, operation and support of aircraft, as well as topics such as infrastructure operations and developments in research and technology.

Aeroservoelasticity (ASE) concerns the interaction of flexible aeroelastic structures with active control systems and is a crucial topic for modern and future aircraft, where such systems can be used to reduce loads due to gusts and manoeuvres and also to extend the flutter stability boundaries. The presence of nonlinearities and uncertainties in the structure, aerodynamics and control system makes an already complex problem even more challenging.

This book, *Adaptive Aeroservoelastic Control*, considers ASE from the control design viewpoint, using a range of adaptive control approaches to solve practical ASE design problems developed by using a consistent theoretical methodology. It fills a significant gap in the current literature and will be of most interest to practicing engineers and researchers working in the fields of aeroelasticity and control.

Preface

Aeroservoelasticity (ASE) lies at the interface of aerodynamics, control and structural dynamics, and by its very nature, it is a difficult topic to deal with. However, it is also an important subject, crucial to the design of modern aircraft, and can be ignored only at the peril of the designer. Unfortunately, there are not many books available that deal with the control design aspects of ASE. It is precisely this gap in the literature that the present book aims to fill. The present work can be regarded as a treatise on adaptive ASE control. While many illustrative examples are offered to the reader, the focus is on the methods and mathematics of essentially nonlinear feedback strategies, which are necessary for deriving a stable, closed-loop ASE system in the presence of modelling uncertainties.

The control challenge for the aeroservoelastician is twofold. There are practical limitations in the plant that prevent a continuous and smooth change of the dynamic variables at all space points. This could be regarded as the natural uncontrollability (or unreachability) of an infinite dimensional system, which is attempted to be controlled by only a finite number of imperfectly modulated control inputs. On the other hand, even if the designer had a large army of control input variables at his disposal, it would still be difficult to devise a sound principle (control law) governing each one of them. This is the other inherent limitation, which arises due to an imperfect knowledge of the plant dynamics, and leads to a deficient mathematical model of the plant. The ASE control design process is thus a perpetual struggle with the combined problem of underactuated and uncertain plant dynamics.

The attempts to control an uncertain ASE system are also twofold:

(a) Devising an accurate mathematical model of the plant by faithfully representing every important physical process, and then designing a controller based upon the plant model.
(b) Using an online identification of the actual plant from its measured input–output record, in order to adapt the controller parameters with the changing plant behaviour.

While method (a) is an effort at achieving modelling precision through sophisticated mathematical models that may not be implementable in real time, its alternative is the adaptive control approach highlighted as method (b). This book underlines the adaptive control approach to solving practical ASE design problems, whereas a previous monograph by the author (*Aeroservoelasticity – Modeling and Control*, Birkhaüser, Boston, 2015) details the modelling approach. The modelling details are hence deferred to the earlier book – which can be regarded as a companion text – and only those aeroelastic principles are highlighted here that are relevant to adaptive control design.

The unsteady aerodynamic behaviour of an aircraft wing is very often uncertain, in so far as both magnitudes and signs of the forces and moments arising out of the aeroelastic motion could be in doubt. This is especially true when simple linear aerodynamic models are applied to problems wherein flow separation and/or shock waves cause an uncertain nonlinear aeroelastic response, typically in the transonic regime. The designer then has the option to either improve the plant model through computational fluid dynamics (CFD) techniques that require iterative and online solution of partial differential equations, or to use an adaptive control scheme, which automatically senses the aeroelastic behaviour and applies a corrective action. While CFD modelling has not arrived at a stage where practical, dynamic aeroelastic computations of separated and shock-induced flows could be performed in the real time, the alternative of adaptive control appears to be more promising due to its relative simplicity.

Adaptive control has reached maturity in the last two decades due to active research in the area of nonlinear control systems design. In the classical sense, adaptive control can be understood to ensure closed-loop, input–output stability via tuning (or describing) functions that automatically adjust the controller gains in accordance with a changing plant dynamics. In the modern sense, adaptive, state-space based techniques are applied to a plant with unknown parameters in order that closed-loop stability exists in the sense of Lyapunov. Such techniques can be either direct – being based upon comparison with a reference model, or indirect – requiring a closed-loop estimation of the unknown (or uncertain) plant dynamics via input–output identification. In either case, closed-loop stability is the primary objective, and neither the reference nor the estimated plant dynamics need be the 'true' representation of the actual behaviour of the aeroelastic plant. In effect, modelling of the true plant behaviour, which is necessary in traditional control design, is bypassed by the adaptive control loop. Herein lie both the strength and the weakness of the adaptive control strategy: while it may not be necessary to have a highly accurate plant model for a successful implementation, large perturbations in the plant's parameters could have unpredictable (usually undesirable) consequences on the closed-loop performance. The control engineer must balance the two opposing tendencies by aiming at a suitable adaptive mechanism that is robust with respect to parametric variations. However, it must be examined whether design robustness can be achieved only if the identified (or reference) plant model is closer to the actual behaviour. In other words, one asks: is it really important for closed-loop stability to have a model that faithfully represents the plant characteristics in every way, or whether a simpler (perhaps highly 'unrepresentative') model might do a better job? This question lies at the heart of robust and adaptive control, and its resolution is an active research area.

A word here is appropriate about the basic difference between the adjectives 'adaptive' and 'robust'. When we consider adaptive control, we have in mind the ultimate adaptation mechanism, viz the human mind, which can almost instantly produce a wide ranging behaviour in response to a drastically changed environment. Such a control system is often said to be 'intelligent' (or even 'smart') – although I dislike such a terminology applied to an artificial controller, because the latter can only respond in very limited manner, entirely depending on the sophistication of the algorithms it has been programmed with. A property of the adaptive controller is the ability to 'learn' (or detect) the changing plant behaviour with operating conditions, and then respond accordingly in order to maintain a desired performance level. An example of such an application is a violinist playing a complicated concerto, when the air-conditioning system of the concert hall breaks down. The player must quickly change the length and pressure of the bow strokes, as well as the spacing of the notes on the fingerboard, in order to adapt to the temperature-induced changes in the strings' natural frequencies, and the expansion or contraction of the wooden body by variations in the humidity. Such an adaptation comes naturally to a good violinist, who has a good ear for the changing notes and tones. Of course, one cannot expect a similar level of adaptive behaviour in an artificial control system, because the level of complexity increases manifold with each parametric variation (temperature, humidity, etc.).

At the other extreme to adaptation (or a fine sensitivity to the changing operating conditions) lies the property of robustness. In order to have a robust control system, there must be the ability to absorb small external disturbances around a specified (or nominal) operating condition, without having a noticeable effect on the performance. In other words, the control system must be quite insensitive to disturbance inputs. The effectiveness of an artificial robust controller thus entirely depends upon how well a nominal performance is achieved in the presence of disturbances, and to design such a controller usually requires meeting a set of conflicting performance objectives. The robust solution is based upon the 'worst-case scenario' (i.e. for the largest expected disturbance measures), and thus can be overly conservative in its performance, applying much larger control inputs than actually required. Furthermore, many performance measures are difficult to quantify as robustness measures (e.g. tonal sound quality in the violin player example). A neglect of important qualitative behaviour in the design process can lead to a stable, but totally unacceptable performance of an automaton playing the violin to a music connoisseur. At the control design level, the difference between adaptive and robust control lies in whether the perturbation variables are considered to be the systemic parameters varying with operating conditions, or external, random inputs about a nominal operating condition. In each case, there exists a design framework evolved over many decades, and which will be explored here in the context of ASE systems.

This book is primarily intended to be a reference for practicing engineers, researchers and academicians in aerospace engineering, whose primary interest lies in flight mechanics and control, especially aeroelasticity. The reader is assumed to have taken a basic undergraduate course in control systems that covers the transfer function and frequency response methods applied to single-input, single-output systems. It is however suggested that the introductory material be supplemented by basic examples and exercises from a textbook on linear control systems, especially if the reader has not had a fundamental course on linear systems theory.

A research monograph on adaptive aeroservoelasticity is an enormous task, as it must access topics ranging in a spectrum as wide as structural dynamics, unsteady aerodynamics and control systems. There are two possible approaches that can be adopted in writing such a book: (i) detailing of the work carried out on the subject by citing and describing various research articles and (ii) offering a fresh insight from the author's perspective by presenting a systematic framework into which the research carried out until now can fit neatly. While the former method (common to survey articles) can give glimpses into the field from the individual viewpoints of the respective researchers, it is only the latter that can add something to the already existing literature, and is hence adopted here. Emphasis is laid on presenting a consistent and unbroken theoretical methodology for adaptive ASE. While many important contributions have been highlighted in the chapter references, they are by no means exhaustive of the developments in ASE. The reader is referred to survey articles for a thorough review of the literature. As mentioned earlier, the companion book on this topic (Aeroservoelasticity – Modeling and Control) can assist the reader in understanding the essential modelling concepts of ASE.

I would like to thank the editorial and production staff of Wiley, Chichester, for their constructive suggestions and valuable insights during the preparation of the manuscript. I also thank my family members for their patience while this book was being prepared.

Ashish Tewari
May 2015

1

Introduction

1.1 Aeroservoelasticity

Aeroservoelasticity (ASE) is the study of interactions among structural dynamics, unsteady aerodynamics and flight control systems of aircraft (Fig. 1.1), and an active research topic in aerospace engineering. The relevance of ASE to modern airplane design has greatly increased with the advent of flexible, lightweight structures, higher airspeeds and large-bandwidth, automatic flight control systems. The latter trend assumes a greater significance in the modern age, as many of the flight tasks that were earlier performed by a much slower human interface, must now be carried out by high-speed, closed-loop digital controllers, resulting in an increased encroachment into the aeroelastic frequency spectrum. Inadvertent ASE couplings can arise between an automatic flight controller and the aeroelastic modes, resulting in signals becoming unbounded in the closed-loop system. Hence, every new aircraft prototype must be carefully flight-tested to evaluate the ever expanding aeroservoelastic interactions domain, and the higher aeroelastic modes that could be safely neglected in the past must now be fully investigated. Furthermore, favourable ASE interactions can be designed by suitably modifying the feedback control laws, such that certain aeroelastic instabilities are avoided in the operating envelope of the aircraft.

Consider the block representation of the typical ASE system shown in Fig. 1.2. Here, an automatic flight control system is designed to fulfil the pilot commands by actuating control inputs applied to the aircraft. It is seldom possible to model all aspects of an aircraft's dynamics by well-defined mathematical representations. The unmodelled dynamics of the system can be treated as unknown external disturbances applied at various points, such as the atmospheric gust inputs acting on the aircraft and the measurement noise present in the sensors. If such disturbances were absent, one could design an open-loop controller to fulfil all the required tasks. However, the presence of random disturbance inputs necessitates a closed-loop system shown by the feedback loop in Fig. 1.2, where the control inputs are continuously updated based on measured outputs. Such a closed-loop system must be stable and should perform well by following the pilot's commands with alacrity and accuracy. Ensuring the stability and good performance of the closed-loop system in the presence of unknown disturbances is the primary task of the control engineer.

Adaptive Aeroservoelastic Control, First Edition. Ashish Tewari.

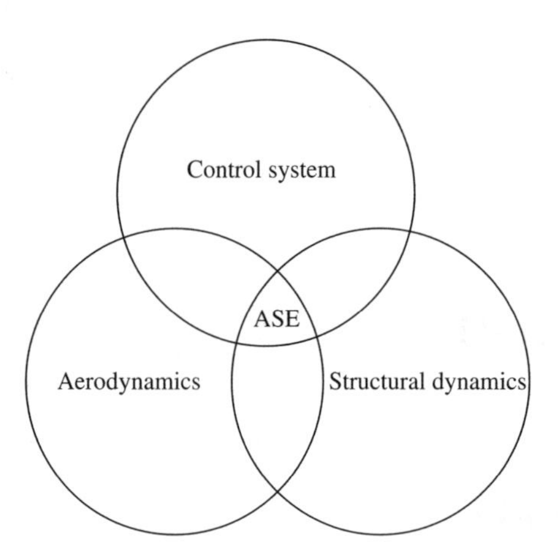

Figure 1.1 Venn diagram showing that aeroservoelasticity (ASE) lies at the intersection of aerodynamics, structural dynamics and flight control systems

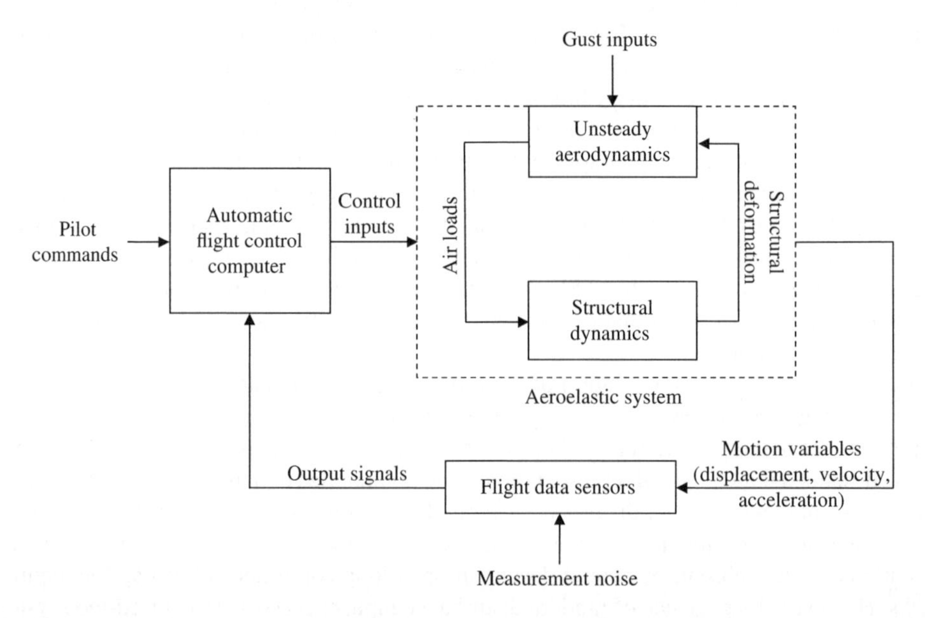

Figure 1.2 Block diagram of a typical flight control system, highlighting the importance of aeroservoelastic analysis

The flight control system is usually designed either without regard to the aeroelastic interactions, or with only the primary, in vacuo structural modes taken into account. When applied to the actual vehicle, such a control system could therefore cause unpredicted consequences due to unmodelled dynamic interaction between the flexible structure and the aerodynamic loads, often leading to instability and structural failure. It is usually left to the flight-test engineers to identify and iron out the problematic ASE coupling of a flying prototype through either a redesign of the structural members, or reprogramming the flight control computer. This process is time consuming, expensive and very often fraught with danger. However, if the ASE analysis is introduced as a systematic procedure into the basic airframe and flight control design from the conceptual stage, such difficulties can be avoided at a more advanced stage. The focus of the present book is to devise such a systematic procedure in the form of an adaptive design of the flight control system.

The most important ASE topic is the catastrophic phenomenon of flutter, which is an unstable dynamic coupling between the elastic motion of the wings (or tails) and the unsteady aerodynamic loading that generally begins at a small amplitude, and grows to large amplitudes thereby causing structural failure. The classical flutter mechanism consists of an interaction between two (or more) natural aeroelastic modes at a critical dynamic pressure, and can be excited by either atmospheric gusts or control surface movement. While traditional method of avoiding flutter consists of stiffening the structure such that the natural modes causing flutter occur outside the normal operating envelope of the aircraft, such a method is not always reliable, and requires many design iterations based on expensive, cumbersome and dangerous flight-tests of actual prototypes. The main problem lies in accurately predicting the critical dynamic pressure, because of a drastic change in aerodynamic characteristics due to Mach number and the equilibrium angle of attack. Such a bifurcation typically occurs at transonic speeds and requires a nonlinear stability analysis. For example, the flutter dynamic pressure computed by linearized subsonic aerodynamics is often much higher than that actually encountered at transonic Mach numbers. Since the non-conservative dip in the flutter dynamic pressure due to transonic effects can be extremely treacherous, either accurate computational fluid dynamics (CFD) modelling or precise wind-tunnel experiments are necessary for predicting transonic flutter modes. However, both CFD modelling and wind-tunnel testing are complicated by the sensitivity of nonlinear transonic aerodynamics to transition and turbulence, for which no CFD model or experimental technique, however advanced, can be entirely relied upon. Even an extremely sophisticated Navier–Stokes computation with tens of million of grid points is unable to resolve the fine turbulence scales of an unsteady transonic flowfield on a complete aircraft configuration. Furthermore, these same aeroelastic phenomena have large-scale effects (Edwards 2008), which make an extrapolation of wind-tunnel data to the full-scale aircraft highly uncertain. The inadequacies of aerodynamic modelling can be practically overcome only by an adaptive, closed-loop identification and control of unsteady aerodynamics, which is the topic of the present book.

Actively suppressing flutter through a feedback control system is an attractive alternative to passive flutter avoidance by haphazard redesign and flight testing. The concept of active flutter suppression began to be explored in the 1970s (Abel 1979), wherein an automatic control system actuated a control surface on the wing, in response to the structural motion sensed by an accelerometer. This modified the aeroelastic coupling between critical modes, such that the closed-loop flutter occurred at a higher dynamic pressure. Linear feedback control design for active flutter suppression requires an accurate knowledge of the aeroelastic modes that cause flutter. Although the classical flutter of a high aspect-ratio wing of a transport type aircraft is caused by an interaction between the primary bending and torsion aeroelastic modes, the flutter mechanism of a low aspect-ratio wing of a fighter-type airplane involves a coupling of several aeroelastic modes. Despite extensive research (Abel and Noll 1988, Perry *et al.* 1995), active flutter suppression has yet to reach operational status. This shortcoming is due to the inability of designing a feedback control system that can be considered sufficiently robust with respect to the parametric uncertainties caused by nonlinear transonic effects which, as mentioned earlier, are difficult to predict. Routine implementation of active flutter suppression must wait until suitably accurate transonic ASE design methods are available. Hence, development of practical adaptive control techniques for transonic flutter suppression will be a revolutionary step in the design of automatic flight control systems.

The process of adaptive aeroservoelastic design is briefly introduced in this chapter, although full explanations will follow in the subsequent chapters. ASE applications require designing an underlying feedback control system (Chapter 2) in order to ensure closed-loop stability in a range of operating conditions. Such a design is typically based upon a linearized model of the underlying aeroelastic system, which is discussed in Chapter 3. The aircraft has a continuous structure, but for computational considerations it is approximated by finite degrees of freedom using a process such as the finite element method (FEM). In a complete wing–fuselage–tail combination, this approximation may require several thousand degrees of freedom for an accurate representation. However, as most aeroelastic phenomena of interest involve only about a dozen structural modes, the structural displacement vector $\{z(t)\}$ [1] can be represented as a linear combination of a few structural vibration modes given by the vector of modal degrees of freedom $\{q(t)\}$ (also called the generalized coordinates), and result in the following generalized equations of motion:

$$[M]\{\ddot{q}\} + [C_d]\{\dot{q}\} + [K]\{q\} = \{Q\}(\{q\}, \{\dot{q}\}, \{\ddot{q}\}), \tag{1.1}$$

where $[M]$, $[C_d]$, $[K]$ are the generalized mass, damping and stiffness matrices representing the individual masses, viscous damping factors and moments of inertia corresponding to the various modal degrees of freedom, and $\{Q(t)\}$ is the generalized aerodynamic force vector, whose dependence upon the modal degrees of freedom (and their time derivatives) requires separate modelling.

1.2 Unsteady Aerodynamics

The computation of unsteady aerodynamic forces $\{Q(t)\}$ from structural degrees of freedom is the main problem in aeroelastic modelling. The fluid dynamics principles upon which such an aerodynamic model is based require a conservation of mass, momentum and energy of fluid flowing through a control volume surrounding the aircraft. As in the case of the structural model, a CFD model necessitates the approximation of the continuous fluid flow by a finite number of cells (called a grid), within each of which the conservation laws can be applied, and then summed over the entire flowfield. The grid can either have a well-defined shape (called structured grid) or could be entirely unstructured in order to give flexibility in accurately modelling the moving, solid boundary. The spatial summation from individual grid points to the entire flowfield can be carried out by finite difference, finite volume or finite element methods, each requiring a definite discretization process. There is also the possibility of using simplifying assumptions in applying the conservation laws. For example, the airflow about a wing $(x, y) \in S, z_\ell \leq z \leq z_u$ at a sufficiently large Reynolds number can be regarded to be largely inviscid, with the viscous effects confined to a thin region close to the wing (boundary layer) and in its wake. This affords a major simplification, wherein $\{Q(t)\}$ is computed from

[1] Vectors and matrices in this chapter are denoted by braces and brackets, respectively. A more compact notation follows in the next chapter.

continuity, inviscid momentum and energy conservation (Euler equations) applied outside the boundary layer and wake, and integrated in space (x, y, z) subject to suitable unsteady boundary conditions. These latter include the solid boundary condition of no flow across the moving wing surface, and tangential velocity continuity at its trailing edge (the Kutta condition) due to the presence of viscosity in the boundary layer and wake.

The unsteady Euler equations can be written in the conservation form as follows:

$$\frac{\partial\{F\}}{\partial t} + \frac{\partial\{f_x\}}{\partial x} + \frac{\partial\{f_y\}}{\partial y} + \frac{\partial\{f_z\}}{\partial z} = \{0\}, \tag{1.2}$$

where

$$\{F\} = \{\rho, \rho u, \rho v, \rho w, \rho e\}, \tag{1.3}$$

is the independent flow variables vector, with ρ being the density, e the specific internal energy,

$$\{V\} = \{u, v, w\}, \tag{1.4}$$

the velocity vector with (u, v, w) being the velocity components along (x, y, z), respectively, and

$$\begin{aligned}\{f_x\} &= u\{F\} + \{0, p, 0, 0, pu\}\\ \{f_y\} &= v\{F\} + \{0, 0, p, 0, pv\}\\ \{f_z\} &= w\{F\} + \{0, 0, 0, p, pw\},\end{aligned} \tag{1.5}$$

are the flux vectors along x, y and z directions, respectively. The flux gradients, $\frac{\partial f_x}{\partial x}, \frac{\partial f_y}{\partial y}, \frac{\partial f_z}{\partial z}$, are required to be modelled differently according to the local direction of the infinitesimal pressure waves. Clearly, even the Euler equations are inherently nonlinear, requiring an iterative solution procedure, which is further complicated by having to model an entropy condition for a unique solution, usually by introducing artificial viscosity into the solution procedure. An artificial viscosity model can lead to spurious frequency spectra in unsteady flow computations. Alternatively, a solution by flux direction biasing or splitting algorithms in finite-element (or finite-volume) methods is employed, which can have further problems of non-physical oscillations when the sonic condition is encountered in the flowfield. Dealing with non-unique and physical solutions is a major problem associated with Euler equations, often requiring sophisticated computational procedures that add to the computational time.

An additional approximation is invariably necessary for modelling purposes, namely that of potential flow with small perturbations. However, even the full-potential (FP) and small-disturbance solutions for the transonic regime are inherently nonlinear and iterative and fraught with non-unicity and non-physical nature, such as the prediction of expansion shock waves. As in the case of Euler solvers, the closure of the inviscid, potential computational problem necessitates the addition of an entropy condition in the form of either artificial viscosity or flux biasing/splitting. Consequently, little is gained in terms of computational complexity by making the potential approximation of unsteady transonic flows. Owing to their iterative nature and high computational times, the unsteady CFD computations of nonlinear governing equations are infeasible to carry out in a real time adaptive control scheme, which may require several evaluations of $\{Q(t)\}$ per time step. Only in the subsonic and supersonic regimes can the small-disturbance potential equation be linearized. In such a case, the unsteady aerodynamic computation involves an integration of pressure distribution $p(x, y, z, t)$ on the wing surfaces, subject to the flow velocity normal to the wing (normalwash) $\{V\} \cdot \{n\}$ created by the structural vibration modes. If the wing is thin ($z_\ell \simeq z_u$), the vibration amplitudes are small, and there are no aerodynamic dissipation mechanisms present (such as viscous flow separation and shock waves), the pressure–normalwash relationship is rendered linear, and is given by the following integral equation:

$$w(x, y, t) = \int_S K[(x, y : \xi, \eta), t]\Delta p(\xi, \eta, t)\mathrm{d}\xi\mathrm{d}\eta, \tag{1.6}$$

where $\Delta p = p_\ell - p_u$ is the pressure difference between the lower and upper faces of the essentially flat wing's mean surface at a given point (ξ, η), and $w(x, y, t)$ the flow component normal to the mean surface (z-component) called the upwash (or its opposite in sign, the downwash). Such a simple relationship is enabled by the process of linear superposition of elementary, flat plate (or panel) solutions to the governing partial differential equation. However, while Eq. (1.6) can be applied in subsonic and supersonic flows about thin wings with small vibrations, it is invalid in the transonic regime, where nearly normal shock waves are always present and cause viscous separation in the boundary layer and wake. Furthermore, even in subsonic and supersonic regimes, the linear superposition cannot be applied around thick wings undergoing large amplitude vibration, as flow separation or strong shock waves could be present.

A linear aerodynamic model Eq. (1.6) combined with the linear structural dynamics Eq. (1.1) yields the following linear aeroelastic state equations that can be used as a baseline plant of the adaptive ASE control system:

$$\{\dot{X}\} = [A]\{X\} + [B]\{u\} + [F]\{p\}, \tag{1.7}$$

where $\{X(t)\} = [\{q(t)\}^T, \{\dot{q}(t)\}^T]^T$ is the state vector of the aeroelastic system, $\{u(t)\}$ the vector of generalized control forces generated by a set of control surfaces and $\{p\}$ the vector of random disturbances called the process noise. In order to derive the constant coefficient matrices $[A], [B], [F]$, an additional step is necessary, even if the generalized aerodynamic forces $\{Q(t)\}$ are linearly related as follows to the modal displacements $\{q(t)\}$ and their time derivatives by virtue of Eq. (1.6):

$$\{Q(s)\} = [G(s)]\{q(s)\}, \tag{1.8}$$

where s is the Laplace operator and $[G(s)]$ denotes the unsteady aerodynamics transfer matrix. The essential step is modelling of $[G(s)]$ by a suitable rational-function approximation (RFA), such as the following:

$$[G(s)] = [A_0] + [A_1]s + [A_2]s^2 + \sum_{j=1}^{N}[A_{j+2}]\frac{s}{s + b_j}, \tag{1.9}$$

where the numerator coefficient matrices, $[A_0], [A_1] + [A_2], [A_{j+2}], j = 1, \cdots, N$ are determined by curve fitting $[G(i\omega)]$ to the simple harmonic aerodynamics data ($s = i\omega$) at a discrete set of frequencies ω, and for each flight condition (speed and altitude). Additionally, the denominator coefficients $b_j, j = 1, \cdots, N$ may be selected by a nonlinear optimization process, whereby the curve fit error in a range of frequencies is minimized. Such an optimized curve fitting is not a trivial matter, and is by itself an area of major research with the objective of deriving an accurate RFA, which is also of the minimum possible order. The order of the state space model (the dimensions of $[A]$) increases rapidly with the order of the RFA, and the computational effort in optimizing the denominator coefficients could be significant especially if a large range of flight conditions is involved. For this reason, several different RFA techniques have been proposed in the literature. However, in keeping with the present objective of designing an adaptive control system, RFA optimization must be carried out offline and its results stored in order to derive the baseline aeroelastic plant model in the flight conditions of interest. The frequency domain (simple harmonic) data to be used for RFA derivation is also pre-computed by a suitable linearized small-disturbance, potential aerodynamic model, such as that based on the integral equation, Eq. (1.6). After the RFA for the aerodynamic transfer-matrix, Eq. (1.9), is derived, a linear, time-invariant, state-space model, Eq. (1.7), for the aeroelastic system – perhaps also including the control-surface actuators model – is obtained.

1.3 Linear Feedback Design

Consider the basic automatic control system shown in Fig. 1.3, where the automatic controller is designed as a generic device to exercise control over the plant, in order that the entire control system meets a certain set of desired objectives, and follows a desired trajectory, $\{x_d(t)\}$. For the purposes of this book, the desired trajectory is taken to be a constant equilibrium state, $\{x_d(t)\} = \{0\}$, wherein the control strategy to be evolved becomes a regulator problem. If the plant can be described precisely by a set of fixed mathematical relationships between the input, $\{u(t)\}$, and output, $\{y(t)\}$, variables, then the controller can usually be designed fairly easily in order to meet the performance requirements in a narrow range of operating conditions (Tewari 2002). Such a controller would have a fixed structure (often linear) and constant parameters. However, a physical plant almost never conforms exactly to any deterministic mathematical description due to either improperly understood physical laws, or unpredictable external disturbances treated as stochastic signals (the process noise vector), which is shown in Fig. 1.3 as the externally applied random vector signal, $\{p(t)\}$.

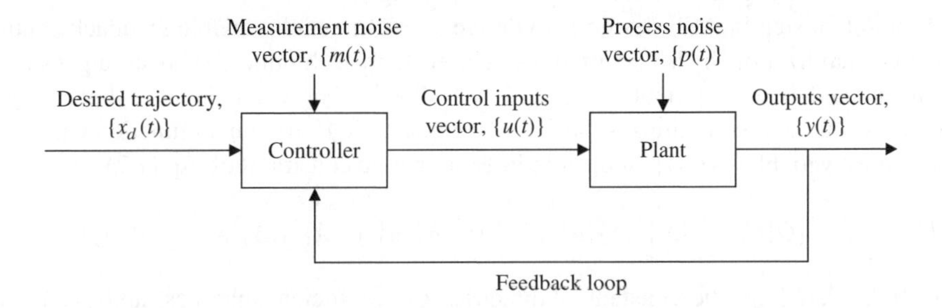

Figure 1.3 Basic automatic control system with a feedback control loop

Similarly, the controller, when physically implemented, has its own imperfections that defy precise mathematical description. For the feedback controller, such a departure from a deterministic controller model is shown as the measurement noise vector, $\{m(t)\}$, appearing in the feedback loop. The success of the automatic controller in performing its task of tracking the reference signals with any accuracy depends upon how sensitive the control system is to the unmodelled noise signals, $\{p(t)\}, \{m(t)\}$. If no regard is given to the noise signals while designing the controller, there is a real possibility that the control system will either break down completely, or have a poor performance when actually implemented. The controller design is therefore carried out to ensure adequate robustness with respect to the noise signals. A feedback loop by itself provides a certain degree of robustness with respect to unmodelled process and measurement noise. If the feedback control parameters are suitably adjusted (fine-tuned), the sensitivity to noise inputs can be further reduced. Such a design is called loop shaping (Chapter 2). For a plant with a linear input–output behaviour and a fair statistical description of the noise inputs that are small in magnitude, the robust control theory (Maciejowski 1989) can be applied to design a linear feedback controller with constant parameters, which will produce an acceptable performance in many applications. However, constant controller gains may either fail to stabilize the system if the plant behaviour is highly uncertain or may have unacceptable performance in the presence of noise inputs. In such cases, the alternative strategy of sensing the actual plant behaviour and to adapt the controller gains to suit a certain minimum performance level in a range of operating conditions is the only answer. Such a strategy where the controller parameters are functions of the sensed plant state vector is called adaptive control, and is nonlinear by definition. In summary, design of an automatic controller can be alternatively based on ensuring a high level of robustness with respect to unmodelled dynamics with constant controller parameters by a design process called robust control or by making the controller parameters adapt to a changing plant behaviour through an adaptation mechanism. The two design techniques of robust control and adaptive control may appear to be contradictory in nature, because in one case the controller is deliberately made impervious to process and measurement noise, while in the other, the controller is asked to change itself with a changing plant dynamics. However, if a compromise can be carried out in the two methods of synthesis, the result can be a synergistic fusion of robust and adaptive control. In such a case, the high-frequency noise (which is typically of small magnitude) is sought to be rejected by an inbuilt control robustness, while the much slower but larger amplitude variations in the plant dynamics are sensed and carefully adapted to. Such an ideal combination of robustness and adaptation is the goal of most control system designers.

An important step in ASE design is to derive a baseline multivariable feedback controller for active stabilization by standard linear closed-loop techniques, such as eigenstructure assignment and linear optimal control (Tewari 2002). For example, if a linear optimal regulator is sought, one minimizes the following quadratic Hamiltonian function with respect to the control variables, $\{u(t)\}$, subject to linear dynamic constraint of Eq. (1.7):

$$H = \frac{1}{2}\{X\}^T[Q]\{X\} + \{X\}^T[S]\{u\} + \frac{1}{2}\{u\}^T[R]\{u\} + \{\lambda\}^T([A]\{X\} + [B]\{u\}), \qquad (1.10)$$

where $[Q], [S], [R]$ are the constant, symmetric cost coefficient matrices, and $\{\lambda(t)\}$ is the vector of co-state variables. The necessary conditions for optimality with an infinite control interval are then given by the following Euler–Lagrange equations:

$$\{\dot{\lambda}\} = -\left(\frac{\partial H}{\partial \{X\}}\right)^T = -[Q]\{X\} - [S]\{u\} - [A]^T\{\lambda\}, \tag{1.11}$$

$$\{\lambda\}(\infty) = \{0\}, \tag{1.12}$$

$$\frac{\partial H}{\partial \{u\}} = \{0\} = [S]^T\{X\} + [R]\{u\} + [B]^T\{\lambda\}, \tag{1.13}$$

the last of which is solved for the optimal control vector to yield the following:

$$\{u\} = -[R]^{-1}\left([S]^T\{X\} + [B]^T\{\lambda\}\right). \tag{1.14}$$

Substitution of Eq. (1.14) into Eqs. (1.7) and (1.11) results in the following set of linear state and co-state equations:

$$\{\dot{X}\} = \left([A] - [B][R]^{-1}[S]^T\right)\{X\} - [B][R]^{-1}[B]^T\{\lambda\}, \tag{1.15}$$

$$\{\dot{\lambda}\} = -\left([A]^T - [S][R]^{-1}[B]^T\right)\{\lambda\} + \left([S][R]^{-1}[S]^T - [Q]\right)\{X\}, \tag{1.16}$$

which have to be solved subject to the following two-point boundary conditions,

$$\{X\}(0) = \{X_0\}; \quad \{\lambda\}(\infty) = \{0\}. \tag{1.17}$$

The simultaneous forward and backward time-marching required for the solution of Eqs. (1.15) and (1.16) is commonly expressed as the following linear feedback control law with a constant gain matrix, $[K]$:

$$\{u\} = -[K]\{X\}, \tag{1.18}$$

where

$$[K] = [R]^{-1}\left([B]^T[P] + [S]^T\right), \tag{1.19}$$

and the constant matrix $[P]$ is the solution to the following algebraic Riccati equation (ARE):

$$\{0\} = [Q] + \left([A] - [B][R]^{-1}[S]^T\right)^T[P] + [P]\left([A] - [B][R]^{-1}[S]^T\right)$$
$$-[P][B][R]^{-1}[B]^T[P] - [S][R]^{-1}[S]^T. \tag{1.20}$$

The ARE is a nonlinear algebraic equation and necessitates an iterative solution procedure, which must be carried out for each set of coefficient matrices $[A], [B], [Q], [R], [S]$. This is, in a nutshell, the linear, quadratic regulator (LQR) problem with a quadratic cost and an infinite control horizon. The cost coefficients $[Q], [R], [S]$ must be selected such that the regulator is an asymptotically stable system, which requires that all the eigenvalues of the closed-loop dynamics matrix, $[A] - [B][K]$, should lie in the left-half side of the Laplace domain. Alternatively, the eigenvalues and eigenvectors of the dynamics matrix, $[A] - [B][K]$, can be directly specified in order to determine $[K]$, which is termed an eigenstructure assignment.

The state feedback regulator cannot be directly implemented because the state variables of the aeroelastic plant, $\{X\}$, are unavailable for direct measurement. What one can measure are the output variables, $\{y\}$, detected by a set of sensors placed strategically on the aircraft. The state vector, $\{X\}$, which is required by the linear feedback control law, Eq. (1.18), must then be constructed by an additional system called an observer (or state estimator). The linear output equation,

$$\{y\} = [C]\{X\} + [D]\{u\} + \{m\}, \tag{1.21}$$

where $\{m(t)\}$ is the vector of random disturbances (the measurement noise), can be used to design a full-order observer, whose dynamics is governed by the following state equation:

$$\{\dot{\hat{X}}\} = ([A] - [L][C])\{\hat{X}\} + ([B] - [L][D])\{u\} + [L]\{y\}, \tag{1.22}$$

where $\{\hat{X}\}$ is the estimated state vector, and $[L]$, the observer gain matrix. Such an observer requires that the plant must be observable with the outputs given by Eq. (1.21). The observer gain matrix, $[L]$, can be selected in a manner similar to (but separately from) the regulator gain, $[K]$, by either eigenstructure assignment for the observer dynamics matrix, $[A] - [L][C]$, or via linear, quadratic, optimal control where $[A]$ is replaced by $[A]^T$, and $[B]$ by $[C]^T$. The optimal observer is also known as the Kalman filter and is guaranteed to minimize the covariance matrix, $[R_e]$, of the estimation error, $\{e\} = \{X\} - \{\hat{X}\}$, in the presence of zero-mean, Gaussian process and measurement noise signals, $\{p\}, \{m\}$. In the infinite horizon case, the Kalman filter gain is determined by the following ARE similar to Eq. (1.20), and hence the Kalman filter is regarded as the dual of the state feedback regulator.

$$\begin{aligned}\{0\} &= [A_G][R_e] + [R_e][A_G]^T - [R_e][C]^T[S_m]^{-1}[C][R_e] \\ &\quad + [F]\left([S_p] - [S_{pm}][S_m]^{-1}[S_{pm}]^T\right)[F]^T,\end{aligned} \tag{1.23}$$

where $[S_m], [S_p], [S_{pm}]$ are the matrices of power spectral density of the measurement noise, $\{m\}$, that of the process noise, $\{p\}$, and the cross-spectral density of $\{p\}, \{m\}$, respectively, and

$$[A_G] = [A] - [F][S_{pm}][S_m]^{-1}[C]. \tag{1.24}$$

The Kalman filter gain matrix is then given by

$$[L] = \left([R_e][C]^T + [F][S_{pm}]\right)[S_m]^{-1}. \tag{1.25}$$

Clearly, the matrices $[S_m], [S_p], [S_{pm}]$ act as the cost coefficients of a quadratic objective function for determining $[L]$ in a manner similar to $[Q], [R], [S]$ for the LQR regulator. These should be suitably selected in the observer design process.

The observer's dynamics must be designed to be stable and much faster than the regulator. It is crucial for practical considerations that the derived control laws must be robust with respect to modelling uncertainties (process noise) and sensor (measurement) noise at a selected range of operating conditions. The procedure by which an LQR and a Kalman filter (also called linear, quadratic estimator (LQE)) are designed separately for a linear, time-invariant plant, and then put together to form a compensator is referred to as the linear, quadratic, Gaussian (LQG) method. Here, the Kalman filter supplies the estimated state for feedback to the LQR regulator. The design of the LQG compensator – specified by the gain matrices, $[K], [L]$ – depends upon the chosen cost parameters, $[Q], [R], [S], [S_m], [S_p], [S_{pm}]$. Suitable performance and robustness requirements of the overall ASE system restrict the choice of the cost parameters to a specific range. Being based upon optimal control, an LQG compensator has excellent performance features for a given set of cost parameters, but its robustness depends upon how much the performance is degraded by state estimation through the Kalman filter. If the observer gains, $[L]$, are too small, the estimation error, $\{e\}$, does not tend to zero fast enough for the feedback to be accurate. On the other hand, if the observer has very large gains, there is an amplification of process and measurement noise by feedback, thereby reducing the overall robustness of the control system. Clearly, a balance must be struck in selecting the Kalman filter design parameters, such that a good robustness is obtained without unduly sacrificing performance. Several linear feedback strategies are in use for striking a compromise between robustness with respect to plant uncertainty, and noise rejection properties. These include LQG compensation with loop-transfer recovery (LTR) (Maciejowski 1989), H_2/H_∞ control (Glover and Doyle 1988) and structured singular value (SSV) (or μ-) synthesis (Packard and Doyle 1992). Chapter 2 is a brief compilation of the basic linear feedback design methods for achieving a robust control system with constant parameters.

1.4 Parametric Uncertainty and Variation

Any aeroelastic model employed in ASE design is likely to have modelling uncertainties in its parameters, $[A], [B], [C], [D]$. These can be either due to errors in the linear aeroelastic plant or due to a part of the dynamics which is entirely unmodelled. The parametric errors in the linear plant are due to inadequacies of the structural dynamics model, as well as those in evaluating the frequency domain aerodynamics and its transfer matrix (RFA) representation. The unmodelled dynamics include nonlinear structural and aerodynamic effects, which are difficult to account for. While linear parametric uncertainties are easier to handle in a control system design, it is the presence of unmodelled dynamics that causes a greater anxiety. Of these, the nonlinear aerodynamic phenomena are the most critical as they can cause unforeseen aeroelastic instabilities, and whose model requires iterative and complex CFD computations which (as noted above) are infeasible to carry out in real time. Aerodynamic nonlinearities encountered in aeroservoelastic systems are divided into two classes: (i) unsteady behaviour involving normal shock waves and (ii) largely separated or vortex-dominated flows. While type (i) is only present at the transonic speeds, nonlinearities of type (ii) occur at high angle-of-attack flight. A fighter-type aircraft manoeuvring at transonic speeds will encounter both the effects. The unsteady flow separation (type (ii)) causes a buffeting of the airframe at low frequencies, and can result in rigid dynamic instabilities, such as wing-rock, nose-slice and coning motions, but rarely cause an aeroelastic coupling. This is due to the fact that the structural dynamics of the airframe simply acts as a stable, linear filter of the nonlinear buffeting forces and moments, allowing the peaks of the spectrum to occur only at the in vacuo structural frequencies. Consequently, notch-filters can be designed to suppress the buffet at well-identified frequencies. Such a nonlinear dynamic characteristic can be analysed by the Popov stability criterion (Chapter 7). However, the shock-wave effects (type (i)) are more interesting; they cause dynamic aeroelastic instabilities, leading to a sharp reduction in the flutter dynamic pressure and a sustained limit-cycle oscillation (LCO), often ending in a catastrophic structural failure. An accurate transonic aerodynamic model is necessary to account for unsteady shock wave effects and an absence of such a model renders the unsteady aerodynamic forces and moments highly uncertain.

In addition to modelling uncertainties, there are significant variations in the aeroelastic characteristics due to changing operating conditions (flight speed and altitude). For example, as the flight Mach number is increased from subsonic to supersonic, the variation of the lift, pitching moment and control-surface hinge moment with angle-of-attack and control deflections vary drastically. Some steady-state aerodynamic derivatives can even change in sign as the transonic regime is crossed. The transonic flight regime is especially critical as it is characterized by different, simultaneously occurring flow regions (subsonic/sonic/supersonic). However, the unsteady transonic variations are more crucial, as they often lead to markedly different aeroelastic behaviour depending upon the flow geometry (airfoil shape, angle-of-attack, control-surface deflections). The unsteady mixing of the different flow regions creates complex, time-dependent flow patterns, and causes interesting aeroelastic interactions. These include a significant dip in the flutter dynamic pressure, transonic buffet, LCO and control surface oscillation (buzz) caused by unsteady shock wave and boundary-layer interaction. Any of these phenomena can cause a catastrophic structural failure, if not properly addressed in airframe (open-loop) and active flight control (closed-loop) designs. In fact, these very transonic aeroelastic instabilities were historically dreaded as the 'sound barrier' which prevented safe supersonic flight in the first half of the 20th century. In the present age, nearly all airline transport aircraft fly in the high subsonic/transonic regime. Furthermore, fighter-type aircraft must not only cross the sonic speed but also perform high-g manoeuvres at transonic speeds in their mission. Hence, transonic ASE is even more important now than at any other time in the history of aviation.

Since transonic ASE applications involve unsteady shock motions, as well as periodic boundary-layer separation and reattachment induced by shock waves, advanced CFD modelling techniques are required for such inherently nonlinear effects (Silva *et al.* 2006). The inviscid, unsteady transonic equations required to capture shock waves are inherently nonlinear, even in their small-disturbance potential form. Furthermore, the presence of normal shock waves in the transonic flow exacerbates the transient (unsteady) flow behaviour by introducing nonlinear shock-induced flow oscillations, which can interact with the viscous boundary layer, thereby causing unsteady flow separation. The ASE plant for such a case is further complicated by the separated wake, or the leading-edge vortex from the wing interacting with the tail, resulting in irregular and often catastrophic deformation of the tail – either on its own or driven by a rapid and large deflections of the elevator. Such a wing–tail–elevator coupling of a post-stall buffet or a shock–vortex interaction requires a fully viscous flow modelling that is only possible by a Navier–Stokes method (Obayashi 1993). Another example of transonic ASE is the control of unsteady control-surface buzz and shock-induced buffet encountered by an aircraft manoeuvring in the transonic regime (Huttsell *et al.* 2001), leading to nonlinear flutter or LCO (Bendiksen 2004). An appropriate CFD model in such a case would require a FP, Euler or Navier–Stokes method, depending upon the geometry, structural stiffness, Mach number and Reynolds number. Sometimes, semi-empirical models are devised from wind-tunnel data for separated and shock-induced flows (Edwards 2008), because they do not require unsteady CFD computations to be performed in loop with structural dynamic and control-law calculations. However, the veracity of such a correlation must be checked carefully before being deployed in ASE design and analysis. An alternative method is to employ flight-test data for deriving an ASE model, such as the neural-network identification by Boëly and Botez (2010).

Any flow model that fully accounts for the unsteady transonic effects over an oscillating wing must necessarily be very complex, hence difficult to solve in real time. Owing to the inherent uncertainty of an unsteady aerodynamic model, a closed-loop controller for ASE application must be quite robust to modelling errors. Furthermore, such a controller must also adapt to changing flight conditions, which renders it mathematically nonlinear even for a linear aeroelastic plant. This implies that as an accurate unsteady aerodynamics plant model is infeasible for aeroservoelastic design, adaptive plant identification in closed loop is the only practical alternative.

1.5 Adaptive Control Design

Following the above discussion, it is logical that the final step in aeroservoelastic system design should be the derivation of adaptive control laws that can fully account and compensate for the parametric uncertainties and variations in the characteristics of the aeroelastic system. Such control laws allow variation of the controller parameters in order to adapt to uncertain and changing plant characteristics. For this, an adaptation mechanism based upon the sensed (identified) input–output behaviour of the plant must be devised. Various adaptation mechanisms that can be applied to adaptive ASE design are now explored.

Design of a control system generally requires a plant model. The ability of a control system in achieving its desired performance depends upon how accurately the plant modelling is carried out. For example, the resolution of a digital camera depends upon how precisely the dynamics of the sensor, aperture and diaphragm are modelled. Similarly, the tolerance of a robotic positioning device largely depends upon the number of structural vibration modes considered in modelling the robotic arm. Most mechanical and electrical systems can be modelled to a very high accuracy because their dynamics are well understood, and hence controller design for the systems can be carried out by traditional methods. The same, however, cannot be said of an ASE system, wherein achieving high accuracy may result in the aeroelastic model becoming too unwieldy and complex to be of any benefit in control system design. For example, accurate modelling of a viscous, unsteady flow over a deforming wing surface would require unsteady, turbulent, Navier–Stokes solutions involving several thousands of grid points and hundreds of hours of computation time. The past several decades have seen significant advancement in CFD, but only at the cost of increasing complexity of modelling, which cannot be practical for closed-loop design and analysis. Rather than pursuing the course of increasingly accurate plant models, which seems to have reached a dead end, it is more profitable to look for simpler models that can capture the fundamental physical aspects of the aeroelastic plant. Therefore, accuracy is sacrificed in the interest of simplicity for a practical ASE design. Simplifying assumptions are usually made by neglecting some aspects of the plant characteristics, such as high-frequency dynamics, and structural and aerodynamic nonlinearities, thereby producing a mathematical model which is more amenable to control system design with either constant, or well-defined controller gains.

Consider an aircraft wing experiencing multimodal vibration in the presence of unsteady airloads. While in vacuo structural modelling of the wing can be accurately performed by a high-order finite-element method, the unsteady air loads acting on the wing are quite another matter. Depending upon the airspeed and altitude, the aerodynamic characteristics can range from low-subsonic, through transonic, to supersonic, each of which is dramatically and fundamentally different from the other. Furthermore, even in a given speed regime, a part of the flow on the wing could be laminar and another part turbulent, attached or separated, subsonic or supersonic, thereby creating almost infinite variation in the magnitude and phase of the dynamic loading. Since the structural deformations (elasticity) and air loads (flowfield) are strongly coupled, each can cause a large change in the other at any given time, and this picture keeps on changing with time in an unpredictable manner. It can be said that an accurate aeroelastic model of the wing must take into account a large number of mutually coupled, randomly varying, local phenomena, an exact accounting of which is impossible. Even the most sophisticated aerodynamic model (Navier–Stokes equations with a statistical turbulence model) falls well short of faithfully capturing the complex flowfield around a flexing wing. Furthermore, such models are cumbersome in terms of computational efficiency, hence cannot be used in control system design. Therefore, the best available aeroelastic plant model is often an inaccurate and uncertain one.

How does one go about designing a good aeroservoelastic system if the plant is not well modelled? This question takes us to adaptive control design. A practical ASE system must operate at different design conditions representative of an aircraft flying at various speeds, altitudes and loadings. In many cases, the aerodynamic behaviour of the aircraft changes drastically when going from one flight regime to another, such as from subsonic to supersonic speeds. If not properly compensated for, the resulting aerodynamic changes (such as appearance of shock waves) can cause a large reduction in aeroelastic stability margin, perhaps leading to a catastrophic condition such as flutter. In order to maintain stability in the presence of varying flight conditions, one has two options: (i) meticulously redesign the control system at a large number of expected conditions and store the design points for a smooth interpolation of controller parameters in a given flight regime. This approach is called gain scheduling and is one of the first adaptive flight control strategies implemented in aircraft. (ii) Render the control system *self-adaptive* with respect to changing flight parameters through an extra feedback loop, which automatically compensates for loss of stability margin. While the former approach relies upon accurate plant modelling, the latter requires updating a 'workable' plant model by actual flight data in real time. Since it is only option (ii) that can be called adaptive in a true sense, it will be the main thrust of the present chapter. While most of the literature on ASE is concerned with accurate plant modelling by sophisticated structural and aerodynamic techniques that are necessary for the gain scheduling approach, we depart from this traditional approach and instead concentrate on developing good adaptive control algorithms that can achieve the closed-loop performance even in the face of a mathematically uncertain plant model. It can be appreciated by an aerodynamicist that even the best possible flow model may fail to capture many essential features of a flowfield, such as turbulent, separated and shock-dominated flows. Unfortunately, it is precisely such flow phenomena that are the most troublesome to an aeroservoelastician. Thus uncertainty in the plant model is unavoidable, and even becomes amplified as one approaches the transonic regime where the majority of modern aircraft operate. Furthermore, even if a high degree of modelling accuracy can been achieved at a particular design condition, the off-design operation usually becomes very sensitive to initial conditions and flow parameters, such as in the nonlinear buffet and limit cycle behaviour caused by separated flows. Owing to its inability to provide a reliable plant model across the flight regimes, the gain scheduling approach has proved to be inadequate for ASE purposes, and has not achieved flight certification status even more than 50 years after it was first devised. Clearly, the answer to a practical implementation of an ASE system lies in the alternative approach, namely self-adaptive control.

The ultimate example of self-adaptive control is the natural flight of birds, where a multitude of muscles move a group of feathers to produce a graceful flight. This is also a fine example of the juggling act involved in multivariable control, such as the symphony generated by the concerted sounds of an orchestra. Such examples of nature provide a motivation and a challenge for the control engineer. Of course, it may be argued that the luxury provided by a plethora of natural control input variables – individually capable of modulation – is unavailable in the average engineering problem. This largely explains why the graceful, quiet and highly manoeuvrable trajectories of birds and insects are only in the realm of dreams of an aircraft designer. The control engineer has to work with a small number of control inputs, each of which is limited in magnitude and rate, often resulting in an underactuated plant. Thus the success in achieving a control objective relies solely upon the sophistication of the control laws employed for the purpose. The ASE designer is acutely aware of this limitation, and has to devote his energy in mathematically deriving a clever control strategy that could compensate for the deficiencies of his plant, which are both physical and mathematical.

1.5.1 Adaptive Control Laws

Adaptive control becomes necessary whenever the plant has either an unknown structure, unknown parameters or changing operating conditions, which imply an absence of any fixed description of input–output relationships. In such a case, an adaptive mechanism becomes necessary for the controller. For simplicity, we focus the discussion to state-feedback regulators, whose parameters are defined by the changing regulator gain matrix, $[K(t)]$. If an output-feedback scheme is used, the observer gains, $[L(t)]$, are also a part of the controller parameters. The adaptive controller is a self-adjusting system that can modify its parameters, $[K(t)]$, based upon the actual inputs, $\{u(t)\}$, applied to the plant, and the measurement of the actual outputs, $\{y(t)\}$, produced by the plant. In essence, an adaptive controller compensates for the lack of knowledge (or a change) of the plant's mathematical model by employing the measured plant characteristics. Owing to the dependence of the controller parameters on the plant's inputs and outputs, the adaptive controller is a nonlinear system, as depicted by the block diagram of Fig. 1.4. On comparison with the basic, fixed gain control system of Fig. 1.3, the presence of the additional adaptation mechanism is evident as the outer feedback loop, which allows for a change in controller parameters, $[K(t)]$, by a set of adaptation laws. Such a mapping of the plant's input and output vectors, $\{u(t)\}$, $\{y(t)\}$, onto the controller parameter vector space, $[K(t)]$, is the hallmark of an adaptive control system.

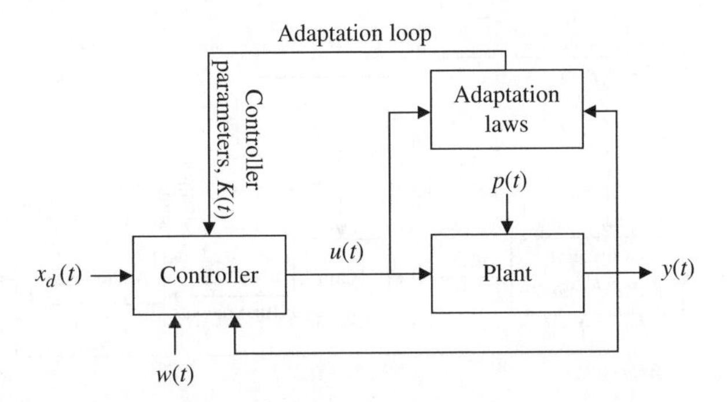

Figure 1.4 Generic adaptive control system with an adaptation mechanism for controller parameters

Adaptive control (Aström and Wittenmark 1995) arose as a discipline especially for designing the flight control systems of high-speed aircraft, which encounter large parametric variations in their operating envelope. Active research in the last several decades has produced many useful adaptive control techniques that can be applied to a wide range of problems. Unfortunately, these are specific to the application, rather than general, and can often be ad hoc procedures. The selection of an adaptive ASE control law is thus a challenging task, because an accurate aeroelastic model is unavailable in most cases. Hence, ASE has remained a formidable technological problem.

A part of the plant's output vector, $\{y(t)\}$, specifies the operating conditions. For example, a flight vehicle's operating (flight) conditions are the airspeed, altitude and Mach number. Often a good mathematical model of the plant can be derived for different sets of flight conditions (flight points), each having a set of linear controller parameters (gains) specially designed for it. In such a case, the adaptation mechanism is simply a table look-up of stored data points and the controller gains can be scheduled with the flight point. The resulting adaptive controller is called a gain scheduler. A schematic diagram of the gain schedule adaptation is shown in Fig. 1.5, where the inner feedback loop is the underlying linear control law for achieving stability for a given set of plant parameters, while the outer feedback loop determines the variation of the underlying controller parameters based upon a pre-set interpolation schedule. The gain scheduling approach was the earliest example of adaptive controllers designed for high-speed aircraft, rockets and spacecraft in the 1950s. As the name implies, most flight applications of gain scheduling involve an adjustment of linear feedback gains, but a more general application can also be envisaged where the controller parameters, $[K(t)]$, appear in a nonlinear relationship with the desired states, $\{x_d(t)\}$, and the outputs, $\{y(t)\}$, as shown in Fig. 1.5. Gain scheduling is thus regarded as a functional mapping method to vary the controller parameters $[K]$, $[L]$ according to the identified operating conditions. This requires solving for and storing the different sets of $[K]$, $[L]$ at various flight conditions. It can be expected that having to design controllers for a wide range of operating conditions requires a massive effort. Furthermore, as there is no possibility of taking into account either modelling errors or unmodelled (nonlinear) plant behaviour, gain scheduling is not regarded as an adaptive controller in the true sense.

A detailed and accurate model of the plant for various operating points is necessary before a gain scheduler can be designed for it, which is not always possible, especially for ASE plants in which we are presently interested. In a typical ASE application, the change in the plant's behaviour can be dramatic and not entirely predictable, such as in the case of transonic

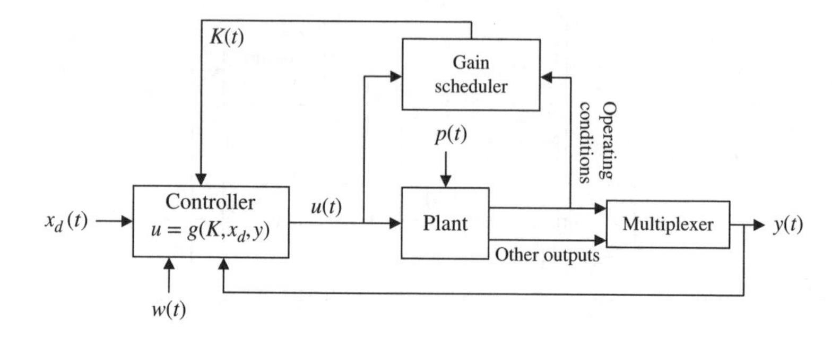

Figure 1.5 Schematic block diagram of a gain scheduling system

flutter and high angle-of-attack (stall) flutter. While an extensive research database exists on the transonic and high angle-of-attack flight, there are no efficient and reliable techniques available at present for modelling the essentially nonlinear characteristics of the aeroelastic plant under such conditions. In such a case, the gain-scheduling approach is not viable and recourse must be made to what is called a 'self-adaptive' control system.

Traditional self-adaptive systems for the regulation of an uncertain plant dynamics can be broadly classified into (i) self-tuning regulators (STRs) and (ii) model-reference adaptation systems (MRASs). An STR is an adaptive controller based on online parameter estimation of the unknown plant dynamics. A MRAS uses a predefined plant model (often linear and time-invariant) as its reference, in order to compare the actual behaviour, and to adapt the controller parameters accordingly. Since the desired behaviour is known a priori, such an adaptive mechanism is said to be *direct*. In contrast, an STR must first estimate the plant behaviour by sensing its input–output relationship in a closed loop and then apply an adaptation (or update) law for the controller parameters. Owing to the online identification of the plant's unknown behaviour, the self-tuning approach is called an *indirect* adaptation method. The two strategies can be further classified depending upon the types of adaptation laws and parameter identification algorithms they employ. We will consider STR for ASE systems in Chapter 5, while MRAS techniques will be the topic of Chapter 8.

A true adaptive controller must detect the actual plant behaviour, and apply a suitable correction to the underlying controller parameters in order to produce a stable closed-loop system. The most formal interpretation of this task is the STR whose schematic diagram is depicted in Fig. 1.6. Note the outer feedback loop for an online identification of the plant parameters, $[A]$, $[B]$, $[C]$, $[D]$, based upon a measurement of the plant's output vector, y, and a knowledge of the applied inputs, u. The slanted arrows in Fig. 1.6 indicate adaptation of the

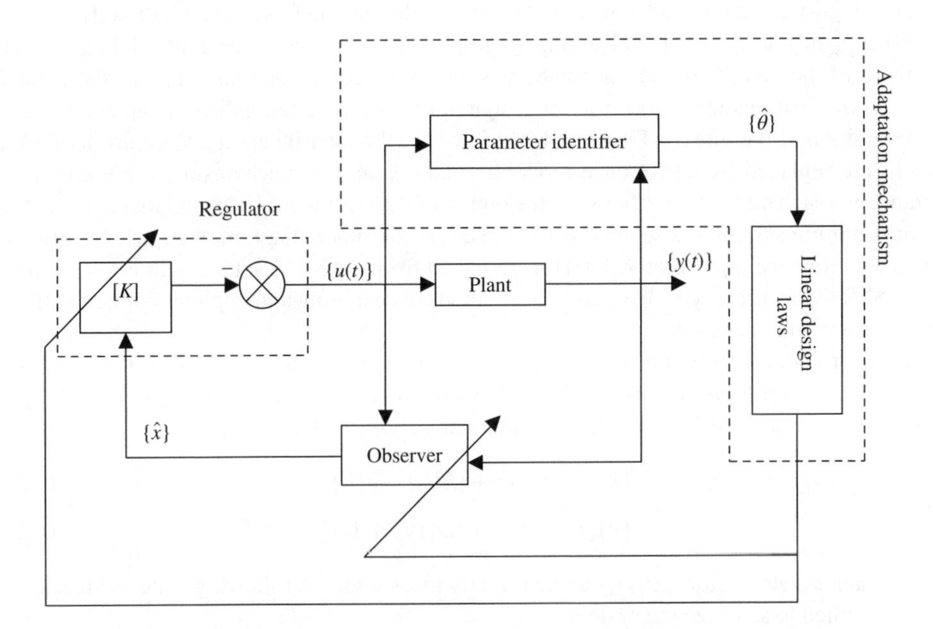

Figure 1.6 Schematic block diagram of a self-tuning regulator (STR)

parameters of the observer and the regulator. Consider a system with the following output equation:

$$\{y(t)\} = [\Phi(t)]\{\theta\}, \tag{1.26}$$

where $\{\theta\}$ are the unknown plant parameters arranged in a column vector and $[\Phi(t)]$ is a regressor matrix of functions of the known inputs and outputs. A parameter estimation scheme derives an estimate $\{\hat{\theta}\}$ by minimizing a positive cost function of the estimation error, $\{e\} = \{y\} - [\Phi(t)]\{\hat{\theta}\}$ at a given time t.

The parameter identification must be carried out with a finite record of the inputs and outputs. As the updated plant parameters become available, they are used to determine the new regulator and observer gains by solving the underlying linear control problem (such as the nonlinear AREs). The identification process is generally based on the solution to a set of linear algebraic equations, and hence the online controller updation is a much less complex task than that of accurately modelling the plant behaviour through a set of nonlinear partial differential equations. The success of the self-tuning approach depends upon active stabilization, rather than on how accurately the plant behaviour can be identified at any given instant. Therefore, guaranteeing closed-loop stability of the adaptation scheme is the primary objective. In some cases, it is even likely that a set of constant controller parameters are found to be stabilizing, albeit the plant parameters may be varying in time. The identified plant parameters are directly used in the underlying controller design, regardless of whether they are the 'true' parameters. Hence, the STR design approach is based upon the *certainty equivalence principle*, which disregards the uncertainty (or error) in plant identification.

The computation of the controller parameters from the underlying control design process of the STR can be transformed into a mapping from the plant's input–output record to the controller parameters space. The plant's parameter identification is then implicit in the adaptation mechanism, and it would appear that the controller parameters are being directly updated from the plant's input–output behaviour. An implicit dependence of the controller parameters on those of the identified plant is sometimes termed direct adaptation, whereas the explicit modules of identification and controller design in Fig. 1.6 is called indirect adaptation.

A variation of the direct STR is the MRAS, where the identification and controller design blocks are replaced by a reference model and an adaptation mechanism for the controller parameters, such that the error between the output of the reference model and that of the actual plant is minimized. Such a scheme is illustrated by the block diagram of Fig. 1.7. Note that when the reference input vector, $\{r(t)\}$, is removed from the MRAS, the result is very similar to the STR of the direct type. However, the methods of designing and implementing the MRAS and STR are quite different.

Consider a linear, time-invariant plant with a state-space model given by the following state-space equations with constant (but unknown) coefficient matrices $[A]$, $[B]$, $[C]$, $[D]$, and unknown process noise, $\{p(t)\}$, and measurement noise, $\{m(t)\}$:

$$\{\dot{x}\} = [A]\{x\} + [B]\{u\} + \{p\}, \tag{1.27}$$

$$\{y\} = [C]\{x\} + [D]\{u\} + \{m\}. \tag{1.28}$$

If the plant's state vector, $\{x(t)\}$, can be directly measured, the following state-feedback law can be applied to stabilize the system:

$$\{u\} = -[K]\{x\} - [K_r]\{r\}, \tag{1.29}$$

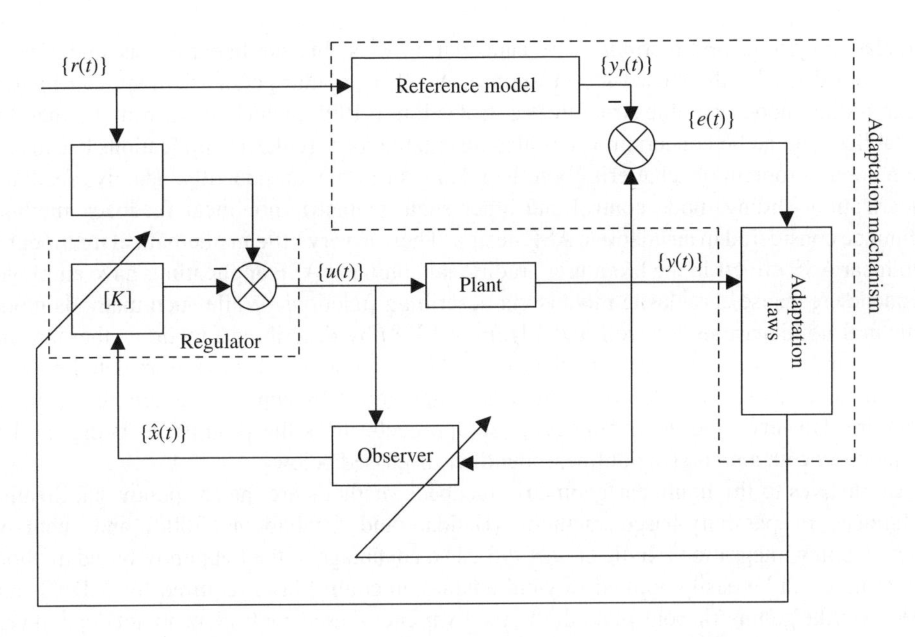

Figure 1.7 A model-reference adaptation system (MRAS)

such that the plant's state tracks a reference input vector, $\{r\}(t)$. Here $[K_r]$ is a feedforward gain matrix, and $[K]$ the regulator gain matrix. A reference model is defined by the following linear, time-invariant state-space representation with known coefficient matrices $[A_r]$, $[B_r]$, $[C_r]$, $[D_r]$, and the reference input vector, $\{r(t)\}$:

$$\{\dot{x}_r\} = [A_r]\{x_r\} + [B_r]\{r\}, \tag{1.30}$$

$$\{y_r\} = [C_r]\{x_r\} + [D_r]\{r\}. \tag{1.31}$$

The estimated state, $\hat{x}$, required by the regulator is supplied by an observer (such as that given by Eq. (1.22)) whose gain matrix, $[L]$, must be designed on the basis of the plant's parameters, $[A]$, $[B]$, $[C]$, $[D]$. Since the plant's parameters are uncertain, an exact set of stabilizing controller parameters, $[K]$, $[K_r]$, $[L]$, is unknown. Thus beginning from an initial guess of stabilizing controller gains, $[K(0)]$, $[K_r(0)]$, $[L(0)]$, the controller parameters must be evolved in time such that the error between the plant output and that of a reference model, $\{e\} = \{y(t)\} - \{y_r(t)\}$, is minimized in the limit $t \to \infty$. This is the broad philosophy behind MRAS schemes.

When the plant is inherently nonlinear, and cannot be linearized about an operating condition, a nonlinear feedback control-with an adaptation mechanism for its parameters – becomes necessary. A possible design strategy for such a controller is the geometric nonlinear feedback approach (Slotine 1995), such as adaptive feedback linearization, wherein the adaptive feedback renders the control system linear by cancelling its nonlinearities. However, a major shortcoming of feedback linearization is its inability to handle parametric uncertainties that are not matched by the control inputs (i.e., the uncertainties occurring in those state equations that do not contain the control inputs). Therefore, while the rigid body motion can be stabilized by adaptive feedback linearization, the same cannot be said for flexible structures or dynamic aeroelastic systems. Furthermore, the cancellation of stable nonlinearities is undesirable, because it degrades the closed-loop response. Another popular geometric nonlinear method is the sliding mode (variable structure) control (Slotine 1995), which apart from the inability to stabilize unmatched uncertainties, is also unsuitable for aeroelastic applications because of the inherent problem of 'chattering' on the sliding surface. Consequently, adaptive feedback linearization, sliding mode control and other such geometric nonlinear feedback methods cannot be considered in an adaptive ASE design. There is very little mathematical treatment of nonlinear ASE effects in the literature. Traditional nonlinear ASE applications have employed frequency response aeroelastic models via describing functions. While such methods model structural nonlinearities (Dowell and I'lgamov 1988) by describing functions, they are not easily found for the nonlinear behaviour caused by separated and shock-dominated flows in the transonic regime. In addition, nonlinear adaptation ASE applications are absent in the literature. However, the describing function approach offers the promise of being used in conjunction with a recursive nonlinear identifier discussed below.

Alternatives to the nonlinear geometric feedback methods are the Lyapunov stabilization techniques of passivity-based methods (Haddad and Chellaboina 2008) and recursive back-stepping integration (Krstic *et al.* 1995). The advantage of the Lyapunov-based methods is that they can be easily applied to yield adaptation control laws required for MRASs and STRs. Unlike geometric control methods, the Lyapunov-based controllers do not depend very much on the plant characteristics, which offer a great flexibility in their design. This book mainly utilizes the Lyapunov-based methods for adaptive controller derivation. However, it is necessary to highlight the important theoretical concepts before applying them in the design process.

1.6 Organization

The treatment of all possible adaptive control techniques that could be applied to the design and analysis of ASE systems is a formidable task. This book attempts to do so by focussing on the important features and concepts. Chapter 2 details the feedback design methods applied to design the underlying controller, whose parameters are to be adjusted by a separate adaptation mechanism. Chapter 3 covers the basic principles and techniques used to derive an aeroelastic plant model that is suitable for use in controller design. Chapter 4 introduces the active suppression of the primary ASE instability, namely flutter, and presents examples of both typical-section (two-dimensional) and lifting-surface (three-dimensional) flutter. Chapter 5 introduces STRs for adaptive ASE systems based upon online plant identification. Chapter 6 details the essential concepts used in analysing the stability, and designing stabilizing controllers for nonlinear systems, of which adaptive ASE systems are the target. Chapter 7 presents the methodology of describing functions analysis, and Nyquist-like techniques based upon Circle and Popov criteria, which can be applied to model LCOs associated with nonlinear aeroelastic behaviour caused by shock-induced and separated flows. Chapter 8 focuses on MRAS techniques, with applications to ASE systems. Chapter 9 presents the essentials of the powerful adaptive control method via backstepping integration as an alternative to the traditional methods (MRAS and STR). Chapter 10 considers robust design of adaptive systems in the presence of nonlinearities and noise inputs. Finally, Chapter 11 covers the ultimate end of adaptive ASE design, namely the possible handling of transonic flutter and LCOs by adaptive control methods.

References

Abel I 1979 An analytical design technique for predicting the characteristics of a flexible wing equipped with an active flutter suppression system and comparison with wind-tunnel data. *NASA Technical Publication* **TP-1367**.

Abel I and Noll TE 1988 Research and applications in aeroservoelasticity at NASA Langley Research Center. *Proceedings of the 16th ICAS Congress*, Tel Aviv, Israel.

Aström KJ and Wittenmark B 1995 *Adaptive Control*. 2nd ed., Addison-Wesley, New York.

Bendiksen OO 2004 Transonic limit-cycle flutter/LCO. *AIAA Paper* **2004-2694**.

Boëly N and Botez RM 2010 New methodologies for the identification and validation of a nonlinear F/A-18 model by use of neural networks. *Proceedings AIAA Atmospheric Flight Mechanics Conference*, Toronto, Canada.

Dowell EH and I'lgamov M 1988 *Studies in Nonlinear Aeroelasticity*. Springer-Verlag, New York.

Edwards JW 2008 Calculated viscous and scale effects on transonic aeroelasticity. *J. Aircr.* **45**, 1863–1871.

Glover K and Doyle JC 1988 State space formulae for all stabilizing controllers that satisfy an H_∞ norm bound and relations to risk sensitivity. *Syst. Control Lett.* **11**, 167–172.

Haddad WM and Chellaboina V 2008 *Nonlinear Dynamical Systems and Control*. Princeton University Press, Princeton, NJ.

Huttsell L, Shuster D, Vol J, Giesing J, and Love M 2001 Evaluation of computational codes for loads and flutter. *AIAA Paper* **2001-569**.

Krstiè M, Kanellakopoulos I, and Kokotoviè PV 1995 *Nonlinear and Adaptive Control Design*. Wiley-Interscience, New York.

Maciejowski JM 1989 *Multivariable Feedback Design*. Addison-Wesley, Reading, MA.

Obayashi S 1993 Algorithm and code development for unsteady three-dimensional Navier-Stokes equations. *NASA Contractor Report* **CR-192760**.

Packard A and Doyle JC 1992 A complex structured singular value. *Automatica* **29**, 71–109.

Perry B, Cole S, and Miller GD 1995 A summary of the active flexible wing program. *J. Aircr.* **32**, 10–15.

Silva WA, Mello OAF, and Azevedo JLF 2006 Sensitivity study of downwash weighting methods for transonic aeroelastic stability analysis. *J. Aircr.* **43**, 1506–1515.

Slotine JJE and Li W 1995 *Applied Nonlinear Control*. Prentice-Hall, Englewood Cliffs, NJ.

Tewari A 2002 *Modern Control Design with MATLAB and Simulink*. John Wiley & Sons, Ltd, Chichester.

References

2

Linear Control Systems

This chapter introduces the basic notations and concepts of linear control theory that are necessary in the design and analysis of aeroservoelastic (ASE) systems. The discussion can be supplemented by textbooks on linear control systems design (Kailath 1980, Kwakernaak and Sivan 1972, Maciejowski 1989, Tewari 2002).

2.1 Notation

Conforming to the advanced treatment presented in the book, the notation is simplified from that found in explanatory textbooks on control systems. There is no attempt made to indicate vectors and matrices by bold symbols, but lower case letters and symbols (a, α) are employed for scalar and vector quantities and capital letters and symbols for matrices. Sometimes, in order to highlight certain modelling features, braces may be used to distinguish vectors from scalars and brackets for matrices. The orders of vector and matrix variables are not indicated separately, but understood to conform to the linear algebraic multiplication rules. The overdot on a letter or symbol represents the time derivative operator of the individual elements (scalar, vector or matrix). The nomenclature for scalar, vector and matrix algebra is given in Table 2.1. Any departure from this labelling scheme, if necessary, is noted. Standard aerospace symbols define relevant flight parameters and variables, as and when used.

2.2 Basic Control Concepts

A system is a self-contained set of physical processes, which can be represented by a set of time-dependent variables (called signals). The externally applied signals are the inputs and the system's variables arising internally are the outputs, which can be measured. Control is the general task of achieving a desired result from a target system, called the plant, by an appropriate manipulation of its inputs. For an ASE system, the plant to be controlled is the aeroelastic system, while the system that exercises the control is the controller. In modelling a system, one must account for the inherent relationships prevailing between the input and output signals. If the system is governed by known physical laws, such a relationship generally takes the form of a set of mathematical (differential, integro-differential and algebraic) equations, and the system is said to be deterministic. In contrast, a system with unknown (or partially

Adaptive Aeroservoelastic Control, First Edition. Ashish Tewari.

Table 2.1 Basic linear algebraic norms

Notation	Mathematical expression	Nomenclature
$\overline{a+ib}$	$a-ib$	Complex conjugate
$\mid a \mid$	$\sqrt{\bar{a}a}$	Magnitude of a complex scalar, a
a^H	$\bar{a}^T$	Hermitian of a complex vector, a
$\mid a \mid$	$\sqrt{\sum_{i=1}^{n} \mid a_i \mid^2} = \sqrt{a^H a}$	Euclidean (or ℓ_2) norm of a vector, a
$\mid a \mid_p$	$\{\sum_{i=1}^{n} \mid a_i \mid^p\}^{1/p}$ $(1 \leq p < \infty)$	Hölder (or p) norm of a vector, a
$\mid A \mid_p$	$\left\{\sum_{i=1}^{n} \sum_{j=1}^{m} \mid A_{ij} \mid^p\right\}^{1/p}$ $(1 \leq p < \infty)$	Hölder (or p) norm of a matrix, A
$\det(A)$		Determinant of a square matrix, A
A^H	$\bar{A}^T$	Hermitian of a matrix, A
$\mathrm{tr}(A)$	$\sum_{i=1}^{n} a_{ii}$	Trace of a square matrix, A
$\mid A \mid_F$	$\sqrt{\mathrm{tr}(A^H A)}$	Frobenius norm of a matrix, A
$\lambda_i(A)$		Eigenvalues of a square matrix, A
$\rho(A)$	$\max_i \mid \lambda_i(A) \mid$	Spectral radius of a square matrix, A
$\sigma_i(A)$	$\sqrt{\lambda_i(A^H A)}$	Singular values (principal gains) of a matrix, A
$\bar{\sigma}(A)$	$\sqrt{\max_i\{\sigma_i(A)\}} = \sup_{z\neq 0} \frac{\mid Az \mid}{\mid z \mid}$	Largest singular value of a matrix, A
$\mid A \mid_S$	$\bar{\sigma}(A)$	Hilbert (or spectral) norm of a matrix, A
$\underline{\sigma}(A)$	$\sqrt{\min_i\{\sigma_i(A)\}} = \inf_{z\neq 0} \frac{\mid Az \mid}{\mid z \mid}$	Smallest singular value of a matrix, A
$\Vert f \Vert_2$	$\sqrt{\int_{-\infty}^{\infty} f(x)^T f(x)\,\mathrm{d}x}$	H_2 norm of a vector function, $f(x)$
$\Vert F \Vert_2$	$\sqrt{\int_{-\infty}^{\infty} \mid F(x) \mid_2^2\,\mathrm{d}x}$	H_2 norm of a matrix function, $F(x)$
$\Vert F \Vert_\infty$	$\sup_x \bar{\sigma}\{F(x)\}$	H_∞ norm of a matrix function, $F(x)$

known) physical laws is called non-deterministic. Most non-deterministic systems are such that certain fixed statistical laws can be applied to them. Such systems are said to be stochastic. If no statistical analysis can be conducted to study a system, it is said to be a completely random system. A stochastic system can be modelled as a set of external input signals – called disturbances – acting upon a deterministic system. Disturbances are generally of two types: (i) process noise that can arise either externally because of unknown inputs or internally because of uncertainty in modelling the system; (ii) a measurement noise that results from the uncertainty in measuring the output signals. The presence of such external and internal modelling errors renders all physical systems stochastic.

A system comprising the plant, the controller and disturbance inputs is called a control system. The controller manipulates the plant through a control input vector, which is an input vector to the plant, but an output of the controller. In a flight control system, the control

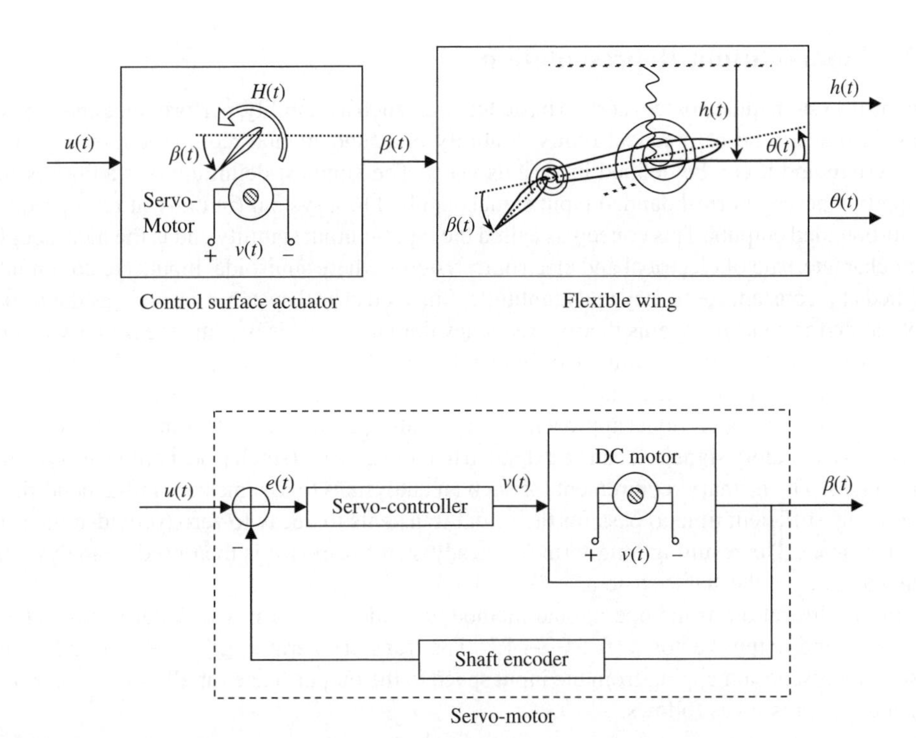

Figure 2.1 Servo-motor-based control surface actuator for a flexible wing equipped with a trailing-edge control surface

inputs are the forces and torques applied to the aircraft by moving certain control surfaces. In actuality, the controller only generates electrical or mechanical signals through wires or cables/hydraulic-lines; these signals must be converted into physical inputs for the plant by separate subsystems of the plant, called the actuators. A feedback controller requires the measurement of the output variables of the plant through separate subsystems of the plant called sensors. For example, the plunge displacement, $h(t)$, and pitch rotation, $\theta(t)$, of an aircraft wing are controlled by deflecting a trailing-edge control surface by an angle, $\beta(t)$, through an actuating torque, $H(t)$, on the control surface's hinge-line, as shown in Fig. 2.1. Here, the actuator is a servo-motor, which is itself a self-contained control system that takes the electrical command (voltage input), $u(t)$, from the flight control computer and produces a driving voltage, $v(t)$, for the DC motor, based upon the measurement of the shaft rotation angle, $\beta(t)$, via an angle encoder (Fig. 2.1). The driving voltage is generated by a servo-controller based upon the error signal, $e = u - v$, between the commanded and actual positions of the shaft.

Although the plant is a stochastic system, the design of a control system is based upon a mathematical model. Such a mathematical model is called a representation of the system. There are two basic kinds of representations: (i) the input–output model and (ii) the state-space model. While an input–output representation describes the relationships between the input and output signals, the state-space representation involves a mathematical model of the system in the time domain based upon other internal signals that are not necessarily the inputs and outputs.

2.3 Input–Output Representation

A control system must possess certain basic features, such as stability, performance and robustness, in order to be of practical utility. Stability is a fundamental requirement of any control system and it can be defined in various ways. The simplest definition of stability is the property whereby norm-bounded input signals applied to a system (initially at rest) produce norm-bounded outputs. This concept is called the input–output stability and is the most sought after characteristic of electrical and electronic systems where sinusoidal inputs are commonly applied at a constant frequency and amplitude. Since electrical systems design was the initial motivation for control systems theory, frequency domain analysis became the primary means of studying input–output stability. Consequently, the use of operational methods (Heaviside operator, Fourier and Laplace transforms) historically took the centre stage in classical systems analysis. While the operational approach was originally developed for deterministic linear systems with oscillatory signals, it can be extended to more general (stochastic, nonlinear) systems and signals. The primary requirement for such an analysis is to assume zero initial conditions or to allow sufficient time to pass for the initial transients to decay to zero (provided the system is stable). The resulting long-term (or steady-state) response is then used to analyse the characteristics of the stable system.

For an illustration of the operational method, consider a system with input vector, $u(t) : \mathbb{R} \to \mathbb{R}^m$, and output vector, $y(t) : \mathbb{R} \to \mathbb{R}^p$. The transfer operator, $\mathcal{G}(.) : \mathbb{R}^m \to \mathbb{R}^p$, of the system represents a mapping from the input space to the output space for all times, $-\infty < t < \infty$, and is expressed as follows:

$$y = \mathcal{G}(u) . \tag{2.1}$$

If the output vector at a time τ, $y(\tau)$, depends only upon the prior input record, $u(t)$, $-\infty < t < \tau$, then the system is said to be causal. Most systems of practical interest are causal. The symbol H used in the vector and matrix norms H_2 and H_∞ (see Table 2.1) of input and output signals of a causal system refers to the Hardy space. If the mapping given by Eq. (2.1) is independent of time, then the system is said to be time invariant (or autonomous).

2.3.1 *Gain and Stability*

Definition 2.3.1 *The system described by Eq. (2.1) is said to have a gain defined in terms of the H_2 norms of the input and output vectors by the following expression:*

$$||\mathcal{G}|| = \sup_{u \neq 0} \frac{||y||_2}{||u||_2},$$

where

$$||f||_2 = \sqrt{\int_{-\infty}^{\infty} f(t)^T f(t) \, \mathrm{d}t}$$

is the H_2 norm[1] of a vector function, $f(t)$, defined over all times, $-\infty < t < \infty$. The supremum (or maximum) is to be taken over all inputs with a finite H_2 norm. However, the outputs need not have a finite H_2 norm. Thus a system can have an infinite gain. Furthermore, for this definition to be valid, the system must be causal, that is, $y(t) = 0$ if $u(t) = 0$, for all times $-\infty < t$.

---

[1] A vector function $f(t)$ whose $\mathcal{H}_2$ norm exists is said to be a square integrable function.

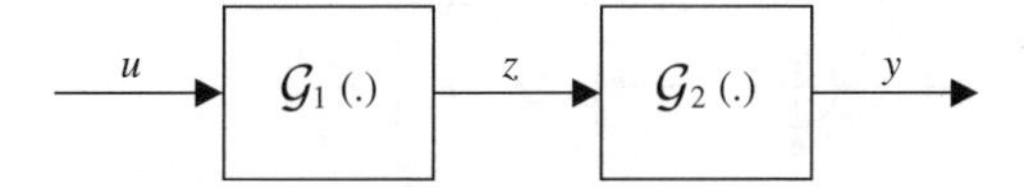

Figure 2.2 Two systems in a series connection

An alternative definition of the gain can be given with respect to the H_∞ (rather than the H_2) norm. In this book, the gain of a multivariable system is defined with respect to the H_2 norm of its signals.

Two subsystems, $\mathcal{G}_1, \mathcal{G}_2$, connected in cascade (series) as shown in Fig. 2.2, have an overall gain less than or equal to the products of their individual gains:

$$\|\mathcal{G}_2\left(\mathcal{G}_1\right)\| \leq \|\mathcal{G}_2\| \cdot \|\mathcal{G}_1\|. \tag{2.2}$$

This can be seen by applying the definition of gain to the subsystems,

$$\begin{aligned} z &= \mathcal{G}_1(u) \\ y &= \mathcal{G}_2(z), \end{aligned} \tag{2.3}$$

or $y = \mathcal{G}_2\left(\mathcal{G}_1(u)\right)$, as well as the inequality,

$$\sup_{z \neq 0} \|z\|_2 \geq \|z\|_2. \tag{2.4}$$

The gain of a system is an indicator of its input–output stability. A system with a finite gain will produce bounded outputs in response to non-zero and bounded inputs. On the other hand, a system with infinite gain will have at least one of the outputs tending to infinity in response to a bounded input, thereby indicating instability.

Definition 2.3.2 *If a causal system has a finite gain, then the system is input–output stable.*

For a feedback control system, the concept of gain as an indicator of input–output stability can be derived from the Small Gain theorem.

2.3.2 Small Gain Theorem

Theorem 2.3.3 *Consider two causal subsystems $\mathcal{G}_1, \mathcal{G}_2$, connected in a feedback loop as shown in Fig. 2.3. The overall closed-loop system is stable if the product of the gains of the two subsystems is less than unity, that is,*

$$\|\mathcal{G}_2\| \cdot \|\mathcal{G}_1\| < 1.$$

Proof. The input–output stability of the closed-loop system requires that the gain of all signals appearing within the loop with respect to each of the two inputs, r_1, r_2, must be finite. Consider the input to the first subsystem, given by the block diagram of Fig. 2.3 as

$$u_1 = r_1 + y_2 = r_1 + \mathcal{G}_2\left(u_2\right) = r_1 + \mathcal{G}_2\left(r_2 + \mathcal{G}_1\left(u_1\right)\right).$$

Using Definition 2.3.1 and the triangle inequality, it can be informally shown that[2]

$$\|u_1\|_2 \leq \|r_1\|_2 + \|\mathcal{G}_2\|\left(\|r_2\|_2 + \|\mathcal{G}_1\| \cdot \|u_1\|_2\right)$$

[2] A more formal proof can be found in Haddad and Chellaboina (2008).

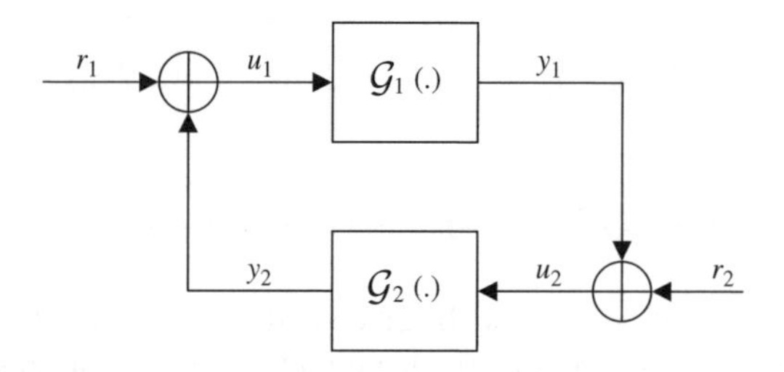

Figure 2.3 Two systems in a feedback connection

or

$$\|u_1\|_2 \leq \frac{\|r_1\|_2 + \|\mathcal{G}_2\| \cdot \|r_2\|_2}{1 - \|\mathcal{G}_2\| \cdot \|\mathcal{G}_1\|},$$

which implies that if $\|\mathcal{G}_2\| \cdot \|\mathcal{G}_1\| < 1$, then

$$\sup_{r_1 \neq 0} \frac{\|u_1\|_2}{\|r_1\|_2},$$

and

$$\sup_{r_2 \neq 0} \frac{\|u_1\|_2}{\|r_2\|_2},$$

are both finite. Similarly, loop gains of the other signals (y_1, u_2, y_2) with respect to r_1, r_2 can also be shown to be finite. Thus, by Definition 2.3.2, the closed-loop system is stable.

2.4 Input–Output Linear Systems

If the input–output behaviour of a system, Eq. (2.1), is such that

$$ay_1 + by_2 = \mathcal{G}\left(au_1 + bu_2\right), \tag{2.5}$$

where y_1, y_2 are the output vectors corresponding to input vectors u_1, u_2, respectively, and a, b are scalar constants, then the system is said to be linear, and can be represented by an impulse response matrix, $G\left(t, t_0\right)$, relating the output $y(t), t \geq t_0$, to the input vector, $u(t)$, which begins acting at a time, t_0, and is zero at all previous times, $t < t_0$. The response of a causal, linear system, initially at rest, when the inputs are unit impulse (Dirac delta) functions, $\delta\left(t - t_0\right)$, applied at $t = t_0$, is called the impulse response vector, and is given by

$$g\left(t, t_0\right) = \int_{-\infty}^{t} G(t, \tau)\, \delta\left(\tau - t_0\right) \mathrm{d}\tau = \int_{-\infty}^{t} G\left(\tau, t_0\right) \delta(\tau) \mathrm{d}\tau, \quad t \geq t_0, \tag{2.6}$$

where the impulse response matrix, $G\left(t, t_0\right)$, now assumes an implicit mathematical expression. The element (i, j) of the impulse response matrix is seen to be the value of the ith output variable at time t when the jth input variable is a unit impulse function applied at time t_0, before which the system was at rest. For an arbitrary, piecewise continuous input vector, $u(t)$, beginning to act at a time, t_0, and zero at all previous times, $t < t_0$, the response of a linear, causal system initially at rest can thus be given by the following convolution integral:

$$y(t) = \int_{-\infty}^{t} G\left(\tau, t_0\right) u(\tau)\,\mathrm{d}\tau, \quad t \geq t_0. \tag{2.7}$$

The impulse response of a linear system is a useful mathematical construct, as it helps in deriving the system's response to an arbitrary input signal by linear superposition. For understanding the concept of impulse response, consider a piecewise-continuous, arbitrary input, $u(t)$, represented by series of impulse inputs, $\delta(t-\tau)$, applied to a linear, time-invariant (LTI) single-input, single-output (SISO) system at various times, $-\infty < \tau \leq t$, and scaled by the current input magnitude, $u(\tau)$. The unit impulse function, $\delta(t)$, can be visualized as a rectangular pulse of width, ϵ, and height, $1/\epsilon$, centred at $t = 0$, by taking the limit, $\epsilon \to 0$, which implies that

$$\int_{-\infty}^{\infty} \delta(t)\,\mathrm{d}t = 1. \tag{2.8}$$

By the mean-value theorem of integral calculus Kreyszig (1998), Eq. (2.8) yields the following expression for input magnitude at time t, called the sampling property of the Dirac delta function:

$$u(t) = \int_{-\infty}^{\infty} u(\tau)\,\delta(t-\tau)\,\mathrm{d}\tau. \tag{2.9}$$

From the definition of the unit impulse function, it is clear that the limits of integration in Eq. (2.9) need not be infinite, but should only bracket the time instant, τ, at which the impulse is applied.

If the system is initially at rest (i.e. the output and all its time derivatives are zero at $t = 0_-$), then the system's response at a subsequent time, t, is simply given by the summation of the individual impulse responses, scaled by the current input magnitude, $u(\tau)$:

$$y(t) = \int_{-\infty}^{\infty} u(\tau)\,g(t-\tau)\,\mathrm{d}\tau, \tag{2.10}$$

which for a multi-input, multi-output (MIMO) system becomes Eq. (2.7). The integral on the right-hand side of Eq. (2.10) is the convolution integral, which is denoted by the operation $(u * g)(t)$. It is symmetric in $u(.)$ and $g(.)$, so we can write

$$y(t) = (u * g)(t) = \int_{-\infty}^{\infty} u(t-\tau)\,g(\tau)\,\mathrm{d}\tau. \tag{2.11}$$

The input, $u(t)$, begins acting at $t = 0$, and as $g(t-\tau) = 0$ for $t < \tau$, we have

$$y(t) = (u * g)(t) = \int_{0}^{t} u(\tau)\,g(t-\tau)\,\mathrm{d}\tau = \int_{0}^{t} u(t-\tau)\,g(\tau)\,\mathrm{d}\tau. \tag{2.12}$$

Numerical evaluation of Eq. (2.12) is carried out by quadrature Kreyszig (1998) for an arbitrary, piecewise continuous input, $u(t)$. The concept of convolution can be extended beyond impulse response.

Working with the impulse response vector and convolution integrals is a cumbersome procedure where integrals are to be evaluated in the time domain. Instead, the input–output relationships of linear systems are expressed in terms of Fourier or Laplace transforms, leading to linear algebraic equations.

2.4.1 Laplace Transform and Transfer Function

Consider a causal, LTI, SISO system, with input, $u(t)$, and output, $y(t)$. Initially, the system is at rest and without any input, that is, $u(t) = 0, y(t) = 0, (t < 0)$. All the time derivatives of the input and output are also zeros for $t < 0$.

Definition 2.4.1 *The transfer function, $G(s)$, of a causal, LTI, SISO system is defined as the ratio of the Laplace transform of the output, $Y(s)$, to that of the input, $U(s)$, subject to the zero initial condition,*

$$G(s) = \frac{Y(s)}{U(s)}.$$

The zero initial condition is denoted by $u\left(0_{-}\right) = 0, y\left(0_{-}\right) = 0$, where $t = 0_{-}$ refers to the time immediately before the application of the input. By the definition of Laplace transform, and noting the causality of the system, we write

$$U(s) = \mathcal{L}\{u(t)\} = \int_0^\infty e^{-st}u(t)\,\mathrm{d}t; \quad Y(s) = \mathcal{L}\{y(t)\} = \int_0^\infty e^{-st}y(t)\,\mathrm{d}t. \tag{2.13}$$

The existence of the Laplace transforms requires that the integrals in Eq. (2.13) should converge to finite values for a given complex variable, $s = \sigma + i\omega$. If $U(s)$ and $Y(s)$ exist, then they are unique. It can be shown Kreyszig (1998) that a Laplace integral, $\mathcal{L}\{f(t)\}$, is finite if and only if the function, $f(t)$, is piecewise continuous (i.e. any time interval, however large, can be broken up into a finite number of sub-intervals over each of which $f(t)$ is continuous, and at the either end of each sub-interval, $f(t)$ is finite) and bounded by an exponential (i.e. there exists a positive, real constant, a, such that $\mid e^{-at}f(t) \mid$ is finite at all times). Laplace transforms of some commonly encountered functions are listed in Table 2.2, along with some important properties. Practically all inputs that can be applied to a physical system are Laplace transformable, and the definition of the transfer function by Definition 2.4.1 requires that the output of a causal LTI system to such an input is also Laplace transformable.

A transfer function is often a ratio of polynomials, called a rational function, in the Laplace variable, s. Let an LTI, SISO system have the governing differential equation,

$$\frac{\mathrm{d}^n y}{\mathrm{d}t^n} + a_n\frac{\mathrm{d}^{n-1}y}{\mathrm{d}t^{n-1}} + \cdots + a_2\frac{\mathrm{d}y}{\mathrm{d}t} + a_1 y$$
$$= b_{m+1}\frac{\mathrm{d}^m u}{\mathrm{d}t^m} + b_m\frac{\mathrm{d}^{m-1}u}{\mathrm{d}t^{m-1}} + \cdots + b_2\frac{\mathrm{d}u}{\mathrm{d}t} + b_1 u, \tag{2.14}$$

then the transfer function is the following:

$$G(s) = \frac{b_{m+1}s^m + b_m s^{m-1} + \cdots + b_2 s + b_1}{s^n + a_n s^{n-1} + \cdots + a_2 s + a_1} = \frac{N(s)}{D(s)}. \tag{2.15}$$

Here, n is the order of the system. If the degree of the numerator polynomial, $N(s)$, is either less than, or equal to that of the denominator polynomial, $D(s)$, $(m \leq n)$, then the LTI system is said to be proper. If $m < n$, then there is no direct connection between input and output, and the system is called strictly proper. The roots of the denominator polynomial, $D(s) = s^n + a_n s^{n-1} + \cdots + a_2 s + a_1 = 0$, are called the poles of the transfer function and they play a crucial role in the system's characteristics, such as stability, response to desired inputs and

Table 2.2 Basic Laplace transforms

Function	Laplace Transform
$f(t)$	$F(s) = \mathcal{L}\{f(t)\}$
$\delta(t) = \begin{cases} \infty & (t = 0) \\ 0 & (t \neq 0) \end{cases}$	1
$u_s(t) = \begin{cases} 1 & (t \geq 0) \\ 0 & (t < 0) \end{cases}$	$\frac{1}{s}$
$tu_s(t)$	$\frac{1}{s^2}$
$t^n u_s(t)$, $(n = 1, 2, \cdots)$	$\frac{n!}{s^{n+1}}$
$e^{at} u_s(t)$	$\frac{1}{s-a}$
$\sin(\omega t) u_s(t)$	$\frac{\omega}{s^2 + \omega^2}$
$\cos(\omega t) u_s(t)$	$\frac{s}{s^2 + \omega^2}$
$e^{at} h(t)$	$H(s-a)$; $H(s) = \mathcal{L}\{h(t)\}$
$h(t-a) u_s(t-a)$	$e^{-as} H(s)$; $H(s) = \mathcal{L}\{h(t)\}$
$-th(t)$	$\frac{dH(s)}{ds}$; $H(s) = \mathcal{L}\{h(t)\}$
$\frac{h(t)}{t}$	$\int_s^{\infty} H(p)\,dp$; $H(s) = \mathcal{L}\{h(t)\}$
$\frac{dh(t)}{dt}$	$sH(s) - h(0_-)$; $H(s) = \mathcal{L}\{h(t)\}$
$\int_0^t h(\tau)\,d\tau$	$\frac{H(s)}{s}$; $H(s) = \mathcal{L}\{h(t)\}$

robustness. The roots of the numerator polynomial of the transfer function, $N(s) = b_{m+1}s^m + b_m s^{m-1} + \cdots + b_2 s + b_1 = 0$, are called the zeros of the system and they have an influence on the system's response to applied inputs.

The impulse response of an LTI, SISO system, $g(t)$, is the inverse Laplace transform of the transfer function with zero initial condition,

$$g(t) = \mathcal{L}^{-1}\{G(s)\} \quad (t \geq 0). \tag{2.16}$$

Any two signals, $u(t), g(t)$ that satisfy the convolution property given by Eq. (2.11) are said to convolve with one another and the Laplace transform of the convolution integral, $y(t) = (u * g)(t)$, is a product of the Laplace transforms of the two functions, $Y(s) = G(s)U(s)$. The following partial-fraction expansion of the transfer function is used to give the impulse response of an LTI, SISO system through Table 2.2:

$$\begin{aligned} G(s) = {} & \frac{r_1}{s-p_1} + \frac{r_2}{s-p_2} + \cdots + \frac{r_n}{s-p_n} \\ & + \frac{r_{m1}}{s-p_m} + \frac{r_{m2}}{(s-p_m)^2} + \cdots + \frac{r_{m(k-1)}}{(s-p_m)^{k-1}} + \frac{r_{mk}}{(s-p_m)^k}, \end{aligned} \tag{2.17}$$

where $p_1, p_2, \ldots, p_n$ are n distinct poles and p_m is a pole of multiplicity k, of $G(s)$. Here, the numerator coefficients of the series, $r_1, r_2, \ldots, r_n$, and $r_{m1}, \ldots, r_{mk}$, are called the residues of

$G(s)$ and are determined as follows:

$$\begin{aligned}
r_1 &= (s - p_1)\, G(s)|_{s=p_1} \\
r_2 &= (s - p_2)\, G(s)|_{s=p_2} \\
r_n &= (s - p_n)\, G(s)|_{s=p_n} \\
r_{mk} &= (s - p_m)^k G(s)|_{s=p_m} \\
r_{m(k-1)} &= \left.\frac{\mathrm{d}(s - p_m)^k G(s)}{\mathrm{d}s}\right|_{s=p_m} \\
r_{m2} &= \left.\frac{\mathrm{d}^{k-2}(s - p_m)^k G(s)}{\mathrm{d}s^{k-2}}\right|_{s=p_m} \\
r_{m1} &= \left.\frac{\mathrm{d}^{k-1}(s - p_m)^k G(s)}{\mathrm{d}s^{k-1}}\right|_{s=p_m} .
\end{aligned} \tag{2.18}$$

In case of complex poles (which always occur in conjugate pairs), the residues are also complex conjugates; hence, the partial fraction expansion involving a complex conjugate pair of poles can be combined to produce a quadratic (or second-order) subsystem.

The response to the unit step and unit ramp functions is related to the transfer function. From the integral property of Laplace transform (Table 2.2), it follows that the unit step function, u_s, is the time integral of the unit impulse function, $\delta(t)$,

$$u_s(t) = \mathcal{L}^{-1}\left\{\frac{1}{s}\right\} = \int_0^t \delta(\tau)\,\mathrm{d}\tau \quad (t > 0). \tag{2.19}$$

Similarly, the indicial (or step) response, $y_s(t)$, of an LTI system, $G(s)$, defined as the output when the input is a unit step function applied at $t = 0$ with a zero initial condition, is the time integral of the impulse response:

$$y_s(t) = \mathcal{L}^{-1}\left\{\frac{G(s)}{s}\right\} = \int_0^t g(\tau)\,\mathrm{d}\tau \quad (t > 0). \tag{2.20}$$

The indicial response is valuable in studying a stable LTI system's performance when a sudden change is desired in the output, and can be derived in a closed-form by partial fraction expansion. The time integral of the indicial response is the ramp response, which is useful in such applications as tracking an object moving with a constant velocity.

For MIMO systems, the concept of transfer function is extended to the transfer matrix.

Definition 2.4.2 *If the Laplace transform of the output vector of an LTI system exists and is denoted by $Y(s)$, and that of the input vector is $U(s)$, then the transfer matrix, $G(s)$, of the system is given by the relationship*

$$Y(s) = G(s)\,U(s).$$

Clearly, the transfer matrix is the Laplace transform of the impulse response matrix, with zero initial conditions.

2.5 Loop Shaping of Linear Control Systems

Consider a LTI control system shown in the block-diagram of Fig. 2.4, with the plant transfer matrix, $G(s)$. The reference signal vector, $r(s)$, acts as the input to the control system, which automatically generates the plant input vector, $u(s)$, in response to the plant output vector, $y(s)$. The connection between the output back to the plant input (control input) is the feedback path (or feedback loop). A pre-filter of transfer matrix $F(s)$ is placed before the feedback loop, while a controller of transfer matrix $H(s)$ is part of the feedback loop. Sometimes, the controller is placed in the path between the output and the summing junction. Owing to the linearity of the control system, this (and other similar) variation in the block-diagram is easily handled by suitably modifying the linear relationships between the various signals, leading to equivalent transfer matrices of the overall system. Such linear operations are called block transformations. The control system is subject to unknown disturbance signals, $p(s)$ and $m(s)$, as shown in Fig. 2.4. The process noise vector, $p(s)$, occurs due to unknown parametric variations, or modelling errors of the plant dynamics, while the measurement noise, $m(s)$, refers to the vector of unknown errors in measuring the output signals. If these unknown disturbances were absent, there would be no need to have a feedback loop. However, as the disturbances are always present in a real system, a feedback loop is necessary for continuously updating the control input signals on the basis of the sensed outputs. The determination of the transfer matrices $F(s), H(s)$ in order to meet desirable properties of stability, performance and robustness of the overall control system in the presence of disturbance inputs is called loop shaping.

From the block diagram of Fig. 2.4, it is evident that

$$y(s) = G(s) H(s) [F(s) r(s) - y(s) - m(s)] + p(s), \tag{2.21}$$

or

$$y(s) = J(s) r(s) - T(s) m(s) + S(s) p(s), \tag{2.22}$$

where

$$S(s) = [I + G(s) H(s)]^{-1} = R_o^{-1}(s) \tag{2.23}$$

is called the sensitivity matrix,

$$R_o(s) = [I + G(s) H(s)] \tag{2.24}$$

is the output return-difference matrix,

$$T(s) = [I + G(s) H(s)]^{-1} G(s) H(s) \tag{2.25}$$

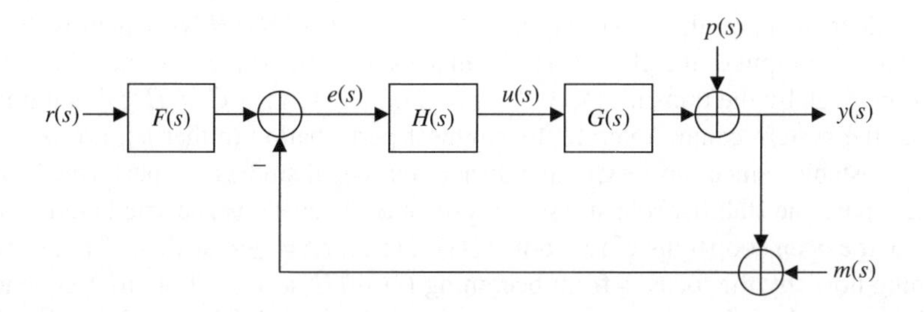

Figure 2.4 A linear control system with plant, $G(s)$, controller, $H(s)$, and pre-filter, $F(s)$

is the closed-loop transfer matrix (or the transmission matrix), and

$$J(s) = [I + G(s)H(s)]^{-1}G(s)H(s)F(s) = T(s)F(s) \tag{2.26}$$

is the overall transfer matrix of the control system. The control input vector can be expressed as follows:

$$u(s) = R_i^{-1}(s)H(s)[F(s)r(s) - m(s) - p(s)], \tag{2.27}$$

where

$$R_i(s) = [I + H(s)G(s)] \tag{2.28}$$

is called the input return-difference matrix. Since the effect of the disturbance inputs, $p(s), m(s)$, on the control input is proportional to the size of $R_i^{-1}(s)$, the latter is sometimes called the input sensitivity matrix. For SISO systems, there is no difference between $R_i(s)$ and $R_o(s)$.

The ability of the system's outputs to quickly and accurately track the reference signals is determined by the overall transfer matrix $J(s)$, which is the product of $T(s)$ and $F(s)$. The sensitivity of the system to process noise is determined by the matrix $S(s)$, while $T(s)$ influences the transmission of the measurement noise to the system's outputs (hence its performance). For robustness with respect to process noise and the rejection of measurement noise, the gains of $S(s)$ and $T(s)$ must both be small. This is difficult to achieve, because of the following relationship:

$$S(s) + T(s) = I, \tag{2.29}$$

which implies a reduction of sensitivity gain would automatically lead to a larger transmission gain, and vice versa. Because of this complementarity condition, $T(s)$ is referred to as the complementary sensitivity matrix. Achieving a trade-off between sensitivity and transmission is one of the major loop-shaping problems. Since both $S(s)$ and $T(s)$ depend upon the subsystems $G(s), H(s)$, which are connected by the feedback loop, and are independent of the pre-filter, $F(s)$, the tasks of designing $H(s)$ and $F(s)$ can be carried out separately. The selection of the pre-filter $F(s)$ is based upon the speed with which a reference signal, $r(t)$, can be tracked with a small error in the presence of the disturbance signals, $p(t), m(t)$. However, as all ASE systems are of the regulator type (i.e. having $r(t) = 0$), only the feedback design is considered here, and the main interest is in studying the stability margin (or stability robustness) of such a system with respect to the unmodelled perturbations, $p(s)$ and $m(s)$.

2.5.1 Nyquist Theorem

For a nominal SISO linear system with a feedback loop as shown in Fig. 2.5, the stability robustness depends upon the sensitivity matrix, $R(s) = I + G(s)H(s)$, which is the same at the input and the output of the plant. If $R(s)$ vanishes for some value of s, then the disturbance propagation given by the transmissivity, $T(s) = G(s)H(s)/[1 + G(s)H(s)]$ is infinite, indicating that the system cannot tolerate the slightest perturbation (either $p(s)$ or $m(s)$) before becoming unstable. Since most experiments and analytical studies are performed for simple harmonic inputs, the stability robustness analysis of SISO systems is carried out by analysing the locus of the open-loop transfer function, $G(s)H(s)$, for $s = \pm i\omega$ with the frequency, ω, and determining how shy the locus is from becoming $G(i\omega)H(i\omega) = -1$ at any frequency. Such a plot of the open-loop frequency response function $G(i\omega)H(i\omega)$ is called the Nyquist locus and is the mapping of the imaginary axis of the s-plane (Laplace domain) on to the $G(s)H(s)$

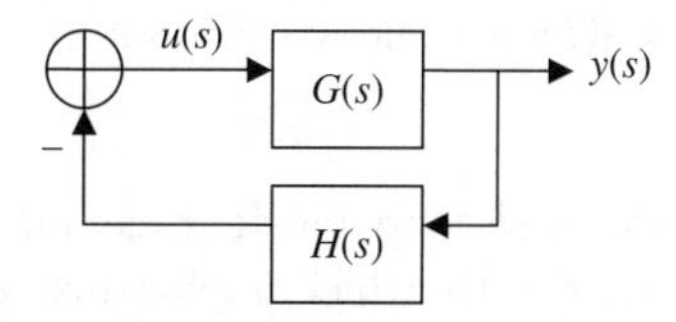

Figure 2.5 A linear, SISO control system with plant $G(s)$ and controller $H(s)$

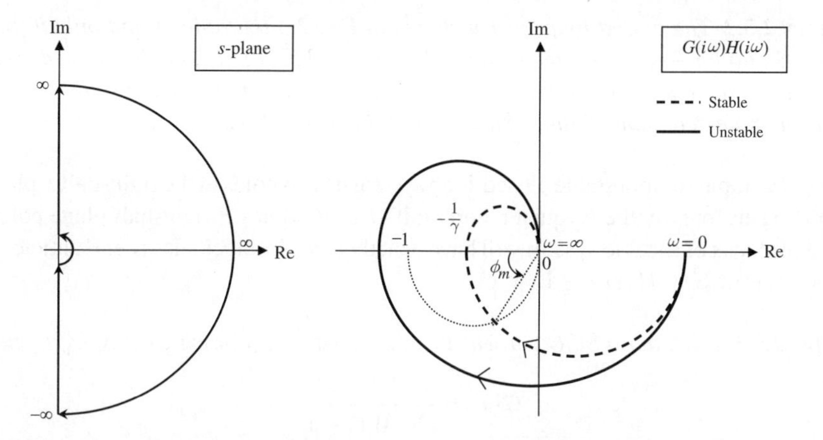

Figure 2.6 Nyquist locii of a linear SISO control system in a feedback configuration of Fig 2.5, with no poles of the open-loop transfer function, $G(s)H(s)$, in the right-half s-plane

plane. The region enclosed by the Nyquist locus is the mapping of the entire right-half s-plane (Fig. 2.6), which can be regarded as the entire region covered by the curve $s = i\omega$ as ω is varied from $-\infty$ to ∞. This is the region falling to the right of the entire imaginary axis, and excluding the origin by a small circle, $s = \sigma + j\sigma$, where σ is a vanishingly small positive number. A practical open-loop transfer function, $G(s)H(s)$, is proper, hence its Nyquist locus is symmetrical about the real axis, and one has to consider only half of the locus (i.e. ω increasing from 0 to ∞), with the direction of increasing frequency marked by an arrow (Fig. 2.6). If there are no cancellations of the poles and zeros in the open-loop transfer function, then closed-loop stability is determined by the characteristic equation,

$$1 + G(s)H(s) = 0. \tag{2.30}$$

If any of the roots of the characteristic equation lie in the right-half s-plane, then the closed-loop system is unstable. Therefore, the point $G(s)H(s) = -1$ assumes a special significance in the closed-loop stability analysis. For determining the range of stability (stability robustness) of the closed-loop system as the open-loop transfer is varied, one can apply the following Nyquist stability theorem.

Theorem 2.5.1 *The closed-loop system of Fig. 2.5 has Z unstable poles, if and only if the locus of* $G(i\omega)H(i\omega)$ *for* $-\infty < \omega < \omega$ *encircles the point* $(-1, 0)$ *in the clockwise direction exactly* $N = Z - P$ *times, where P is the number of poles of the open-loop transfer function* $G(s)H(s)$ *in the right-half s-plane, provided no pole-zero cancellations have occurred in* $G(s)H(s)$.

Proof. The proof can be derived from Cauchy's theorem of complex analysis (D'Azzo and Houpis 1966).

For many practical systems, the open-loop transfer function, $G(s)H(s)$, does not have any poles in the right-half s-plane (i.e. $P = 0$), which implies that any clockwise encirclement of the point $G(s)H(s) = -1$ by the Nyquist locus means an unstable closed-loop system. This fact is illustrated by Fig. 2.6.

Corollary 2.5.2 *The closed-loop system shown in Fig. 2.5 is stable, if and only if the locus of $G(i\omega)H(i\omega)$ for $-\infty < \omega < \omega$ encircles the point $(-1, 0)$ in the anticlockwise direction as many times as the open-loop transfer function $G(s)H(s)$ has poles in the right-half s-plane, provided no pole-zero cancellations have occurred in $G(s)H(s)$.*

Proof. An input–output stable closed-loop system has no poles in the right-half s-plane, that is, $Z = 0$. Therefore, by the Nyquist theorem, if $G(s)H(s)$ has P right-half plane poles, then the Nyquist locus of a stable system will have exactly $N = -P$ clockwise (i.e. P anticlockwise) encirclements of $G(s)H(s) = -1$.

Example 2.5.3 *Consider a SISO aeroelastic system with the following transfer function:*

$$G(s) = \frac{1}{s^2 + 0.1s - 1},$$

which is input–output unstable (a pole in the right-half s-plane). In order to stabilize the system, a controller $H(s) = 2$ is added in the configration of Fig. 2.5, resulting in the open-loop transfer function

$$G(s)H(s) = \frac{2}{s^2 + 0.1s - 1}.$$

Now, while $G(s)H(s)$ has a pole in the right-half s-plane ($s = 0.9512$), its Nyquist plot (Fig. 2.7) has exactly one anticlockwise encirclement ($N = -1$) of the point $(-1, 0)$. Therefore, by the Nyquist theorem, the closed-loop system is stable ($Z = P + N = 1 - 1 = 0$) This can be verified from the roots of the characteristic equation,

$$1 + G(s)H(s) = s^2 + 0.1s + 1,$$

which are $s_{1,2} = -0.0500 \pm j0.9987$.

2.5.2 Gain and Phase Margins

The Nyquist locus of a LTI, SISO feedback control system is determined by the following open-loop frequency response function:

$$G(i\omega)H(i\omega) = \mid G(i\omega)H(i\omega) \mid \; e^{j\phi}. \tag{2.31}$$

The main utility of the Nyquist stability theorem is for finding out how much the loop can gain, $\mid G(i\omega)H(i\omega) \mid$, and the loop phase, ϕ, be varied before the closed-loop system becomes unstable. These stability margins are referred to as the gain and phase margins, respectively, and their sizes give a measure of the stability robustness of the control system. Referring to

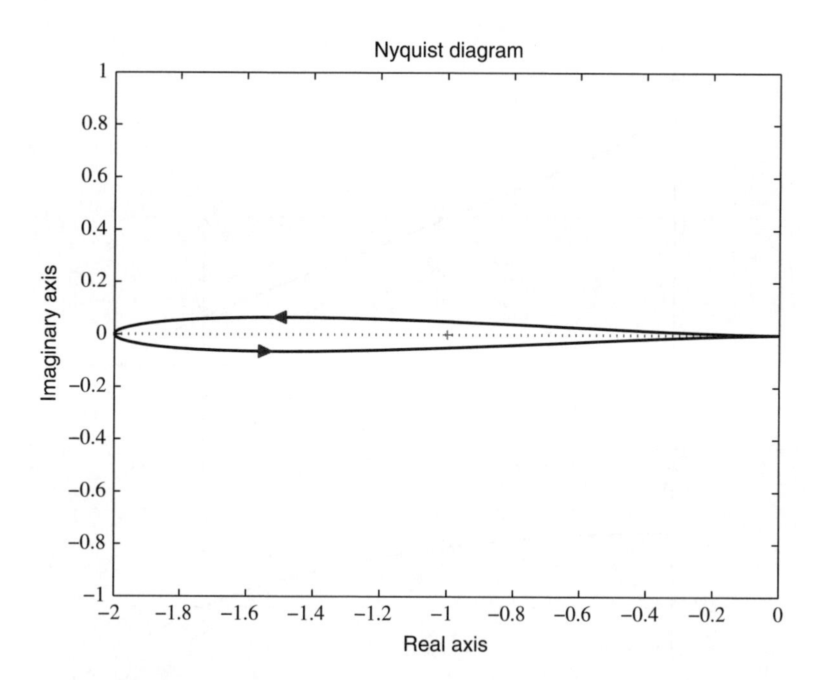

Figure 2.7 The Nyquist locus of $G(s)H(s) = 2/\left(s^2 + 0.1s - 1\right)$, showing one anticlockwise encirclement of the point $G(s)H(s) = -1$

the stable system of Fig. 2.6 (dashed line), it is evident that the gain margin is the factor, γ, by which $\mid G(i\omega)H(i\omega)\mid$ can be increased at the point of crossing the negative real axis (i.e. $\phi = -180°$) before it hits the point $(-1, 0)$. Similarly, the phase margin, ϕ_m, is the angle by which the phase, ϕ, can be increased before reaching $\phi = -180°$ (negative real axis). As shown in Fig. 2.6, ϕ_m is obtained by subtracting $-180°$ from the phase at the crossover frequency when the Nyquist locus crosses the unit circle (dotted line), $\mid G(i\omega)H(i\omega)\mid = 1$ centred at the origin. The gain and phase margins are easily identified on a Bode plot (Fig. 2.8) of the loop frequency response function $G(i\omega)H(i\omega)$. Recall that owing to the linearity of the closed-loop system, the controller can be placed at any point inside the feedback loop (before or after the summing junction in Fig. 2.5), and the Nyquist stability analysis would still be valid.

For a given plant $G(s)$, the controller transfer function, $H(s)$, must be adjusted until it leads to acceptable gain and phase margins for the control system. A practical approach to the problem of loop shaping of SISO systems is to take the following controller transfer function, called a compensator (Tewari 2002):

$$H(s) = K\frac{\alpha(1 + \tau s)}{1 + \alpha\tau s}, \tag{2.32}$$

where the real positive constants, K, α, τ, are the design parameters to be determined from the desired robustness properties. For example, by choosing $\alpha < 1$, the compensator speeds up the closed-loop dynamics, thereby increasing the phase margin, while for $\alpha > 1$, the phase is lagged by slowing down the closed-loop dynamics, thereby increasing the gain margin. The change of phase occurs between the corner frequencies, $\omega = 1/\tau$ and $\omega = 1/(\alpha\tau)$, with the maximum phase change occurring at $\omega = 1/\left(\tau\sqrt{\alpha}\right)$ (which appears at the middle of the two

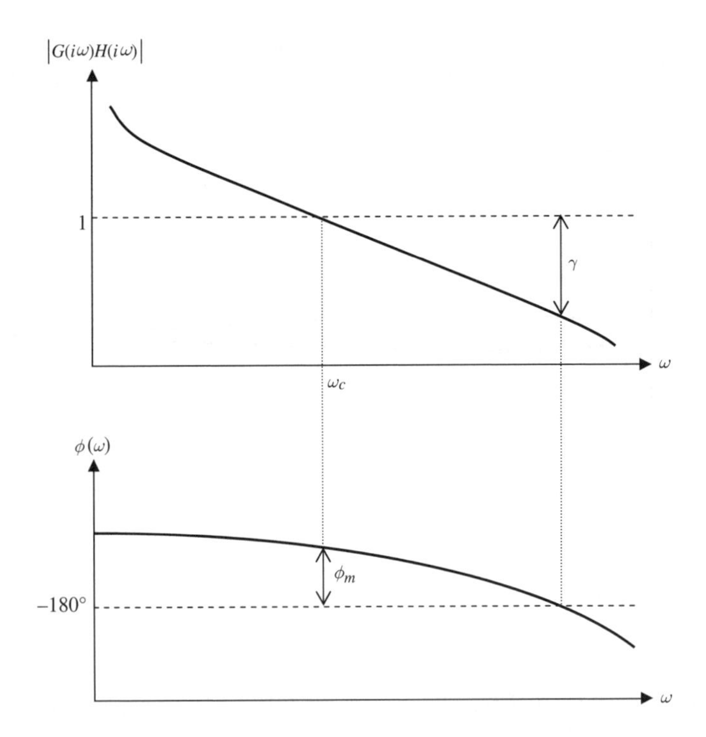

Figure 2.8 Bode plot of the loop transfer function $G(s)H(s)$ showing gain margin, γ, and phase margin, ϕ_m, of a closed-loop system

corner frequencies in a logarithmic Bode plot). The phase lead and lag effects are mutually exclusive due to a single possible choice of α. However, the phase margin in the lag compensator ($\alpha > 1$) can be increased by suitably increasing the controller gain, K, thereby offering some compromise between low-frequency and high-frequency behaviour. This is depicted by the dashed curve in Fig. 2.9.

2.5.3 Loop Shaping for Single Variable Systems

The sensitivity and complementary sensitivity of a SISO system are respectively the following:

$$S(s) = \frac{1}{1 + G(s)H(s)}, \quad T(s) = \frac{G(s)H(s)}{1 + G(s)H(s)}, \tag{2.33}$$

which implies $S(s) + T(s) = 1$. By having phase-lead compensation, the sensitivity to process noise, $|S(s)|$, is reduced, while the transmission of measurement noise frequencies, $|T(s)|$, is increased. The converse is true for a phase-lag compensator. By choosing a combination of phase-lead and phase-lag compensators, it is possible to reduce both sensitivity, $|S(s)|$, and complementary sensitivity, $|T(s)|$, in different frequency ranges, as required (Tewari 2002). In all practical systems, the process noise has its spectrum concentrated at low frequencies, while the measurement noise occurs in a much wider frequency spectrum. Therefore, it is possible to reduce noise sensitivity by choosing $|S(i\omega)|$ to be small in a certain frequency

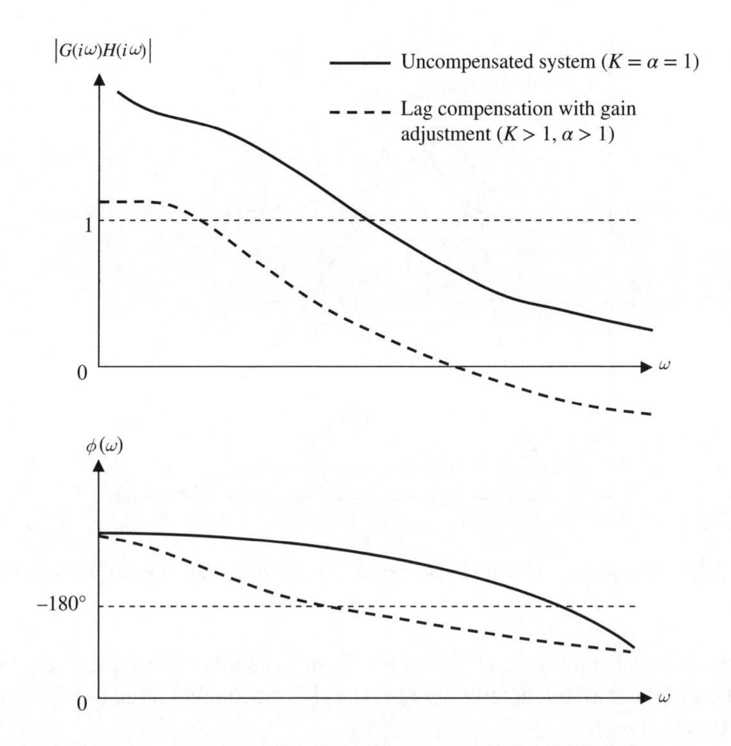

Figure 2.9 Bode plot of the loop transfer function $G(s)H(s)$ with (and without) lag compensation and gain adjustment

range, $0 \leq \omega \leq \omega_0$, while having $|T(i\omega)|$ small at higher frequencies, $\omega > \omega_b$. Usually, ω_b is the system's bandwidth defined by

$$|T(i\omega)| = \frac{|T(0)|}{\sqrt{2}}. \tag{2.34}$$

For making $|S(i\omega)|$ small at low frequencies, it is required that $|G(i\omega)H(i\omega)|$ is large; this can be ensured by making $|H(i\omega)|$ large at low frequencies. However, as all physical plants have the property of declining gain at large frequencies, $|G(i\omega)| \rightarrow 0$ as $\omega \rightarrow \infty$, the requirement of keeping $|T(i\omega)|$ small at high frequencies is easily satisfied by having $|H(i\omega)|$ declining at a given rate (or rolling off) with the frequency. Such a design is illustrated by the gain plot of Fig. 2.10.

The minimization of the complementary sensitivity in a large frequency range gives the additional advantage of minimization of the net control effort. By Parseval's theorem, the net control effort (or energy) of a stable feedback control system in response to an initial disturbance (either $p(t)$ or $m(t)$ in Fig. 2.4) is proportional to the following integral:

$$\int_0^\infty u^2(t)\,\mathrm{d}t = \frac{1}{2\pi}\int_{-\infty}^\infty |U(i\omega)|^2 \mathrm{d}\omega, \tag{2.35}$$

where $U(s)$ is the Laplace transform of $u(t)$. Since $|U(i\omega)|$ is proportional to

$$\left|\frac{H(i\omega)}{1 + G(i\omega)H(i\omega)}\right| = \frac{|T(i\omega)|}{|G(i\omega)|}, \tag{2.36}$$

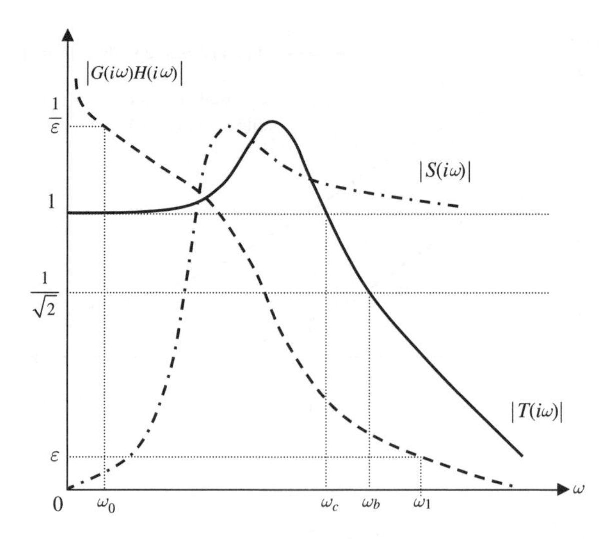

Figure 2.10 Sensitivity and complementary sensitivity of a stable feedback control system

the net control energy is minimized by keeping $\mid T(i\omega) \mid$ small at all frequencies, with the possible exception of those frequencies where $\mid G(i\omega) \mid$ is reasonably large (such as the system's bandwidth, $0 \leq \omega \leq \omega_b$).

The choice of the loop gain, $\mid G(i\omega) H(i\omega) \mid$, for satisfying the conflicting sensitivity, complementary sensitivity and stability robustness requirements can be summarized as follows:

(a) $\mid S(i\omega) \mid < \epsilon$, $(\epsilon \ll 1)$, implies $\mid G(i\omega) H(i\omega) \mid > 1/\epsilon$, for $0 \leq \omega < \omega_0$.
(b) $\mid T(i\omega) \mid < \epsilon$, $(\epsilon \ll 1)$, implies $\mid G(i\omega) H(i\omega) \mid < \epsilon$, for $\omega > \omega_1$.
(c) Certain minimum gain and phase margins must be achieved by the graph of $G(i\omega) H(i\omega)$.

The frequencies ω_0, ω_1 for requirements (a) and (b) are chosen such that the gap between them is the smallest (the ideal gap would be zero, but this is practically impossible to achieve). The result is a compromise between (a) and (b) where the two frequencies are close to, and on either side of the crossover frequency, ω_c, implying that if $\mid S(i\omega) \mid$ cannot be made small, it should be at least close to unity so that $\mid T(i\omega) \mid$ can be made sufficiently small. An alternative to (c) is the requirement that the Nyquist locus $G(i\omega) H(i\omega)$ must remain outside a certain neighbourhood of the point $-1 + i0$, which is specified by an M circle of a certain radius centred at $-1 + i0$. An M circle is the locus of points in the complex z plane such that

$$\left| \frac{z}{z+1} \right| = M.$$

2.5.4 *Singular Values*

For multi-variable systems, the direct application of frequency domain methods (such as the Nyquist stability theorem) is possible only for square systems (those with equal number of inputs and outputs) (Maciejowski 1989). Clearly, such an approach cannot be applied to general ASE systems. However, the concepts of loop-shaping are readily extended to non-square MIMO systems through the concept of principal gains (singular values).

Definition 2.5.4 *Singular values (or principal gains) of the frequency response matrix, $G(i\omega)$, of a LTI system, $y = Gu$, at a given frequency, ω, are the positive square roots of the eigenvalues of the following square matrix:*

$$G^T(-i\omega)\,G(i\omega).$$

The computation of singular values is carried out by singular value decomposition (SVD) as follows:

$$G(i\omega) = U(i\omega)\,\Sigma V^T(-i\omega), \tag{2.37}$$

where U, V are unitary complex matrices satisfying

$$U(i\omega)\,U^T(-i\omega) = I; \qquad V(i\omega)\,V^T(-i\omega) = I \tag{2.38}$$

and the diagonal matrix Σ contains the singular values of $G(i\omega)$, as its diagonal elements, denoted by

$$\sigma_k\{G(i\omega)\}; \qquad k = 1, 2, \dots, n.$$

The largest among all the singular values is denoted by $\bar{\sigma}$, while the smallest is indicated by σ.

Definition 2.5.5 *The spectral (or Hilbert) gain of a LTI system, $y = Gu$, is the largest singular value of the transfer matrix, given by the Hilbert norm*

$$\mid G|_S = \bar{\sigma}(G) = \max_i \sqrt{\lambda_i\left(G^H G\right)} = \sup_{u\neq 0}\frac{\mid Gu\mid}{\mid u\mid}.$$

Definition 2.5.6 *The smallest singular value of the transfer matrix, G, where $y = Gu$, is given by*

$$\sigma(G) = \min_i \sqrt{\lambda_i\left(G^H G\right)} = \inf_{u\neq 0}\frac{\mid Gu\mid}{\mid u\mid}.$$

The spectral gain, $|A|_S = \bar{\sigma}(A)$, indicates the largest possible 'size' (*not* the dimensions) of a square matrix signal, A, while the smallest singular value, $\sigma(A)$, indicates how close the square matrix, A, is to being singular ($\sigma(A) = 0$ implies a singular matrix A). The ratio $\mathrm{cond}(A) = \bar{\sigma}(A)/\sigma(A)$, which is always greater than unity, indicates the condition number of the square matrix, A. The larger the condition number, the closer is the matrix A to being singular.

A frequency spectrum of the singular values indicates the variation of the 'magnitude' of the frequency response matrix, which is supposed to lie between $\sigma(\omega)$ and $\bar{\sigma}(\omega)$. Similarly to SISO systems, the range of frequencies $\left(0 \leq \omega \leq \omega_b\right)$ over which $\bar{\sigma}(\omega)$ stays above $0.707\bar{\sigma}(0)$ is called the system's bandwidth and is denoted by the frequency, ω_b. The bandwidth indicates the highest frequency signal to which the system has an appreciable response. For higher frequencies $\left(\omega > \omega_b\right)$, the singular values of a strictly proper system generally decline rapidly with increasing frequency (roll off). A system with a steep roll off of the spectral gain, $\bar{\sigma}$, has a reduced sensitivity to high-frequency noise, which is a desirable property. The choice of the loop principal gains, $\bar{\sigma}\{G(i\omega)H(i\omega)\}$ and $\bar{\sigma}\{G(i\omega)H(i\omega)\}$, for satisfying the conflicting

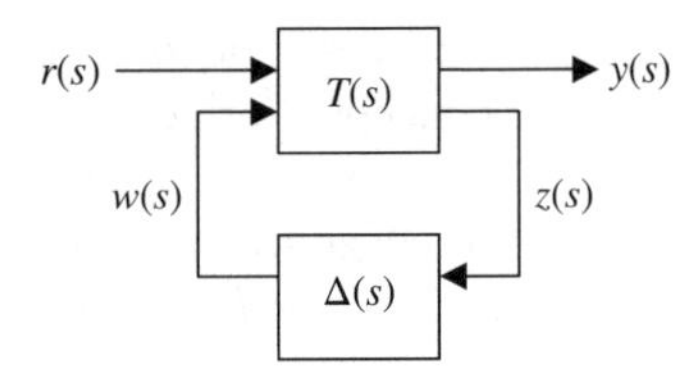

Figure 2.11 Block uncertainty representation of a feedback control system

sensitivity, complementary sensitivity and control energy requirements, can be summarized in terms of the singular values as follows:

(a) $\bar{\sigma}\{S(i\omega)\} < \epsilon$, ($\epsilon \ll 1$), implies $\sigma\{G(i\omega)H(i\omega)\} > 1/\epsilon$, for $0 \le \omega < \omega_0$.
(b) $\bar{\sigma}\{T(i\omega)\} < \epsilon$, ($\epsilon \ll 1$), implies $\bar{\sigma}\{G(i\omega)H(i\omega)\} < \epsilon$, for $\omega > \omega_1$.
(c) For minimum control energy, $\bar{\sigma}\{H(i\omega)\}$ must be as small as possible in the controller bandwidth.

The highest magnitude achieved by the largest singular value (spectral gain) over the system bandwidth is the H_∞ norm of the transfer matrix, given by

$$||\mathbf{G}||_\infty = \sup_\omega[\bar{\sigma}\{\mathbf{G}(i\omega)\}], \tag{2.39}$$

where $\sup_\omega(.)$ is the supremum (the maximum value) with respect to the frequency. We note that the different norms ($| G|_S$, $||\mathbf{G}||$, $||\mathbf{G}||_\infty$) give alternative scalar measures of the transfer matrix, each of which could be applied to determine a system's stability margins.

2.5.5 *Multi-variable Robustness Analysis: Input–Output Model*

While the stability robustness of SISO systems is indicated by their gain and phase margins, a multi-variable MIMO system requires a more sophisticated treatment due to the matrix operations inherent in its input–output behaviour. The unknown variations (or uncertainty) in the plant transfer matrix, $G(s)$, and controller transfer matrix, $H(s)$, give rise to uncertainty in the overall transfer matrix, $T(s)$, of the control system, which can be schematically represented as shown in Fig. 2.11. Here, a block uncertainty matrix, $\Delta(s)$, appears in closed-loop with the control system through certain output variables, $z(s)$, and reference inputs vector, $w(s)$. It is as if all parametric uncertainties of the control system have been pulled out of it, and collected into a feedback transfer matrix, $\Delta(s)$. The loop transfer matrix is thus given by $\Delta(s)T(s)$ and the return difference matrix at control system's input is $I + \Delta(s)T(s)$. Similarly to the Nyquist analysis of SISO systems, the stability of the uncertain system depends upon the return difference matrix.

Further discussion of multi-variable robust design is taken up later in the chapter.

2.6 State-Space Representation

Often the input–output relationships do not give a complete description of a system's behaviour, which is actually based upon the internal structure of the system. An example of incomplete description by input–output representation is when there is cancellation of a pole with a zero in the transfer function, which gives misleading information about the system's characteristics. The condition, or internal status, of a system at a given time is specified by a set of real signals, called state variables, which are collected into the state vector. The vector space spanned by the state vector is called a state-space. The state of a system is thus defined as a collection of the smallest number of variables necessary to completely specify the system's evolution in time, in the absence of external inputs. The number of state variables necessary to represent a system is called order of the system, because it is equal to the net order of differential equations governing the system. While the size of the state-space (i.e. order of a system) is unique, any given system can be described by infinitely many, alternative state-space representations. A general model of a system can be expressed in terms of a set of first-order, ordinary differential equations called the state equations, such as the following:

$$\dot{x} = f(x, u, p, t), \tag{2.40}$$

where $x(t) : \mathbb{R} \to \mathbb{R}^n$ is the state vector, $u(t) : \mathbb{R} \to \mathbb{R}^m$ is the control input vector, and $p(t) : \mathbb{R} \to \mathbb{R}^\ell$ is the process noise vector. The vector function, $f(.) : \mathbb{R}^n \times \mathbb{R}^m \times \mathbb{R}^\ell \times \mathbb{R} \to \mathbb{R}^n$, is assumed to possess partial derivatives with respect to x, p and u in the neighbourhood of a nominal trajectory, $x_n(t), t_i \le t \le t_f$, which is a special solution of the state equation. A practical choice of the nominal trajectory is the one which satisfies Eq. (2.40) for the unforced case, that is, for $u(t) = 0, p(t) = 0$:

$$\dot{x}_n = f\left(x_n, 0, 0, t\right) \quad t_i \le t \le t_f, \tag{2.41}$$

where $t_i \le t \le t_f$ is called the control interval with initial time, t_i, and final time, t_f.

The output variables of a plant, $y(t) : \mathbb{R} \to \mathbb{R}^r$, result from either direct or indirect measurements related to the state variables and control inputs through sensors. Certain errors can arise in the measurement process due to sensor imperfections (the measurement noise), which are denoted by $m(t) : \mathbb{R} \to \mathbb{R}^q$. Therefore, the output vector is related to the state vector, the control input vector and the measurement noise vector by the following output equation:

$$y = h(x, u, m, t), \tag{2.42}$$

where $h(.) : \mathbb{R}^n \times \mathbb{R}^m \times \mathbb{R}^q \times \mathbb{R} \to \mathbb{R}^n$ is a functional possessing continuous derivatives with respect to x, m and u in the neighbourhood of the nominal trajectory, $x_n(t), t_i \le t \le t_f$. Most practical systems have $q = r$. The disturbances, $p(t), m(t)$, are often treated as outputs of separate systems having a stochastic behaviour, which cannot be modelled by state equations. When the disturbance inputs are absent, the system's state, $x(t)$, and output variables, $y(t)$, are uniquely determined from the initial conditions, $x\left(t_i\right) = x_0$, and a prescribed control input history, $u(t), t_i \le t \le t_f$. Such a system is said to be deterministic in nature. When $f(.)$ and $h(.)$ do not explicitly depend upon the time, the system is said to be time-invariant (or autonomous) and without loss of generality, the control interval can be taken as $0 \le t < \infty$. The ASE systems are generally time-invariant, hence our discussion will be focussed on time-invariant systems.

2.6.1 State-Space Theory of Linear Systems

The theory of linear systems refers to the mathematical framework of a discretized (finite dimensional) system linearized about a particular solution (Kailath 1980).

Definition 2.6.1 *A deterministic, time-invariant system with state vector, $x(t)$, and input vector, $u(t)$, governed by state equations,*

$$\dot{x} = f(x, u), \tag{2.43}$$

initial conditions,

$$x(0) = x_0, \tag{2.44}$$

and output equation,

$$y = h(x, u), \tag{2.45}$$

is said to be linear if its output vector resulting from the applied input vector,

$$u(t) = c_1 u_1(t) + c_2 u_2(t), \tag{2.46}$$

is given by

$$y(t) = c_1 y_1(t) + c_2 y_2(t), \tag{2.47}$$

where $y_1(t)$ and $y_2(t)$ are the output vectors of the system to the inputs $u_1(t)$ and $u_2(t)$, respectively, and c_1, c_2 are arbitrary scalar constants.

By inspecting the governing equations of a system, Eqs.(2.43) and (2.45), it is possible to determine whether it is linear. If the functions, $f(.), h(.)$, are continuous, and do not contain nonlinear functions of the state and input variables, then the system is linear.

Let a nominal state vector, $x_n(t)$, and a corresponding reference input vector, $u_n(t)$, satisfy the system's governing vector state equation, Eq. (2.43),

$$\dot{x}_n = f\left(x_n, u_n\right), \tag{2.48}$$

subject to the initial condition,

$$x_n(0) = x_{n0}. \tag{2.49}$$

Let $\Delta x(t)$ and $\Delta u(t)$ be deviations in state and control input vectors, respectively, from the reference solution, $\left(x_n, u_n\right)$, such that the perturbed solution is given by

$$\begin{aligned} x(t) &= x_n(t) + \Delta x(t) \\ u(t) &= u_n(t) + \Delta u(t), \end{aligned} \tag{2.50}$$

subject to initial conditions, Eqs. (2.44) and (2.49). If the vector function, $f(.)$, possesses continuous derivatives with respect to state and control variables up to an infinite order at the reference solution, $\left(x_n, u_n\right)$, then the state equation can be expanded about the reference solution by neglecting the quadratic and higher order terms as follows:

$$\dot{x} - \dot{x}_n = \Delta\dot{x} = f\left(x_n + \Delta x, u_n + \Delta u\right) - f\left(x_n, u_n\right), \tag{2.51}$$

where

$$f\left(x_n + \Delta x, u_n + \Delta u\right) \simeq f\left(x_n, u_n\right) + \frac{\partial f}{\partial x} f\left(x_n, u_n\right) \Delta x + \frac{\partial f}{\partial u}\left(x_n, u_n\right) \Delta u.$$

Substitution of Eq. (2.52) into Eq. (2.51) yields the following linearized state equation about the reference solution:

$$\Delta\dot{x} = A\Delta x + B\Delta u, \tag{2.52}$$

where A and B are the following constant Jacobian matrices:

$$A = \frac{\partial f}{\partial x}\left(x_n, u_n\right)$$
$$B = \frac{\partial f}{\partial u}\left(x_n, u_n\right). \quad (2.53)$$

The solution to the LTI state equation,

$$\dot{x} = Ax + Bu, \quad (2.54)$$

subject to initial condition,

$$x(0) = x_0, \quad (2.55)$$

is expressed as the sum of homogeneous and particular solutions. The homogeneous solution is derived by writing for the case $u(t) = 0$,

$$\dot{x}(t) = Ax(t), \quad (2.56)$$

and

$$x(t) = \Phi(t)x_0, \qquad t \geq 0, \quad (2.57)$$

where $\Phi(t)$ is the state transition matrix for the evolution of the state from $t = 0$ to the time t, with the following properties:

Inversion:

$$\Phi(t) = \Phi^{-1}(-t). \quad (2.58)$$

Association:

$$\Phi(t) = \Phi\left(t - t_0\right)\Phi\left(t_0\right). \quad (2.59)$$

Differentiation:

$$\frac{\mathrm{d}\Phi(t)}{\mathrm{d}t} = A(t)\Phi(t). \quad (2.60)$$

The general solution to the non-homogeneous state equation, Eq. (2.54), subject to initial condition, $x_0 = x(0)$, is expressed as follows:

$$x(t) = \Phi(t)x_0 + \int_0^t \Phi(t-\tau)Bu(\tau)\,\mathrm{d}\tau, \qquad t \geq 0, \quad (2.61)$$

which can be verified by substituting into Eq. (2.54), along with the properties of $\Phi(t)$.

Definition 2.6.2 *The output (or response) of an LTI system is given by*

$$y(t) = Cx(t) + Du(t), \quad (2.62)$$

where C is called the output coefficient matrix and D the direct transmission matrix. If $D = 0$, the system is said to be strictly proper.

Substitution of Eq. (2.61) into Eq. (2.62) yields the following expression for the system's response:

$$y(t) = C\Phi(t)x_0 + \int_0^t \left[C\Phi(t-\tau)B + D\delta(t-\tau)\right]u(\tau)\,\mathrm{d}\tau, \qquad t \geq 0, \quad (2.63)$$

where $\delta(t-\tau)$ is the Dirac delta function representing a unit impulse applied at $t=\tau$. The first term on the right-hand side of Eq. (2.63) is called the initial response, while the integral term is the convolution integral, which was encountered earlier while discussing input–output models of a linear system. The convolution integral gives the system's response when the initial condition is zero ($x_0 = 0$), and is denoted by

$$y(t) = \int_0^t G(t-\tau)u(\tau)\mathrm{d}\tau, \qquad t \geq 0, \tag{2.64}$$

where $G(t-\tau)$ is the impulse response matrix, which can now be represented as follows:

$$G(t-\tau) = C\Phi(t,\tau)B + D\delta(t-\tau), \qquad t \geq \tau. \tag{2.65}$$

The indicial (step) response matrix is defined as the integral of $G(t-\tau)$ and given by

$$\mathcal{S}(t) = \int_0^t G(t-\tau)\mathrm{d}\tau, \qquad t \geq 0. \tag{2.66}$$

The element (i,j) of the step response matrix, $\mathcal{S}(t)$, is the value of the ith output variable at time t when the jth input variable is a unit step applied at time $t=0$, subject to zero initial condition, $x_0 = 0$.

The derivation of the state transition, impulse response and step response matrices for a time-varying system is usually impossible in a closed-form, except for some special cases. Generally, a linear, time-varying system's state equations are integrated by a numerical procedure in a manner similar to that employed for nonlinear systems. Thus, much of the utility of linear systems analysis is lost if the state-space coefficient matrices are time-varying.

The state transition matrix of an LTI system is often denoted by the matrix exponential as follows:

$$\Phi(t) = e^{At}, \tag{2.67}$$

where the matrix exponential, e^M, of a square matrix, M, is defined by an infinite series in a manner similar to the scalar exponential:

$$e^M \doteq I + M + \frac{1}{2}M^2 + \cdots + \frac{1}{n!}M^n + \cdots \tag{2.68}$$

Computation of e^{At} by the infinite series is impossible. Instead, either a numerical approximation, or an analytical derivation is required. The numerical approximation is carried out by breaking up the time duration, t, into many smaller intervals over each of which a finite series approximation is performed by neglecting terms of higher power. The final state transition matrix is then obtained by multiplication, using the associative property (Tewari 2002). Alternatively, for a system of small order, the Laplace transform provides a means of analytical evaluation of e^{At}. Upon taking the Laplace transform of Eq. (2.54) for an LTI system, subject to the initial condition, $x_0 = x(0)$, we have

$$sX(s) - x_0 = AX(s) + BU(s), \tag{2.69}$$

where $X(s)$ and $U(s)$ are the Laplace transforms of $x(t)$ and $u(t)$, respectively. The state-transition matrix is then derived for the homogeneous system by taking the inverse Laplace transform as follows:

$$x(t) = \mathcal{L}^{-1}(sI - A)^{-1}x_0, \tag{2.70}$$

or

$$e^{At} = \mathcal{L}^{-1}(sI - A)^{-1}. \tag{2.71}$$

The general solution of an LTI system to an arbitrary, Laplace transformable input $u(t)$, which begins to act at time $t = 0$ when the system's state is $x(0) = x_0$ is thus given by

$$x(t) = e^{At}x_0 + \int_0^t e^{A(t-\tau)}B(\tau)\,u(\tau)\,d\tau. \tag{2.72}$$

The first term on the right-hand side (initial response) decays to zero for an asymptotically stable system (defined below) in the limit $t \to \infty$. However, in the same limit, the integral term either converges to a finite value (called the steady state) or assumes the same functional form as that of the input (called forced response).

The transfer matrix, $G(s)$, of an LTI system is the linear relationship between the output's Laplace transform, $Y(s)$, and that of the input vector, $U(s)$, subject to zero initial conditions, $y(0) = \dot{y}(0) = \ddot{y}(0) = \cdots = 0$: $Y(s) = G(s)\,U(s)$. Clearly, the transfer matrix is the Laplace transform of the impulse response matrix subject to zero initial conditions. The roots of the lowest common denominator polynomial of the transfer matrix (the poles) are the same as the eigenvalues of the system's state dynamics matrix, A. By taking the Laplace transform of the state and output equations, the transfer matrix of an LTI system can be expressed in terms of its state-space coefficients as follows:

$$G(s) = C(sI - A)^{-1}B + D. \tag{2.73}$$

For $s = i\omega$, the transfer matrix becomes the frequency response matrix, $G(i\omega)$, whose elements denote the steady-state response of an output variable to a simple harmonic input variable, subject to zero initial conditions, all other inputs being zero.

Equation (2.69) for $U(s) = 0$ represents an eigenvalue problem, whose solution yields the eigenvalues, s, and eigenvectors, $\mathbf{X}(s)$. The eigenvalues of the linear system are obtained by solving the following characteristic equation:

$$\det(sI - A) = 0. \tag{2.74}$$

The n generally complex roots of the characteristic equation (eigenvalues of A) signify an important system property, called stability. Considering that an eigenvalue is generally complex, its imaginary part denotes the frequency of oscillation of the characteristic vector about the equilibrium point, and the real part signifies the growth (or decay) of its amplitude with time. The criteria for the stability of an LTI system are defined as follows:

Definition 2.6.3

(a) *If all eigenvalues have negative real parts, the system is asymptotically stable and regains its equilibrium in the steady state.*

(b) *A system having complex eigenvalues with zero real parts (and all other eigenvalues with negative real parts) displays oscillatory behaviour of a constant amplitude and is said to be stable (but not asymptotically stable).*

(c) *If at least one eigenvalue has a positive real part, its contribution to the system's state is an exponentially growing amplitude, and the system is said to be unstable.*

(d) *If a multiple eigenvalue of multiplicity* p *is at the origin (i.e. has both real and imaginary parts zero), its contribution to the system's state has terms containing the factors* $t^i, i = 0 \cdots (p-1)$*, which signify an unbounded behaviour with time. Hence, the system is unstable.*

Definition 2.6.4 *Controllability is defined as the property of a system where it is possible to move it from* any *initial state,* $x(0)$*, to* any *final state,* $x(t)$*, solely by the application of the input vector,* $u(t)$*, in a* finite *time,* t*.*

The words 'any' and 'finite' are highlighted in the definition of controllability, because it may be possible to move an uncontrollable system only between some specific states by applying the control input, or to require an infinite time for moving an uncontrallable system between arbitrary states. For a system to be controllable, all its state variables must be influenced, either directly or indirectly, by control inputs. If there is a subsystem that is unaffected by the control inputs, then the entire system is uncontrollable.

Theorem 2.6.5 *A LTI system* (A, B) *is controllable if and only if the following test matrix is of the rank* n*, the order of the system:*

$$P = (B, AB, A^2B, \ldots, A^{n-1}B).$$

Proof. See (Kailath 1980).

If a system is unstable but controllable, it can be stabilized by a feedback control system. It is often possible to decompose an uncontrollable LTI system into controllable and uncontrollable subsystems. A system that is both unstable and uncontrollable could also be stabilized, provided its uncontrollable subsystem is stable. In such a case, the system is said to be stabilizable.

Definition 2.6.6 *Observability is the property of an unforced (homogeneous) system where it is possible to estimate* any *initial state,* $x(0)$*, of the system solely by a* finite *record,* $t \geq 0$*, of the output vector,* $y(t)$*.*[3]

For a system to be observable, all of its state variables must contribute, either directly or indirectly, to the output vector. If there is a subsystem that leaves the output vector unaffected, then the entire system is unobservable.

Theorem 2.6.7 *An unforced LTI system,* $\dot{x} = Ax$*, whose output is related to the state vector by*

$$y = Cx,$$

is observable if and only if the following test matrix has rank n*, the order of the system:*

$$N = \left(C^T, A^T C^T, \left(A^T\right)^2 C^T, \ldots, \left(A^T\right)^{n-1} C^T\right).$$

Proof. See (Kailath 1980).

It is often possible to decompose an unobservable LTI system into observable and unobservable subsystems. A system whose unobservability is caused by a stable subsystem is said to be detectable.

[3] The definition of observability is extended to a forced linear system by requiring in addition that the applied input vector, $u(t)$, is known in the period of observation, $t \geq 0$.

2.6.2 State Feedback by Eigenstructure Assignment

Consider the plant dynamics expressed in an LTI state-space form as follows:

$$\dot{x} = Ax + Bu, \tag{2.75}$$

where $x(t)$ is the state vector, $u(t)$ the control input vector, and A, B are the constant coefficient matrices. Design of a linear, state feedback regulator for the LTI plant of Eq. (2.75) with the control law,

$$u = -Kx, \tag{2.76}$$

is possible by assigning a structure for the eigenvalues and eigenvectors of the closed-loop dynamics matrix, $A - BK$. In case of single-input plants, this merely involves selecting the locations for the closed-loop poles (pole placement) by the following Ackermann's formula that yields the desired closed-loop characteristics (Tewari 2002):

$$K = \left(a_d - a\right)\left(PP'\right)^{-1}, \tag{2.77}$$

where a is the row vector formed by the coefficients, a_i, of the plant's characteristic polynomial in *descending order* $[a = \left(a_n, a_{n-1}, \dots, a_2, a_1\right)]$:

$$\det(sI - A) = s^n + a_n s^{n-1} + a_{n-1} s^{n-2} + \cdots + a_2 s + a_1, \tag{2.78}$$

a_d is the row vector formed by the characteristic coefficients of the closed-loop system in descending order $\left[a_d = \left(a_{dn}, a_{d(n-1)}, \dots, a_{d2}, a_{d1}\right)\right]$:

$$\det(sI - A + BK) = s^n + a_{dn} s^{n-1} + a_{d(n-1)} s^{n-2} + \cdots + a_{d2} s + a_{d1}, \tag{2.79}$$

P is the controllability test matrix of the plant and P' is the following upper triangular matrix:

$$P' = \begin{pmatrix} 1 & a_n & a_{n-1} & \cdots & a_3 & a_2 \\ 0 & 1 & a_n & \cdots & a_4 & a_3 \\ 0 & 0 & 1 & \cdots & a_5 & a_4 \\ \cdots & \cdots & \cdots & \cdots & \cdots & \cdots \\ 0 & 0 & 0 & \cdots & 1 & a_n \\ 0 & 0 & 0 & \cdots & 0 & 1 \end{pmatrix}. \tag{2.80}$$

Of course, this requires that the plant must be controllable, $|P| \neq 0$. A popular choice of the closed-loop poles is the Butterworth pattern (Tewari 2002) wherein all the poles are equidistant from the origin, $s = 0$. Such a pattern generally requires the least control effort for a given bandwidth and is thus considered to be optimal placement of the poles. The pole-placement method is inapplicable to multi-input plants, which have many more controller gains to be found than the number of equations available from the pole locations. In such a case, we need additional equations that can be derived from the shape of the eigenvectors using the method of eigenstructure assignment. A popular method in this regard is the robust pole assignment method of (Kautsky *et al.* 1985) wherein the eigenvectors, $v_i, i = 1, 2, \dots, n$, corresponding to the eigenvalues, λ_i, respectively, and satisfying the eigenvalue problem,

$$(A - BK)\, v_i = \lambda_i v_i; \tag{2.81}$$

are chosen such that the modal matrix,

$$V = \left(v_1, v_2, \dots, v_n\right); \tag{2.82}$$

is as well-conditioned as possible. An alternative method of state-feedback regulator design is the linear, quadratic regulator (LQR), which is discussed later.

2.6.3 Linear Observers and Output Feedback Compensators

Control systems with output (rather than state) feedback require observers that can reconstruct the missing information about the system's states from the input applied to the plant and the output fed back from the plant. An observer mimics the plant by generating an estimated state vector, $\hat{x}$, instead of the actual plant state vector, x, and supplies it to the regulator. A control system that contains both an observer and a regulator is called a compensator. Owing to a decoupling of the observer and plant states in the control system, it is possible to design the regulator and observer separately from each other by what is known as the separation principle. The separation principle states that the regulator design can be carried out exactly in the same manner as if the estimated state were the true plant state. As we shall see later, the separation principle can also be applied to the design of adaptive controllers based upon the estimation of plant parameters (rather than plant states), which in turn are used to determine the controller parameters (rather than the control inputs). In such an application, the separation principle is termed 'certainty equivalence'.

The output equation,

$$y = Cx + Du, \tag{2.83}$$

is used in the design of a full-order observer with the following state equation:

$$\dot{\hat{x}} = (A - LC)\hat{x} + (B - LD)u + Ly, \tag{2.84}$$

where $\hat{x}(t)$ is the estimated state vector and L the observer gain matrix, provided the plant (A, C) is observable. The observer gain matrix, L, is selected in a manner similar to the regulator gain, K, by either eigenstructure assignment for the observer dynamics matrix, $A - LC$, or linear, quadratic, optimal control where A is replaced by A^T, and B by C^T.

The closed-loop control system dynamics with a desired state, $\bar{x}(t)$, and linear feedforward/feedback control with output feedback,

$$u = \bar{K}\bar{x} + K(\bar{x} - \hat{x}), \tag{2.85}$$

is given by the state equation

$$\begin{Bmatrix} \dot{x} \\ \dot{\hat{x}} \end{Bmatrix} = \begin{pmatrix} A & -BK \\ LC & A - BK - LC \end{pmatrix} \begin{Bmatrix} x \\ \hat{x} \end{Bmatrix} + \begin{pmatrix} B\left(K + \bar{K}\right) \\ B\left(K + \bar{K}\right) \end{pmatrix} \bar{x}, \tag{2.86}$$

where K, L are separately designed and the feedforward gain matrix, $\bar{K}$, is selected to ensure that the closed-loop error dynamics is independent of the desired state, $\bar{x}(t)$, with a given state equation

$$\dot{\bar{x}} = \bar{f}(\bar{x}), \tag{2.87}$$

such that

$$(A + B\bar{K}\bar{x}) - \bar{f}(\bar{x}) = 0. \tag{2.88}$$

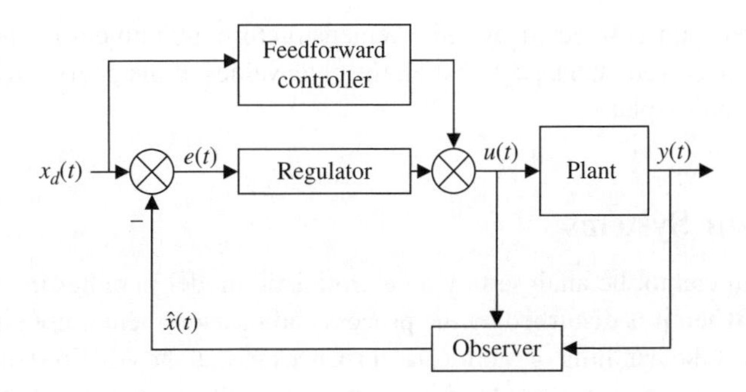

Figure 2.12 Schematic block-diagram of an observer-based, output feedback compensator

Thus, one can design a tracking system for a plant that is both controllable and observable with the available inputs and outputs, respectively, as well as satisfying Eq. (2.88) with its desired state vector. An observer-based output feedback compensator is depicted in a schematic form in Fig. 2.12.

When a part of the plant's state vector can be directly obtained from the output vector, it is unnecessary to estimate the entire state vector by a full-order observer. Consider a plant whose state vector is partitioned as follows:

$$x = \left(x_1^T, x_2^T\right)^T, \tag{2.89}$$

such that

$$\begin{aligned} \dot{x}_1 &= A_{11}x_1 + A_{12}x_2 + B_1 u \\ \dot{x}_2 &= A_{21}x_1 + A_{22}x_2 + B_2 u. \end{aligned} \tag{2.90}$$

The measurable part of the state vector, x_1, can be directly obtained by inversion of the output equation with a square coefficient matrix, C:

$$y = Cx_1. \tag{2.91}$$

The unmeasurable part, x_2, needs estimation by a reduced-order observer and can be expressed as follows:

$$\bar{x}_2 = Ly + z, \tag{2.92}$$

where z is the state vector of the reduced-order observer with the following state equation:

$$\dot{z} = Fz + Hu + Gy, \tag{2.93}$$

whose coefficient matrices are determined from the requirement that the estimation error, $\bar{e} = x_2 - \bar{x}_2$, should go to zero in the steady-state, irrespective of the control input and the output (Tewari 2002):

$$\begin{aligned} F &= A_{22} - LCA_{12} \\ G &= FL + (A_{21} - LCA_{11})C^{-1} \\ H &= B_2 - LCB_1, \end{aligned} \tag{2.94}$$

with the observer gain L selected by either eigenstructure assignment or the Kalman filter approach (to be discussed later), such that all the eigenvalues of the oberver dynamics matrix, F, are in the left-half s-plane.

2.7 Stochastic Systems

A control system cannot be analysed by a deterministic model in either the time or the frequency domain when it is disturbed by the process and measurement noise signals. A crucial task of this book is the handling of random disturbances so that the ASE system remains stable in their presence. In a more general tracking system, it is also required that the tracking error should remain small in the presence of disturbances. However, it is not one of the objectives in ASE design. If the random signals were the outputs of deterministic systems, a pre-filter can be designed to completely block out such signals to prevent them from affecting the closed-loop system's stability. Since noise signals are not governed by deterministic processes, a feedback control loop is especially necessary to suppress the system's sensitivity to noise inputs. Such a design is then said to be robust. Before a robust controller can be designed, a statistical estimate of the expected disturbances is necessary via the theory of probability, wherein future outcomes of an event are predicted based upon its past observation. In this section, we review the essential probability concepts applied to stochastic signals and systems.

The probability of occurrence of a specific outcome of an event, x, is represented as the discrete probability, $p(x)$, such that $0 \leq p(x) \leq 1$, defined as the ratio of the number of events n in which the outcome is x to the total number N of observed events:

$$p(x) = \frac{n}{N}. \tag{2.95}$$

The accuracy of predicting an outcome is increased by increasing the number of observations, N (called the sample size). For an event with M possible multiple outcomes, the sum of probabilities of all possible individual outcomes must be unity:

$$\sum_{i=1}^{M} p(x_i) = \frac{1}{N} \sum_{i=1}^{M} n_i = 1. \tag{2.96}$$

If x is a continuous scalar variable assuming a random value between x_1 and x_2, a probability density function, $p(x)$ can be defined, which must satisfy

$$\int_{x_1}^{x_2} p(x)\,\mathrm{d}x = 1. \tag{2.97}$$

Equation (2.97) is made more general by having the integration performed from $-\infty$ to ∞, and noting that $p(x) = 0$ whenever x lies outside its feasible range, $x_1 \leq x \leq x_2$:

$$\int_{-\infty}^{\infty} p(x)\,\mathrm{d}x = 1. \tag{2.98}$$

The expected value of a continuous, random variable, x, with a probability density function, $p(x)$, is given by

$$E(x) = \int_{-\infty}^{\infty} x p(x)\,\mathrm{d}x = \bar{x}, \tag{2.99}$$

and is also known as, $\bar{x}$, called the mean value of x. From the definition of expected value it follows that

$$E(x - \bar{x}) = 0. \tag{2.100}$$

The variance of a continuous random variable from its mean value is defined by

$$\sigma^2 = E\left[(x - \bar{x})^2\right] = \int_{-\infty}^{\infty} (x - \bar{x})^2 p(x)\,\mathrm{d}x. \tag{2.101}$$

The square-root of variance, σ, is called the standard deviation of x from its mean value.

When groups of random variables, such as a pair (x, y), are concerned, a joint probability of both x and y occurring simultaneously is expressed as $p(x, y)$. If x and y are unrelated outcomes of separate, random events, then we have

$$p(x, y) = p(x)\,p(y), \tag{2.102}$$

which can exceed neither $p(x)$ nor $p(y)$. If x and y are continuous random variables (which may or may not be related), then we have

$$\int_{-\infty}^{\infty}\int_{-\infty}^{\infty} p(x, y)\,\mathrm{d}x\mathrm{d}y = 1. \tag{2.103}$$

A conditional probability of x given y is defined by

$$p(x|y) = \frac{p(x, y)}{p(y)}, \tag{2.104}$$

from which it follows that both conditional and unconditional probabilities of x given y are the same, if x and y are unrelated variables. The relationship between the conditional probabilities of x given y, and y given x, is provided by following Bayes's rule:

$$p(y|x) = \frac{p(x|y)\,p(y)}{p(x)}. \tag{2.105}$$

An arithmetic sum of scalar random variables, x and y,

$$z = x + y, \tag{2.106}$$

has the following probability descriptions:

$$p(z|x) = p(z - x) = p(y), \tag{2.107}$$

and

$$p(z) = \int_{-\infty}^{\infty} p(z|x)\,p(x)\,\mathrm{d}x = \int_{-\infty}^{\infty} p(z - x)\,p(x)\,\mathrm{d}x. \tag{2.108}$$

The mean and variance of the sum are given by

$$E(z) = \bar{z} = E(x + y) = E(x) + E(y) = \bar{x} + \bar{y} \tag{2.109}$$

and

$$\sigma_z^2 = E\left[(z - \bar{z})^2\right] = E\left[(x - \bar{x})^2\right] + E\left[(y - \bar{y})^2\right] = \sigma_x^2 + \sigma_y^2, \tag{2.110}$$

respectively. The definitions given above can be extended to a sum of any number of random variables, or even to their linearly weighted sums. The central limit theorem states that a sum of a large number of random variables (having various unconditional probability density functions) approaches a specific probability density function, called the normal (or Gaussian) distribution, given by

$$p(x) = \frac{1}{\sqrt{2\pi}\sigma} e^{-\frac{(x-\bar{x})^2}{2\sigma^2}}. \tag{2.111}$$

Here $\bar{x}$ and σ denote the mean and standard deviation of the Gaussian distribution, respectively. From the central limit theorem, it follows that a Gaussian signal can be regarded as a purely random signal, because it is the sum of a large number of stochastic signals, each with a different probability distribution. Since Eq. (2.111) gives a simple probability function, the effect of random noise inputs on control system stability can be easily analysed by assuming a Gaussian model.

For a Gaussian process, the probability that x lies in an error band of $\pm\xi$ about the mean is given by the following integral expression:

$$\int_{\bar{x}-\xi}^{\bar{x}+\xi} p(x)\,\mathrm{d}x = \frac{1}{\sqrt{2\pi}\sigma} \int_{-\xi}^{\xi} e^{-\frac{\eta^2}{2\sigma^2}}\,\mathrm{d}\eta = \frac{2}{\sqrt{\pi}} \int_{0}^{\frac{\xi}{\sigma\sqrt{2}}} e^{-u^2}\,\mathrm{d}u = \mathrm{erf}\left(\frac{\xi}{\sigma\sqrt{2}}\right), \tag{2.112}$$

which is called the error function of x.

The discussion up to this point is confined to statistical measures of variables that can assume random values at any given instant. This is like taking a snapshot of a time-varying process and results in the measures called the ensemble properties of the process. However, most processes require studying how the statistical properties change over time. There are special random processes called stationary processes, whose ensemble properties are constants with time. There are some random processes – called ergodic processes – whose properties sampled over time give rise to exactly the same probability density functions as those obtained by taking the ensemble average. Stationarity and ergodicity are very useful assumptions in deriving an unknown system's statistical measures.

The correlation function of two signals, $x(t)$ and $y(t)$, is defined as the expected value of the product of the two signals evaluated at different times,

$$\psi_{xy}(t,\tau) = E\left[x(t)\,y(\tau)\right]; \qquad t \neq \tau. \tag{2.113}$$

If $E\left[x(t)\,y(\tau)\right] = 0$ for all arbitrary times, (t,τ), then $x(t)$ and $y(t)$ are said to be uncorrelated. The degree of correlation of a signal with itself at different time instants is defined by the autocorrelation function,

$$\psi_{xx}(t,\tau) = E\left[x(t)\,x(\tau)\right]; \qquad t \neq \tau. \tag{2.114}$$

A random signal, $x(t)$, which is totally uncorrelated with itself at different values of time ($\psi_{xx}(t,\tau) = 0, t \neq \tau$) is called a white noise, and has an important place in control theory. For a white noise signal, the conditional probability at the present time, given its value at some other time, is no different from the unconditional probability of the signal at the present time

$$p\left[x(t)\right] = p\left[x(t)\,|x(\tau)\right]; \qquad t \neq \tau \tag{2.115}$$

Therefore, knowing the value of the signal at some time does not help in predicting its value at any other time. In order to find the joint probability of a white noise process evolving from $t = t_0$ with a given probability, $p\left[x\left(t_0\right)\right]$, to the present time, t, one must know the following unconditional probabilities at all intermediate times:

$$\begin{aligned} p \quad & \left[x(t), x(t-T), x(t-2T), \ldots, x\left(t_0+T\right), x\left(t_0\right)\right] \\ & = p\left[x(t)\right] p\left[x(t-T)\right] p\left[x(t-2T)\right] \cdots p\left[x\left(t_0+T\right)\right] p\left[x\left(t_0\right)\right], \end{aligned} \tag{2.116}$$

where T is a sampling interval. Since evaluating the probability function at an infinite number of time instants is not a possible task, a white noise process is said to be completely unpredictable.

A signal that is correlated with itself at different values of time ($\psi_{xx}(t, \tau) \neq 0, t \neq \tau$) is called coloured noise. An example of coloured noise is the Markov process, $x(t)$, beginning at $t = t_0$ with probability, $p\left[x\left(t_0\right)\right]$, and defined by

$$\begin{aligned} p \quad & \left[x(t), x(t-T), x(t-2T), \ldots, x\left(t_0+T\right), x\left(t_0\right)\right] \\ & = p\left[x(t) \,|x(t-T)\right] p\left[x(t-T) \,|x(t-2T)\right] \cdots p\left[x\left(t_0+T\right) |x\left(t_0\right)\right] p\left[x\left(t_0\right)\right]. \end{aligned} \tag{2.117}$$

The joint probability of a Markov process evolving from $t = t_0$ with a given probability, $p\left[x\left(t_0\right)\right]$, to the present time, t, thus depends only upon the product of conditional probabilities, $p\left[x(\tau) \,|x(\tau-T)\right], t_0 + T \leq \tau \leq t$, of evolving over one sampling interval (called the transitional probability). The value of a Markov signal at a given time, $x(t)$, can be predicted from the transitional probabilities at previous times, $p\left[x(\tau) \,|x(\tau-T)\right], t_0 + T \leq \tau \leq t$, and the initial probability, $p\left[x\left(t_0\right)\right]$. The simplest Markov process is obtained by passing a white noise through a linear, time-varying system (called a linear filter). The state equation of a scalar, Markov process can thus be written as follows:

$$\dot{x}(t) = a(t) x(t) + b(t) w(t), \tag{2.118}$$

where $w(t)$ is a white noise process and $a(t), b(t)$ are the time-varying filter coefficients. A Markov process that has a Gaussian probability distribution is termed a Gauss–Markov process, which can be generated by passing a Gaussian white noise through a linear filter [Eq. (2.118)] with a Gaussian initial state, $x\left(t_0\right)$.

By subtracting the (ensemble) mean values of the signals in the correlation and autocorrelation functions, the cross-covariance and auto-covariance functions are defined as follows:

$$\phi_{xy}(t, \tau) = E\left[\{x(t) - \bar{x}(t)\}\{y(\tau) - \bar{y}(\tau)\}\right]; \qquad t \neq \tau \tag{2.119}$$

and

$$\phi_{xx}(t, \tau) = E\left[\{x(t) - \bar{x}(t)\}\{x(\tau) - \bar{x}(\tau)\}\right]; \qquad t \neq \tau, \tag{2.120}$$

respectively. Hence, the variance of a signal is obtained by substituting $t = \tau$:

$$\phi_{xx}(t, t) = E\left[\{x(t) - \bar{x}(t)\}^2\right] = \sigma_x^2. \tag{2.121}$$

The covariance of two signals can be similarly defined by substituting $t = \tau$ in the cross-covariance function:

$$\phi_{xy}(t,t) = E\left[\{x(t) - \bar{x}(t)\}\{y(t) - \bar{y}(t)\}\right] = \sigma_{xy}. \tag{2.122}$$

If a process is stationary, the actual time, t, is immaterial. Then the results depend only upon the time shift, α, where $\tau = t + \alpha$:

$$\begin{aligned}
\psi_{xy}(\alpha) &= E\left[x(t)\,y(t+\alpha)\right] \\
\psi_{xx}(\alpha) &= E\left[x(t)\,x(t+\alpha)\right] \\
\phi_{xy}(\alpha) &= E\left[\{x(t) - \bar{x}(t)\}\{y(t+\alpha) - \bar{y}(t+\alpha)\}\right] \\
\phi_{xx}(\alpha) &= E\left[\{x(t) - \bar{x}(t)\}\{x(t+\alpha) - \bar{x}(t+\alpha)\}\right].
\end{aligned} \tag{2.123}$$

The concepts of probability can be extended to vectors of random variables. However, there is a qualitative difference between a scalar random variable and a vector random variable in that the joint probability, $p(x)$, of a vector, $x(t) : \mathbb{R} \to \mathbb{R}^n$, taking up a specific value, $x = x_0$, is very much smaller than the unconditional probability of any of its elements, $x_i(t)$, individually assuming a given value, x_{i0}, irrespective of the values taken up by the remaining elements. A vector of continuous, random variables, $x = \left(x_1, x_2, \ldots, x_n\right)^T$, is thus said to have a joint probability density function, $p(x)$, which satisfies the identity

$$\int_{-\infty}^{\infty} \cdots \int_{-\infty}^{\infty} \int_{-\infty}^{\infty} p(x)\,\mathrm{d}x_1 \mathrm{d}x_2 \cdots \mathrm{d}x_n = 1 \tag{2.124}$$

and has a mean value given by

$$E(x) = \bar{x} = \int_{-\infty}^{\infty} \cdots \int_{-\infty}^{\infty} \int_{-\infty}^{\infty} x p(x)\,\mathrm{d}x_1 \mathrm{d}x_2 \cdots \mathrm{d}x_n. \tag{2.125}$$

The state vector of any physical, stochastic process, $x(t)$, can be regarded as a random vector whose elements continuously vary with time. Its mean at a given time is given by

$$\bar{x}(t) = E\left[x(t)\right], \tag{2.126}$$

while the statistical correlation among the elements of a stochastic vector signal, $x(t)$, is measured by the correlation matrix defined by

$$R_x(t,\tau) = E\left[x(t)\,x^T(\tau)\right]. \tag{2.127}$$

A diagonal correlation matrix indicates that the state variables of a stochastic process are uncorrelated with one another, although they may be autocorrelated with themselves at different times. Substituting $t = \tau$ in the correlation matrix produces the following covariance matrix:

$$R_x(t,t) = E\left[x(t)\,x^T(t)\right], \tag{2.128}$$

which, by definition, is symmetric. The trace of the covariance matrix is therefore the square of the H_2 norm of the signal:

$$\mathrm{tr}\{R_x(t,t)\} = \|x(t)\|_2^2. \tag{2.129}$$

2.7.1 Ergodic Processes

If a stationary process is assumed to be also ergodic, the mean and autocorrelation matrix are derived by taking the time averages (rather than ensemble averages) as follows:

$$\bar{x} = \lim_{T\to\infty} \frac{1}{T} \int_{-\frac{T}{2}}^{\frac{T}{2}} x(t)\,\mathrm{d}t \tag{2.130}$$

$$R_x(\alpha) = \lim_{T\to\infty} \frac{1}{T} \int_{-\frac{T}{2}}^{\frac{T}{2}} x(t)\,x^T(t+\alpha)\,\mathrm{d}t, \tag{2.131}$$

where α is the time shift. If a power spectral density matrix is defined by taking the Fourier transform (Kreyszig 1998) of the correlation matrix, the time domain, stationary signal is transformed to the frequency domain, as follows:

$$S_x(\omega) = \int_{-\infty}^{\infty} R_x(\alpha)\,e^{-i\omega\alpha}\mathrm{d}\alpha. \tag{2.132}$$

The power spectral density (PSD) matrix, $S_x(\omega)$, represents the energy content of the ergodic, vector random signal, $x(t)$, distributed over the frequency, ω. The presence of peaks in the spectral norm of the PSD matrix, $|\,S_x(\omega)\,|_S$, at some frequencies indicates that the system can be excited to a large response by applying harmonic inputs at those particular frequencies. The Fourier transform of the signal is given by

$$X(i\omega) = \int_{-\infty}^{\infty} x(t)\,e^{-i\omega t}\mathrm{d}t. \tag{2.133}$$

From the Eqs. (2.131)–(2.133), the following relationship between $S_x(\omega)$ and $X(i\omega)$ is derived:

$$S_x(\omega) = X(i\omega)\,X^T(-i\omega). \tag{2.134}$$

An inverse Fourier transform produces the correlation matrix from the PSD matrix as follows:

$$R_x(\alpha) = \frac{1}{2\pi} \int_{-\infty}^{\infty} S_x(\omega)\,e^{i\omega\alpha}\mathrm{d}\omega, \tag{2.135}$$

whereas the covariance matrix is simply obtained by substituting $\alpha = 0$:

$$R_x(0) = \frac{1}{2\pi} \int_{-\infty}^{\infty} S_x(\omega)\,e^{i\omega}\mathrm{d}\omega. \tag{2.136}$$

The cross-correlation matrix of two ergodic vector signals, $x(t), y(t)$, is defined by

$$R_{xy}(\alpha) = \lim_{T\to\infty} \frac{1}{T} \int_{-\frac{T}{2}}^{\frac{T}{2}} x(t)\,y^T(t+\alpha)\,\mathrm{d}t, \tag{2.137}$$

and the cross-spectral density by

$$S_{xy}(\omega) = \int_{-\infty}^{\infty} R_{xy}(\alpha)\,e^{-i\omega\alpha}\mathrm{d}\alpha. \tag{2.138}$$

The Fourier transforms of the two signals are given by

$$X(i\omega) = \int_{-\infty}^{\infty} x(t)\, e^{-i\omega t} \mathrm{d}t, \quad Y(i\omega) = \int_{-\infty}^{\infty} y(t)\, e^{-i\omega t} \mathrm{d}t, \tag{2.139}$$

hence the following relationship exists between $S_{xy}(\omega)$ and $S_{yx}(\omega)$:

$$S_{xy}(\omega) = X(i\omega)\, Y^T(-i\omega) = \left[Y(-i\omega)\, X^T(i\omega)\right]^T = S_{yx}^T(-\omega) = S_{yx}^H(\omega). \tag{2.140}$$

If the two signals are totally uncorrelated, then $R_{xy}(\alpha) = 0$, and hence $S_{xy}(\omega) = S_{yx}(i\omega) = 0$.

2.7.1.1 Zero-Mean Gaussian White Noise

It was earlier commented that the central limit theorem predicts that a linear superimposition of a large number of random signals produces a stationary, Gaussian signal. If it is further specified that all the superimposed signals are uncorrelated with one another and by themselves in time, and also have zero mean values, then the resulting signal is called a zero-mean, Gaussian white noise (ZMGWN). While it may not be possible to find a ZMGWN signal in practice, its assumption greatly simplifies control system design. Let $w(t)$ be a continuous time, vector ZMGWN signal, whose mean is given by

$$\bar{w} = \lim_{T\to\infty} \frac{1}{T} \int_{-\frac{T}{2}}^{\frac{T}{2}} w(t)\, \mathrm{d}t = 0 \tag{2.141}$$

and

$$R_w(\alpha) = \lim_{T\to\infty} \frac{1}{T} \int_{-\frac{T}{2}}^{\frac{T}{2}} w(t)\, w^T(t+\alpha)\, \mathrm{d}t = 0; \qquad \alpha \neq 0. \tag{2.142}$$

Furthermore, because we have

$$W(i\omega) = \int_{-\infty}^{\infty} w(t)\, e^{-i\omega t} \mathrm{d}t = c = \text{const.} \tag{2.143}$$

we have

$$S_w(\omega) = W(i\omega)\, W^T(-i\omega) = cc^T = \text{const.} \tag{2.144}$$

Therefore, the PSD matrix of a ZMGWN is a constant matrix. However, if one takes the inverse Fourier transform of the constant $\left[S_w\right]$, the result is a covariance matrix that has all elements tending to infinity:

$$R_w(0) = \frac{1}{2\pi} cc^T \int_{-\infty}^{\infty} e^{i\omega} \mathrm{d}\omega \to \infty \tag{2.145}$$

This difficulty is resolved by writing

$$R_w(\alpha) = cc^T \delta(\alpha), \tag{2.146}$$

where $\delta(\alpha)$ is the following Dirac delta function:

$$\delta(\alpha) = \begin{cases} \infty & (\alpha = 0) \\ 0 & (\alpha \neq 0) \end{cases} \tag{2.147}$$

with the property,

$$\int_{-\infty}^{\infty} \delta(\theta)\, d\theta = 1. \tag{2.148}$$

2.7.2 Filtering of Random Noise

When a signal is passed through a linear filter, its statistical measures are modified. A white noise signal (i.e. a purely random signal) when passed through a linear, time-varying filter becomes coloured (e.g. Markov process). Consider a linear, stable system with control input vector, $u(t)$, output vector $y(t)$, and an ergodic noise vector, $w(t)$, described by the transfer matrices, $G(s)$ and $F(s)$, as follows:

$$y(s) = G(s)\, u(s) + F(s)\, w(s). \tag{2.149}$$

The power spectral densities of the control inputs and the noise are $S_u(\omega)$ and $S_w(\omega)$, respectively. Let the noise be totally uncorrelated with control inputs, that is, $S_{uw}(\omega) = 0$. The cross-spectral density of the output with the control input is the following:

$$S_{yu}(\omega) = G(i\omega)\, S_u(\omega). \tag{2.150}$$

If the noise were absent, the power spectral density of the output would be given by

$$S_y(\omega) = G(i\omega)\, S_u(\omega)\, G^H(i\omega), \tag{2.151}$$

In the presence of the noise, the spectral density is modified to

$$S_y(\omega) = G(i\omega)\, S_u(\omega)\, G^H(i\omega) + F(i\omega)\, S_w(\omega)\, F^H(i\omega). \tag{2.152}$$

In the special case of the control being absent, and $w(t)$ being a white noise signal ($S_w = cc^T =$ const.), the output is a coloured signal (i.e. it has a frequency-dependent spectrum). The most general description of random noise is the zero mean, Gaussian white noise (ZMGWN), which by definition, is generated by an infinite bandwidth process. An objective of robust control design is to select a feedback law, $u = -H(s)\, y$, such that the overall dependence of the output on the noise signal is minimized. Since a linear control system has a finite bandwidth in which it can effectively respond to applied inputs, it is possible to select a compensator, $H(s)$, such that the sensitivity to noise is reduced in a given range of frequencies. Of course, the entire noise spectrum cannot be blocked by a linear stabilizing compensator, because of the latter's finite bandwidth. However, traditional methods of low-pass filtering have been successfully employed in most practical cases, because the actual noise has predominantly high-frequency content. The output spectrum of a low-pass filter contains peaks at only the lower frequencies, implying a smoothening of the raw noise signal. Often a band-pass filter is used to suppress both high- and low-frequency contents of a noisy signal. The magnitude of a filtered signal above (or below) a given frequency can be made to decay rapidly with frequency by suitably selecting the feedback gain, $\|H(s)\|$. Such a decay of signal magnitudes with frequency is called attenuation, or roll off.

When a feedback control law, $u(s) = -H(s)\, y(s)$, is applied, the resulting closed-loop system is represented as follows:

$$y(s) = [I + G(s)\, H(s)]^{-1} F(s)\, w(s). \tag{2.153}$$

The sensitivity of the system to noise depends directly on the power spectral density,

$$S_y(\omega) = [I + G(i\omega) H(i\omega)]^{-1} F(i\omega) S_w(\omega) F^H(i\omega) \{[I + G(i\omega) H(i\omega)]^{-1}\}^H, \tag{2.154}$$

which must be reduced in size by minimizing the spectral norm of the noise transmission matrix, $N(s) = [I + G(s) H(s)]^{-1} F(s)$, for a white noise disturbance, $S_w = cc^T$. As discussed earlier, such a minimization is not possible across a large frequency range. The general problem of finding a stabilizing controller, $H(s)$, for a desired noise transmission spectrum and white noise intensity is called spectral factorization.

In trying to find a stabilizing solution for an open-loop unstable plant in the presence of disturbances, a crucial intermediate task of the controller is to estimate the state variables from the output vector. Thus every practical solution to the stabilization problem includes an observer for state estimation. The observer must perform its task by minimizing the sensitivity of the estimation error to noise and is therefore inherently a filter. Estimation theory begins with the important result of the Wiener filter.

2.7.3 *Wiener Filter*

Consider a linear system with transfer matrix, $G(s)$. If $G(s)$ does not have any poles on the imaginary axis, $s = i\omega$, then $G(s)$ can be expressed as follows:

$$G(s) = \hat{G}(s) + \overline{G}(s), \tag{2.155}$$

where $\hat{G}$, called the stable part, has all poles in the left-half s-plane, and $\overline{G}$ is the anti-stable part having all its poles in the right-half s-plane.

Let a stochastic signal, $x(t)$, be estimated from the measurement of another signal, $y(\tau)$, for some time $\tau \le t$, and let the cross-spectral density of the two signals be $S_{xy}(\omega)$. The Wiener filter is a stable, linear, strictly proper filter that gives the state estimate, $\hat{x}$, as follows:

$$\hat{x}(s) = W(s) y(s), \tag{2.156}$$

with the filter's transfer matrix given by

$$W(s) = \hat{L}(s) Z^{-1}(s), \tag{2.157}$$

where

$$S_{xy}(\omega) = Z(i\omega) Z^H(i\omega), \tag{2.158}$$

with Z and Z^{-1} being stable and proper, and $\hat{L}(s)$ being the stable part of the following matrix:

$$L(i\omega) = S_{xy}(\omega) \left[Z^H(i\omega)\right]^{-1}, \tag{2.159}$$

expressed as a sum of stable, $\hat{L}$, and anti-stable, $\overline{L}$, parts:

$$L(s) = \hat{L}(s) + \overline{L}(s). \tag{2.160}$$

The Wiener filter minimizes the error covariance matrix, $R_e(0) = E\left[e(t) e^T(t)\right]$, of the state estimation error, $e(t) = x(t) - \hat{x}(t)$, and is therefore an optimal filter. This can be seen by writing the error spectral density as follows:

$$S_e(\omega) = S_x(\omega) - W(i\omega) S_{xy}^H(\omega) - S_{xy}(\omega) W^H(i\omega) + W(i\omega) Z(i\omega) Z^H(i\omega) W^H(i\omega), \tag{2.161}$$

or

$$S_e(\omega) = \left(WZ - S_{xy}[Z^H]^{-1}\right)\left(WZ - S_{xy}[Z^H]^{-1}\right)^H + S_x - S_{yx}[Z^H]^{-1}Z^{-1}S_{xy}^H, \tag{2.162}$$

substituting Eqs. (2.158)–(2.160), and carrying out the complex integrals for inverse Fourier transform over a path that includes the imaginary axis and a large semicircle of radius r enclosing the right-half s-plane. The only non-zero contribution of the terms on the right-hand side of Eq. (2.162) in the limit $r \to \infty$ results in the following:

$$R_e(0) = \frac{1}{2\pi}\int_{-\infty}^{\infty}\left(WZ - \hat{L}\right)\left(WZ - \hat{L}\right)^H \mathrm{d}\omega + \text{Terms independent of } W(s) \tag{2.163}$$

This implies that every error covariance matrix (being by definition positive semi-definite), for which the first term on the right-hand side of Eq. (2.163) is non-zero, is non-minimal. Hence, $W = \hat{L}Z^{-1}$ is the optimal solution.

We note that the Wiener filter bases its state estimate, $\hat{x}(t)$, from a record of the output signal, $y(t)$, at all prior times. While the Wiener filter ensures the existence of an optimal solution to the state estimation problem, it requires an infinite record of the output for the state estimate, which is not practical. State estimation from a finite record of the output requires a further assumption. If we assume the linear system to be a Markov process, then a knowledge of its state at any given time, $x(t)$, is sufficient to predict the state at all future times. This is a major assumption, which requires that all the disturbance signals driving the linear system must be Gaussian white noises. State estimation for a Markov process requires only a finite record of the output vector.

2.7.4 Kalman Filter

Consider a LTI plant with the state vector, $x(t)$, output vector, $y(t)$, process noise, $p(t)$, measurement noise, $m(t)$, and the coefficient matrices, A, B, F, C, D, of appropriate dimensions, resulting in the following state-space representation:

$$\dot{x} = Ax + Bu + Fp$$

$$y = Cx + Du + m. \tag{2.164}$$

By assuming $p(t)$ and $m(t)$ to be ZMGWN, we can greatly simplify the model of the stochastic plant. The correlation matrices of the white noises, $p(t)$ and $m(t)$, are expressed as follows:

$$R_p(t) = S_p\delta(t)$$

$$R_m(t) = S_m\delta(t) \tag{2.165}$$

$$R_{pm}(t) = S_{pm}\delta(t), \tag{2.166}$$

where S_p, S_m and S_{pm} are the constant power spectral density matrices of the signals $p(t)$ and $m(t)$, with the corresponding infinite covariance matrices, $R_p(0), R_m(0)$ and $R_{pm}(0)$, respectively. For convenience, we shall represent all covariance matrices without the (0) notation, that is, as R_p, R_m, and so on. A state-feedback control system cannot be designed for a stochastic plant, because its state vector, $x(t)$, is unknown at any given time. Instead, the feedback from an estimated state vector, $\hat{x}(t)$, is employed, which in turn is derived from the measurement of the output vector, $y(\tau)$, over a previous, finite time interval, $t_0 \leq \tau \leq t$. Hence an observer must be present as a part of the feedback controller. However, unlike a deterministic observer discussed previously, an observer that estimates the state-vector on the basis of the statistical description of the vector output and plant state is required. The Kalman filter is one such observer. Before considering the design of the Kalman filter, an important result from estimation theory must be covered.

Consider an LTI system in the absence of control inputs, ($u = 0$), driven by a ZMGWN noise, $w(t)$, of covariance matrix, R_w, and represented by the following state equation:

$$\dot{\hat{x}} = \hat{A}\hat{x} + Fw. \tag{2.167}$$

Let the state dynamics matrix $\hat{A}$ be Hurwitz (i.e. with all eigenvalues in the left-half plane). The state solution is then expressed as follows:

$$\hat{x}(s) = \left(sI - \hat{A}\right)^{-1} Fw(s), \tag{2.168}$$

with the power spectral density given by

$$S_{\hat{x}}(\omega) = \left(i\omega I - \hat{A}\right)^{-1} FR_w F^T \left[\left(-i\omega I - \hat{A}\right)^{-1}\right]^T, \tag{2.169}$$

and the covariance matrix by

$$R_{\hat{x}} = \frac{1}{2\pi} \int_{-\infty}^{\infty} S_{\hat{x}}(\omega)\, \mathrm{d}\omega. \tag{2.170}$$

Theorem 2.7.1 *The observer represented by Eqs. (2.167)–(2.170), having a Hurwitz dynamics matrix, $\hat{A}$, has its covariance matrix, $R_{\hat{x}}$, given by the unique solution to the following Lyapunov equation:*

$$\hat{A}R_{\hat{x}} + R_{\hat{x}}\hat{A}^T + FR_w F^T = 0. \tag{2.171}$$

Proof. The proof is derived by substituting Eq. (2.169) into Eq. (2.170), taking the inverse Fourier transform, and integrating the result by parts. Since $\hat{A}$ has all eigenvalues in the left-half plane, the resolvent, $\left(sI - \hat{A}\right)^{-1}$ is of full rank, hence the solution $R_{\hat{x}}$ is unique.

The Kalman filter is a special observer designed to minimize the covariance matrix of the state estimation error,

$$\hat{e}(t) = x(t) - \hat{x}(t), \tag{2.172}$$

where $\hat{x}$ is the estimated state vector. For the time-varying Kalman filter, the covariance of estimation error, being a non-stationary signal, is the conditional covariance matrix based on a finite record of the output and can be written as follows:

$$\begin{aligned} R_e(t,t) &= E\left[\hat{e}(t)\hat{e}^T(t)\,|y(\tau), t_0 \leq \tau \leq t\right] \\ &= E\left[\{x(t) - \hat{x}(t)\}\{x^T(t) - \hat{x}^T(t)\}|y(\tau), t_0 \leq \tau \leq t\right], \end{aligned} \tag{2.173}$$

which is simplified (Tewari 2011) to the following:

$$\begin{aligned} R_e(t,t) &= E\left[x(t)x^T(t)\right] - \hat{x}(t)\bar{x}^T(t) - \hat{x}^T(t)\bar{x}(t) + \hat{x}(t)\hat{x}^T(t) \\ &= E\left[x(t)x^T(t)\right] - \bar{x}(t)\bar{x}^T(t) + \Delta x(t)\,\Delta x^T(t), \end{aligned} \tag{2.174}$$

where $\Delta x = \hat{x} - \bar{x}$ is the deviation of the estimated state from the conditional mean, $\bar{x}(t)$. Therefore, the best estimate of the state-vector, that is, $\Delta x(t) = 0$, would result in a minimization of the conditional covariance matrix, $R_e(t, t)$, and the process leading to the minimization is an 'optimal' observer (Kalman filter). The same argument applies to the LTI Kalman filter, with the difference that now the signals driven by white noise processes are stationary, and hence lead to a unique, constant covariance matrix of the estimation error, R_e.

The state equation of the LTI Kalman filter as a full-order observer is given by

$$\dot{\hat{x}} = \hat{A}\hat{x} + \hat{B}u + Ly, \tag{2.175}$$

where L is the Kalman filter gain matrix and $\hat{A}, \hat{B}$ the constant coefficient matrices. The state equation for the estimation error dynamics is thus obtained to be the following:

$$\dot{\hat{e}} = \hat{A}\hat{e} + \left(A - LC - \hat{A}\right)x + \left(B - LD - \hat{B}\right)u + Fp - Lm. \tag{2.176}$$

In order that the estimation error dynamics be independent of the state and control variables, it must be true that

$$\begin{aligned} \hat{A} &= A - LC \\ \hat{B} &= B - LD, \end{aligned} \tag{2.177}$$

which yields

$$\dot{\hat{e}} = (A - LC)\hat{e} + Fp - Lm. \tag{2.178}$$

Since $p(t)$ and $m(t)$ are ZMGWN processes, their linear combination,

$$w = Fp - Lm, \tag{2.179}$$

is also a ZMGWN signal. Therefore, we are ready to apply the result of Theorem 2.7.1 by expressing the estimation error dynamics of the Kalman filter as follows:

$$\dot{\hat{e}} = \hat{A}\hat{e} + w, \tag{2.180}$$

where $\hat{A} = A - LC$ must be Hurwitz by a suitable choice of the Kalman filter gain matrix, L. This is carried out by either eigenstructure assignment (covered previously), or linear, optimal control (to be discussed in the following section). While it is tempting to quickly drive an initial estimation error, $\hat{e}(0)$, to zero by selecting large Kalman filter gains, L, which places all the eigenvalues of $\hat{A}$ deep in the left-half plane, this is not an ideal solution, because the feedback from the measurement noise, $m(t)$, increases with L. Therefore, the Kalman filter design is a balance between the conflicting requirements of a low filter gain, L, for robustness with respect to the measurement noise, and a high filter gain for moving the eigenvalues of $\hat{A} = A - LC$ sufficiently deep into the left-half plane so that the estimation error is quickly driven to small values.

By substituting Eqs. (2.177) and (2.179) into the result of Theorem 2.7.1, we have the following equation for the minimum error covariance, $R_{\hat{e}}$:

$$\hat{A}R_{\hat{e}} + R_{\hat{e}}\hat{A}^T + R_w = 0 \tag{2.181}$$

with R_w being the following covariance matrix of the combined white noise, $w = Fp - Lm$:

$$R_w = FR_pF^T + LR_mL^T - FR_{pm}L^T - LR_{pm}^TF^T. \tag{2.182}$$

Equations (2.181) and (2.182) yield the following algebraic Riccati equation (ARE):

$$\hat{A}R_{\hat{e}} + R_{\hat{e}}\hat{A}^T + FR_pF^T + LR_mL^T - FR_{pm}L^T - LR_{pm}^TF^T = 0, \tag{2.183}$$

which can also be expressed as follows,

$$AR_{\hat{e}} + R_{\hat{e}}A^T + FR_pF^T - LR_mL^T = 0, \tag{2.184}$$

where

$$L = (R_{\hat{e}}C^T + FR_{pm})R_m^{-1}. \tag{2.185}$$

Hence, a unique, positive semi-definite solution, $R_{\hat{e}}$, to the ARE yields a stabilizing solution for the Kalman filter gain. The following lemma establishes the sufficient condition for the existence of such a solution.

Lemma 2.7.2 *The algebraic Riccati equation, Eq. (2.183), has a unique, positive semi-definite solution, $R_{\hat{e}}$, if the matrix R_m is symmetric and positive definite, the pair (A, C) is detectable, and the pair $\left(A - FR_{pm}R_m^{-1}C, FR_pF^T - FR_{pm}R_m^{-1}R_{pm}^TF^T\right)$ is stabilizable.*

Proof. See (Glad and Ljung 2002).

By specifying the noise covariances matrices, R_p, R_m, R_{pm}, as the cost parameters of the minimization problem (see the next section), a stabilizing solution for $(A - LC)$ with eigenvalues at desired locations can be obtained. However, as the ARE is a nonlinear equation, an iterative numerical solution must be sought.

A Kalman filter estimate, $\hat{x}(t) = x(t) - \hat{e}(t)$, has the following property:

$$E\left[x(t)x(t)^T\right] = E\left[\hat{x}(t)\hat{x}^T(t)\right] + E\left[\hat{e}(t)\hat{e}^T(t)\right], \tag{2.186}$$

which implies that the two processes, $\hat{x}(t)$ and $\hat{e}(t)$, are totally uncorrelated. Thus $\hat{e}(t+\tau), \tau > 0$ is uncorrelated with all past estimates, $\hat{x}(t)$, hence the Kalman filter is not only an optimal observer that minimizes the error covariance but it is also the only such causal observer. Therefore, the Kalman filter estimate can be derived from a finite record of the outputs, $y(t), 0 < t < \tau$.

A major simplification in the Kalman filter occurs if the process and measurement noise are uncorrelated with each other, that is, $R_{pm} = 0$, which is a commmon situation. In such a case, the filter gain simplifies to the following expression:

$$L = R_{\hat{e}}C^TR_m^{-1}, \tag{2.187}$$

where $R_{\hat{e}}$ is the unique, positive semi-definite solution to the following ARE:

$$AR_{\hat{e}} + R_{\hat{e}}A^T - R_{\hat{e}}C^TR_m^{-1}CR_{\hat{e}} + FS_pF^T = 0. \tag{2.188}$$

for a time-varying Kalman filter, see (Tewari 2011). The Kalman filter updates its estimation error by the linear state equation. However, in cases where the error dynamics are essentially nonlinear, a much more accurate state estimate can be obtained using the nonlinear plant state equation solved for the nominal case, rather than its linearized version. However, the ARE is still based upon the linear Kalman filter. Such an implementation of the Kalman filter is called an extended Kalman filter (EKF).

2.8 Optimal Control

Consider a dynamic system given by the following state equation with a known initial condition:

$$\dot{x} = f(x, u, t), \quad x(t_0) = x_0, \tag{2.189}$$

where $u(t)$ is the control input vector bounded by constraints in a given interval, $t_0 \leq t \leq t_f$ (called admissible control input), and $f(.)$ is a continuous functional and has a continuous partial derivative with respect to state, $\partial f/\partial x$, in the given interval. The random disturbance inputs, $p(t), m(t)$, are excluded from the optimal control problem, but are taken into account in estimating the state vector from measurements. Let the transient performance objectives be specified in terms of a scalar function of control and state variables, $L[x(t), u(t), t]$, called the Lagrangian. For an acceptable performance, the system's response, $x(t)$, to the applied control input, $u(t)$, should be such that the Lagrangian is minimized with respect to the control input in a control interval, $t_0 \leq t \leq t_f$. Furthermore, the performance at the final time, t_f, is prescribed by another scalar function, $\varphi\left[x\left(t_f\right), t_f\right]$, called the terminal cost that must also be minimized. Hence, both transient and terminal performance objectives are combined into the following scalar objective function to be minimized with respect to the control input, $u(t)$:

$$J = \varphi\left[x\left(t_f\right), t_f\right] + \int_{t_0}^{t_f} L[x(t), u(t), t]\,\mathrm{d}t. \tag{2.190}$$

The optimization is subject to the equality constraint of Eq. (2.189), which must be satisfied by $x(t), u(t)$ at all times. This is ensured by adjoining the constraint equation to the Lagrangian in an augmented objective function, $\mathcal{J}$, as follows:

$$\begin{aligned}\mathcal{J} &= J + \lambda^T(t) f[x(t), u(t), t] \\ &= \varphi\left[x\left(t_f\right), t_f\right] + \int_{t_0}^{t_f} \{L[x(t), u(t), t] + \lambda^T(t)(f[x(t), u(t), t] - \dot{x})\}\mathrm{d}t,\end{aligned} \tag{2.191}$$

where $\lambda(t)$ is a vector of Lagrange multipliers (or co-state vector) of the same size as the order of system. The co-state vector must be determined from the optimization process and is related to the partial derivative of L with respect to f, when u is held constant:

$$\lambda^T = -\left(\frac{\partial L}{\partial f}\right)_u. \tag{2.192}$$

In addition to the dynamic state equation, there could be other equality and inequality constraints on the state vector, $x(t)$, and the control vector, $u(t)$, which must be satisfied during minimization of the objective function, $\mathcal{J}$. For simplicity of formulation, such constraints are being excluded here.

2.8.1 Euler–Lagrange Equations

The necessary conditions for the existence of a unique solution to the optimal control problem are called Euler–Lagrange equations, which can be stated for a fixed control interval as the following theorem.

Theorem 2.8.1 *If a finite, piecewise-continuous, admissible control vector, $u(t)$, of the system,*

$$\dot{x} = f(x, u, t),$$

minimizes the Hamiltonian,

$$H\left[x(t), u(t), t\right] = L\left[x(t), u(t), t\right] + \lambda^T(t) f\left[x(t), u(t), t\right],$$

where $f(.)$ is a continuous functional, and has a continuous partial derivative with respect to state, $\partial f / \partial x$, then there exists an absolutely continuous, co-state vector, $\lambda(t)$, non-zero in the fixed control interval, $t_0 \leq t \leq t_f$, such that the following conditions are satisfied:

$$\begin{aligned} \frac{\partial H}{\partial u} &= 0, \\ \dot{x} &= \frac{\partial H}{\partial \lambda}, \\ \dot{\lambda} &= -\left(\frac{\partial H}{\partial x}\right)^T, \\ \lambda^T\left(t_f\right) &= \left(\frac{\partial \varphi}{\partial x}\right)_{t=t_f}. \end{aligned}$$

Proof. See (Bryson and Ho 1975).

Since they are valid on a specific extremal trajectory, $x(t)$, the Euler-Lagrange equations guarantee minimization of the Hamiltonian with respect to only small variations in $u(t)$. Hence, there could exist other extremal trajectories that also satisfy the Euler–Lagrange equations. Since they are the necessary conditions for optimal control, the Euler–Lagrange equations can be solved to produce a specific extremal trajectory and the corresponding optimal control history, depending upon the conditions imposed upon the state and co-state variables. A major simplification occurs for time-invariant systems, that is, systems whose dynamic state equations as well as the Lagrangian do not explicitly depend upon time, t. For a time-invariant system, we can write

$$\dot{x} = f(x, u), \quad x\left(t_0\right) = x_0, \tag{2.193}$$

and

$$H\left[x(t), u(t)\right] = L\left[x(t), u(t)\right] + \lambda^T(t) f\left[x(t), u(t)\right]. \tag{2.194}$$

Differentiating Eq. (2.194) with time, we have

$$\begin{aligned} \dot{H} &= \frac{\partial L}{\partial t} + \dot{\lambda}^T f + \lambda^T\left(\frac{\partial f}{\partial x}\dot{x} + \frac{\partial f}{\partial u}\dot{u}\right) \\ &= \left(\frac{\partial L}{\partial x} + \lambda^T \frac{\partial f}{\partial x}\right)\dot{x} + \left(\frac{\partial L}{\partial u} + \lambda^T \frac{\partial f}{\partial u}\right)\dot{u} + \dot{\lambda}^T f \end{aligned} \tag{2.195}$$

or

$$\dot{H} = H_x \dot{x} + H_u \dot{u} + \dot{\lambda}^T f. \tag{2.196}$$

All the derivatives are evaluated at the optimal point, for which the Euler–Lagrange equations dictate

$$\begin{aligned} H_u &= 0 \\ -H_x &= \dot{\lambda}^T. \end{aligned}$$

Therefore, we have

$$\dot{H} = \dot{\lambda}^T (f - \dot{x}) = 0 \tag{2.197}$$

or $H = \text{const}$. Thus, the Hamiltonian remains constant along the optimal trajectory for a time-invariant problem. The most commonly used terminal cost and the Lagrangian functions for a time-invariant problem are of the following quadratic forms:

$$\varphi\left[x\left(t_f\right)\right] = \left[x\left(t_f\right) - x\right]^T Q_f \left[x\left(t_f\right) - x\right] \tag{2.198}$$

and

$$L\left[x(t), u(t)\right] = \left(x^T, u^T\right) \begin{pmatrix} Q & S \\ S^T & R \end{pmatrix} \begin{pmatrix} x \\ u \end{pmatrix}. \tag{2.199}$$

Here, Q_f, Q, S and R are constant matrices known as cost coefficients that specify the relative penalties (weightages) in minimizing the deviations in terminal-state, state and control variables. Since a quadratic objective function results from a second-order Taylor series expansion about the optimal trajectory and penalizes large deviations more than the small ones, it is a logical choice in any practical optimal control problem.

2.8.2 *Linear, Quadratic Optimal Control*

The basis of robust control systems design is the optimal control theory applied to linear systems. Here, we shall confine the treatment to the regulator problem, which is of interest in ASE systems. Consider an aeroelastic plant with state vector, $\xi(t)$, and control input vector, $\eta(t)$, governed by the following state equation:

$$\dot{\xi} = f(\xi, \eta, t). \tag{2.200}$$

Let an extremal trajectory, $\xi_d(t)$, and the corresponding extremal control history, $\eta_d(t)$, be available from the necessary conditions for the solution of the optimal control problem minimizing an objective function,

$$J(\xi, \eta) = \varphi\left[\xi\left(t_f\right), t_f\right] + \int_{t_0}^{t_f} L\left[\xi(t), \eta(t), t\right] \mathrm{d}t, \tag{2.201}$$

subject to certain specific constraints. The extremal control and trajectory are the nominal functions satisfying Eq. (2.200),

$$\dot{\xi}_d = f\left(\xi_d, \eta_d, t\right), \tag{2.202}$$

about which small control and state deviations,

$$u(t) = \eta(t) - \eta_d(t); \qquad t_0 \le t \le t_f \tag{2.203}$$

and

$$x(t) = \xi(t) - \xi_d(t); \qquad t_0 \le t \le t_f, \tag{2.204}$$

are to be minimized. Employing a first-order Taylor series expansion about the nominal trajectory, we have the following linear, state equation governing small, off-nominal deviations:

$$\dot{x}(t) = A(t)x(t) + B(t)u(t); \qquad x\left(t_0\right) = x_0, \tag{2.205}$$

where

$$A(t) = \left.\frac{\partial f}{\partial \xi}\right|_{\xi_d, \eta_d} ; \quad B(t) = \left.\frac{\partial f}{\partial \eta}\right|_{\xi_d, \eta_d} \tag{2.206}$$

are the Jacobian matrices of the expansion. In a similar manner, the objective function can be expanded about the extremal solution, (ξ_d, η_d), up to the second-order terms

$$J(\xi_d + x, \eta_d + u) \simeq J(\xi_d, \eta_d) + \Delta^2 J(x, u). \tag{2.207}$$

The first variation of J about the extremal trajectory is identically zero,

$$\Delta J(x, u) = \left.\frac{\partial J}{\partial \xi}\right|_{\xi_d, \eta_d} x + \left.\frac{\partial J}{\partial \eta}\right|_{\xi_d, \eta_d} u = 0. \tag{2.208}$$

The second variation of J about the extremal trajectory is given by

$$\begin{aligned} \Delta^2 J(x, u) = & \frac{1}{2} x^T(t_f) Q_f x(t_f) \\ & + \frac{1}{2} \int_{t_0}^{t_f} \{x^T(t), u^T(t)\} \begin{bmatrix} Q(t) & S(t) \\ S^T(t) & R(t) \end{bmatrix} \begin{Bmatrix} x(t) \\ u(t) \end{Bmatrix}, \end{aligned} \tag{2.209}$$

which is a quadratic form with the following cost coefficient matrices:

$$Q_f = \left.\frac{\partial^2 \varphi}{\partial \xi^2}\right|_{\xi_d(t_f)} ; \quad Q(t) = \left.\frac{\partial^2 L}{\partial \xi^2}\right|_{\xi_d, \eta_d} \tag{2.210}$$

and

$$S(t) = \left.\frac{\partial^2 L}{\partial \xi \partial \eta}\right|_{\xi_d, \eta_d} ; \quad R(t) = \left.\frac{\partial^2 L}{\partial \eta^2}\right|_{\xi_d, \eta_d}. \tag{2.211}$$

Hence, the second variation of the objective function, $\Delta^2 J$, about the extremal trajectory is a quadratic cost function, which must be minimized, subject to a linearized dynamic equation for a neighbouring extremal trajectory. This forms the basis of the linear, quadratic, optimal control problem for neighbouring extremal trajectories.

The Hamiltonian corresponding to the quadratic cost function, $\Delta^2 J$, subject to linear dynamic constraint of Eq. (2.205) is the following:

$$\begin{aligned} H = & \frac{1}{2} x^T(t) Q(t) x(t) + x^T(t) S(t) u(t) + \frac{1}{2} u^T(t) R(t) u(t) \\ & + \lambda^T(t) [A(t) x(t) + B(t) u(t)]. \end{aligned} \tag{2.212}$$

The necessary conditions for optimality with a fixed terminal time, t_f, are then given by Euler–Lagrange equations (Theorem 2.8.1):

$$\dot{\lambda} = -\left(\frac{\partial H}{\partial x}\right)^T = -Q(t) x(t) - S(t) u(t) - A^T(t) \lambda(t), \tag{2.213}$$

$$\lambda(t_f) = \left(\frac{\partial \varphi}{\partial x}\right)^T_{t=t_f} = Q_f x(t_f), \tag{2.214}$$

$$\frac{\partial H}{\partial u} = 0 = S^T(t) x(t) + R(t) u(t) + B^T(t) \lambda(t). \tag{2.215}$$

Equation (2.215) is solved for the optimal control as follows:

$$u(t) = -R^{-1}(t)\,[S^T(t)\,x(t) + B^T(t)\,\lambda(t)]. \tag{2.216}$$

Substitution of Eq. (2.216) into Eqs. (2.205) and (2.213) results in the following set of linear state and co-state equations:

$$\dot{x} = [A(t) - B(t)R^{-1}(t)S^T(t)]x(t) - B(t)R^{-1}(t)B^T(t)\,\lambda(t), \tag{2.217}$$

$$\dot{\lambda} = -[A^T(t) - S(t)R^{-1}(t)B^T(t)]\lambda(t) + [S(t)R^{-1}(t)S^T(t) - Q(t)]x(t), \tag{2.218}$$

which must be solved subject to the boundary conditions,

$$x(t_0) = x_0; \qquad \lambda(t_f) = Q_f x(t_f). \tag{2.219}$$

The linear, two-point boundary value problem (TPBVP) given by Eqs. (2.217)–(2.219) must be integrated in time, such that the boundary conditions are satisfied. However, as the state and co-state vectors are related by Eq. (2.214) at the final time, a solution is sought to Eqs. (2.217)–(2.219), which satisfies a state transition matrix ensuring the linear independence of solutions, $x(t)$ and $\lambda(t)$. To this end, the state and co-state equations are expressed as follows:

$$\begin{Bmatrix} \dot{x} \\ \dot{\lambda} \end{Bmatrix} = \begin{pmatrix} A - BR^{-1}S^T & -BR^{-1}B^T \\ SR^{-1}S^T - Q & -A^T + SR^{-1}B^T \end{pmatrix} \begin{Bmatrix} x \\ \lambda \end{Bmatrix}. \tag{2.220}$$

These must satisfy the boundary conditions of Eq. (2.219). The solution is obtained by integrating backward in time from $t = t_f$, for which we have

$$\begin{Bmatrix} x(t) \\ \lambda(t) \end{Bmatrix} = \begin{pmatrix} X(t) & 0 \\ 0 & \Lambda(t) \end{pmatrix} \begin{Bmatrix} x(t_f) \\ \lambda(t_f) \end{Bmatrix}, \tag{2.221}$$

where $x(t)$ and $\lambda(t)$ are the transition matrices corresponding to the backward evolution in time of $x(t)$ and $\lambda(t)$, respectively. Clearly, the following must be satisfied:

$$\lambda(t) = \Lambda(t)\,\lambda(t_f) = \Lambda(t)\,Q_f x(t_f), \tag{2.222}$$

$$x(t) = X(t)\,x(t_f), \tag{2.223}$$

and

$$x(t_0) = X(t_0)\,x(t). \tag{2.224}$$

Inversion of Eq. (2.223) and substitution into Eq. (2.222) yields

$$\lambda(t) = \Lambda(t)\,Q_f X^{-1}(t)\,x(t). \tag{2.225}$$

Since both the transition matrices must satisfy

$$X(t_f) = I; \qquad \Lambda(t_f) = I, \tag{2.226}$$

Eq. (2.225) represents the adjoint relationship between the solutions, $x(t)$ and $\lambda(t)$, written as follows:

$$\lambda(t) = P(t)\,x(t), \tag{2.227}$$

where $P(t) = \lambda(t) Q_f x^{-1}(t)$. Substituting Eq. (2.227) into Eqs. (2.216), we have the following linear, optimal feedback control law:

$$u(t) = -R^{-1}(t)\left[B^T(t)P(t) + S^T(t)\right]x(t). \tag{2.228}$$

Taking the time derivative of Eq. (2.227) and substituting into Eqs. (2.217), (2.218) and (2.228), we have the following matrix Riccati equation (MRE) to be satisfied by the matrix, $P(t)$:

$$\begin{aligned} -\dot{P} &= Q + \left(A - BR^{-1}S^T\right)^T P + P\left(A - BR^{-1}S^T\right) \\ &\quad - PBR^{-1}B^T P - SR^{-1}S^T, \end{aligned} \tag{2.229}$$

which must be solved subject to the boundary condition,

$$P\left(t_f\right) = Q_f. \tag{2.230}$$

A sufficient condition for optimality is the existence of a positive definite solution to MRE, $P(t)$, for all times in the control interval, $t_0 \leq t \leq t_f$. Equations (2.228)–(2.230) give the solution to the optimal LQR problem based on state feedback for tracking a nominal, optimal trajectory, $\xi_d(t)$, and is guaranteed to result in a neighbouring optimal trajectory, $\xi_d(t) + x(t)$, and is derived here from the necessary conditions of optimality (the Euler–Lagrange equations). An alternative derivation of the MRE is possible from the sufficient condition of optimality (the Hamilton–Jacobi–Bellman equation) (Athans and Falb 2007). Therefore, the MRE reflects both necessary and sufficient conditions for the existence of an optimal control law for linear systems with a quadratic performance index.

The MRE requires iterative numerical solution methods. Although simple iterative schemes based on repeated linear, algebraic solutions are usually applied when the coefficient matrices are either slowly or periodically varying, the convergence to a positive definite (or even positive semi-definite) solution is not always guaranteed. Given the complexity of an MRE solution, it is often much easier to directly solve the linear state and co-state equations, Eqs. (2.217)–(2.219), by either the shooting or collocation methods (Tewari 2011). Linearity of the adjoint system of equations, Eq. (2.220), assures the existence of a transition matrix, $\Phi\left(t, t_0\right)$, such that

$$\left\{\begin{matrix} x(t) \\ \lambda(t) \end{matrix}\right\} = \Phi\left(t, t_0\right)\left\{\begin{matrix} x\left(t_0\right) \\ \lambda\left(t_0\right) \end{matrix}\right\}, \tag{2.231}$$

with the boundary conditions

$$x\left(t_0\right) = x_0; \qquad \lambda\left(t_f\right) = Q_f x\left(t_f\right). \tag{2.232}$$

The transition matrix has the properties

$$\Phi\left(t_0, t_0\right) = I; \qquad \Phi\left(t_0, t\right) = \Phi^{-1}\left(t, t_0\right) \tag{2.233}$$

and

$$\dot{\Phi}\left(t, t_0\right) = \begin{pmatrix} A - BR^{-1}S^T & -BR^{-1}B^T \\ SR^{-1}S^T - Q & -A^T + SR^{-1}B^T \end{pmatrix}\Phi\left(t, t_0\right), \tag{2.234}$$

as well as the special property of being a symplectic matrix, that is,

$$\Phi^T\left(t, t_0\right)\begin{pmatrix} 0 & I \\ -I & 0 \end{pmatrix}\Phi\left(t, t_0\right)=\begin{pmatrix} 0 & I \\ -I & 0 \end{pmatrix}. \tag{2.235}$$

Partitioning $\Phi\left(t, t_0\right)$ as follows:

$$\begin{Bmatrix} x(t) \\ \lambda(t) \end{Bmatrix}=\begin{pmatrix} \Phi_{xx}\left(t, t_0\right) & \Phi_{x\lambda}\left(t, t_0\right) \\ \Phi_{\lambda x}\left(t, t_0\right) & \Phi_{\lambda\lambda}\left(t, t_0\right) \end{pmatrix}\begin{Bmatrix} x\left(t_0\right) \\ \lambda\left(t_0\right) \end{Bmatrix}, \tag{2.236}$$

the symplectic nature of the transition matrix implies that

$$\Phi^{-1}\left(t, t_0\right)=\begin{pmatrix} \Phi^T_{\lambda\lambda}\left(t, t_0\right) & -\Phi^T_{x\lambda}\left(t, t_0\right) \\ -\Phi^T_{\lambda x}\left(t, t_0\right) & \Phi^T_{xx}\left(t, t_0\right) \end{pmatrix}, \tag{2.237}$$

which is very useful in carrying out the matrix operations required for the solution of the boundary-value problem.

2.9 Robust Control Design by LQG/LTR Synthesis

Most tracking control problems require dissipation of all errors to zero when the time becomes large compared with the time-scale of plant dynamics. Since the control interval is quite large, it is unimportant to account for the relatively much faster variation of plant parameters, which can – in many cases – be averaged out over a long period, and essentially approximated by LTI systems where the plant coefficient matrices, A, B, C, D, are constants. Such a problem with the objective of a zero, steady-state (i.e. as $t \to \infty$) error is referred to as infinite-horizon control, because the control interval, t_f, is taken to be infinite. In such cases, both LQR and Kalman filter designs are greatly simplified by having an infinite control interval, for which the corresponding solutions, $P(t), R_e(t, t)$, approach steady-state values given by, $P(\infty), R_e(\infty, \infty)$, expressed simply as the constants, P, R_e. The governing equation for a time-invariant LQR problem is derived simply by substituting $\dot{P}=0$ in MRE, Eq. (2.229), resulting in the following ARE:

$$\begin{aligned} 0 &= (A - BR^{-1}S^T)^T P + P(A - BR^{-1}S^T) \\ &\quad - PBR^{-1}B^T P + Q - SR^{-1}S^T. \end{aligned} \tag{2.238}$$

The optimal feedback control law is obtained from the algebraic Riccati solution,

$$u(t) = -R^{-1}(B^T P + S^T)x(t), \tag{2.239}$$

where the cost coefficient matrices, Q, R, S, are constants. For asymptotic stability of the regulated system, all the eigenvalues of the closed-loop dynamics matrix,

$$A - BR^{-1}(B^T P + S^T),$$

must be in the left-half s-plane, which requires that the ARE solution, P must be a symmetric, positive semi-definite matrix (Tewari 2002), that is, a matrix with all eigenvalues being either greater than, or equal to zero. There may not always be a positive semi-definite solution; on the other hand, there could be several such solutions of which it cannot be determined which one is to be regarded as the best one. However, it can be proved (Glad and Ljung 2002) that if the following sufficient conditions are satisfied, there exists a unique, symmetric, positive semi-definite solution to the ARE:

- The control cost coefficient matrix, R, is symmetric and positive definite, the matrix $(Q - SR^{-1}S^T)$ is symmetric and positive semi-definite and the pair $(A - BR^{-1}S^T, Q - SR^{-1}S^T)$ is detectable (if not observable).
- The pair (A, B) is either controllable or, at least, stabilizable.

It is interesting to note the equivalence between the Kalman filter and the LQR, in that both the LQR gain and the Kalman filter gain are based on the solution to the same ARE. This duality is shown in Table 2.3.

The time-invariant LQR regulator has extremely nice robustness properties. Consider the LTI plant with an LQR regulator of constant gain matrix, K. The transfer matrix representation of the regulated system is depicted in Fig. 2.13. The system is described by

$$\dot{x} = Ax + Bu, \tag{2.240}$$

with optimal control-law,

$$u(t) = -R^{-1}(B^T P + S^T)x = -Kx. \tag{2.241}$$

The loop gain at plant input (point marked by '(1)' in Fig. 2.13) is given by

$$H(s)G(s) = K(sI - A)^{-1}B, \tag{2.242}$$

Table 2.3 Duality between the LQR regulator and the Kalman filter

Kalman Filter	$\Longleftrightarrow$	LQR
A^T	$\Longleftrightarrow$	A
C^T	$\Longleftrightarrow$	B
S_m	$\Longleftrightarrow$	R
FS_pF^T	$\Longleftrightarrow$	Q
FS_{pm}	$\Longleftrightarrow$	S

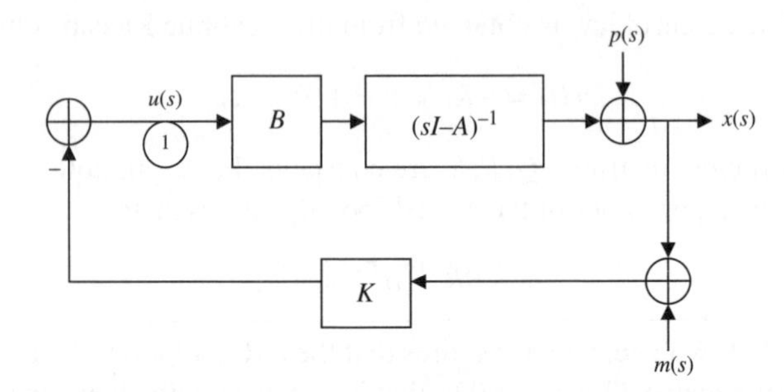

Figure 2.13 Transfer matrix representation of the linear, quadratic regulator (LQR)

while the return difference matrix at the same point is

$$R_i(s) = I + H(s)G(s) = I + K(sI - A)^{-1}B. \tag{2.243}$$

Since the sensitivity of the plant input $u(t)$ with respect to the noise inputs, $p(t), m(t)$, is proportional to the magnitude of $R_i^{-1}(s)$, large singular values of $R_i(s)$ would imply high robustness with respect to both process and measurement noise. By substituting Eqs. (2.243) and Eq. (2.241) into the ARE Eq. (2.238) and simplifying, it can be shown that for a positive semi-definite matrix Q and a positive definite matrix R, the following must be satisfied:

$$R_i^H(s)RR_i(s) \geq R, \tag{2.244}$$

from which it follows that if $R = \rho I$, the following must be true:

$$R_i^H(s)R_i(s) \geq I, \tag{2.245}$$

or

$$\sigma\{R_i(s)\} \geq 1. \tag{2.246}$$

Hence, the smallest singular value of the input return difference matrix never drops below unity. For a SISO system, this implies that the Nyquist locus never enters the unit circle centred at $(-1, 0)$, which implies a phase margin of at least 60 deg, with gain margin ranging from -6 dB to infinity. Such high levels of robustness are greatly valued in practical implementation, which is the main reason why LQR controllers are a common choice in a wide range of applications. However, the high level of robustness comes at the cost of high regulator gains, which has associated control magnitude (energy) and saturation issues. With increased controller bandwidth, there is also the possibility of high-frequency noise amplification due to increased complementary sensitivity. Thus the theoretically high robustness levels are almost never achieved in practice with high feedback gains. Instead, other means of increasing robustness must be sought, which do not rely upon high controller bandwidth.

In the stochastic sense, a constant error covariance matrix, R_e, implies a stationary white noise process. If the estimation error of a linear system is stationary, the system must be driven by stationary processes. Therefore, an LTI Kalman filter essentially involves the assumption of stationary, zero-mean, Gaussian white (ZMGWN) models for both process noise, $p(t)$, and measurement noise, $m(t)$. Hence, the error covariance matrix must now satisfy the following ARE:

$$0 = A_G R_e + R_e A_G^T - R_e C^T S_m^{-1} C R_e + F(S_p - S_{pm} S_m^{-1} S_{pm}^T)F^T, \tag{2.247}$$

where S_m, S_p, S_{pm} are constant spectral density matrices and

$$A_G = A - F S_{pm} S_m^{-1} C. \tag{2.248}$$

The constant Kalman filter gain matrix is the following:

$$L = (R_e C^T + F S_{pm}) S_m^{-1}. \tag{2.249}$$

Clearly, the ARE for the Kalman filter must also have a symmetric, positive semi-definite solution, R_e, for an asymptotically stable Kalman filter dynamics. Furthermore, by satisfying sufficient conditions that are dual to those stated above for the LQR problem, a unique, positive semi-definite solution to the ARE can be found. The ARE is thus at the heart of both LQR and Kalman filter design for LTI systems. Being a nonlinear algebraic equation, it must be solved numerically, such as by iteration of the following Lyapunov equation for a symmetric matrix, P:

$$AP + PA^T + Q = 0. \tag{2.250}$$

There are several efficient algorithms for iteratively solving the ARE, which are programmed into commercially available software, such as MATLAB®.

The procedure by which an LQR and a Kalman filter are designed separately for an LTI plant, and then put together to form a feedback compensator is referred to as the linear, quadratic, Gaussian (LQG) method. The resulting feedback compensator is called an LQG compensator. Figure 2.14 depicts a general case where the output vector, $y(t)$, is to match a desired output (also called a reference or commanded output), $\mathbf{y_d}(t)$, in the steady state. Such a reference output is usually commanded by the terminal controller (not shown in the figure). Clearly, the measured signal given to Kalman filter is $\left[y(t) - \mathbf{y_d}(t)\right]$, based on which (as well as the known input vector, $u(t)$) it supplies the estimated state for feedback to the LQR regulator. Since the design of the LQG compensator – specified by the gain matrices, $(\mathbf{K}, L)$ – depends upon the chosen LQR cost parameters, $\mathbf{Q}, R, S$, and the selected Gaussian white-noise spectral densities, S_m, S_p, S_{pm}, it is possible to design infinitely many compensators for a given plant. Usually, there are certain performance and robustness requirements specified for the closed-loop system that indirectly restrict the choice of the cost parameters to a given range. Being based on optimal control, an LQG compensator has excellent performance features for a given set of cost parameters, but its robustness is subject to the extent the performance is degraded by state estimation through the Kalman filter. If the filter gains are too small, the estimation error does not tend to zero fast enough for the feedback to be accurate. On the other hand, if the Kalman filter has very large gains, there is an amplification of process and measurement noise by feedback, thereby reducing the overall robustness of the control system. Clearly, a balance must be struck in selecting the Kalman filter design parameters, S_m, S_p, S_{pm}, such that a good robustness is obtained without unduly sacrificing performance.

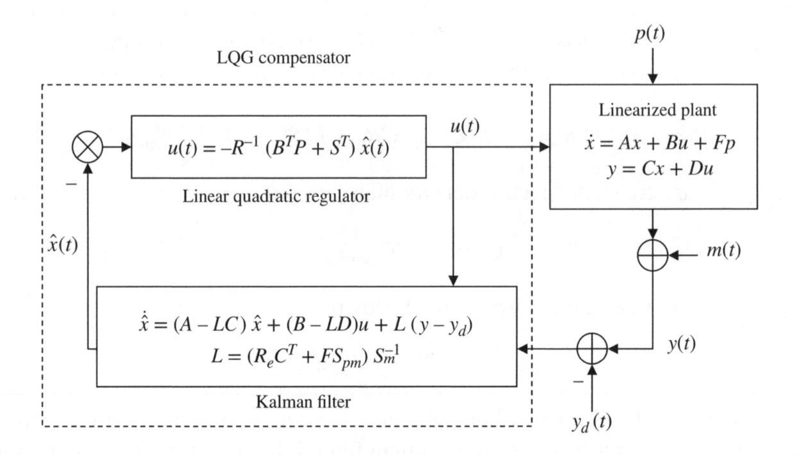

Figure 2.14 Linear, quadratic, Gaussian (LQG) compensator for linearized plant dynamics

To study the robustness of an LQG compensated system, refer to the block diagram of the control system transformed to the Laplace domain in a negative feedback configuration, as shown in Fig. 2.15. For simplicity, consider a strictly proper plant (i.e $D = 0$) of order n represented by the transfer matrix,

$$G(s) = C(sI - A)^{-1}B$$

of dimension $\ell \times m$, where ℓ is the number of outputs, and m the number of inputs. An LQG compensator of dimension $m \times \ell$ has the transfer matrix,

$$H(s) = K(sI - A + BK + LC)^{-1}L.$$

The process noise is represented by a ZMGWN disturbance, $p(s)$, appearing at the plant's output, while the ZMGWN measurement noise, $m(s)$, affects the feedback loop as shown. The overall system's transfer matrix, $T(s)$, from the desired output to the actual output, is the transmission matrix (which was discussed previously). On the other hand, the effect of the process noise on the output is given by the transfer matrix, $S(s)$, which is the sensitivity matrix. Both $T(s)$ and $S(s)$ are derived (with reference to Fig. 2.15) as follows:

$$y = p + Gu = p + G[H(y_d - y - m)], \tag{2.251}$$

or

$$(I + GH)y = p + GH(y_d - m), \tag{2.252}$$

thereby implying

$$y = (I + GH)^{-1}p + (I + GH)^{-1}GH(y_d - m), \tag{2.253}$$

or

$$y(s) = S(s)p(s) + T(s)[y_d(s) - m(s)], \tag{2.254}$$

where

$$\begin{aligned} S(s) &= [I + G(s)H(s)]^{-1} \\ T(s) &= [I + G(s)H(s)]^{-1}G(s)H(s). \end{aligned} \tag{2.255}$$

Because it is true that $T(s) = I - S(s)$, the transmission matrix is also called the complementary sensitivity matrix.

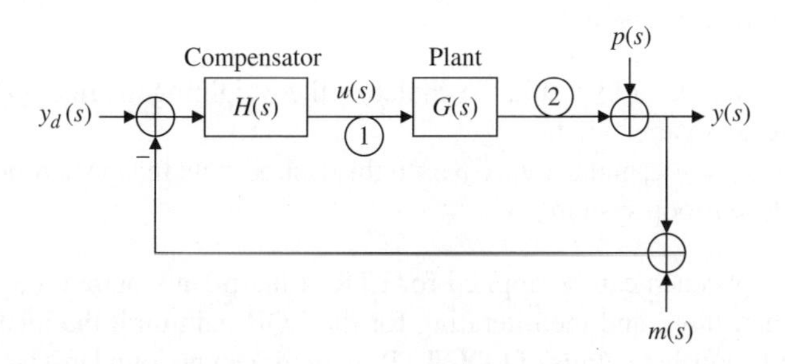

Figure 2.15 Transfer matrix representation of an LQG compensated system

Since the introduction of Kalman filter into the system degrades the overall complementary sensitivity, it is necessary to select the Kalman filter and regulator gains such that a measure of robustness is retained. We have already discussed in the earlier part of this chapter how a good overall robustness can be achieved by loop-shaping, such that the singular values of the sensitivity matrix and the transmission matrix are kept within pre-selected bounds. Since the two requirements are conflicting, a design compromise is obtained by choosing different ranges of frequencies for the minimization of $\bar{\sigma}(S)$ and $\bar{\sigma}(T)$. This is practically achieved by assigning suitable weightages to the cost parameters of the LQG design, and the resulting compensator is said to have recovered the loop-transfer function, $G(s)H(s)$, which would have been present if the state variables were directly fed to the regulator. The systematic design procedure for loop-transfer recovery (LTR), which attempts to regain the robustness of the LQR regulator by an iterative selection of either the regulator or the Kalman filter gains is termed the LQG/LTR method (Maciejowski 1989).

To further indicate the LQG/LTR procedure, refer to Fig. 2.15, and note that the transfer matrix, $H(s)G(s)$, denotes the transfer of the input, $u(s)$, back to itself if the loop is broken at the point marked '(1)'. Hence, $H(s)G(s)$ is the return ratio at the plant's input. If all the states are available for measurement, there would be no need for a Kalman filter, hence the ideal return ratio at input is given by

$$H(s)G(s) = K(sI - A)^{-1}B.$$

On the other hand, if the feedback loop is broken at the point marked '(2)' of Fig. 2.15, then the transfer matrix, $G(s)H(s)$, represents the return ratio at output. If there is no regulator in the system, then the ideal return-ratio at output is given by

$$G(s)H(s) = C(sI - A)^{-1}L.$$

If the ideal return ratio is recovered at either the plant's input, or the output, by suitably designing the LQG compensator, the best possible combination of the LQR regulator and the Kalman filter is achieved. If (for simplicity) the process and measurement noise are uncorrelated, that is, $S_{pm} = 0$, then it can be shown (Maciejowski 1989) that by selecting

$$F = B; \qquad S_p = \rho S_m$$

and making the positive scalar parameter ρ arbitrarily large, the LQG return-ratio at input,

$$H(s)G(s) = K(sI - A + BK + LC)^{-1}LC(sI - A)^{-1}B$$

can be made to approach the ideal return ratio at input. The following procedure for LTR at the input can therefore be applied:

(a) Select an LQR regulator by a suitable choice of the weighting matrices Q, R, S such that a desired robustness is obtained.
(b) Select $F = B, S_p = \rho S_m$ and increase ρ until the desired state feedback robustness is recovered in the closed-loop system.

An alternative approach can be applied for LTR at the plant's output, beginning with the design of a Kalman filter, and then iterating for the LQR gain until the ideal return ratio at output is recovered. Further details of LQG/LTR methods can be found in Maciejowski (1989) and Glad and Ljung (2002).

2.10 H_2/H_∞ Design

The state-space design by the separation principle using a regulator and a Kalman filter, and followed by loop shaping is a commonly used design method for achieving a desired level of stability robustness in multivariable systems. An example of such an approach was given in the previous section via the LQG/LTR synthesis. However, we recall that the stability robustness properties were enforced on the sensitivity and complementary sensitivity matrices of the system in a manner quite similar to the SISO design by Nyquist type techniques in the frequency domain. Therefore, it is logical to ask why is it at all necessary to start with the state-space design, if the robustness properties are to be ensured in the frequency domain. Hence an alternative design procedure can be evolved based on the frequency domain design methodology commonly used for single-variable (SISO) systems. The state feedback regulator and the Kalman filter then follow naturally where the singular value spectra replace the Bode gain plot, and parameters analogous to the gain and phase margins of a SISO system are extended to stability robustness measures of multivariable systems. In deriving the transfer matrix, $H(s)$, of a stabilizing controller, one minimizes a combined, frequency-weighted scalar measure of sensitivity, complementary sensitivity and transfer matrix from disturbance to plant output, over a range of frequencies. This is the basis of the H_2 and H_∞ synthesis procedure.

Consider a strictly proper plant transfer matrix realization, $G(s)$, with control inputs, $u(s)$, and measured outputs, $y(s)$. All exogenous inputs to the feedback control system, namely, the process noise, $p(s)$, appearing at plant's input and the measurement noise, $m(s)$, appearing at the plant's output, are clubbed together in a single vector, $w(s)$, called external disturbances, as shown in Fig. 2.16. The plant's transfer matrix description is thus the following:

$$y(s) = G(s)\,u(s) + w(s)\,. \tag{2.256}$$

The design problem is to find a stabilizing output feedback controller, $H(s)$, with the feedback control law

$$u(s) = -H(s)\,y(s)\,, \tag{2.257}$$

such that the control system has adequate stability robustness with respect to the random disturbance inputs, $w(s)$, which are not modelled in any way. In terms of loop-shaping terminology, this translates into simultaneously minimizing the norms of the following transfer matrices:

$$\begin{aligned} S(s) &= [I + G(s)H(s)]^{-1}, \\ T(s) &= I - S(s) = [I + G(s)H(s)]^{-1}G(s)H(s) = G(s)H(s)[I + G(s)H(s)]^{-1} \\ G_{wu}(s) &= -H(s)S(s) \end{aligned} \tag{2.258}$$

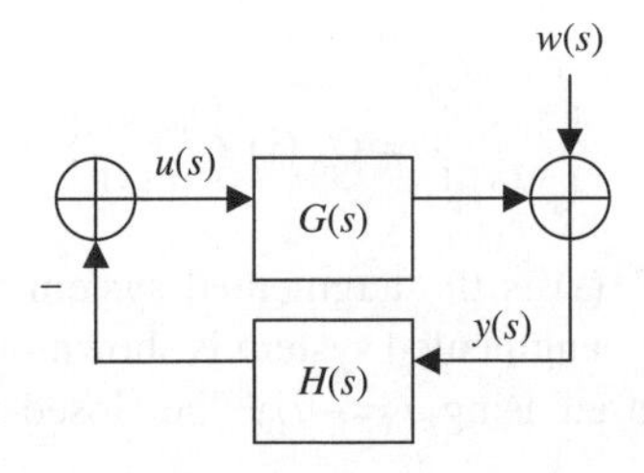

Figure 2.16 Basic stabilization system for H_2/H_∞ design

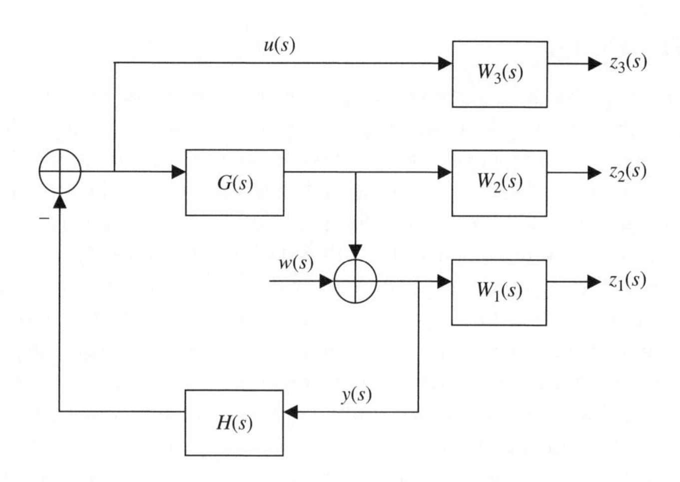

Figure 2.17 Augmented plant and closed-loop system for H_2/H_∞ design

over a given frequency range. However, this is impossible to carry out because the minimization of both sensitivity, $S(s)$, and complementary sensitivity, $T(s)$, are conflicting requirements. Furthermore, the transfer matrix, $G_{wu}(s)$, from the disturbance to the input variables is maximized when $T(s)$ is minimized. The only possibility of achieving small magnitudes of these matrices is to minimize them at different frequencies, which is implemented through pre-multiplying each matrix by a different frequency-weighting matrix, and minimizing the norm of the following array over a specified frequency range:

$$\begin{pmatrix} W_1(s)S(s) \\ W_2(s)T(s) \\ W_3(s)G_{uw}(s) \end{pmatrix}. \tag{2.259}$$

Here, the weighting matrix $W_1(s)$ is strictly proper and square, while $W_2(s), W_3(s)$ are square matrices. Introduction of frequency weights is tantamount to augmenting the original system by additional output variables, $z_1(s), z_2(s), z_3(s)$, called error signals. The transfer matrix description of the augmented system is given by

$$\begin{Bmatrix} z_1(s) \\ z_2(s) \\ z_3(s) \\ y(s) \end{Bmatrix} = \begin{pmatrix} W_3(s) & 0 \\ W_2(s)G(s) & 0 \\ W_1(s)G(s) & W_1(s) \\ G & I \end{pmatrix} \begin{Bmatrix} u(s) \\ w(s) \end{Bmatrix} \tag{2.260}$$

or, in an abbreviated form,

$$\begin{Bmatrix} z(s) \\ y(s) \end{Bmatrix} = G_a(s) \begin{Bmatrix} u(s) \\ w(s) \end{Bmatrix} \tag{2.261}$$

where $z^T = \left(z_1^T, z_2^T, z_3^T\right)$ and $G_a(s)$ is the augmented system's transfer matrix given in Eq. (2.260). A block-diagram of the augmented system is shown in Fig. 2.17.

When the control loop is closed using $u = -Hy$, the closed-loop system has the following description:

$$z(s) = G_c(s)w(s), \tag{2.262}$$

where

$$G_c(s) = \begin{pmatrix} -W_3(s)\,G_{uw}(s) \\ -W_2(s)\,T(s) \\ W_1(s)\,S(s) \end{pmatrix}. \tag{2.263}$$

Thus H_2/H_∞ design problem is to find a stabilizing controller $H(s)$ that minimizes either the H_2 norm or the H_∞ norm of the closed-loop transfer matrix, $G_c(i\omega)$; that is, either

$$\|G_c\| = \sqrt{\frac{1}{2\pi}\int_{-\infty}^{\infty} \mathrm{tr}[G_c(i\omega)\,G_c^T(-i\omega)]\mathrm{d}\omega} \tag{2.264}$$

or

$$\|G_c\|_\infty = \sup_\omega[\bar{\sigma}\{G_c(i\omega)\}], \tag{2.265}$$

must be minimized with respect to $H(s)$ such that all the poles of $G_c(s)$ are in the left-half plane. The success of the design process depends entirely upon the ability to find a suitable set of weighting matrices, $W_1(s), W_2(s), W_3(s)$, for which a feasible solution to the constrained minimization problem exists.

2.10.1 H_2 Design Procedure

The H_2 norm of $G_c(s)$ can be expressed as follows:

$$\|G_c\|_2 = \|W_1 S\|_2 + \|W_2 T\|_2 + \|W_3 G_{uw}\|_2. \tag{2.266}$$

Hence, the minimization of $\|G_c\|_2$ ensures a simultaneous minimization of the H_2 norm of the weighted sensitivity, complementary sensitivity and $G_{uw}(s)$. The power spectral density of the error vector, $z(t)$, for a white noise disturbance, $w(t)$, of unit intensity is the following:

$$S_z(\omega) = Z(i\omega)\,Z^T(-i\omega) = G_c(i\omega)\,G_c^T(-i\omega). \tag{2.267}$$

Therefore, the minimization of $\|G_c\|$ results in a minimization of the error power spectral density, which is an objective quite similar to that of the LQG compensator. Hence the H_2 controller design is quite similar to that in the LQG case.

The augmented plant, Eq. (2.260), can be represented in an LTI state-space form as follows:

$$\begin{aligned} \dot{x} &= Ax + Bu + Fw, \\ y &= Cx + w, \\ z &= Mx + Nu. \end{aligned} \tag{2.268}$$

The controller design, being based on output feedback, requires the following inherent observer dynamics as a part of the augmented plant:

$$\dot{\hat{x}} = (A - FC)\hat{x} + Bu + Fy. \tag{2.269}$$

Since the coefficients A, F, C depend upon the chosen frequency weights, a stable observer requires a judicious selection of the frequency weights. The optimal H_2 synthesis then consists of deriving an LQR regulator with $Q = M^T M, R = I$ and gain, $K = B^T P$ such that

$$H(s) = -K(sI - A + BB^T P + FC)^{-1}F, \tag{2.270}$$

where P is a symmetric, positive semi-definite solution to the following Riccati equation:

$$0 = A^T P + PA - PBB^T P + M^T M. \tag{2.271}$$

For simplicity, it is assumed that

$$N^T M = 0; \qquad N^T N = I. \tag{2.272}$$

A possible method of ensuring a Hurwitz observer matrix, $(A - FC)$ is by replacing F with a Kalman filter gain, L, which requires a particular structure for the frequency weights. The regulator and observer gains can be further modified by loop shaping, as is done in the LQG/LTR process, until a satisfactory singular-value spectrum is obtained for the sensitivity and complementary sensitivity matrices.

2.10.2 H_∞ Design Procedure

While the H_2 design could be carried out in a manner similar to the LQG design, the H_∞ norm minimization is based on an iterative solution. However, there being no direct relationship between H_∞ norm of the closed-loop transfer matrix and the objective function of an LQG design now, it is difficult to know in advance what the minimum value of $\|G_c\|_\infty$ would be for an acceptable design. A practical method (Glover and Doyle 1988) is to choose a positive real number, γ, and then derive a stabilizing compensator by trial and error for achieving the following objective:

$$\|G_c\|_\infty = \sup_\omega[\bar{\sigma}\{G_c(i\omega)\}] \leq \gamma. \tag{2.273}$$

By decreasing γ until the compensator fails to stabilize the system, one can find the limit on the minimum value of $\|G_c\|_\infty$. It is to be noted that if γ is increased to a large value, the design approaches that of the optimal H_2 (or LQG) compensator. Thus, one can begin iterating for γ from a value corresponding to a baseline H_2 (or LQG) design.

The controller derived by satisfying Eq. (2.273) is indeed an optimal solution (Glad and Ljung 2002). This is illustrated by defining the following function for the plant of Eq. (2.268) subject to condition, Eq. (2.272):

$$f(t) = x(t)^T P x(t) + \int_0^t \left[z(\tau)^T z(\tau) - \gamma^2 w(\tau)^T w(\tau)\right] d\tau, \tag{2.274}$$

where P is a symmetric, positive semi-definite matrix and $\gamma > 0$. For the zero initial condition, $x(0) = 0$, we have $f(0) = 0$. The time derivative of $f(t)$ is given by

$$\begin{aligned}
\dot{f} &= \dot{x}^T P x + x^T P \dot{x} + z^T z - \gamma^2 w^T w \\
&= x^T A^T P x + u^T B^T P x + w^T F^T P x + x^T P A x + x^T P B u + x^T P F w \\
&\quad + x^T M^T M x + u^T u - \gamma^2 w^T w \\
&= x^T \left[A^T P + PA + M^T M + P\left(\frac{1}{\gamma^2} F F^T - B B^T\right) P\right] x \\
&\quad + \left(u + B^T P x\right)^T \left(u + B^T P x\right) - \gamma^2 \left(w - \frac{1}{\gamma^2} F^T P x\right)^T \left(w - \frac{1}{\gamma^2} F^T P x\right).
\end{aligned} \tag{2.275}$$

If P is a symmetric, positive semi-definite solution of the following ARE:

$$0 = A^T P + PA + P\left(\frac{1}{\gamma^2}FF^T - BB^T\right)P + M^T M, \tag{2.276}$$

and a feedback control law is selected such that

$$u(t) = -B^T Px(t), \tag{2.277}$$

then we have $\dot{f}(t) \leq 0$, which implies that $f(t) \leq 0$ for all t. Since the first term on the right-hand side of Eq. (2.274) is always non-negative, the integral term is always less than or equal to zero, which implies

$$z(t)^T z(t) \leq \gamma^2 w(t)^T w(t), \tag{2.278}$$

for all t, and for any disturbance signal, $w(t)$. Note that a minimization of γ implies a minimization of the H_2 norm of the error vector for any given disturbance signal. Hence, the optimality of a compensator satisfying Eq. (2.273) is guaranteed, provided there exists a positive semi-definite solution of the Riccati equation, Eq. (2.276).

A problem in the H_∞ approach is that it does not automatically produce a stabilizing controller, $H(s)$. Hence, stability of the H_∞ design must be separately ensured by requiring that the dynamics matrix, $(A - BB^T P)$, must have all the eigenvalues in the left-half s-plane. An optimization procedure which enforces this constraint in every iteration is listed below (Glover and Doyle 1988):

(a) Select a set of frequency weights, $W_1(s), W_2(s), W_3(s)$, and ensure that these yield a Hurwitz observer dynamics matrix, $(\mathbf{A} - FC)$.
(b) Select a value for γ (usually 1).
(c) Solve the ARE, Eq. (2.276), for a symmetric, positive semi-definite matrix, P. If no such solution exists, go back to (b) and increase γ. If a symmetric, positive semi-definite P exists which yields a Hurwitz matrix, $(A - BB^T P)$, try to find a better solution by going back to (b) and decreasing γ.
(d) Repeat steps (a)–(c) until the smallest γ value is obtained.

2.11 μ-Synthesis

Structured singular value (SSV or μ-) synthesis is an alternative method of analysing the robustness of linear feedback systems. The H_2/H_∞ method of the previous section is an application of the small gain theorem to obtain stability margins of uncertain linear feedback systems without assuming any particular structure of the uncertainty. This usually results in an overly conservative design, because the actual perturbations are present in only some of the plant parameters, and not necessarily all of them. By assigning a structure to the uncertainty model, it is possible to have a more realistic and less conservative design. This is the objective of the SSV design method, whose methodology is presented in a tutorial paper (Packard and Doyle 1992).

Consider a square system with transfer matrix, $M(s) \in \mathbb{C}^{m\times m}$, between inputs, $u \in \mathbb{R}^m$ and outputs, $v \in \mathbb{R}^m$, with a multiplicative uncertainty, $\Delta(s) \in \mathbb{C}^{m\times m}$, as shown in Fig. 2.18. This

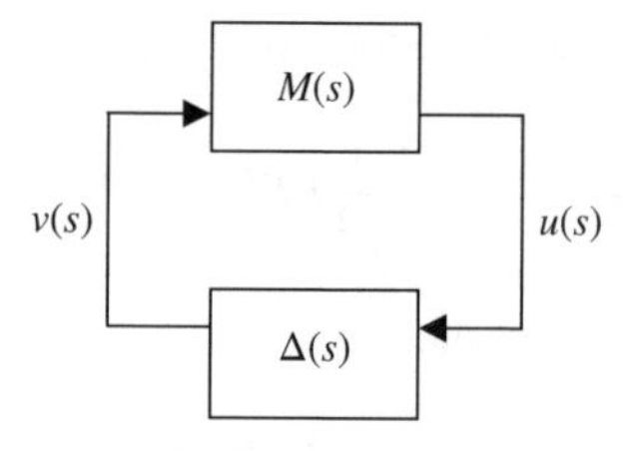

Figure 2.18 A square plant with a multiplicative uncertainty

representation implies the following loop equations:

$$v(s) = \Delta(s)\, u(s)$$
$$u(s) = M(s)\, v(s), \tag{2.279}$$

which requires $\Delta(s)$ to have a block-diagonal structure given by

$$\Delta \subset \mathbf{\Delta} = \left\{ \text{diag}\left[\delta_i I_{r_1}, \ldots, \delta p I_{r_p}, \Delta_{p+1}, \ldots, \Delta_{p+q}\right] : \right.$$
$$\left. \delta_i \in \mathbb{C}, \Delta_{p+j} \in \mathbb{C}^{n_j \times n_j}, 1 \leq i \leq p, 1 \leq j \leq q \right\}, \tag{2.280}$$

where

$$\sum_{i=1}^{p} r_i + \sum_{j=1}^{q} n_j = m.$$

If the matrix $I - M\Delta$ is nonsingular, then Eq. (2.279) has only the trivial solution, $u = v = 0$. For a non-trivial solution, the matrix $I - M\Delta$ must be singular and hence there are infinitely many possible solutions to Eq. (2.279), including the ones in which the signals $u(s), v(s)$ are unbounded at some frequency $s = i\omega$. Thus for stability, we require the singularity of the matrix $I - M\Delta$, and define a structured singular value, $\mu(M)$, as the measure of the smallest structured uncertainty, Δ, which causes the instability of the feedback loop depicted in Fig. 2.18.

Definition 2.11.1 *For a matrix, $M \in \mathbb{C}^{m \times m}$, the structured singular value is defined by*

$$\mu(M) = \frac{1}{\min\{\bar{\sigma}(\Delta) : \Delta \in \mathbf{\Delta}, \det(I - M\Delta) = 0\}}, \tag{2.281}$$

unless there exists no $\Delta \in \mathbf{\Delta}$ that makes $\det(I - M\Delta) = 0$, *for which case* $\mu(M) = 0$.

Definition 2.11.2 *A norm bounded subset of all block-diagonal matrices is defined to be the following:*

$$\mathbf{B}_{\mathbf{\Delta}} = \{\Delta \in \mathbf{\Delta} : \bar{\sigma}(\Delta) \leq 1\}. \tag{2.282}$$

An important property of the structured singular value is the following:

$$\mu(M) = \max_{\Delta \in \mathbf{B}_{\mathbf{\Delta}}} \rho(M\Delta), \tag{2.283}$$

where $\rho(.)$ denotes the spectral radius of a square matrix. Furthermore, we have the following bounds:

$$\rho(M) \leq \mu(M) \leq \bar{\sigma}(M), \tag{2.284}$$

which can be used to possibly bracket $\mu(M)$ in a computation. If $\Delta \in \mathbb{C}^{m\times m}$ (i.e. it has a full-block structure), then we have $\mu(M) = \bar{\sigma}(M)$. However, the upper and lower bounds on $\mu(M)$ provided by Eq. (2.284) can be arbitrarily far apart. A better fixing of the bounds on $\mu(M)$ is provided by the following theorem.

Theorem 2.11.3 *For all $U \in \mathbf{U}$ and $D \in \mathbf{D}$, where*

$$\begin{aligned}
\mathbf{U} &= \{U \in \mathbf{\Delta} : U^H U = I\} \\
\mathbf{D} &= \Big\{ \mathrm{diag}\left(D_1, \dots, D_p, d_{p+1} I_{n_1}, \dots, d_{p+q} I_{n_q}\right) : \\
&\qquad D_i \in \mathbb{C}^{r_i \times r_i}, D_i = D_i^H > 0, d_{p+j} \in \mathbb{R}, d_{p+j} > 0 \Big\},
\end{aligned} \tag{2.285}$$

the structured singular value has the following property:

$$\mu(MU) = \mu(UM) = \mu(M) = \mu\left(D^{1/2} M D^{-1/2}\right). \tag{2.286}$$

Proof. The proof is derived by expanding the determinant of $(I - M\Delta)$ as $\det\left(I - MD^{-1/2}\Delta D^{1/2}\right)$.

Theorem 2.11.3 lets us tighten the bounds on $\mu(M)$ to the following:

$$\max_{U\in\mathbf{U}} \rho(UM) \leq \max_{\Delta\in\mathbf{B}_\Delta} \rho(\Delta UM) = \mu(M) \leq \inf_{D\in\mathbf{D}} \bar{\sigma}\left(D^{1/2} M D^{-1/2}\right). \tag{2.287}$$

In computing the infimum, one of the diagonal elements in $\mathbf{D}$ is assumed to be unity, such as $d_{p+q} = 1$. More information about the bounds and properties of $\mu(M)$ can be found in the tutorial paper (Packard and Doyle 1992). The main difficulty in the μ-synthesis procedure is to evaluate the maximum in Eq. (2.287), which requires some form of optimization. For a transfer matrix, the maximum must be found over a frequency range, which could be a cumbersome process. The prevalent approaches for carrying out the optimization with a frequency sweep include the gradient-based methods (Doyle 1982, Fan and Tits 1986), or a non-gradient search scheme (Tewari and Balakrishnan 1991). As the number of uncertainty blocks increase in the block-diagonal structure, so does the computation time for the SSV, which points towards finding efficient computational algorithms, which possibly do not require a frequency sweep.

2.11.1 Linear Fractional Transformation

The main utility of the structured singular value (SSV) method is in modelling the uncertainty of a linear system described by a transfer-matrix representation, $G(s) \in \mathbb{C}^{p\times m}$, relating the inputs, $u \in \mathbb{R}^m$, and the outputs, $y \in \mathbb{R}^p$. The uncertain part of $G(s)$ is represented by a multiplicative structured uncertainty, $\Delta_2 \in \mathbf{\Delta}_2 \subset \mathbb{C}^{n\times n}$, between the exogenous (disturbance) inputs, $w \in \mathbb{R}^n$, and exogenous outputs, $z \in \mathbb{R}^n$, as shown in Fig. 2.19. Such a representation

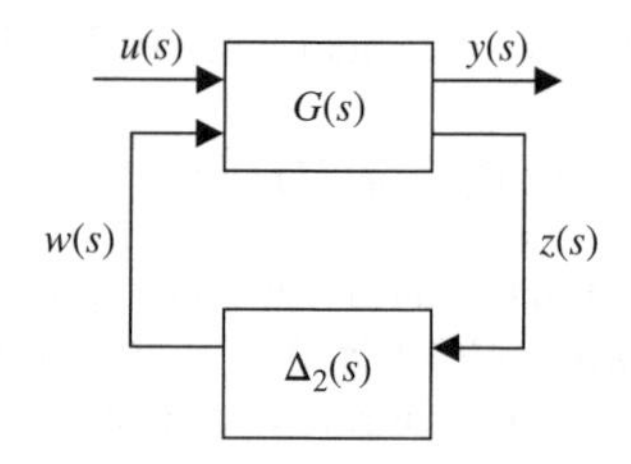

Figure 2.19 Linear fractional transformation (LFT) representation of an uncertain system

of an uncertain linear system is called a linear fractional transformation (LFT), denoted by $y = \mathcal{L}\left(G, \Delta_2\right) u$, and has the following loop equations:

$$\begin{aligned} y(s) &= G_{11}(s)\, u(s) + G_{12}(s)\, w(s), \\ z(s) &= G_{21}(s)\, u(s) + G_{22}(s)\, w(s), \end{aligned} \tag{2.288}$$

$$w(s) = \Delta_2(s)\, z(s). \tag{2.289}$$

The transfer matrix (LFT) of the uncertain system is thus given by

$$\mathcal{L}\left(G, \Delta_2\right) = G_{11} + G_{12}\Delta_2\left(I - G_{22}\Delta_2\right)^{-1} G_{21}. \tag{2.290}$$

If $\Delta_1 \in \mathbf{\Delta}_1 \subset \mathbb{C}^{n \times n}$ is an alternative block uncertainty on G_{11}, then an analogous LFT of the system is given by

$$\mathcal{L}\left(\Delta_1, G\right) = G_{22} + G_{21}\Delta_1\left(I - G_{11}\Delta_1\right)^{-1} G_{12}. \tag{2.291}$$

If one applies the small gain theorem to the uncertain part of the system (i.e. the loop involving z and w), the following stability requirement can be obtained without taking into account any structure of the uncertainty Δ_2:

$$\|G_{22}\| \leq 1, \tag{2.292}$$

where the gain $\|.\|$ can be taken using either the H_2, or the H_∞ norm. This is the robust stability (or stability robustness) requirement underlying the H_2/H_∞ design method (see the previous subsection), and could be unnecessarily restrictive (conservative) in selecting a feedback controller to stabilize the system between u and y. However, when we assume a norm-bounded, block-diagonal structure for $\Delta_2 \in \mathbf{B}_\Delta$, the structured singular value of G_{22} defined by

$$\mu\left(G_{22}\right) = \frac{1}{\min\{\bar{\sigma}\left(\Delta_2\right) : \Delta \in \mathbf{B}_\Delta, \det\left(I - G_{22}\Delta_2\right) = 0\}}. \tag{2.293}$$

forms the basis of a less restrictive (or more efficient) design. Since $G_{22}(s)$ and $\Delta_2(s)$ for $s = i\omega$ vary with the frequency, ω, $\mu(G_{22})$ indicates the measure of the smallest block-diagonal perturbation, $\Delta_2(i\omega)$, which can move a pole of $G_2(s)$ (i.e. the zero of $I - G_{22}\Delta_2$) to the imaginary axis, $s = i\omega$. The maximum allowable perturbation before instability (or stability robustness) is therefore given by the supremum of $\mu(G_{22})$ over a given frequency range, defined by the following μ norm:

$$\|G_{22}\|_\mu = \sup_\omega \left\{\mu\left[G_{22}(i\omega)\right]\right\}. \tag{2.294}$$

The less-restrictive design requirement based on μ is thus given by

$$\|G_{22}\|_{\mu} \leq 1. \tag{2.295}$$

The most common form of structured uncertainty is in the elements of the state-space coefficient matrices, (A, B, C, D), of a linear system, $G(s) = C(sI - A)^{-1}B + D$. Thus we have the following constant real matrix, M, whose uncertainty is represented by the structured singular value, $\mu(M)$, by a norm-bounded, block-diagonal perturbation $\Delta_1 \in \mathbf{B}_{\Delta}$:

$$M = \begin{pmatrix} A & B \\ C & D \end{pmatrix}. \tag{2.296}$$

The LFT of such a representation is therefore the following:

$$\mathcal{L}\left(\frac{1}{s}I, M\right) = D + C(sI - A)^{-1}B, \tag{2.297}$$

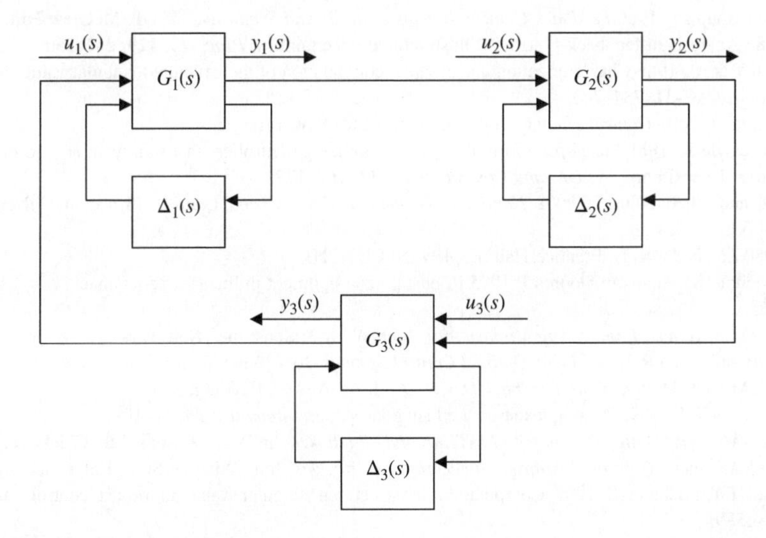

Figure 2.20 Linear fractional transformation (LFT) representation of a system comprising three uncertain subsystems connected by a feedback loop

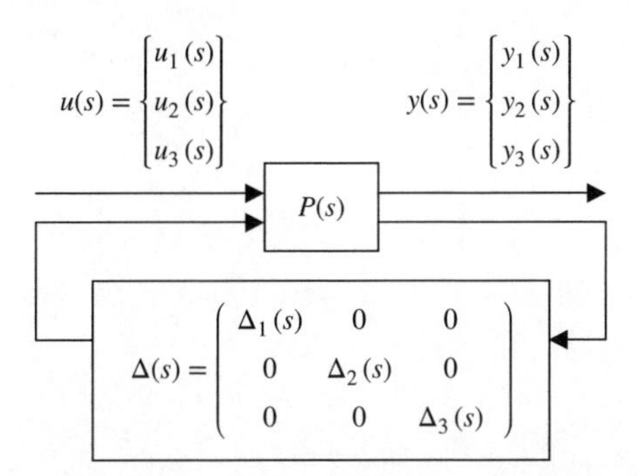

Figure 2.21 Overall linear fractional transformation representation of the feedback system shown in Fig. 2.20

where the block perturbation is taken to be $\Delta_1 = I/s$ on $M_{22} = D$. Such an uncertainty model is very useful for robust ASE analysis, as we shall see in Chapter 7.

The LFT of two (or more) systems connected in a feedback loop can be derived from the individual LFTs. For example, the system shown in the block diagram of Fig. 2.20 consists of three subsystems of individual LFTs $\mathcal{L}(G_1, \Delta_1), \mathcal{L}(G_2, \Delta_2)$ and $\mathcal{L}(G_3, \Delta_3)$. The overall system's LFT is derived in the block diagram of Fig. 2.21, where the overall transfer matrix, P, depends only on G_1, G_2, G_3, and the overall perturbation Δ consists of $\Delta_1, \Delta_2, \Delta_3$ as its diagonal blocks. We shall return to such a representation while discussing uncertain ASE systems in Chapter 7.

References

Athans M and Falb PL 2007 *Optimal Control*. Dover Publications, New York.

Bryson AE Jr., and Ho YC 1975 *Applied Optimal Control*. Hemisphere, Washington, DC.

D'Azzo JJ and Houpis C 1966 *Feedback Control Systems Analysis and Synthesis*, 2nd ed., McGraw-Hill, New York.

Doyle JC 1982 Analysis of feedback systems with structured uncertainties. *Proc. IEE* **129**, 242–250.

Fan MKH and Tits AL 1986 Characterization and efficient computation of the structured singular value. *IEEE Trans. Autom. Control* **AC-31**, 734–743.

Glad T and Ljung L 2002 *Control Theory*. Taylor & Francis Ltd, New York.

Glover K and Doyle JC 1988 State-space formulae for all stabilizing controllers that satisfy an H_∞ norm bound and relations to risk sensitivity. *Systems and Control Letters* **11**, 167–172.

Haddad WM and Chellaboina V 2008 *Nonlinear Dynamical Systems and Control*. Princeton University Press, Princeton, NJ.

Kailath T 1980 *Linear Systems*. Prentice-Hall, Englewood Cliffs, NJ.

Kautsky J, Nichols NK, and van Dooren P 1985 Robust pole assignment in linear state feedback. *Int. J. Control* **41**, 1129–1155.

Kreyszig E 1998 *Advanced Engineering Mathematics*. John Wiley & Sons, Inc., New York.

Kwakernaak H and Sivan R 1972 *Linear Optimal Control Systems*. John Wiley & Sons, Inc., New York.

Maciejowski JM 1989 *Multivariable Feedback Design*. Addison-Wesley, Reading, MA.

Packard A and Doyle JC 1992 A complex structured singular value. *Automatica* **29**, 71–109.

Tewari A 2002 *Modern Control Design with MATLAB and Simulink*. John Wiley & Sons, Ltd, Chichester.

Tewari A 2011 *Advanced Control of Aircraft, Spacecraft, and Rockets*. John Wiley & Sons, Ltd, Chichester.

Tewari A and Balakrishnan S 1991 Computation of structured singular value in robust control. *AIAA Paper* **AIAA-91-0589**.

3

Aeroelastic Modelling

The emphasis in this book is on developing an adaptive control strategy for carrying out effective control of aeroelastic systems. Towards this end, a working basic model of the plant is necessary that can be revised and updated by online, feedback estimation techniques. The model should be such that it captures the essential dynamical features of the system without sacrificing the simplicity, and without detracting from our main thrust, namely adaptive control. In order to build such an aeroservoelastic (ASE) model, we rely upon the fundamental ideas in structures and aerodynamics as applied to lifting surfaces. The simplifying assumptions of causality, linearity and small perturbations are highly valued in building a basic ASE model, and are applied wherever it is possible to do so. Such a baseline model is inapplicable to situations where nonlinear effects become significant, such as at high flow incidence, transonic and hypersonic speed regimes, and inelastic structural deformations. One must go to great lengths in accurately modelling nonlinear aerodynamic and structural behaviour, often requiring sophisticated computational models that are not amenable to repeated online calculations in a closed-loop ASE design process. However, as it is the task of the adaptive control design methodology to compensate for the nonlinear effects, which are treated as uncertain but measurable parts of the plant dynamics, a linearized baseline model is adequate as a building block of the overall ASE system and serves as a benchmark for the synthesis of an adaptive control law.

A companion monograph has been prepared by the author (Tewari 2015) as an introduction to the basic modelling techniques employed in ASE design and analysis. Therefore, the discussion of this chapter only briefly covers the essential methods to be employed in deriving a suitable aeroelastic plant model. The reader will also benefit from a review of classical textbooks on aeroelasticity, such as (Bisplinghoff and Halfman 1955, Fung 1955, Scanlan and Rosenbaum 1951).

ASE analysis and design begins with a plant model that is simple and yet captures the essential dynamical features. Linearity is an important requirement for a working plant model, because it allows a quick closed-loop solution for control-law design. This is even more important in case of adaptive control where iterative, closed-loop simulations are necessary. An ASE plant model is based on linear structural dynamics and unsteady aerodynamics, both of which can be described either by linear operational forms or integral equations. The baseline structural model is derived either by the Newtonian or the Lagrangian method and has aerodynamic

Adaptive Aeroservoelastic Control, First Edition. Ashish Tewari.

forces and moments as the unknown driving functions of the generalized coordinates. Thus, it is important to express unsteady aerodynamics either in a closed form, or by inversion of simple integral equations. While an accurate aerodynamic modelling may require the solution of nonlinear, partial differential equations (Navier–Stokes (N–S), Euler, full-potential or transonic small-perturbation equations), a working baseline model employs simplification of both the governing equations and the boundary conditions. The compressible subsonic and supersonic modelling for the unsteady case requires the inversion of a pressure-upwash integral equation. A simplified solution of the subsonic harmonic problem begins with the incompressible limit, where the analytical solution is available in a closed form. The results of harmonic motion can be extended to arbitrary motion by analytic continuation, and practically carried out by Duhamel's integral of indicial results or by transform methods.

3.1 Structural Model

Modern aerospace structures are of semi-monocoque type where the outer covering (skin) shares a significant load with the internal members (spars, ribs, longerons, frames and bulkheads). Such a construction is not only necessary but also leads to an optimal combination of light weight and high stiffness. Owing to the high stiffness of the skin panels, it is possible to assume that the bending and torsion loads on the lifting surface do not cause an inelastic stress–strain behaviour at any point. In fact, if the elastic limit is deemed to be crossed at any location, the entire structure is assumed to have failed. However, an aircraft structure has a much more stringent failure criterion than even the elastic limit. This is due to the fact that an elastic buckling of the skin would cause an unacceptable deformation of the external shape, leading to an extensive and off-design modification of the aerodynamic loads, and therefore must not be allowed. Under static conditions, the lifting structure has a tensile load on the bottom surface and a compressive load on the top surface, which situation, of course, alternates under a dynamic loading. The structural failure is then typically analysed by testing for the critical stress for elastic buckling of the skin panels under limit loads. This conservative elastic buckling criterion not only leads to a safer structure but also simplifies the task of the aeroservoelastician by assuming a linear load–displacement behaviour that also preserves the cross-sectional shape. The post-buckling behaviour and dynamic aeroelastic analysis of individual skin panels (panel flutter) require nonlinear structural modelling and are thus excluded from our present scope.

3.1.1 Statics

Consider an aircraft wing modelled as a thin, elastic structure, with an unloaded and undeformed, mean surface described by $S(x, y) = 0$. Here (x, y) are coordinates running along the mean surface. If the wing's mean surface is approximated to be flat, these become the Cartesian coordinates. Suppose a concentrated, static load, P, is now applied at a point, (ξ, η), on the structure. Because P could be either a normal force, $P^{(1)}$, a chordwise tangential force, $P^{(2)}$, or any of the two moments, $P^{(3)}, P^{(4)}$, as shown in Fig. 3.1, it is called the *generalized load* (or force). The static load will cause a structural deformation (subject to any geometric constraints) such that equilibrium is achieved. The resulting structural displacement at any given point from the original (undeformed) shape can be represented by a linear combination of the

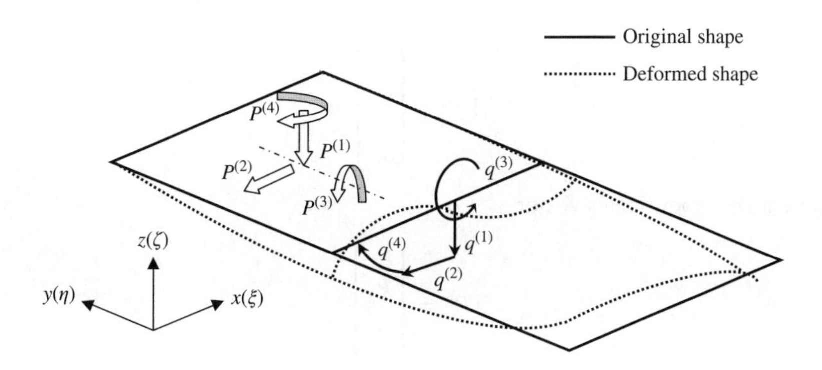

Figure 3.1 Generalized loads and coordinates for structural modelling

normal deflection, $q^{(1)}$, the chordwise deflection, $q^{(2)}$, the twist angle, $q^{(3)}$, and the warp angle, $q^{(4)}$, as depicted in Fig. 3.1. These displacement coordinates are called the *generalized coordinates* (or displacements). Let q be such a generalized coordinate at an arbitrary point, (x, y), on the structure. Since the structure is assumed to be linearly elastic, the generalized load and displacement are linearly related by

$$q(x, y) = R(x, y : \xi, \eta)P(\xi, \eta), \tag{3.1}$$

where $R(x, y : \xi, \eta)$ is the *flexibility influence-coefficient* function (also called *Green's function*). By applying linear superposition, Eq. (3.1) can be extended for the case of a continuous, generalized load per unit area (either pressure or distributed moment), $p(\xi, \eta)$, as follows:

$$q(x, y) = \int_S R(x, y : \xi, \eta)p(\xi, \eta)\mathrm{d}\xi\mathrm{d}\eta. \tag{3.2}$$

For ease of computation, the surface integral in Eq. (3.2) is discretized by considering only a finite number of generalized loads and generalized coordinates. This is tantamount to approximating a continuous (or infinite dimensional) structure by an equivalent finite dimensional form. For example, the mean surface can be thought of consisting of n flat elemental panels of individual dimensions, $(\Delta\xi_i, \Delta\eta_i), i = 1, \ldots, n$. The load distribution on the jth panel is then approximated by an average generalized load, $P_j = p(\xi, \eta)\Delta\xi_j\Delta\eta_j$, acting at a given load point (such as the panel centroid) in each panel. Similarly, the generalized displacement, $q(x, y)$, averaged over the ith panel is taken as q_i at a given collocation point, $(x_i, y_i), i = 1, \ldots, n$. The discretized load–displacement relationship is then given for the ith panel as follows:

$$q_i = \sum_{i=1}^{N} R_{ij}P_j; \quad i = 1, \ldots, N, \tag{3.3}$$

where the upper limit of the indices (i, j) is $N = 4n$, indicating that there are four generalized loads (and coordinates) at any given point (see Fig. 3.1). Then all the generalized displacements at all the collocation points are collected and described by the following vector-matrix equation:

$$\{q\} = [R]\{P\}, \tag{3.4}$$

where

$$\{q\} = \begin{Bmatrix} q_1 \\ q_2 \\ \vdots \\ q_N \end{Bmatrix} \tag{3.5}$$

is the generalized coordinates vector,

$$\{P\} = \begin{Bmatrix} P_1 \\ P_2 \\ \vdots \\ P_N \end{Bmatrix} \tag{3.6}$$

the generalized loads vector, and

$$[R] = \begin{pmatrix} R_{11} & R_{12} & \dots & R_{1N} \\ R_{21} & R_{22} & \dots & R_{2N} \\ \vdots & \vdots & \vdots & \vdots \\ R_{N1} & R_{N2} & \dots & R_{NN} \end{pmatrix} \tag{3.7}$$

the generalized influence-coefficients matrix with $N = 4n$. A way of understanding the discrete influence coefficient, R_{ij}, is by visualizing the ith generalized virtual coordinate, δq_{ij}, caused by an isolated generalized load at a given point, P_j,

$$\delta q_{ij} = R_{ij} P_j. \tag{3.8}$$

A virtual coordinate can be regarded as an arbitrary, infinitesimal variation in any of the four possible directions (Fig. 3.1) at a given point due to an isolated generalized load, and must be compatible with any geometric constraints on the structure. The actual generalized coordinate, q_i, is a sum of all the virtual coordinates, δq_{ij}, corresponding to the generalized load, P_j, and is given by the ith row of Eq. (3.4),

$$q_i = \sum_{j=1}^{N} \delta q_{ij} = \sum_{j=1}^{N} R_{ij} P_j. \tag{3.9}$$

Hence, the discrete influence coefficient, R_{ij}, is the ith virtual generalized coordinate due to the jth *unit* generalized load. The reciprocal principle states that the ith virtual coordinate due to the jth unit load is the same as the jth virtual coordinate caused by the ith generalized load, that is,

$$R_{ij} = R_{ji}, \tag{3.10}$$

which implies that the matrix $[R]$ is symmetric.

Equation (3.4) predicts the generalized displacements caused by a general static loading. However, it should also be possible to determine the generalized loads from the generalized displacements they actually produce. This requires an inversion of Eq. (3.4) as follows:

$$\{P\} = [K]\{q\}, \tag{3.11}$$

where $[K] = [R]^{-1}$ must exist and is called the generalized stiffness matrix of the structure. Thus, both $[R]$ and $[K]$ are nonsingular, symmetric matrices. An element, k_{ij}, of $[K]$ is called the stiffness coefficient and can be considered the ith generalized virtual load due to the jth unit generalized displacement. The work done by a generalized virtual load,

$$\delta P_{ij} = k_{ij} q_j, \tag{3.12}$$

in producing a generalized displacement, q_j, at a point is given by

$$U_{ij} = \int \delta P_i \mathrm{d}q_j = \int k_{ij} q_j \mathrm{d}q_j = \frac{1}{2} k_{ij} q_j^2 = \frac{1}{2} \delta P_i q_j. \tag{3.13}$$

When summed over all points on the structure, the net work done by all the static forces is the total strain energy stored in the structure, derived as follows:

$$U = \sum_{i=1}^{N} \sum_{j=1}^{N} U_{ij} = \frac{1}{2} \{P\}^T \{q\} = \frac{1}{2} \{q\}^T [K] \{q\}. \tag{3.14}$$

The potential (strain) energy is responsible for restoring the structure to its original shape once the loading is removed, and its quadratic form is an important consequence of the linearly elastic behaviour. Since the external forces must be balanced by equal and opposite internal forces for a static equilibrium, one can regard U as the net work done by the internal, restoring (or conservative) forces.

By making simplifying assumptions for a typical aircraft, the number of generalized coordinates (hence the dimension, N, of the matrix $[K]$) can be significantly reduced. For example, most aircraft structures can be assumed to have chordwise rigidity, which results in chordwise cross sections remaining plane. Thus, the warp angle, $q^{(4)}$, can be neglected at all points. Furthermore, for a lifting structure, the chordwise deformation, $q^{(2)}$, can be neglected in comparison with the normal deflection, $q^{(1)}$, because of the following reasons: (i) the chordwise component of the aerodynamic force (drag) is an order of magnitude smaller than the normal force (lift) and (ii) the chordwise bending stiffness of the wing is an order of magnitude larger than that in the normal direction. The overall consequence of the above assumptions is an infinite stiffness in the chordwise direction, which enables discarding the rows and columns in $[K]$ corresponding to the coordinates $q^{(2)}$ and $q^{(4)}$. When a control surface is present, its deflection is modelled simply by a rotation angle about a rigid hinge axis.

Structural modelling is further simplified for a typical aircraft wing of high aspect ratio and a small thickness ratio, wherein shear deformation is neglected and the vertical displacement, $w(x, y)$, can be represented at any point by the normal deflection, $q^{(1)}$, and the twist angle, $q^{(3)}$, both of which are functions of only the spanwise location, y:

$$w(x, y) = q^{(1)}(y) - x q^{(3)}(y). \tag{3.15}$$

3.1.2 Dynamics

For a dynamic loading on the structure, it is necessary to consider not only the potential (internal) energy, U, but also the net kinetic energy, T, as well as the work done by non-conservative forces, W_n. The motion is completely described by the generalized coordinates vector, $\{q(t)\}$, measured in an inertial frame, and its time derivative, $\{\dot{q}(t)\}$, for which we can write

$$T = T(\{q\}, \{\dot{q}\}). \tag{3.16}$$

The generalized coordinates now represent the degrees of freedom (d.o.f.s) of the dynamic motion. Before proceeding further, it is necessary to split the generalized force vector into conservative and non-conservative parts, $\{Q_c(t)\}$ and $\{Q_n(t)\}$, respectively:

$$\{Q\} = \{Q_c\}(\{q\}) + \{Q_n\}(\{q\}, \{\dot{q}\}). \tag{3.17}$$

By definition, the conservative force is a function only of the generalized coordinates, while the non-conservative force can also depend upon the time derivatives of the generalized coordinates. Since the elastic stiffness creates a restoring internal force given by Eq. (3.11), it is a conservative force. For a linear structure, the generalized forces created by viscous (Rayleigh) damping effects are proportional to $\{\dot{q}\}$, and are therefore non-conservative forces. Finally, the generalized, unsteady aerodynamic forces given by the vector $\{Q_a\}$ are non-conservative in nature. Thus we write

$$\begin{aligned}\{Q_c\} &= -[K]\{q\} \\ \{Q_n\} &= -[C]\{\dot{q}\} + \{Q_a\},\end{aligned} \tag{3.18}$$

where the negative sign indicates an internal (opposing) force, $[C]$, is the generalized damping matrix comprising the viscous damping coefficients, and $\{Q_a\}$ can have a nonlinear relationship with the generalized coordinates and their time derivatives. By Newton's second law of motion applied to the discretized structure, we have

$$[M]\{\ddot{q}\} = \{Q\} = -[K]\{q\} - [C]\{\dot{q}\} + \{Q_a\}, \tag{3.19}$$

or

$$[M]\{\ddot{q}\} + [C]\{\dot{q}\} + [K]\{q\} = \{Q_a\}, \tag{3.20}$$

where $[M]$ is the generalized mass matrix representing the individual masses and moments of inertia corresponding to the various d.o.f.s.

One can alternatively adopt an energy approach to derive Eq. (3.20). For achieving an arbitrary, infinitesimal generalized displacement, $\delta q_i(t)$, Hamilton's principle requires that of all possible trajectories, the correct one is that which minimizes the net mechanical energy, $(T - U + W_n)$. This statement is given in the variational form by the necessary condition,

$$\delta \int \left(T - U + W_n\right) \mathrm{d}t = \int \delta\,(T - U)\,\mathrm{d}t + \int \delta W_n \mathrm{d}t = 0, \tag{3.21}$$

where $\delta(.)$ represents the variational operator,

$$\delta T = \sum_{i=1}^{N} \frac{\partial T}{\partial q_i}\delta q_i + \frac{\partial T}{\partial \dot{q}_i}\delta \dot{q}_i, \tag{3.22}$$

$$\delta U = \sum_{i=1}^{N} \frac{\partial U}{\partial \dot{q}_i}\delta \dot{q}_i \tag{3.23}$$

and

$$\delta W_n = \sum_{i=1}^{N} Q_{ni}\delta \dot{q}_i. \tag{3.24}$$

Since the initial virtual displacement must be 0, $\delta q_i(0) = 0$, integrating the first term of kinetic energy variation, Eq. (3.22), by parts yields the following expression:

$$\int \frac{\partial T}{\partial \dot{q}_i} \delta \dot{q}_i \mathrm{d}t = -\int \frac{\mathrm{d}}{\mathrm{d}t}\left(\frac{\partial T}{\partial \dot{q}_i}\right) \delta q_i \mathrm{d}t, \tag{3.25}$$

which substituted into Eq. (3.21) produces the well-known Lagrange's equations:

$$\frac{\mathrm{d}}{\mathrm{d}t}\left(\frac{\partial T}{\partial \dot{q}_i}\right) - \frac{\partial T}{\partial q_i} + \frac{\partial U}{\partial q_i} = Q_{ni} \quad (i = 1, \ldots, N). \tag{3.26}$$

For a linearly elastic structure, the potential and kinetic energies are expressed as follows:

$$U = \frac{1}{2}\{q\}^T[K]\{q\}$$
$$T = \frac{1}{2}\{\dot{q}\}^T[M]\{\dot{q}\}, \tag{3.27}$$

thereby yielding

$$\frac{\mathrm{d}}{\mathrm{d}t}\left(\frac{\partial T}{\partial \{\dot{q}\}}\right) = [M]\{\ddot{q}\}$$
$$\frac{\partial T}{\partial \{q\}} = [0]$$
$$\frac{\partial U}{\partial \{q\}} = [K]\{q\}. \tag{3.28}$$

Thus, substituting Eq. (3.28) along with the second of Eq. (3.18) into Eq. (3.26) yields the equation of motion, Eq. (3.20).

Much of the effort in structural modelling involves the derivation of the generalized mass, stiffness and viscous damping matrices by an appropriate discretization scheme and the separation of the generalized coordinates into spatial and time-dependent parts. The various techniques commonly applied for this purpose are the finite-element method (FEM) (or Galerkin method), the assumed-modes method (also known as either the Ritz or Rayleigh–Ritz method) and the boundary-element method. Of these, FEM is the most popular because of its ease of implementation; it is also quite efficient in terms of model size for a given accuracy. Further details about the FEM can be found in Tewari (2015).

Often, a much simpler structural model can be applied for aeroelastic purposes, that is, the lumped-parameter approach based on a typical section. This is the topic of the next section.

3.1.3 Typical Wing Section

An aircraft wing is a thin structure cantilevered at the root, with the span generally much larger than the average chord (Fig. 3.2). Since our interest is in exercising ASE control, we are mainly concerned with wings of large aspect ratio (span/chord), which can be regarded as more prone to ASE instabilities rather than those of smaller aspectratio. A large aspect-ratio wing has a clearly defined elastic axis (e.a.) as the line joining the shear centres of all cross sections (Fig. 3.2). The line joining the centres of mass (c.m.) at each chordwise section is called the

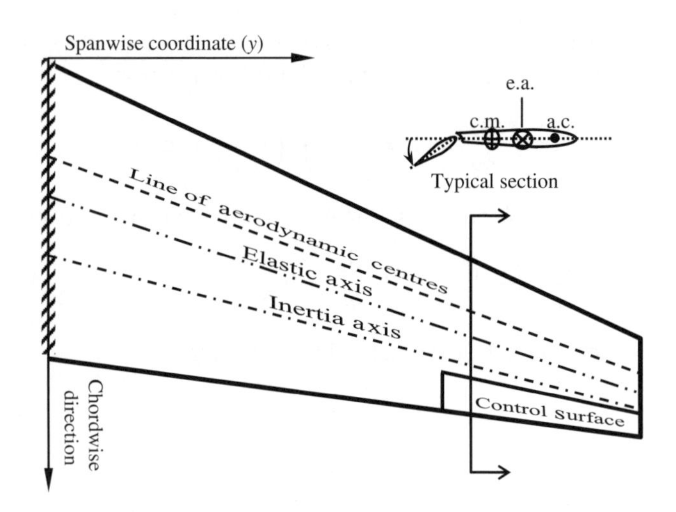

Figure 3.2 Schematic diagram of wing semi-span and the typical chordwise section

inertia line and affects the natural frequencies of the structural modes. The net aerodynamic loads at each chordwise section essentially act at a point different from the shear centre. This point is called the aerodynamic centre (a.c.) and is defined as the location about which the pitching moment is invariant with the angle of attack. Thus, the net effect of aerodynamic loading per unit span of the wing is a concentrated pitching moment, a lift force and a drag force, all applied at the a.c. The drag is much smaller than the lift, and therefore has a negligible aeroelastic contribution as explained above. Owing to the offset of the line joining the a.c.s from the e.a. (see Fig. 3.2), there is a net spanwise bending moment, a vertical shear load and a torsional (or twisting) moment at each spanwise location. The bending and twisting produced at each spanwise location cause a vertical translation (plunge), a rotation (pitch) and a negligible shear deformation. If a control surface is also present, the control-surface rotation is the third d.o.f at a given section. It has been shown (Theodorsen 1942) that for a wing without a large sweep angle or concentrated masses, one can reasonably approximate the aeroelastic motion as consisting of pitch, plunge and control-surface rotation at a reference spanwise location, y. This results in the typical-section approximation of a wing, which is quite popular in classical aeroelasticity (Bisplinghoff and Halfman 1955, Fung 1955) because it allows the calculation of three-dimensional, dynamical properties by considering only two-dimensional aerodynamics.

Since airplanes are constructed to have chordwise rigid wing sections, deformation at any point along the span can be described by a linear combination of spanwise bending of and torsional deflections about the e.a. At a particular spanwise location, the wing deformation is therefore represented by a vertical downward deflection, h, (plunge) of the e.a., and twist, θ, (pitch) about the e.a., defined positive in the nose-up sense. A thin wing of high aspect ratio cantilevered at the root (Fig. 3.2) has the following partial differential equations governing its deformation:

$$m(y)\frac{\partial^2 h}{\partial t^2} + S_y(y)\frac{\partial^2 \theta}{\partial t^2} + EI(y)\frac{\partial^4 h}{\partial y^4} = P(y,t)$$

$$S_y(y)\frac{\partial^2 h}{\partial t^2} + I_y(y)\frac{\partial^2 \theta}{\partial t^2} - GJ(y)\frac{\partial^2 \theta}{\partial y^2} = M_y(y,t), \quad (3.29)$$

where P is the plunge force applied at the e.a., M_y the twisting moment about the e.a., m, S_y, I_y denote the inertial properties per unit span, EI is the bending stiffness and GJ the torsional stiffness at the given station. These equations can be alternatively expressed in terms of a coordinate running along the swept e.a., $y' = y \sec \Gamma$, where Γ is the sweep angle of the e.a. (Fig. 3.2). In such a case, the typical section is to be taken normal to the e.a. and all the quantities in the chordwise direction are similarly transformed.

The structural mounting of a control surface on a hinge results in high chordwise, as well as spanwise rigidity. Hence, its deflection is idealized by a single rotation angle, β, about the hinge line, defined positive for the downward rotation of the trailing edge. The external forces and moments driving the d.o.f.s are the plunge force, P_e, applied downward at the e.a., the nose-up pitching (twisting) moment, M_{ye}, acting about the e.a., and the hinge moment, H_e, causing a downward rotation of the control surface about its hinge line. The d.o.f.s at a given spanwise location can be considered independent of those at another location in the typical-section model for a high-aspect-ratio wing. This implies applying a lumped-parameter approach to Eq. (3.29). Such a model is generally valid, except for very thin structures of small aspect ratio, which have a significant chordwise bending, and for thick wings having a significant shear deformation in bending. The bending stiffness at the given spanwise location is modelled by linear spring stiffness, k_h, while the stiffness in torsion and control-surface d.o.f.s are described by the rotational spring constants, k_θ and k_β, respectively, as shown in Fig. 3.3. These stiffness constants are assumed to be sufficiently large such that the vertical displacement, h, is small relative to the sectional chord, $2b$ and the angular displacements, θ, β are also small.

In order to be a typical section, the reference spanwise location must be carefully selected such that the in vacuo natural frequencies of the wing's predominant structural modes (primary bending, primary torsion and primary control deflection) match with the corresponding modes of the typical section (plunge, pitch and control rotation, respectively). Usually, either the 70% semi-span location from the wing root or the mid-span location of control surface is taken to be the typical section for a straight wing without concentrated masses. Consider the typical section detailed in Fig. 3.4. The three d.o.f.s are considered to be the generalized coordinates h (plunge), θ (pitch), and β (control-surface deflection angle). The wing is originally in a static equilibrium, moving in the horizontal direction[1] with a constant speed U, and then encounters small perturbations (h, θ, β). Thus h is the vertically downward displacement of the e.a. from its equilibrium position, θ gives the rotation of the chord line about the e.a. (considered positive nose-up as shown in Fig. 3.4), whereas β is the angle between the chord lines of the wing

[1] If the flight direction is normal to the span, y, and the typical section is normal to the e.a., y', then the effective freestream speed seen by the section is $U \cos \Gamma$.

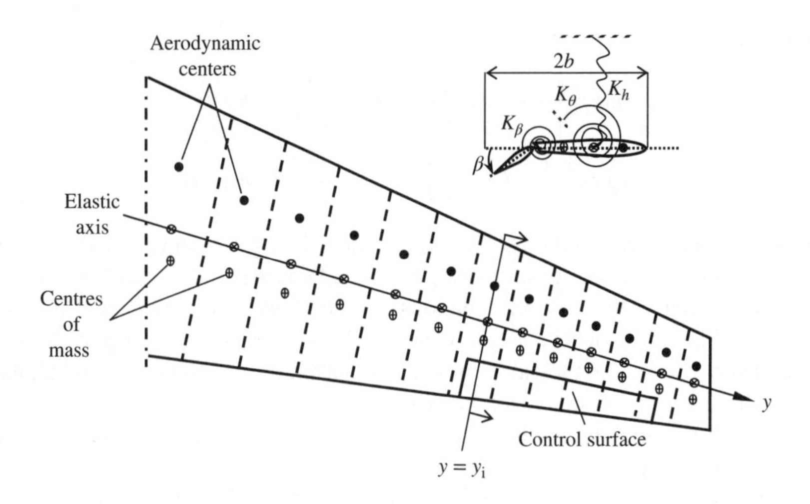

Figure 3.3 Lumped-parameter model of a wing with structural degrees of freedom at a typical spanwise section taken normal to the elastic axis

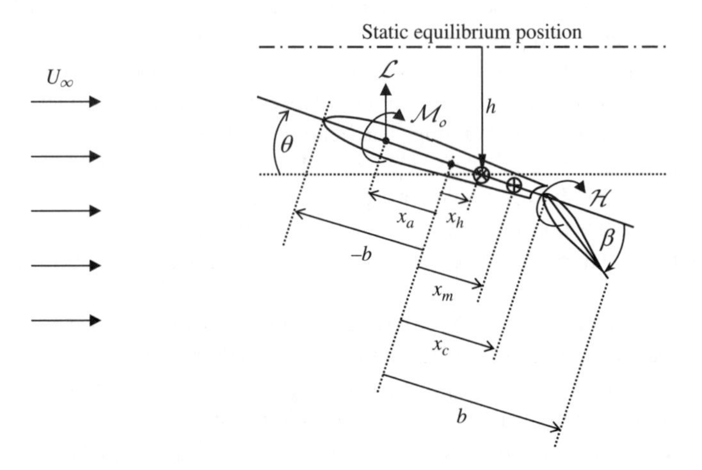

Figure 3.4 Schematic diagram of the typical wing section

and the control surface (positive with the trailing edge down). The mid-chord point of the wing is used as the datum for measuring the chordwise distances, positive towards the trailing edge. The chordwise coordinate x runs along the direction in which the typical section is taken (i.e. normal to either y- or y'-axis) The location of wing leading edge is $x = -b$, and that of the trailing edge, $x = b$, where $b = c/2$, the wing semi-chord. The chordwise locations of the e.a., the a.c., the c.m. and the control-surface hinge-line are $x = x_h$, $x = x_a$, $x = x_m$ and $x = x_c$, respectively.

The structural damping is generally considered negligible in comparison with aerodynamic damping. The generalized forces corresponding to the d.o.f.s must be calculated from the aerodynamic loads shown in Fig. 3.4. These are $\mathcal{L}$ (lift), $\mathcal{M}_0$ (zero-lift pitching moment) concentrated at the a.c. and $\mathcal{H}$ (hinge moment) acting at the hinge-line. The equations of motion are derived by the Lagrange's equations as follows, using the generalized coordinates $\{q\} = (h, \theta, \beta)^T$. The net potential (strain) energy and kinetic energy are given by

$$U = \frac{1}{2}\left(k_h h^2 + k_\theta \theta^2 + k_\beta \beta^2\right)$$

$$T = \frac{1}{2}\int_{-b}^{b} \dot{w}^2 \mathrm{d}m = \frac{1}{2}\int_{-b}^{x_c} \left[\dot{h} + \left(x - x_h\right)\dot{\theta}\right]^2 \mathrm{d}m + \frac{1}{2}\int_{x_c}^{b} \left[\dot{h} + \left(x_c - x_h\right)\dot{\theta} + \left(x - x_c\right)\dot{\beta}\right]^2 \mathrm{d}m$$

$$= \frac{1}{2}\left[m\dot{h}^2 + I_\theta \dot{\theta}^2 + I_\beta \dot{\beta}^2 + 2S_\theta \dot{\theta}\dot{h} + 2S_\beta \dot{\beta}\dot{h} + 2\left(x_c - x_h\right) S_\beta \dot{\theta}\dot{\beta}\right], \tag{3.30}$$

where

$$I_\theta = \int_{-b}^{b} \left(x - x_h\right)^2 \mathrm{d}m \tag{3.31}$$

is the moment of inertia of the wing about the e.a.,

$$I_\beta = \int_{x_c}^{b} \left(x - x_c\right)^2 \mathrm{d}m \tag{3.32}$$

is the moment of inertia of the control surface about its hinge line,

$$S_\theta = \int_{-b}^{b} \left(x - x_h\right) \mathrm{d}m = m\left(x_m - x_h\right)$$

$$S_\beta = \int_{-b}^{x_c} \left(x - x_c\right) \mathrm{d}m = m_c\left(x_{m_c} - x_c\right) \tag{3.33}$$

with m_c being the mass of the control surface and x_{m_c} its c.m. By substituting Eq. (3.30) into Eq. (3.159), we have

$$\frac{\mathrm{d}}{\mathrm{d}t}\left(\frac{\partial T}{\partial\{\dot{q}\}}\right) = [M]\{\ddot{q}\} = \begin{pmatrix} m & S_\theta & S_\beta \\ S_\theta & I_\theta & \left(x_c - x_h\right) S_\beta \\ S_\beta & \left(x_c - x_h\right) S_\beta & I_\beta \end{pmatrix} \begin{Bmatrix} \ddot{h} \\ \ddot{\theta} \\ \ddot{\beta} \end{Bmatrix}$$

$$\frac{\partial T}{\partial\{q\}} = [0]$$

$$\frac{\partial U}{\partial\{q\}} = [K]\{q\} = \begin{pmatrix} k_h & 0 & 0 \\ 0 & k_\theta & 0 \\ 0 & 0 & k_\beta \end{pmatrix} \begin{Bmatrix} h \\ \theta \\ \beta \end{Bmatrix}. \tag{3.34}$$

Thus the mass and stiffness matrices are symmetric, and the structural dynamical system is linear. The generalized loads vector on the structure is due to unsteady aerodynamics and can be expressed as follows:

$$\{Q_a(t)\} = \begin{Bmatrix} -\mathcal{L}(t) \\ \mathcal{M}_0(t) + \left(x_h - x_a\right)\mathcal{L}(t) \\ \mathcal{H}(t) \end{Bmatrix}. \tag{3.35}$$

Since structural damping has been neglected, $\{Q_a\}$ is the only non-conservative force acting on the structure. The aerodynamic loading, $(\mathcal{L}, \mathcal{M}_0, \mathcal{H})$, is a function of the flow speed and density, as well as the control-surface deflection, β. In addition, the lift, $\mathcal{L}$, and hinge moment, $\mathcal{H}$, also depend upon the flow incidence (called the geometric angle of attack), which is defined as the angle made by the chord with the air flow far upstream of the wing (called the freestream). The geometric angle of attack at a given point is related to the freestream flow component seen normal to the wing, w (called the upwash), which, in turn, is a function of the generalized coordinates h, θ, β. For example the local angle of attack at the a.c. is given by[2]

$$\alpha \simeq \frac{w}{U_\infty} = \theta + \frac{\dot{h} + \left(x_a - x_h\right)\dot{\theta}}{U_\infty}, \tag{3.36}$$

where U_∞ is the speed of the uniform, relative air flow far upstream of the wing (freestream speed) (Fig. 3.4). Furthermore, the location of the a.c. can be a function of freestream flow properties. For a thin airfoil, the wing's a.c. in subsonic freestream is near the quarter-chord location ($x_a = -b/2$) and moves to the mid-chord point ($x_a = 0$) for a supersonic freestream. It may thus be appreciated that there is a complicated relationship between the generalized coordinates and the generalized loads on the structure. The rest of the chapter is concerned with how such a relationship can be modelled by the use of aerodynamic concepts.

3.2 Aerodynamic Modelling Concepts

The modelling of generalized unsteady aerodynamic forces, $\{Q_a(t)\}$, resulting from an arbitrary motion, $\{q(t)\}$, is a complex undertaking, generally requiring computational fluid dynamics (CFD) solutions. However, in many cases, it is possible to approximate the dependence of unsteady airloads on the motion variables by a linear operational relationship,

$$\{Q_a\} = [\mathcal{F}](\{q\}), \tag{3.37}$$

where $[\mathcal{F}](.)$ is the aerodynamic operator matrix. Whenever such a relationship exists, the design and analysis of an ASE system can be carried out in a systematic manner. Furthermore, problems of a more complex geometry can be analysed by linear superposition of elementary solutions. In this section, we focus on such flow situations where unsteady aerodynamic modelling is possible by Eq. (3.37), which can be regarded as the baseline aerodynamic model.

As discussed later, the linear relationship of Eq. (3.37) is typically derived for a wing from the following integral equation by applying linear superposition of elementary (flat-plate) solutions to a governing partial differential equation of unsteady aerodynamics:

$$w(x, y, t) = \int_S K\left[(x, y : \xi, \eta), t\right] \Delta p\,(\xi, \eta, t)\, \mathrm{d}\xi \mathrm{d}\eta, \tag{3.38}$$

where Δp is the pressure difference distribution between the upper and lower surfaces of the wing and w denotes the normal (z-component) flow (the upwash) experienced at the wing's mean surface, $S(x, y) = 0$ (Fig. 3.1). The *kernel function* (or Green's function), $K[(x, y : \xi, \eta), t]$, represents the influence coefficient of downwash induced at *collocation*

[2] In a steady flow, we have $\alpha \simeq \theta$, which is why most textbooks on aeroelasticity use the two angles interchangeably. However, we distinguish the angles α and θ here, because we are mainly concerned with unsteady flow.

point (x, y) due to a unit discrete pressure load acting at the *load point* (ξ, η). The similarity between the linear load–deflection relationship of Eq. (3.2) and the pressure-upwash integral equation, Eq. (3.38) is to be carefully noted. This integral equation has to be solved for the unsteady pressure distribution (hence $\{Q_a(t)\}$) on the wing due to a prescribed upwash distribution resulting from the structural motion, $\{q(t)\}$, which is applied as an unsteady boundary condition. The main advantage of linear modelling is that the solution can be superimposed on the steady-state solution of a wing with thickness and camber. Thus the linear coupling between structural dynamics and unsteady aerodynamics [Eq. (3.37)] is clear.

Since our focus is on developing adaptive ASE techniques, we confine the discussion of the aerodynamic model to two-dimensional (2-D) flow over a typical wing section (see Fig. 3.4) in the interest of simplicity. Such an approach of neglecting spanwise (y-component) flow variations at a given cross section is called the strip-theory, and is quite valid for high-aspect-ratio wings, which are the most prone to aeroelastic instabilities. The extension of the approach to three-dimensional (3-D) aerodynamics only adds a spatial variable to the governing equations without essentially changing their character.

3.2.1 Governing Equations for Unsteady Flow

The basic governing equations of unsteady fluid dynamics are the N–S equations (Tewari 2015), which are derived using the principles of conservation of mass, momentum and energy. N–S equations must be solved if one needs to accurately model viscous flow effects, such as flow separation, turbulence and shock-wave/boundary-layer interaction. However, N–S solutions generally require highly sophisticated, memory intensive, iterative numerical techniques, even for steady flows over simple geometries. Their application to aeroservoelasticity involving a rapidly changing flowfield due to a dynamically deforming boundary is practically ruled out. Therefore, one always looks for reasonable approximations by which the unsteady N–S equations can be simplified. Since the viscous effects are of secondary importance in aeroelastic deformation of wings (unless large areas of separated flow originally exist in the undeformed case), one can often drop viscosity and thermal conductivity from the N–S equations, leading to the following inviscid Euler equations:

$$\frac{\partial}{\partial t}\begin{Bmatrix}\rho\\ \rho u\\ \rho w\\ e\end{Bmatrix}+\frac{\partial}{\partial x}\begin{Bmatrix}\rho u\\ \rho u^2+p\\ \rho uw\\ u(e+p)\end{Bmatrix}+\frac{\partial}{\partial z}\begin{Bmatrix}\rho w\\ \rho uw\\ \rho w^2+p\\ w(e+p)\end{Bmatrix}=\{0\} \tag{3.39}$$

Here $e = h - p/\rho$ is the specific internal energy. For the normal aircraft flight, air is assumed to be a perfect gas, which enables the following relationships among the thermodynamic variables:

$$p = \rho RT \quad \text{(State)} \tag{3.40}$$

$$h = c_p T, \tag{3.41}$$

where R is the specific gas constant and c_p the constant-pressure specific heat. Equations (3.40) and (3.41) yield the following thermodynamic relationship:

$$p = (\gamma - 1)\left[e - \frac{1}{2}\left(u^2 + w^2\right)\right], \tag{3.42}$$

where

$$\gamma = \frac{c_p}{c_p - R} \tag{3.43}$$

is the specific heat ratio.

In principle, Euler equations are much simpler in form than the N–S equations, but they retain the nonlinear character of the latter. They can capture the most essential unsteady aerodynamic phenomena that are necessary for practical ASE models. These include pressure distributions due to moving boundaries and shock waves. The isentropic shock wave relations are solutions of Euler equations. Thus it is not necessary to model shock waves separately as they arise naturally out of Euler solutions. However, Euler solvers do require certain degree of care and complexity. The typical unsteady Euler solver is based on a time-marching, finite-difference (or finite-element/finite-volume) procedure, such as explicit (MacCormack, Lax–Wendroff) or implicit (ADI) methods (Tannehill *et al.* 1997). The primitive flow variables, u, w, ρ, e, must be solved in an iterative manner over a computational domain discretized into a large number of grid points for a realistic problem. Furthermore, the convergence and stability of an Euler solution algorithm are problematic because of the non-dissipative nature of the inviscid momentum flux terms, particularly so if strong shock waves are present in the flowfield. This typically requires the introduction of artificial viscosity (or entropy condition) into a solution procedure. Even more importantly, Euler equations have non-unique solutions, which must be resolved by a proper application of the tangential boundary condition on the solid surface. This is termed closure and typically takes the form of Kutta–Joukowski condition at the trailing edge which determines-circulation around an airfoil. Another problem is the generation of body conforming grids for the Euler solver when the solid boundaries and shock waves are moving, as in an unsteady application. Therefore, much of the advantage of simplified governing equations is lost in having to devise a sophisticated numerical scheme. The unsteady Euler equations are thus of limited utility in ASE modelling and must be simplified further.

3.2.2 *Full-Potential Equation*

Further simplification of the governing Euler equations is possible by defining the specific entropy, S, as an additional thermodynamic variable through the following Gibbs' relation:

$$T\mathrm{d}S = \mathrm{d}h - \frac{1}{\rho}\mathrm{d}p. \tag{3.44}$$

Entropy is a measure of disorder in the flowfield, which, by Gibbs' relation, is seen to increase with heat transfer (enthalpy gradient, $\mathrm{d}h$) and the presence of strong shock waves (large pressure gradient, $\mathrm{d}p$). When substituted into the Euler momentum equations, the Gibbs' relation along with the continuity equation, yield the following important result, called the unsteady Crocco's equation:

$$T\{\nabla\}S + \{V\} \times \{\Omega\} = \{\nabla\}h_0 + \frac{\partial\{V\}}{\partial t}, \tag{3.45}$$

where $\{V\} = (u, w)^T$ is the *velocity vector*,

$$\{\Omega\} = -\{\nabla\} \times \{V\} \tag{3.46}$$

is the vorticity vector (a measure of rotation in the flowfield),

$$\{\nabla\}(.) = \begin{Bmatrix} \partial(.)/\partial x \\ \partial(.)/\partial z \end{Bmatrix} \tag{3.47}$$

is the gradient operator and

$$h_0 = h + \frac{1}{2}\left(u^2 + w^2\right) = h + \frac{1}{2}\{V\} \cdot \{V\} \tag{3.48}$$

is the stagnation enthalpy. If the flow is steady, adiabatic ($h_0 = $ const.) and irrotational ($\{\Omega\} = \{0\}$) at all points, then Crocco's equation implies a constant entropy flowfield (isentropic flow). The condition of irrotational flow requires that the velocity vector must be the gradient of a scalar function, Φ, called the velocity potential:

$$\{V\} = \{\nabla\}\Phi = \begin{Bmatrix} \partial\Phi/\partial x \\ \partial\Phi/\partial z \end{Bmatrix} = \begin{Bmatrix} \Phi_x \\ \Phi_z \end{Bmatrix}. \tag{3.49}$$

However, even if the flow is unsteady, it can still be isentropic as long as it is irrotational and the following condition is satisfied by the velocity potential:

$$h_0 = -\frac{\partial\Phi}{\partial t} = -\Phi_t. \tag{3.50}$$

This can be verified by substituting Eqs. (3.48)–(3.50) into Eq. (3.45). For an isentropic flow of a perfect gas, Eqs. (3.42)–(3.44) imply that

$$\frac{p}{\rho^\gamma} = \text{const.} \tag{3.51}$$

Isentropic flow is a special case of barotropic flow in which the pressure is a function of the density only. The momentum equation of such a potential flow can be directly integrated in order to yield the following unsteady Bernoulli equation:

$$\Phi_t + \frac{V^2}{2} + \int \frac{\mathrm{d}p}{\rho} = 0. \tag{3.52}$$

The speed at which infinitesimal pressure disturbances move in an otherwise undisturbed medium is called the speed of sound, which for a perfect gas is given by the following isentropic relation:

$$a = \left.\frac{\partial p}{\partial \rho}\right|_{\text{isentropic}} = \sqrt{\gamma R T}. \tag{3.53}$$

The non-dimensional flow parameter governing the compressible flow is the Mach number, M, defined as the ratio of flow speed, U, and the speed of sound,

$$M = \frac{\sqrt{u^2 + w^2}}{a} = \frac{U}{a}. \tag{3.54}$$

Aeroelastic problems are concerned with an essentially steady flow far upstream of a dynamically flexing wing. Let the subscript ∞ denote the steady-flow conditions far upstream that are unaffected by the unsteady flow in the wing's vicinity. Then the freestream Mach number is given by $M_\infty = U_\infty / a_\infty$, and the isentropic flow conditions produce the following interesting relationship for density variation:

$$\frac{\rho}{\rho_\infty} = \left(\frac{T}{T_\infty}\right)^{1/(\gamma-1)} = \left[\frac{(\gamma-1)}{2a_\infty^2}\left(-2\Phi_t - U^2\right)\right]^{1/(\gamma-1)}$$
$$= \left\{1 + \frac{(\gamma-1)}{2}M_\infty^2\left[1 - \frac{\Phi_t}{U_\infty^2} - \left(\frac{U}{U_\infty}\right)^2\right]\right\}^{1/(\gamma-1)}. \quad (3.55)$$

Of course, Eq. (3.55) requires that the flow far upstream should be steady, that is, $(\Phi_t)_\infty = 0$. By substituting Eq. (3.49) into Eq. (3.55), we have the following form of unsteady Bernoulli equation:

$$\frac{\rho}{\rho_\infty} = \left\{1 + \frac{(\gamma-1)}{2}M_\infty^2\left[1 - \frac{1}{U_\infty^2}\left(\Phi_t + \Phi_x^2 + \Phi_z^2\right)\right]\right\}^{1/(\gamma-1)}. \quad (3.56)$$

Substitution of Eq. (3.49) into the continuity equation, the first equation of the set Eq. (3.39), yields

$$\rho_t + \rho_x\Phi_x + \rho\Phi_{xx} + \rho_z\Phi_z + \rho\Phi_{zz} = 0. \quad (3.57)$$

Then by substituting Eq. (3.56) into Eq. (3.57) and carrying out the partial differentiations of ρ, the following *full-potential equation* (FPE) governing the inviscid, isentropic flow can be derived:

$$\left(a^2 - \Phi_x^2\right)\Phi_{xx} + \left(a^2 - \Phi_z^2\right)\Phi_{zz} = \Phi_{tt} + 2\left(\Phi_x\Phi_{xt} + \Phi_z\Phi_{zt}\right) + 2\Phi_x\Phi_z\Phi_{xz}, \quad (3.58)$$

where a is the local speed of sound. It is interesting to note that the FPE, Eq. (3.58), can be alternatively expressed as follows (Garrick 1957):

$$\nabla^2\Phi = \frac{1}{a^2}\frac{D^2\Phi}{Dt^2}, \quad (3.59)$$

where

$$\nabla^2(.) = \{\nabla\}\cdot\{\nabla\}(.) = \frac{\partial^2}{\partial x^2}(.) + \frac{\partial^2}{\partial z^2}(.) \quad (3.60)$$

is the Laplacian operator and

$$\frac{\mathrm{D}}{\mathrm{D}t}(.) = \frac{\partial}{\partial t}(.) + \{V\}\cdot\{\nabla\}(.) \quad (3.61)$$

is the Eulerian (or material) derivative representing the time derivative seen by a fluid particle convecting with the flow at the local velocity $\{V\}$. Equation (3.59) is the governing *wave equation* of acoustics, with the rate of wave propagation a. Hence, merely by transforming the spatial coordinates[3] from a body-fixed frame to a frame convecting with the flow, we have

[3] It can be verified that Eq. (3.59) is alternatively expressed as follows:

$$\Phi_{\xi\xi} + \Phi_{\zeta\zeta} = \frac{1}{a^2}\Phi_{tt},$$

where the fluid-fixed coordinates (ξ, ζ) are obtained from the body-fixed coordinates (x, z) by the following Galilean transformation:

$$\begin{Bmatrix}\xi\\ \zeta\end{Bmatrix} = \begin{Bmatrix}x\\ z\end{Bmatrix} - \{V\}t.$$

established an equivalence between potential unsteady aerodynamics and acoustics. This is an important result, which is quite useful in analysing and solving the FPE.

Another interesting aspect of potential flow is the acceleration potential, Ψ, which is related to the velocity potential as follows:

$$\Psi = \frac{\mathrm{D}\Phi}{\mathrm{D}t} = \frac{\partial \Phi}{\partial t} + \{V\} \cdot \{\nabla\}\Phi, \tag{3.62}$$

which, by the integrated form of the Euler momentum equation (unsteady Bernoulli equation) can also be expressed as

$$\Psi = -\int \frac{\mathrm{d}p}{\rho}. \tag{3.63}$$

Clearly, the acceleration potential is directly related to the pressure difference and provides an alternative description of the potential flowfield. We shall return to acceleration potential later in the chapter.

The full-potential formulation is a practical alternative to Euler equations because, in order to completely determine the flowfield, one has to solve only for the velocity potential, $\Phi(x, z, t)$, rather than each of the primitive variables, u, w, ρ, e, of Euler equations. However, while Euler equations can be applied to non-isentropic flow caused by strong shock waves, the FPE is valid only for isentropic flow. The nonlinear nature of the FPE makes it almost as formidable to solve as Euler equations, although the number of dependent variables is reduced to one. The absence of viscous dissipation calls for artificial viscosity and tangential flow conditions for closure, as in the case of Euler formulation. The main utility of the FPE formulation is for nearly isentropic, transonic flows where weak, normal shock waves are present, and for which Euler equations are unnecessary. As discussed below, a major simplification of the FPE is possible for transonic small-disturbance (TSD) problems, where the boundary conditions can be applied on a mean surface rather than the actual moving boundary. Furthermore, in subsonic and supersonic regimes, the FPE can be effectively linearized for thin wings undergoing small amplitude motions.

In order to check the applicability of the FPE to transonic speeds, consider the following normal shock relation for entropy gradient at upstream Mach number close to unity, $M_\infty \approx 1$:

$$\frac{\partial S}{\partial n} \simeq \frac{2\gamma}{3(\gamma+1)^2}\left(M_\infty^2 - 1\right)^3. \tag{3.64}$$

In the limit $M_\infty \to 1$, the entropy variation becomes negligible across the shock wave, and an isentropic condition prevails, thereby enabling the application of the FPE. But the momentum is not exactly conserved in the presence of a normal shock wave, however weak it may be. Thus the validity of the FPE for transonic flows with weak shock waves is only approximate, but can give a reasonable model for ASE purposes.

Solution procedures for the FPE are essentially based on an iterative finite-difference approach, although Green's function integral solution of the boundary-value problem is also possible (Tewari 2015). In the unsteady case, the FPE is hyperbolic in nature for all speed regimes, hence a time-marching scheme can be adopted (as in a typical unsteady Euler solver). However, for the steady-state problem, the FPE and Euler equations change their character from being elliptic in the local subsonic region to hyperbolic in the supersonic region. Therefore, in a transonic steady-state application, a special treatment of the mixed elliptic/hyperbolic behaviour is required when locally supersonic regions (normal shock waves) are present in the flowfield. This is either carried out by a switching (or type-dependent) procedure when the coefficient of the Φ_{xx} term changes sign (Murman 1973) or by spatial upwinding techniques (Hafez *et al.* 1979, Holst and Ballhaus 1979) for density computation in a conservative form. As in the case of Euler computations, there is also the need for the introduction of artificial viscosity/entropy for avoiding physically unrealistic solutions (Osher *et al.* 1985). Furthermore, special treatment of circulation at the trailing edge and the wake is also necessary (Steger and Caradonna 1980) for closure, as in the case of Euler equations. Fortunately, a typical ASE application involves small amplitude motion of thin lifting surfaces, which quite significantly simplifies the full-potential model.

3.2.3 Transonic Small-Disturbance Equation

Aircraft geometries are streamlined for dag minimization such that the cross-flow (lateral) variations in the flow variables caused by the body thickness are small compared to those along the freestream (longitudinal) direction. The velocity potential is thus only slightly changed from its freestream value. This fact offers a major simplification in the governing equations called the small-disturbance (or small-perturbation) approximation. Consider a two-dimensional airfoil with the x-axis along the relative freestream of speed U_∞ and the z-axis normal to it. The net velocity potential is then regarded as a linear superposition of the perturbation velocity potential, ϕ, over that of the freestream:

$$\Phi = U_\infty x + \phi, \tag{3.65}$$

which results in the velocity components

$$u = \frac{\partial \Phi}{\partial x} = U_\infty + \phi_x; \quad w = \frac{\partial \Phi}{\partial z} = \phi_z. \tag{3.66}$$

When Eq. (3.65) is substituted into the FPE, Eq. (3.58), with the small-perturbation assumptions,

$$\phi_x \ll U_\infty; \quad \phi_z \ll U_\infty \tag{3.67}$$

and

$$\phi_x \ll a; \quad \phi_z \ll a, \tag{3.68}$$

one can safely neglect second- (and higher) order terms involving ϕ_z and ϕ_t, and third-order terms involving ϕ_x. However, the second-order term involving ϕ_x must be retained for accuracy in the transonic limit, $U_\infty + \phi_x \simeq a$, at which weak normal shock waves may be present (Landahl 1961). With these approximations, the FPE is approximated by the following TSD equation:

$$\left[1 - M_\infty^2 - \frac{(\gamma + 1)M_\infty^2}{U_\infty}\phi_x\right]\phi_{xx} + \phi_{zz} = \frac{2M_\infty^2}{U_\infty}\phi_{xt} + \frac{M_\infty^2}{U_\infty^2}\phi_{tt}. \tag{3.69}$$

The main utility of the TSD equation is its applicability in the unsteady, transonic limit, $M_\infty \simeq 1$, for which its essentially nonlinear character cannot be neglected. However, it is much simpler to solve than the FPE, because the unsteady boundary conditions can be applied on the mean surface (rather than the actual boundary) of the thin wing.

A practical simplification of the TSD equation is called the low-frequency limit (Landahl 1961), for which the term involving ϕ_{tt} can be neglected, leading to the following:

$$\left[1 - M_\infty^2 - \frac{(\gamma+1)M_\infty^2}{U_\infty}\phi_x\right]\phi_{xx} + \phi_{zz} = \frac{2M_\infty^2}{U_\infty}\phi_{xt}. \tag{3.70}$$

The applicability of the low-frequency transonic small-disturbance (LFTSD) approximation, Eq. (3.70), requires that the largest characteristic frequency, ω, governing the unsteady motion must be sufficiently small such that

$$\omega \ll \frac{U_\infty}{b}, \tag{3.71}$$

where b is a characteristic length. Typically, b is taken to be the mean semi-chord of the wing and ω the largest elastic modal frequency (bending, torsion or control-surface rotation) that can influence the flowfield. In terms of these parameters, Eq. (3.70) can be rendered non-dimensional as follows:

$$\left[1 - M_\infty^2 - (\gamma+1)M_\infty^2\bar{\phi}_\xi\right]\bar{\phi}_{\xi\xi} + \bar{\phi}_{\zeta\zeta} = 2kM_\infty^2\bar{\phi}_{\xi\tau}, \tag{3.72}$$

where the non-dimensional variables are given by

$$\xi = x/b,\ \zeta = z/b,\ \tau = t\omega,\ \bar{\phi} = \frac{\phi}{U_\infty b},\ k = \frac{\omega b}{U_\infty}. \tag{3.73}$$

The reduced frequency, k, is a similarity parameter of the unsteady flow in addition to the Mach number, M_∞. In the low-frequency limit, we have $k \ll 1$. Solution of the LFTSD equation is usually carried out by finite-difference techniques which are quite similar to (but much simpler than) those required for the FPE. Being based on an iterative solution of the linearized equation

$$\left(1 - M_\infty^2\right)\bar{\phi}_{\xi\xi} + \bar{\phi}_{\zeta\zeta} = 2kM_\infty^2\bar{\phi}_{\xi\tau}, \tag{3.74}$$

by an approximate factorization approach (Ballhaus and Steger 1975), the LFTSD equation is the simplest nonlinear unsteady aerodynamic model. Its solution can be derived iteratively through a linear governing equation, which is essentially the aim of an adaptive ASE design. Therefore, it is envisaged that the low-frequency TSD equation will be the crux of future developments in transonic ASE modelling and control.

When all the nonlinear terms are dropped from the TSD equation, the result is the linearized compressible flow equation governing the potential subsonic and supersonic flows. The various governing equations of unsteady aerodynamics in the increasing order of approximation are summarized below for a two-dimensional, inviscid flow:

1. *Euler*:

$$\frac{\partial\{F\}}{\partial t} + \frac{\partial\{f_x\}}{\partial x} + \frac{\partial\{f_z\}}{\partial z} = \{0\}$$

with

$$\{F\} = \{\rho, \rho u, \rho w, \rho e\}$$
$$\{f_x\} = u\{F\} + \{0, p, 0, pu\}$$
$$\{f_z\} = w\{F\} + \{0, 0, p, pw\}$$
$$e = \frac{p}{\gamma - 1} + \frac{1}{2}\rho\left(u^2 + w^2\right).$$

2. *Full Potential*:

$$[a^2 - (U_\infty + \Phi_x)^2]\Phi_{xx} + (a^2 - \Phi_z^2)\Phi_{zz} - 2(U_\infty + \Phi_x)(\Phi_z\Phi_{xz} + \Phi_{xt}) - 2\Phi_z\Phi_{zt} - \Phi_{tt} = 0$$

with

$$a^2 = a_\infty^2 - (\gamma - 1)\left[U_\infty\Phi_x + \frac{1}{2}\left(\Phi_x^2 + \Phi_z^2\right) + \Phi_t\right].$$

3. *Transonic Small Disturbance (TSD)*:

$$\left[1 - M_\infty^2\left(1 + \frac{\gamma + 1}{U_\infty}\phi_x\right)\right]\phi_{xx} + \phi_{zz} - 2\frac{M_\infty^2}{U_\infty}\phi_{xt} - \frac{M_\infty^2}{U_\infty^2}\phi_{tt} = 0.$$

4. *Low-Frequency Transonic Small Disturbance (LFTSD)*:

$$\left[1 - M_\infty^2\left(1 + \frac{\gamma + 1}{U_\infty}\phi_x\right)\right]\phi_{xx} + \phi_{zz} - 2\frac{M_\infty^2}{U_\infty}\phi_{xt} = 0.$$

5. *Linearized Subsonic/Supersonic*:

$$\left(1 - M_\infty^2\right)\phi_{xx} + \phi_{zz} - 2\frac{M_\infty^2}{U_\infty}\phi_{xt} - \frac{M_\infty^2}{U_\infty^2}\phi_{tt} = 0.$$

3.3 Baseline Aerodynamic Model

As discussed above, we require a linear unsteady aerodynamic model to serve as a baseline for designing adaptive ASE control laws. In effect, we are interested in linear operator relationship of the kind given by Eq. (3.162). The governing equation of such a model in terms of the disturbance velocity potential is derived from the TSD equation by neglecting the nonlinear term in Eq. (3.69), resulting in the following linearized equation:

$$\left(1 - M_\infty^2\right)\phi_{xx} + \phi_{zz} = \frac{2M_\infty}{a_\infty}\phi_{xt} + \frac{1}{a_\infty^2}\phi_{tt}, \tag{3.75}$$

which can be expressed as the following wave equation:

$$\nabla^2\phi = \frac{1}{a_\infty^2}\left.\frac{\mathrm{D}^2\phi}{\mathrm{D}t^2}\right|_\infty, \tag{3.76}$$

where

$$\left.\frac{\mathrm{D}}{\mathrm{D}t}\right|_\infty (.) = \frac{\partial}{\partial t}(.) + \begin{Bmatrix} U_\infty \\ 0 \end{Bmatrix} \cdot \{\nabla\}(.) = \frac{\partial}{\partial t}(.) + U_\infty\frac{\partial}{\partial x}(.) \tag{3.77}$$

is the Eulerian derivative representing the time derivative seen by an observer moving with the freestream velocity $(U_\infty, 0)^T$. Note that the wave propagation speed in Eq. (3.76) is a_∞, the freestream speed of sound, rather than a, the local speed of sound in Eq. (3.59). Evidently, the effect of linearization on the FPE is to make small disturbances due to the body spread out at a constant speed in all directions relative to the freestream. Of course, this does not allow the presence of any shock waves in the flowfield. The same linear governing equation is satisfied by the disturbance acceleration potential, ψ, which is related to the disturbance velocity potential by virtue of Eq. (3.62) as follows:

$$\psi = \left.\frac{D\phi}{Dt}\right|_{\infty} (.) = \frac{\partial \phi}{\partial t} + U_{\infty}\frac{\partial \phi}{\partial x}. \tag{3.78}$$

Equation (3.75) is accurate at subsonic and supersonic speeds, but cannot be applied to the transonic regime where the nonlinear TSD equation must be solved. Being a linear equation, Eq. (3.75) possesses the important property that its solution to arbitrary boundary conditions is a linear superposition of elementary solutions corresponding to much simpler boundary conditions. Such elementary solutions can be source doublets distributed over a solid boundary on which the flow tangency and Kutta condition at the trailing edge are to be satisfied (Garrick 1957).

The boundary conditions in terms of the disturbance velocity potential can be posed as follows (Garrick 1957):

1. The flow is uniform (undisturbed) far upstream of the body,

$$\phi(-\infty, z, t) = 0, \tag{3.79}$$

   and perturbations remain bounded at infinity.
2. Pressure is continuous across the wake. The small perturbation causes a planar wake that always lies in the mean wing plane. The vorticity of the wake, along with the bound circulation around the wing, satisfies the conservation of vorticity in the potential flowfield.
3. Viscous effects, although unmodelled, are assumed to be just large enough to cause a smooth flow tangential to the mean wing surface at the trailing edge. This assumption is called the Kutta condition, and allows inviscid flow modelling without the attendant problem of non-unique solutions. Its physical validity, however, is questionable when the reduced frequency, $b\omega/U_{\infty}$ becomes large. In a typical aeroelastic application with small amplitude, low-frequency oscillations, the Kutta condition is widely held to be valid.
4. The flow is tangential to the impervious, solid boundary. This implies that a flow particle in contact with the body must follow the instantaneous surface contour of the dynamic body, which can be defined for the two-dimensional case by the functional

$$F(x, z, t) = z - z_b(x, t) = 0.$$

   Hence, the solid boundary condition is given by

$$\frac{\mathrm{D}F}{\mathrm{D}t} = \frac{\partial F}{\partial t} + \{V\} \cdot \{\nabla\}F = 0. \tag{3.80}$$

   For a thin wing, this boundary condition is effectively linearized by assuming that the local normal on the body surface is along the z-axis and the product $w\partial z_b/\partial x$ is negligible. Furthermore, the thickness of the body can be considered to be either negligible or its effect included in the steady flowfield. Thus the unsteady effect of the body can be represented by the mean surface (i.e. the median plane between upper and lower surfaces), $z = 0$, as follows:

$$w(x, 0, t) = \left.\frac{\partial \phi}{\partial z}\right|_{z=0} = \frac{\partial z_b}{\partial t} + U_{\infty}\frac{\partial z_b}{\partial x}. \tag{3.81}$$

Since the upwash, w, cannot be physically discontinuous across the mean surface, it must be the same at the upper and lower surfaces, denoted by $z = +0$ and $z = -0$, respectively. However, ϕ must change sign across the wing, that is, be an antisymmetric (odd) function of z, because its z-derivative, w, is a symmetric (even) function. Conventionally, ϕ is taken to be positive on the upper surface and negative on the lower surface. Implementation of the tangential flow condition is useful in determining the unknown strength of a distribution of elementary solution (source, vortex, doublet, etc.) on the solid boundary.

Generally, a solid boundary has a different pressure distribution on the upper surface, p_u, from that on the lower surface, p_ℓ. However, the approximation of the body by a mean surface of zero thickness, $z = 0$, only allows the pressure distribution to change sign, but not the magnitude, across the surface. The linearization of the unsteady Bernoulli equation, Eq. (3.52), written as

$$p(x, z, t) - p_\infty = -\rho_\infty \left(\frac{\partial \phi}{\partial t} + U_\infty \frac{\partial \phi}{\partial x} \right), \tag{3.82}$$

results in the following relationship between the pressure difference across the wing, $\Delta p = p_u - p_\ell$, and the disturbance velocity potential, ϕ:

$$\Delta p(x) = p(x, +0, t) - p(x, -0, t) = 2p(x, 0, t) = -2\rho_\infty \left(\frac{\partial \phi}{\partial t} + U_\infty \frac{\partial \phi}{\partial x} \right)\bigg|_{z=0}. \tag{3.83}$$

Equation (3.83) can be inverted in order to yield the following integral relationship between the pressure and velocity potential on the wing:

$$\phi(x, 0, t) = -\frac{1}{\rho_\infty U_\infty} \int_{-\infty}^{x} p\left(\xi, 0, t - \frac{x - \xi}{U_\infty} \right) \mathrm{d}\xi. \tag{3.84}$$

The condition of uniform flow far upstream, Eq. (3.79), is implicit in Eq. (3.84). Typically, the integral equation Eq. (3.84) must be solved for pressure difference across the wing, while taking into account how the velocity potential evolves in the wake region. An alternative description to the velocity potential is possible in terms of the disturbance acceleration potential, ψ, defined by

$$\psi = -\frac{p - p_\infty}{\rho_\infty} = \frac{\partial \phi}{\partial t} + U_\infty \frac{\partial \phi}{\partial x}, \tag{3.85}$$

for which the wake region need not be modelled as ψ vanishes identically across the wake.

3.3.1 *Integral Equation Formulation*

The solution to the baseline governing equation, Eq. (3.75), subject to appropriate unsteady boundary conditions, must be sought in a simple, non-iterative manner, in order to be utilized in control-law derivations. Towards this end, an integral equation governing the unsteady, two-dimensional flow can be derived from Eq. (3.84) by taking the z-derivative of both the sides and substituting the expression for upwash,

$$w = \frac{\partial \phi}{\partial z}, \tag{3.86}$$

resulting in

$$w(x,0,t) = -\frac{1}{\rho_\infty U_\infty}\int_{-\infty}^{x}\frac{\partial}{\partial z}p\left(\xi, z, t-\frac{x-\xi}{U_\infty}\right)\bigg|_{z=0} d\xi. \tag{3.87}$$

Since pressure disturbance is related to acceleration potential by Eq. (3.85), the integral equation Eq. (3.87) is referred to as the acceleration potential formulation. This integral equation must be solved for the pressure difference across the mean surface, given a prescribed upwash distribution, Eq. (3.81). The upper-limit of integration, x, is handled differently for subsonic and supersonic freestreams. In the supersonic case, the pressure disturbance cannot travel upstream of the Mach cone, whereas no such restriction exists for the subsonic case. Typically, Eq. (3.87) is amenable to either analytical or numerical solution for the harmonically oscillating wing,

$$w(x,0,t) = \bar{w}e^{i\omega t}, \tag{3.88}$$

which results in the following integral equation for the complex amplitude (magnitude and phase) in the frequency domain:

$$\bar{w}(x) = -\frac{1}{\rho_\infty U_\infty}e^{-i\omega x/U_\infty}\int_{-\infty}^{x}\frac{\partial}{\partial z}\bar{p}\left(\xi, z, t-\frac{x-\xi}{U_\infty}\right)\bigg|_{z=0} d\xi, \tag{3.89}$$

where

$$p(x,z,t) - p_\infty = \bar{p}(x,z)e^{i\omega t}, \tag{3.90}$$

is the harmonic pressure difference with a complex amplitude, $\bar{p}$.

It is interesting to note the equivalence of the integral equation Eq. (3.89) with the Green's identity resulting from the superposition of harmonically pulsating, acceleration potential doublets (Garrick 1957). This equivalence can be exploited in order to obtain an analytical solution in the harmonic limit. Such a formulation for the three-dimensional case is the subsonic doublet-lattice method (Albano and Rodden 1969) and the subsonic/supersonic doublet-point method (Tewari 1994, Ueda and Dowell 1982). For the simplicity of discussion, we will confine ourselves here to the two-dimensional case.

3.3.2 *Subsonic Unsteady Aerodynamics*

A line acceleration potential doublet convecting with the freestream at subsonic velocity, $U_\infty < a_\infty$, corresponds to the following elementary solution of the governing wave equation (Garrick 1957), Eq. (3.76):

$$\frac{\bar{p}(x,z)}{\rho_\infty} = \frac{i}{4\beta}\frac{\partial}{\partial z}e^{-i\mu(x-\xi)}e^{i\omega t}H_0^{(2)}(\kappa R), \tag{3.91}$$

where

$$\beta = \sqrt{1-M_\infty^2}, \tag{3.92}$$

$$\kappa = \frac{kM_\infty}{\beta^2}, \tag{3.93}$$

$$\mu = \kappa M_\infty, \tag{3.94}$$

$$k = \frac{\omega b}{U_\infty}, \tag{3.95}$$

is the *reduced frequency* of harmonic motion,

$$R = \sqrt{(x-\xi)^2 + \beta^2(z-\zeta)^2}, \tag{3.96}$$

and

$$H_n^{(2)}(.) = J_n(.) - iY_n(.) \tag{3.97}$$

represents the Hankel function of the second kind and nth order, expressed as a complex combination of Bessel functions of the first (J_n) and second kind (Y_n) and of the same order. Here, the length dimension is rendered non-dimensional by dividing by the wing semi-chord, $b = c/2$. By substituting Eq. (3.91) into Eq. (3.89) and integrating by parts (Garrick 1957), we have the following integral equation for the subsonic harmonic case:

$$\bar{w}(x) = \int_{-1}^{1} K\left(x-\xi, k, M_\infty\right) \bar{\gamma}(\xi)\mathrm{d}\xi, \tag{3.98}$$

where $\bar{\gamma}(\xi)$ is the complex amplitude of *bound vorticity* on the wing at a non-dimensional chordwise location ξ from the leading edge, related to the local pressure jump across the surface by

$$\Delta p(\xi) = -\rho_\infty U_\infty \bar{\gamma}\,(\xi)\, e^{i\omega t}. \tag{3.99}$$

The limits of integration are from the leading edge, $x = -1$, to the trailing edge, $x = 1$. A form of the integral equation relating the upwash amplitude to that of the pressure difference across the mean surface is due to Possio (1938):

$$\bar{w}(x) = -\frac{1}{\rho_\infty U_\infty} \int_{-1}^{1} K\left(x-\xi, k, M_\infty\right) \Delta\bar{p}(\xi)\mathrm{d}\xi, \tag{3.100}$$

or as follows in terms of the non-dimensional *lift* amplitude per unit span, $\bar{\mathcal{L}}$, and the *angle-of-attack* amplitude, $\bar{\alpha} = \bar{w}/U_\infty$:

$$\bar{\alpha}(x) = \int_{-1}^{1} K\left(x-\xi, k, M_\infty\right) \bar{\mathcal{L}}(\xi)\mathrm{d}\xi. \tag{3.101}$$

Thus the kernel function of the integral equation represents the important aerodynamic influence coefficient between non-dimensional lift and angle of attack. The kernel function is expressed as follows in terms of the non-dimensional variables (Garrick 1957):

$$\begin{aligned} K\left(x-\xi, k, M_\infty\right) = -\frac{k}{4\beta} e^{i\mu(x-\xi)} \Bigg\{ & H_0^{(2)}\left(\kappa \mid x-\xi \mid\right) - iM_\infty \frac{(x-\xi)}{\mid x-\xi \mid} H_1^{(2)}\left(\kappa \mid x-\xi \mid\right) \\ & - i\beta^2 e^{-ik(x-\xi)/\beta^2} \left(\frac{2}{\pi\beta} \ln \frac{1+\beta}{M_\infty} \right. \\ & \left. + \int_0^{k(x-\xi)/\beta^2} e^{i\lambda} H_0^{(2)} M_\infty \mid \lambda \mid \mathrm{d}\lambda \right) \Bigg\} \end{aligned} \tag{3.102}$$

The integral equation must be solved for the complex bound vorticity amplitude, $\bar{\gamma}(\xi)$, given a prescribed upwash amplitude, $\bar{w}(x)$, on the wing. The Kutta condition at the trailing edge requires that $\bar{\gamma}(1) = 0$, which is generally satisfied by selecting the bound vorticity distribution as the following infinite series:

$$\bar{\gamma}(x) = U_\infty \left(2a_0 \cot\frac{\theta}{2} + 4\sum_{1}^{\infty} a_n \frac{\sin n\theta}{n} \right), \tag{3.103}$$

where $x = -\cos\theta$. A numerical scheme called the *collocation method* (cf. Possio 1938) is adopted for determining the unknown, complex coefficients, a_i, in Eq. (3.103) (hence the solution to Eq. (3.102)) by enforcing the upwash boundary condition at several collocation (or control) points. Alternative techniques work with the kernel function, such as the *iterative kernel evaluation method* (Dietze 1947, Turner and Rabinowitz 1950), and *kernel function expansion method* (Fettis 1952, Schade 1944). A more direct solution procedure – that does not employ the integral equation – is the approximate boundary-value solution of the governing wave equation, Eq. (3.76), transformed into elliptical coordinates, and the result expanded as a series of *Mathieu functions* (Küssner 1954, Reissner 1951, Timman and Van de Vooren 1949). However, there are difficulties associated with the numerical evaluation of integrals involving infinite series expansion in Mathieu functions. Other approaches for Possio's integral solution are based on series expansions in compressible reduced frequency, $k/(1 - M_\infty^2)$, such as the work of Runyan (1952) and Timman *et al.* (1951). However, these expansions are often incomplete and give erroneous results, both at low frequencies, and higher subsonic Mach numbers. More recently, Lin and Iliff (2000) have presented an approximate closed-form solution of Possio's wave equation, which offers the promise of implementation in an ASE control-law derivation. Appendix B outlines the iterative solution procedure of Dietze and the analytical expansion by Fettis, which are valuable because they utilize an analytical development for the incompressible limit ($M_\infty \to 0$), thereby offering an additional insight into the physical characteristics of subsonic unsteady aerodynamics. Furthermore, Appendix B also presents the alternative closed-form development by Lin and Iliff as the likely baseline scheme for an active flutter suppression system (which is the topic of the next chapter).

3.3.2.1 Analytical Solution in Incompressible Limit

The incompressible flow past an oscillating airfoil was the first analytical development in unsteady aerodynamics (Birnbaum 1924, Theodorsen 1935, Wagner 1925). The problem was posed as a Laplace equation (derived from Eq. (3.75) by putting $M_\infty = 0$) and developed by using conformal mapping of the complex velocity potential of a source distribution on the wing's chord plane (flat plate) taken to be the mean surface. Thus the effects of thickness and camber are neglected and a simple model relating the velocity potential to the unsteady pressure difference, Δp, is derived. This approach is briefly outlined in Appendix A. We especially refer to the method by Theodorsen (1935), which separately integrates the circulatory (or wake-dependent) and non-circulatory (wake-independent) harmonic potentials, in order to give the following analytical expression for pressure difference across the airfoil:

$$\frac{\Delta\bar{p}(x)}{\frac{1}{2}\rho_\infty U_\infty^2} = \mathcal{F}(\bar{\alpha}). \tag{3.104}$$

Here $\mathcal{F}(.)$ is the *aerodynamic operator*, which operates on the local angle-of-attack amplitude, $\bar{\alpha}(\xi) = \bar{w}(\xi)/U_\infty$, and is given by

$$\mathcal{F}(.) = \mathcal{F}_c(.) + \mathcal{F}_{nc}(.), \tag{3.105}$$

where the non-circulatory part is the following (Bisplinghoff and Halfman 1955):

$$\mathcal{F}_{nc}(.) = \frac{4}{\pi}\int_{-1}^{1}\left[\frac{\sqrt{1-\xi^2}}{r\sqrt{1-x^2}} - \frac{ik}{2}\log\left(\frac{1-x\xi+\sqrt{1-\xi^2}\sqrt{1-x^2}}{1-x\xi-\sqrt{1-\xi^2}\sqrt{1-x^2}}\right)\right](.)\mathrm{d}\xi \tag{3.106}$$

with $r = x - \xi$. The circulatory part is represented as follows:

$$\mathcal{F}_c(.) = \frac{4}{\pi}[1 - C(k)]\sqrt{\frac{1-x}{1+x}}\int_{-1}^{1}\sqrt{\frac{1+\xi}{1-\xi}}(.)\mathrm{d}\xi, \tag{3.107}$$

where

$$C(k) = \frac{\int_1^\infty \frac{x}{\sqrt{x^2-1}}e^{-ikx}\mathrm{d}x}{\int_1^\infty \frac{x+1}{\sqrt{x^2-1}}e^{-ikx}\mathrm{d}x} = \frac{H_1^{(2)}(k)}{H_1^{(2)}(k) + iH_0^{(2)}(k)} \tag{3.108}$$

is *Theodorsen's function*, which physically represents the deficiency in lift due to the unsteady wake. In the steady flow limit ($k = 0$), the wake is fully developed and thus $C(k)$ approaches unity, whereby the circulatory part vanishes. This is tantamount to a wake vortex that has convected downstream to infinity, and corresponds to the maximum lift magnitude. The chief achievement of Theodorsen (1935) is in the evaluation of the two improper integrals of Eq. (3.108) by Hankel functions of the reduced frequency. The integration considers k to be complex, with a small real part (resulting in slightly damped oscillation) and applies *analytic continuation* by making the real part vanish in the harmonic limit (real k). Such a technique would appear to be a purely mathematical device without a sound physical argument, but others (Küssner and Schwarz 1940, Söhngen 1940) obtained the same result by Cauchy principal value and Fourier series, thereby lending credence to Theodorsen's method. Alternatively, if one employs an integral equation formulation using velocity potential doublets, then Eq. (3.104) is derived as a solution by separation of variables (Garrick 1957). Tabulated values of Theodorsen's function give unsteady aerodynamic loads for a prescribed harmonic motion (Smilg and Wasserman 1942).

Another practical application of the incompressible solution is in transient aerodynamics, which involves non-harmonic (indicial, impulsive or oscillating with variable amplitude) motion. Clearly, a general aerodynamic theory must model the transient response to suddenly applied inputs, and is also necessary for ASE applications where an arbitrary, time-dependent input can be applied by the control system. The first step in this direction was taken by Wagner (1925) who proposed a simple function for the lift of an airfoil suddenly put into motion at a given geometric angle of attack. The indicial aerodynamics represented by Wagner's function is related to Theodorsen's harmonic aerodynamics by Fourier transform (Garrick 1938). This is now a basic result of linear systems theory, but in the 1930s and 1940s it was a novel idea explored by Garrick. Consider the *pressure gradient* amplitude due to circulation obtained from Eq. (3.104) by taking the derivative with x:

$$\left.\frac{\mathrm{d}\Delta p}{\mathrm{d}x}\right|_c = \frac{2\rho_\infty U_\infty}{\pi}C(k)\sqrt{\frac{1-x}{1+x}}\sqrt{\frac{1+\xi}{1-\xi}}\bar{w}(\xi)e^{i\omega t}, \tag{3.109}$$

which is integrated chordwise to yield the circulatory lift per unit span:

$$\mathcal{L}_c = 2b\pi\rho_\infty U_\infty C(k)\bar{w}e^{i\omega t}. \tag{3.110}$$

This is the lift at large times, that is, when all the stable transients have decayed to zero, and the airfoil is oscillating harmonically at a constant frequency ω with a time-independent amplitude and phase $\bar{w}$. Now consider a situation when there is a unit-step change in the upwash at $t = 0$, which is expressed as follows:

$$w(\xi, t) = w_0(\xi)u_s(t), \tag{3.111}$$

where

$$u_s(t) = \begin{cases} 0 & (t < 0) \\ 1 & (t \geq 0) \end{cases} \tag{3.112}$$

is the unit-step function. Wagner's function, $\Phi(\hat{t})$, gives the circulatory (indicial) lift in this case by

$$\mathcal{L} = 2b\pi\rho_\infty U_\infty w_0 \Phi(\hat{t}), \tag{3.113}$$

where $\hat{t} = U_\infty t/b$ is the non-dimensional time, or $\omega t = k\hat{t}$. From a comparison of Eqs. (3.110) and (3.113), it is clear that the following transform relationship holds between $\Phi(\hat{t})$ and $C(k)$:

$$\frac{C(k) - 1}{ik} = \int_0^\infty [\Phi(\hat{t}) - 1]e^{-ik\hat{t}}\mathrm{d}\hat{t}, \tag{3.114}$$

which is Fourier transform in k (or Laplace transform with Laplace variable ik). Being an indicial function, Wagner's function is 0 for negative values of time, and has the property $\Phi(\infty) = 1$.

Let us examine the pitching moment contribution of the circulatory lift,

$$\mathcal{M}_c = b\int_{-1}^{1} (\Delta p)_c(x - x_h)\mathrm{d}x = 2\pi\rho_\infty U_\infty b\left(x_h + \frac{b}{2}\right) C(k)w_{3c/4}, \tag{3.115}$$

where

$$w_{3c/4} = \theta + \dot{h} + \left(\frac{b}{2} - x_h\right)\dot{\theta} \tag{3.116}$$

is the geometric (or resultant) upwash at the 3/4-chord point ($x = b/2$) from the leading edge (see Eq. (3.36) for the definition of geometric angle of attack and upwash). Therefore, while the pitching moment due to wake vanishes at the 1/4-chord position ($x = -b/2$) – thereby yielding the a.c. – it is directly proportional to the geometric upwash at the 3/4-chord point. This information is extremely useful while trying to satisfy the tangential flow condition at the trailing edge (the Kutta condition) by a distribution of discrete elementary solutions of the Laplace equation, that is, velocity potential doublets and vortices. One selects the strengths of discrete vortices (or doublets) such that flow becomes tangential to the surface at the 3/4-chord location, which is tantamount to having an *induced upwash* cancel the geometric upwash given by Eq. (3.116). The Kutta condition is thus satisfied by having the correct bound and wake vorticity which induces an upwash equal in magnitude, but opposite in sign, to that of the geometric upwash at the 3/4-chord point. Hence, the 3/4-chord location is called the control (or upwash collocation) point of a discrete element, panel method.

For a general transient motion, Wagner's function can give the unsteady lift by the following Duhamel's integral by linear superposition of the prescribed geometric upwash:

$$\mathcal{L}(\hat{t}) = 2b\pi\rho_\infty U_\infty \left[\Phi(\hat{t})w(\xi,0) + \int_0^s \Phi(\hat{t}-\tau)\frac{\partial}{\partial\tau}w(\xi,\tau)\mathrm{d}\tau\right]. \quad (3.117)$$

Such a technique is adopted by Leishman (1994) in order to generate unsteady lift and moment for an airfoil with an arbitrary flap motion, and is seen to have a reasonable comparison with experimental results for compressible subsonic case. While exploring indicial concepts, we shall develop in a later chapter an alternative approach for arbitrary motion through approximate rational functions in the Laplace domain, fitted to the harmonic data by replacing the Laplace variable with *ik* by analytic continuation. This method is termed rational function approximation (RFA) and can be applied to both subsonic and supersonic regimes (Eversman and Tewari 1991).

3.3.3 Supersonic Unsteady Aerodynamics

While our interest in ASE modelling practically lies in the compressible subsonic and transonic regimes (because that is where the most challenging ASE problems appear), we consider the linearized supersonic flow for the sake of completeness. Another reason for studying supersonic flow is to analyse the flutter of missile/rocket fins, which typically occurs at supersonic speeds. For a thin airfoil oscillating in supersonic flow, it is quite possible to use a modified form of the acceleration potential integral equation [Eq. (3.100)] that was employed by Possio for subsonic flow. The modification involves limiting the integration to the area inside the Mach cone emanating from the given load point (flow perturbations cannot travel upstream of the Mach cone), (ξ, ζ), and changing β to $\beta = \sqrt{M_\infty^2 - 1}$ in the kernel function. However, it is much simpler to employ the distribution of pulsating velocity potential sources first proposed by Possio (1937) and developed into a numerical procedure by Garrick and Rubinow (1946).

Consider a thin airfoil with mid-plane $\zeta = 0$, approximated by a distribution of infinitesimal strength, line source pulses spread out in ξ-direction, with the leading edge at $\xi = 0$. The net effect of the airfoil at a location (x, z) is given by the velocity potential at that point as follows (Garrick 1957):

$$\phi(x,z,t) = -\frac{a_\infty}{\pi\beta}\int_0^{\xi_1}\int_{\tau_1}^{\tau_2}\frac{w(\xi,t-\tau)}{\sqrt{a_\infty^2\tau^2-(x-\xi-U_\infty t)^2-z^2}}\mathrm{d}\tau\mathrm{d}\xi, \quad (3.118)$$

where $\xi_1 = x - \beta z$ is the point of intersection of the Mach line passing through (x, z) and the ξ-axis, and τ_1, τ_2 are the roots of the quadratic expression inside the square-root (denominator). There is no change in the flow tangency condition from the subsonic case applied on the mean surface,

$$w(x,t) = \frac{\partial z_b}{\partial t} + U_\infty\frac{\partial z_b}{\partial x}. \quad (3.119)$$

On the mean surface, the potential is given by

$$\phi(x,0,t) = -\frac{1}{\pi\beta}\int_0^{x}\int_{\tau_1'}^{\tau_2'}\frac{w(\xi,t-\tau)}{\sqrt{\left(\tau-\tau_1'\right)\left(\tau_2'-\tau\right)}}\mathrm{d}\tau\mathrm{d}\xi, \quad (3.120)$$

where

$$\tau_1' = \frac{1}{M_\infty + 1}\frac{x-\xi}{a_\infty}$$
$$\tau_2' = \frac{1}{M_\infty - 1}\frac{x-\xi}{a_\infty}. \tag{3.121}$$

A major distinction between supersonic and subsonic integral formulations is that the former is valid for arbitrary motions. Stewartson (1950) shows that the harmonic solution for the velocity potential can be derived by taking the Laplace transform of the governing differential (wave) equation. The reader is referred to Bisplinghoff and Halfman (1955) for a detailed derivation of the supersonic kernel function and its computation in a tabular form. Here we present the final result as the following linear operator relationship:

$$\frac{\Delta\bar{p}(x)}{\frac{1}{2}\rho_\infty U_\infty^2} = \mathcal{F}(\bar{\alpha}), \tag{3.122}$$

where

$$\mathcal{F}(.) = -\frac{4}{\beta}\left(\frac{\partial}{\partial x} + ik\right)\int_{-1}^{x} e^{-ikM_\infty^2 r/\beta^2} J_0\left(kM_\infty^2 r/\beta^2\right)(.)\mathrm{d}\xi \tag{3.123}$$

with $r = x - \xi$ and $J_0(.)$ being the Bessel function of the first kind and zero order.

It is to be noted that the linearized supersonic aerodynamics is valid only for thin airfoils undergoing infinitesimal oscillations. For a thick airfoil (or finite amplitude oscillation) the Mach waves are replaced by oblique shock waves whose analysis is essentially nonlinear. However, at large reduced frequencies, a linear approximation called the piston theory becomes valid at reasonably large Mach numbers even for thick airfoils (Hayes and Probstein 1959).

3.4 Preliminary Aeroelastic Modelling Concepts

Before attempting to devise ASE systems, it is necessary to understand how structural and aerodynamic subsystems interact with each other. Towards this end, the typical-section model will be employed. Structural motion causes aerodynamic forces and moments which depend upon the flow parameters (speed, density, Mach number, Reynolds number), as well as on the time rate of structural deformation. The aerodynamic force acting normal to the freestream direction is the lift, L, acting at the a.c., the aerodynamic pitching moment about the pitch axis is denoted, M_y, while the aerodynamic hinge moment of the control surface is H. The aerodynamic drag acting along the freestream direction is unimportant in causing aeroelastic effects, and is thus neglected. Since the aerodynamic forces and moments arise owing to structural motion, and in turn affect the structural response, they are considered to be internal reactions of the aeroelastic system. The functional dependence of aerodynamic forces on structural motion must be analysed at every possible flow condition that can cause an instability. When the structural motion takes place slowly (at very low natural frequencies), the aerodynamic forces and moments can sometimes be approximated to functions of the instantaneous angle of attack (and its time derivatives) taken as if it were constant, and their time history (or transient behaviour) can be entirely neglected. Such an assumption, called the quasi-steady approximation, leads to an intuitive and straightforward aeroelastic analysis. Quasi-steady aerodynamic forces and

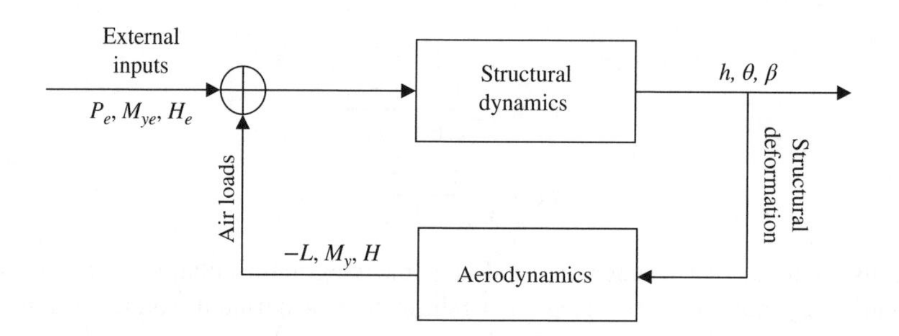

Figure 3.5 Aeroelastic system for a typical section

moments result in a natural damping in each d.o.f. The plunging motion, $h(t)$, at a constant rate, results in aerodynamic damping (a positive plunge causes an increase in the instantaneous angle of attack and therefore increases the quasi-steady lift in proportion to the constant plunge rate, $\mathrm{d}h/\mathrm{d}t$). Furthermore, a pure pitching rotation about any axis is naturally damped owing to a negative pitching moment caused by the steady pitch rate, $\mathrm{d}\theta/\mathrm{d}t$, (a constant nose-up pitch rate results in a nose-down pitching moment). Similarly, control-surface rotation, $\beta(t)$, is an aerodynamically damped motion (a constant nose-up rotation rate $\mathrm{d}\beta/\mathrm{d}t$ causes a nose-down hinge moment). Before 1935, quasi-steady approximation of the aerodynamic forces saw wide usage for carrying out an intuitive analysis and is also employed in aeroelasticity textbooks for illustrative purposes. But such an approximation is inaccurate for even moderate frequencies, where an interaction of the flow over the wing takes place with the wake in a time-dependent manner.

Unsteady aerodynamic behaviour of a wing is markedly different from the quasi-steady case, because the forces and moments can have significant differences in phase from that of the structural d.o.f.s, mainly due to the effects of a vortical (circulatory) wake. Under an unsteady aerodynamic coupling, the structural deformations, $h(t)$, $\theta(t)$, $\beta(t)$, assume an oscillatory behaviour in time with phase differences. If the amplitudes of the oscillations are sufficiently small, flow separation can be neglected, and the resulting motion can be considered to be independent of viscous effects. Furthermore, if the flow speeds do not approach the speed of sound, shock waves are assumed to be absent. Flutter phenomenon in the absence of viscous separation and shock waves is called classical flutter, and is easier to analyse because it is driven by linear aerodynamic characteristics. In contrast, flutter under viscous and shock wave effects – called stall flutter, and transonic flutter, respectively – require a nonlinear analysis of the flutter mechanism. In this chapter, we consider the basic linear analysis of the classical flutter mechanism. Since the aerodynamics of classical flutter is linearized, the methods of linear analysis can be applied to it. Frequency fidelity is a basic property of linear systems, wherein a forcing function at a particular excitation frequency, ω, excites all the system variables only at that frequency. In a classical flutter mechanism, a harmonic excitation results in all the d.o.f.s oscillating at a common frequency. Since the aerodynamic and structural systems are coupled in a feedback loop as shown in Fig. 3.5, all the signals sustain themselves at the excitation frequency, but can have differences in magnitude and phase.

The simplest unsteady aerodynamic model is for the inviscid (infinite Reynolds number)[4] and incompressible (zero Mach number) flow. Such a flow is termed the ideal flow, because it can yield closed-form solutions. In the ideal flow case, the aerodynamic forces and moments can be divided into non-circulatory and circulatory parts. For example, the lift can be expressed as $L = L_{nc} + L_c$, where $L_{nc}(t)$ is the non-circulatory part, and $L_c(t)$ the circulatory part. The non-circulatory lift, L_{nc}, is due to the vertical acceleration of a cylindrical mass of fluid with diameter equalling the chord, and is thus termed the apparent inertia effect. Similarly, there are the non-circulatory pitching and hinge moment contributions. The non-circulatory effects are independent of the flow speed and merely add to the structural mass and moments of inertia. The circulatory aerodynamic effects are caused by the vortical flow pattern (or circulation) around the airfoil, which is necessary for enforcing the flow tangency condition at the trailing edge. For example, the lift's operational dependence upon a small angle of attack, α, is regarded to be linear and can be described by the following convolution integral:

$$L_c(t) = \int_0^\infty \alpha(\tau) g(t-\tau) \mathrm{d}\tau, \tag{3.124}$$

where

$$g(t) = \mathcal{L}^{-1}\{G_a(s)\} = \mathcal{L}^{-1}\left\{\frac{L_c(s)}{\alpha(s)}\right\} \tag{3.125}$$

is the lift response for a unit impulsive change in the angle of attack, $\alpha = \delta(t)$, and the linear dependence of the circulatory lift on the angle of attack (subject to zero initial condition) is described by the following aerodynamic transfer function:

$$G_a(s) = \frac{L_c(s)}{\alpha(s)}. \tag{3.126}$$

In terms of the lift indicial admittance (step response function), $\mathbb{A}(t)$, where

$$\mathbb{A}(t) = \int_0^t g(\tau)\mathrm{d}\tau, \tag{3.127}$$

we write the circulatory lift response as the following Duhamel's integral:

$$L_c(t) = \alpha(0)\mathbb{A}(t) + \int_0^t \alpha(\tau)\frac{\mathrm{d}\mathbb{A}(t-\tau)}{\mathrm{d}t}\mathrm{d}\tau = \int_0^t \mathbb{A}(t-\tau)\frac{\mathrm{d}\alpha(\tau)}{\mathrm{d}\tau}\mathrm{d}\tau. \tag{3.128}$$

Here $\alpha(0)$ is the step change in the angle of attack at $t = 0$, derived by approaching the limit $t = 0$ from the positive side. Both $g(t)$ and $\mathbb{A}(t)$ are subject to the initial condition,

$$\alpha(0-) = \dot{\alpha}(0-) = \frac{\mathrm{d}^{n-1}\alpha}{\mathrm{d}t^{n-1}}(0-) = 0$$

[4] The fluid flow everywhere is regarded as inviscid, except in the case of the fluid layer in contact with the airfoil surface, which is assumed to have just enough viscosity to create a circulation around the airfoil, whose strength should be sufficient to satisfy the flow tangency (Kutta–Joukowski) condition at the trailing edge.

for an nth-order aerodynamic system, implying $g(t) = \mathbb{A}(t) = 0$ for $t < 0$. The relationship between lift indicial admittance and the aerodynamic transfer function is the following:

$$\mathcal{L}\{\mathbb{A}(t)\} = \frac{G_a(s)}{s}. \tag{3.129}$$

Other aerodynamic transfer functions (and indicial admittances) describing the changes in the lift, pitching moment and hinge moment with respect to angle of attack, α, and control-surface rotation, β, respectively, can be derived in a similar manner.

In the ideal flow case, the aeroelastic system characteristics depend only upon the density, ρ, flight (or freestream) speed, U, and oscillation frequency, ω. Flutter analysis requires an investigation of the characteristic poles of the aeroelastic system at each possible flow condition described by the combination (ρ, U, ω). Alternatively, a frequency response analysis $(s = i\omega)$ via Bode magnitude and phase plots of the aeroelastic transfer functions can reveal the excitation frequency at which a resonance type, self-sustained oscillation can take place for a given flow condition when the source of external excitation is removed ($P_e = M_{ye} = H_e = 0$). Such an analysis requires that suitable aerodynamic transfer functions such as $G_a(s)$, (or the respective frequency response functions, such as $G_a(i\omega)$) be available at various combinations of (ρ, U). A dimensional analysis reveals that the independent parameters (ρ, U, ω) can be combined to yield a non-dimensional reduced frequency,

$$k = \omega b / U$$

and the non-dimensional time,

$$\hat{t} = Ut/b,$$

where b is the airfoil's semi-chord. Other non-dimensional parameters are the ratio of fluid and structural densities, ρ/σ, and the non-dimensional stiffness ratios,

$$\frac{k_h}{\sigma U^2 b^2}, \quad \frac{k_\theta}{\sigma U^2 b^3}, \quad \frac{k_\beta}{\sigma U^2 b^3}.$$

If a flutter condition exists, it is described by the critical frequency, ω_f, and speed, U_f, and the corresponding reduced frequency,

$$k_f = \frac{\omega_f b}{U_f}.$$

Appendix A presents the analytical solution of the unsteady aerodynamic transfer function in the ideal flow case. The results are expressed in terms of Wagner's indicial admittance function, $\Phi(\hat{t})$, and its close relation, Theodorsen's circulation function, $C(k)$. Wagner's function describes the lift of an airfoil suddenly started from rest, such that it sees a unit step change in the angle of attack at $t = 0$, $\alpha(t) = u_s(t)$. In non-dimensional time, Wagner's function describes the aerodynamic indicial admittance as follows:

$$\mathbb{A}(\hat{t}) = 2\pi b \rho U^2 \Phi(\hat{t}) \tag{3.130}$$

Substitution of Eq. (3.130) into Eq. (3.128) yields

$$L_c(\hat{t}) = 2\pi b \rho U^2 \left[\Phi(\hat{t})\alpha(0) + \int_0^{\hat{t}} \Phi(\hat{t} - \tau)\frac{\mathrm{d}\alpha(\tau)}{\mathrm{d}\tau}\mathrm{d}\tau \right]. \tag{3.131}$$

Equations (3.130) and (3.129) relate Wagner's function to the aerodynamic transfer function as follows:

$$2\pi b\rho U^2 \mathcal{L}\{\Phi(\hat{t})\} = G_a(p)/p, \tag{3.132}$$

where $p = sb/U$ is the non-dimensional Laplace variable. In the steady-state harmonic limit, $p = ik$, where $k = \omega b/U$ is the reduced frequency of oscillation, the aerodynamic frequency response function, called Theodorsen's function, $C(k)$, is defined as follows (Appendix A):

$$\mathcal{L}\{\Phi(\hat{t})\} = C(-ip)/p = \frac{C(k)}{ik}. \tag{3.133}$$

Clearly, Theodorsen's function is related to the aerodynamic transfer function for circulatory lift by

$$G_a(ik) = 2\pi b\rho U^2 C(k), \tag{3.134}$$

and hence $C(k)$ is also called the circulation function. Wagner's function represents the deficiency in the indicial circulatory lift response from its steady-state value,

$$L_\infty = \lim_{\hat{t}\to\infty} L_c(\hat{t}), \tag{3.135}$$

with the property that $\Phi(\infty) = 1$, that is, lift deficiency vanishes in the steady state. By the application of the final-value theorem of Laplace transform, we have

$$\lim_{\hat{t}\to\infty} \Phi(\hat{t}) = \lim_{p\to 0} p\mathcal{L}\{\Phi(\hat{t})\} = \lim_{k\to 0}(ik)\frac{C(k)}{ik} = C(0) = 1. \tag{3.136}$$

Theodorsen's function describes the phase difference, ϕ, between the harmonic (complex) amplitude of the circulatory lift, $\bar{L}$, and its static value,

$$L_0 = \lim_{k\to 0} L_c(t), \tag{3.137}$$

caused by the vertical velocity component (downwash) induced by the oscillating wake on the airfoil. The unsteady lift is expressed as follows:

$$L_c(\hat{t}) = \bar{L}(k)e^{ik\hat{t}} = L_0 e^{i[k\hat{t}+\phi(k)]} = L_0 e^{i\phi(k)} e^{ik\hat{t}} \tag{3.138}$$

or

$$\bar{L}(k) = L_0 e^{i\phi(k)}. \tag{3.139}$$

Note that $\phi(0) = 0$ (which follows from the property $C(0) = 1$). The phase difference, ϕ, in the lift due to circulation is crucial in understanding the flutter mechanism. For this purpose, consider the following expression for Theodorsen's function, $C(k)$ (Appendix A):

$$C(k) = \frac{\int_1^\infty \frac{x}{\sqrt{x^2-1}} e^{-ikx} \mathrm{d}x}{\int_1^\infty \frac{x+1}{\sqrt{x^2-1}} e^{-ikx} \mathrm{d}x}, \tag{3.140}$$

where the improper integrals, assumed to be convergent for a general oscillation (see Appendix A), are evaluated by Hankel functions of the second kind and order n, $H_n^{(2)}(k)$:

$$C(k) = \frac{H_1^{(2)}(k)}{H_1^{(2)}(k) + iH_0^{(2)}(k)} = F(k) + iG(k). \tag{3.141}$$

Substitution of Eq. (3.141) into Eq. (3.134) results in the following expression for the phase difference in the lift due to angle of attack:

$$\phi = \tan^{-1}\frac{\mathrm{Im}[\bar{\alpha}C(k)]}{\mathrm{Re}[\bar{\alpha}C(k)]}, \tag{3.142}$$

where $\bar{\alpha}$ is the complex amplitude for a harmonic oscillation in the angle of attack.

3.5 Ideal Flow Model for Typical Section

As mentioned previously, the typical-section model based on incompressible, inviscid, irrotational (ideal) flow with small perturbations over a freestream is the simplest one that can be used as the basis of flutter suppression. Two equivalent frequency-domain formulations in the ideal flow case for an airfoil oscillating in pitch, plunge and control-surface modes are provided by Theodorsen (1935) and Küssner and Schwarz (1940), using alternatively the conformal mapping of complex upwash due to harmonically pulsating sources and sinks (Appendix A) and the Fourier series representation of the induced upwash, respectively. The formulation by Theodorsen (1935) is universally recognized as the conventional one, and hence is employed here. Consider a thin airfoil of semi-chord, b, idealized as a flat plate, placed in uniform, incompressible flow of speed, U, and density, ρ, oscillating with reduced frequency, $k = \omega b/U$, in plunge, h, and pitch, θ, both of and about an axis located at a distance ab behind the mid-chord point. In addition, there is an unbalanced trailing-edge control surface, whose deflection, β, has an oscillation at the same frequency, ω, about the hinge line located at a distance bc aft of the mid-chord point (Fig. 3.6).

Since the plunge displacement, h, is the vertical deflection of the pitch axis from the static equilibrium position due to pure bending, the net angle of attack at a given point x aft of the mid-chord point is the following[5]:

$$\alpha = \theta + \frac{\dot{h}}{U} + (x - ab)\frac{\dot{\theta}}{U}. \tag{3.143}$$

The two-dimensional lift, L, pitching moment, M_y, and control-surface hinge moment, H, in the case of simple harmonic oscillation in all the d.o.f.s are given as follows (Theodorsen 1935):

$$L = \pi\rho b^2\left(\ddot{h} + U\dot{\theta} - ab\ddot{\theta} - \frac{U}{\pi}T_4\dot{\beta} - \frac{b}{\pi}T_1\ddot{\beta}\right) + 2\pi\rho UbQ \tag{3.144}$$

$$\begin{aligned} M_y = &-\rho b^2\left[-\pi ab\ddot{h} + \pi Ub\left(\frac{1}{2} - a\right)\dot{\theta} + \pi b^2\left(\frac{1}{8} + a^2\right)\ddot{\theta} + (T_4 + T_{10})U^2\beta\right. \\ &\left. + Ub\left\{T_1 - T_8 - (c-a)T_4 + \frac{1}{2}T_{11}\right\}\dot{\beta} - b^2\{T_7 + (c-a)T_1\}\ddot{\beta}\right] \\ &+ 2\pi\rho Ub^2\left(a + \frac{1}{2}\right)Q \end{aligned} \tag{3.145}$$

[5] There is a difference in our notation for pitch angle compared to Theodorsen's (1935). The rotation about the e.a. is termed here as the pitch angle, θ, whereas Theodorsen defines the same angle as α. In our notation, α is the net change in the angle of attack caused by the plunging and pitching motions.

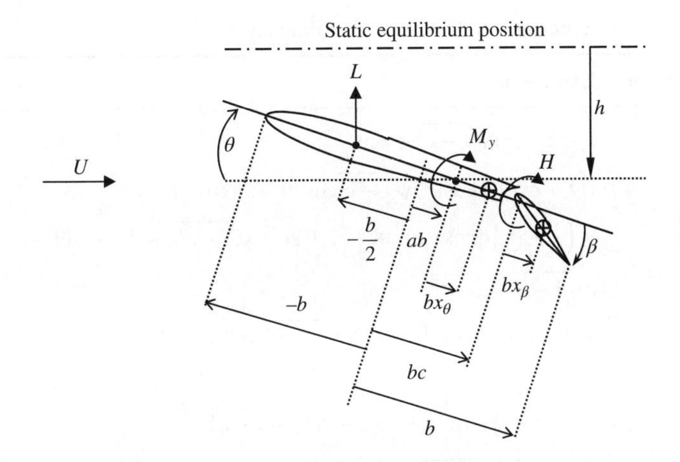

Figure 3.6 Typical section for Theodorsen's model

$$H = -\rho b^2 \left[-T_1 b\ddot{h} + Ub \left\{ -2T_9 - T_1 + T_4 \left(a - \frac{1}{2} \right) \right\} \dot{\theta} + 2T_{13}\ddot{\theta} + \frac{1}{\pi} \left(T_5 - T_4 T_{10} \right) U^2 \beta \right.$$
$$\left. - \frac{1}{2\pi} UbT_4 T_{11} \dot{\beta} - \frac{1}{\pi} b^2 T_3 \ddot{\beta} \right] + 2\pi\rho Ub^2 \left(a + \frac{1}{2} \right) - \rho Ub^2 T_{12} Q. \tag{3.146}$$

The coefficients T_i, $i = 1, \ldots, 14$, employed in Eqs. (3.144)–(3.147) are listed in Table 3.1. In the aerodynamic load expressions, Q is a common term arising out of circulatory lag due to the wake, and can be regarded as the frequency response of the upwash w induced by the wake at the 3/4-chord location ($x = ab/2$) as follows:

$$Q(k) = C(k)w(k)$$
$$= C(k) \left\{ \dot{h} + U\theta + b \left(\frac{1}{2} - a \right) \dot{\theta} + \frac{U}{\pi} T_{10} \beta + \frac{b}{2\pi} T_{11} \dot{\beta} \right\}. \tag{3.147}$$

The non-circulatory lift is the term in parentheses on the right-hand side of Eq. (3.144) and does not involve Q. By comparison with Eq. (3.143), the first three terms of the non-circulatory lift are seen to be proportional to the angle of attack at the mid-chord point ($x = 0$), while the remaining two terms are due to the control-surface rotation.

The total mass of the wing per unit span is m, mass of the control surface per unit span is m_c, the distance of the c.m. of the wing behind the pitch axis is bx_θ, while the distance of the control-surface c.m. behind its own hinge line is bx_β (Fig. 3.6). The structural stiffness constants in plunge, pitch and control-surface rotation are k_h, k_θ, k_β, respectively, which denote the bending, torsion and control stiffness, respectively.

However, before proceeding further, we would like to simplify our notation by dropping the braces and brackets signifying the vectors and matrices, respectively, and instead adopt the following notation introduced in Chapter 2 for control systems analysis:

$$M\ddot{q} + C_d \dot{q} + Kq = Q_a + Q_c, \tag{3.148}$$

Hence, $q(t) = (h, \theta, \beta)^T : \mathbb{R} \to \mathbb{R}^3$ is to be understood to represent the generalized coordinates vector corresponding to the 3 d.o.f.s, and $M \in \mathbb{R}^{3\times3}$, $C_d \in \mathbb{R}^{3\times3}$ and $K \in \mathbb{R}^{3\times3}$ are the constant

Table 3.1 Coefficients for Theodorsen's formulation

Coefficient	Expression
T_1	$-\frac{1}{3}(2+c^2)\sqrt{1-c^2}+c\cos^{-1}c$
T_2	$c(1-c^2)-(1+c^2)\sqrt{1-c^2}\cos^{-1}c+c\left(\cos^{-1}c\right)^2$
T_3	$-\left(\frac{1}{8}+c^2\right)\left(\cos^{-1}c\right)^2+\frac{c}{4}\left(7+2c^2\right)\sqrt{1-c^2}\cos^{-1}c-\frac{1}{8}\left(1-c^2\right)\left(5c^2+4\right)$
T_4	$c\sqrt{1-c^2}-\cos^{-1}c$
T_5	$-(1-c^2)-\left(\cos^{-1}c\right)^2+2c\sqrt{1-c^2}\cos^{-1}c$
T_6	T_2
T_7	$-\left(\frac{1}{8}+c^2\right)\cos^{-1}c+\frac{c}{8}\left(7+2c^2\right)\sqrt{1-c^2}$
T_8	$\frac{1}{3}(2c^2+1)\sqrt{1-c^2}+c\cos^{-1}c$
T_9	$\frac{1}{2}\left[\frac{1}{3}(1-c^2)^{3/2}+aT_4\right]$
T_{10}	$\sqrt{1-c^2}+\cos^{-1}c$
T_{11}	$(2-c)\sqrt{1-c^2}+(1-2c)\cos^{-1}c$
T_{12}	$(2+c)\sqrt{1-c^2}-(2c+1)\cos^{-1}c$
T_{13}	$-\frac{1}{2}[(c-a)T_1+T_7]$
T_{14}	$\frac{1}{16}+\frac{1}{2}ac$

generalized mass, damping and stiffness matrices, respectively, of the structure. For further development, $C_d = 0$ will be assumed for a conservative flutter analysis (also because structural damping is very difficult to model for an actual structure). The generalized aerodynamic force vector, $Q_a(t) = (-L, M_y, H)^T : \mathbb{R} \to \mathbb{R}^3$ is assumed to be linearly related to $q(t)$, $\dot{q}(t)$ and $\ddot{q}(t)$, as well as to certain additional state variables collected into the aerodynamic state vector, $x_a(t) : \mathbb{R} \to \mathbb{R}^\ell$, which is necessary for modelling the aerodynamic lag caused by a circulatory wake by an RFA (to be discussed in the next section). Since the only applied control input is the control torque, $u(t)$, acting about the control-surface hinge line, the generalized control force vector is given by $Q_c(t) = (0, 0, u)^T : \mathbb{R} \to \mathbb{R}^3$. The generalized mass and stiffness, matrices of the structural system with the generalized coordinates q are given as follows:

$$M = \begin{pmatrix} m & S_\theta & S_\beta \\ S_\theta & I_\theta & b(c-a)S_\beta \\ S_\beta & b(c-a)S_\beta & I_\beta \end{pmatrix}$$

$$K = \begin{pmatrix} k_h & 0 & 0 \\ 0 & k_\theta & 0 \\ 0 & 0 & k_\beta \end{pmatrix}, \tag{3.149}$$

where the inertial coupling parameters are the following:

$$S_\theta = mbx_\theta; \quad S_\beta = m_c bx_\beta. \tag{3.150}$$

The generalized loads vector Q_a contains both circulatory and non-circulatory terms. It is to be noted here that the non-circulatory terms in Eqs. (3.144)–(3.147) (i.e. those that do not involve Q) are simply clubbed together with the mass, stiffness and damping coefficients of the structure, resulting in the generalized mass, damping and stiffness matrices of the modified structural system, and are expressed in a non-dimensional form as follows:

$$\bar{M} = \begin{pmatrix} \bar{M}_{11} & \bar{M}_{12} & \bar{M}_{13} \\ \bar{M}_{21} & \bar{M}_{22} & \bar{M}_{23} \\ \bar{M}_{31} & \bar{M}_{32} & \bar{M}_{33} \end{pmatrix}$$

where

$$\begin{aligned}
\bar{M}_{11} &= \kappa + 1 \\
\bar{M}_{12} &= x_\theta - a\kappa \\
\bar{M}_{13} &= x_\beta - T_1 \frac{\kappa}{\pi} \\
\bar{M}_{21} &= x_\theta - a\kappa \\
\bar{M}_{22} &= r_\theta^2 + \kappa \left\{ \frac{1}{8} + a^2 \right\} \\
\bar{M}_{23} &= r_\beta^2 + (c - a)x_\beta - \frac{\kappa}{\pi} \{T_7 + T_1(c - a)\} \\
\bar{M}_{31} &= x_\beta - T_1 \frac{\kappa}{\pi} \\
\bar{M}_{32} &= r_\beta^2 + (c - a)x_\beta - \frac{\kappa}{\pi} \{T_7 + T_1(c - a)\} \\
\bar{M}_{33} &= r_\beta^2 - \frac{\kappa}{\pi^2} T_3
\end{aligned}$$

and

$$\bar{C} = \begin{pmatrix} 0 & \kappa & -\frac{\kappa}{\pi} T_4 \\ 0 & \kappa\left(\frac{1}{2} - a\right) & \frac{\kappa}{\pi}\left\{4T_9 - \left(a + \frac{1}{2}\right)T_4\right\} \\ 0 & -\frac{\kappa}{\pi}\left\{T_1 + 2T_9 - \left(a - \frac{1}{2}\right)T_4\right\} & -\frac{\kappa}{2\pi^2} T_4 T_{11} \end{pmatrix}$$

$$\bar{K} = \begin{pmatrix} \omega_h^2 & 0 & 0 \\ 0 & \omega_\theta^2 r_\theta^2 & \frac{\kappa}{\pi}\{T_4 + T_{10}\} \\ 0 & 0 & \omega_\beta^2 r_\beta^2 + \frac{\kappa}{\pi^2}\{T_5 - T_4 T_{10}\} \end{pmatrix},$$

where

$$\kappa = \frac{\pi \rho b^2}{m}$$

is the non-dimensional mass parameter representing the ratio of mass of a cylinder of air per unit span with diameter equalling the chord to the wing mass per unit span. The non-dimensional radii of gyration in pitch and control rotation are given by

$$r_\theta = \sqrt{\frac{I_\theta}{mb^2}}, \quad r_\beta = \sqrt{\frac{I_\beta}{mb^2}},$$

respectively, while the non-dimensional structural frequencies of plunge, pitch and control-surface modes are respectively the following:

$$\omega_h = \frac{b}{U}\sqrt{\frac{k_h}{m}}, \quad \omega_\theta = \frac{b}{U}\sqrt{\frac{k_\theta}{I_\theta}}, \quad \omega_\beta = \frac{b}{U}\sqrt{\frac{k_\beta}{I_\beta}}.$$

The symmetry of the modified structural mass matrix $\bar{M}$ and its independence from the airspeed U is to be noted.

The aeroelastic equations of motion in a non-dimensional form are thus the following:

$$\bar{M}\hat{q}'' + \bar{C}\hat{q}' + \bar{K}\hat{q} = \hat{Q}_a + (0, 0, 1)^T \hat{u}, \tag{3.151}$$

where the prime represents differentiation with respect to the non-dimensional time, $\hat{t} = tU/b$, $\hat{q} = (h/b, \theta, \beta)^T$ is the non-dimensional generalized coordinates vector, $\hat{u}$ the non-dimensional torque input acting on the control surface and $\hat{Q}_a$ is the non-dimensional, circulatory part of the generalized force vector, which can be expressed as follows:

$$\hat{Q}_a = 2\kappa C(k) \begin{Bmatrix} -1 \\ \frac{1}{2} + a \\ -\frac{1}{2\pi} T_{12} \end{Bmatrix} \hat{w}, \tag{3.152}$$

with the non-dimensional upwash induced by the wake given by

$$\hat{w} = \frac{h'}{b} + \theta + \left(\frac{1}{2} - a\right)\theta' + \frac{1}{\pi} T_{10}\beta + \frac{1}{2\pi} T_{11}\beta'. \tag{3.153}$$

Equation (3.152) provides the following linear relationship between the circulatory generalized forces and generalized coordinates:

$$\hat{Q}_a = G(ik)\hat{q}, \tag{3.154}$$

where $G(ik) \in \mathbb{R}^{3\times3}$ is the following aerodynamic frequency response matrix that operates on the generalized coordinates:

$$G(ik) = 2\kappa C(k) \begin{Bmatrix} -1 \\ \frac{1}{2} + a \\ -\frac{1}{2\pi} T_{12} \end{Bmatrix} \left\{ ik, \quad 1 + ik\left(\frac{1}{2} - a\right), \quad \frac{1}{\pi}\left(T_{10} + \frac{ik}{2} T_{11}\right) \right\}. \tag{3.155}$$

3.6 Transient Aerodynamics of Typical Section

Before proceeding further, it is necessary to consider a more general dynamics than simple harmonic oscillations. Since the frequency response matrix $G(ik)$ arises owing to Theodorsen's function representing the frequency response of a circulatory wake-induced upwash, it can be generalized as a linear aerodynamic operator for general motions, provided that one can do the same for the Theodorsen's function. Such an extension from the simple harmonic limit to a general transient motion – termed analytic continuation – is justified because a step change of the upwash results in a stable (decaying) pressure magnitude response, which is related to the inverse Fourier transform of Theodorsen's function as follows:

$$\Phi(\tau) = 1 + \frac{2}{\pi}\int_0^{\infty} \frac{C(ik) - 1}{ik} e^{ik\tau} dk, \tag{3.156}$$

where $\tau = Ut/b > 0$ is the non-dimensional time and $\Phi(\tau)$ is Wagner's function (Wagner 1925). The process of analytic continuation is brought into effect by adopting a stable transfer function representation of $C(k)$ wherein ik is replaced by the non-dimensional Laplace variable $s = \sigma + ik$, and results in the following RFA:

$$C(s) = \frac{N(s)}{D(s)}. \tag{3.157}$$

For a proper transfer function $C(s)$, the degree of $N(s)$ must be no greater than that of $D(s)$. A useful representation of $C(s)$ is by a series of first-order, stable poles given by

$$C(s) = a_0 + \sum_{n=1}^{\ell} \frac{a_n s}{s + b_n}, \tag{3.158}$$

where $b_n > 0, n = 1, \dots, \ell$ are the aerodynamic poles. The real coefficients a_0, a_n, b_n can be determined by fitting $C(k) = C(-is)$ at a discrete set of reduced frequencies k. The curve-fitting is practically carried out by using a least-squares fit error minimization (Tewari 2015). It has been shown (Eversman and Tewari 1991) that in a selected range of reduced frequencies, the optimum fit accuracy with the harmonic Theodorsen function can be remarkably improved by using a multiple-pole approximation:

$$C(s) = a_0 + \sum_{n=1}^{r}\sum_{p=1}^{m_n} \frac{a_{n_p} s}{(s + b_n)^p}, \tag{3.159}$$

where

$$\sum_{n=1}^{r} m_n = \ell,$$

which has the advantage of reducing the total number of poles ℓ required for a given curve-fit accuracy. The substitution of either Eq. (3.158) or Eq. (3.159) into Eq. (3.155) produces the following unsteady aerodynamic transfer matrix for general transient motions:

$$G(s) = 2\kappa C(s) \begin{Bmatrix} -1 \\ \frac{1}{2} + a \\ -\frac{1}{2\pi} T_{12} \end{Bmatrix} \left\{ s, \quad 1 + \left(\frac{1}{2} - a\right) s, \quad \frac{1}{\pi}\left(T_{10} + \frac{1}{2} T_{11} s\right) \right\}. \tag{3.160}$$

Equation (3.160) is quite useful in deriving the ASE plant, such as that given by the state-space representation of Eqs. (3.135)–(3.137).

3.7 State-Space Model of the Typical Section

Once the aerodynamic transfer function is obtained by RFA, the equations of motion of the aeroelastic plant, Eq. (3.151), can be expressed in the Laplace domain as follows, assuming zero initial conditions:

$$(s^2\bar{M} + s\bar{C} + \bar{K})\bar{q}(s) = G(s)\bar{q}(s) + (0, 0, 1)^T u(s). \tag{3.161}$$

Clearly, the order of the plant is $6 + \ell$, where ℓ is the total number of aerodynamic states (the number of poles of $C(-is)$). A possible choice of the augmented state vector $x(t) : \mathbb{R} \to \mathbb{R}^{6+\ell}$ is the following:

$$x = \begin{Bmatrix} \bar{q} \\ \dot{\bar{q}} \\ x_a \end{Bmatrix}, \tag{3.162}$$

where the aerodynamic states $x_a(t)$ are derived from the transfer function relationship derived from non-dimensionalized, analytic continuation of Eq. (3.147):

$$\bar{Q}(s) = C(s)\bar{w} = C(s)\left\{s,\quad 1 + \left(\frac{1}{2} - a\right)s,\quad \frac{1}{\pi}\left(T_{10} + \frac{1}{2}T_{11}s\right)\right\}\bar{q}$$

and either Eq. (3.158) or Eq. (3.159). The aerodynamic state equations can thus be expressed as follows:

$$\dot{x}_a = F_a x_a + \Gamma_a \begin{Bmatrix} \bar{q} \\ \dot{\bar{q}} \end{Bmatrix}, \tag{3.163}$$

where $F_a \in \mathbb{R}^{\ell \times \ell}$, $\Gamma_a \in \mathbb{R}^{\ell \times 6}$ are the aerodynamic coefficient matrices. The structural equations of motion in non-dimensional time are then written as follows:

$$\bar{M}\ddot{\bar{q}} + \bar{C}\dot{\bar{q}} + \bar{K}\bar{q} = C_\ell \dot{\bar{q}} + K_\ell \bar{q} + N_a x_a + (0, 0, 1)^T u. \tag{3.164}$$

Here, the aerodynamic states arising from the aerodynamic transfer matrix $G(s)$ contribute to the structural state equations through an aerodynamic coefficient matrix $N_a \in \mathbb{R}^{3\times\ell}$, as well as the circulatory damping and stiffness matrices, $C_\ell \in \mathbb{R}^{3\times3}$ and $K_\ell \in \mathbb{R}^{3\times3}$, respectively. Since the matrices C_ℓ, K_ℓ have been thus far excluded from the structural damping and stiffness matrices, they must now be added to the structural equations of motion. For example, if the simple pole series approximation of Eq. (3.158) is adopted for $C(s)$, we have

$$N_a = -2\kappa \begin{Bmatrix} -1 \\ \frac{1}{2} + a \\ -\frac{1}{2\pi}T_{12} \end{Bmatrix} \left\{a_1 b_1 \quad a_2 b_2 \quad \ldots \quad a_\ell b_\ell\right\}, \tag{3.165}$$

$$K_\ell = 2\kappa\left(a_0 + a_1 + \ldots + a_\ell\right) \begin{Bmatrix} -1 \\ \frac{1}{2} + a \\ -\frac{1}{2\pi}T_{12} \end{Bmatrix} \left\{0 \quad 1 \quad \frac{1}{\pi}T_{10}\right\}, \tag{3.166}$$

$$C_\ell = 2\kappa\left(a_0 + a_1 + \ldots + a_\ell\right)\begin{Bmatrix} -1 \\ \frac{1}{2}+a \\ -\frac{1}{2\pi}T_{12} \end{Bmatrix}\left\{1 \quad \left(\frac{1}{2}-a\right) \quad \frac{1}{2\pi}T_{11}\right\}. \tag{3.167}$$

Finally, the state equations of the aeroelastic system represented by Eqs. (3.163) and (3.164) can be expressed as follows:

$$\dot{x}_e = A_e x_e + B_e u, \tag{3.168}$$

where

$$A_e = \begin{pmatrix} 0 & I & 0 \\ -\bar{M}^{-1}\hat{K} & -\bar{M}^{-1}\hat{C} & -\bar{M}^{-1}N_a \\ \Gamma_a & & F_a \end{pmatrix} \tag{3.169}$$

and

$$B_e = \left(0_{1\times 5}, 1, 0_{1\times \ell}\right)^T, \tag{3.170}$$

where $\hat{C} = \bar{C} - C_\ell$, $\hat{K} = \bar{K} - K_\ell$. The controllability of the plant with the control-surface torque input $u(t)$ can be easily verified by the rank of the following controllability test matrix (Chapter 2):

$$P = \left(B_e, A_e B_e, A_e^2 B_e, A_e^3 B_e, A_e^4 B_e, A_e^5 B_e, A_e^6 B_e\right). \tag{3.171}$$

Since B_e is a column vector, P is square. The determinant $\mid P \mid$ is always non-zero, which signifies an unconditionally controllable plant.

Example 3.7.1 *Let us derive a state-space representation of Theodorsen's aeroelastic model by taking only a single pole in the RFA of Theodorsen's function. Hence, $\ell = 1$, the plant is of seventh order, and Theodorsen's function is approximated in the non-dimensional Laplace domain by*

$$C(s) = a_0 + \frac{a_1 s}{s + b_1}.$$

Selecting the single aerodynamic state as follows:

$$x_a(s) = \frac{\bar{w}(s)}{s + b_1},$$

the aerodynamic state equation is obtained as

$$\dot{x}_a = -b_1 x_a + \Gamma_a \begin{Bmatrix} q \\ \dot{q} \end{Bmatrix},$$

where

$$\Gamma_a = \left\{0 \quad 1 \quad \frac{1}{\pi}T_{10} \quad 1 \quad \left(\frac{1}{2}-a\right) \quad \frac{1}{2\pi}T_{11}\right\}$$

and $F_a = -b_1$ for Eq. (3.169). The aerodynamic coefficient matrices for the structural state equations are the following:

$$N_a = -2\kappa a_1 b_1 \begin{Bmatrix} -1 \\ \frac{1}{2}+a \\ -\frac{1}{2\pi}T_{12} \end{Bmatrix},$$

$$\hat{K} = \bar{K} - 2\kappa(a_0 + a_1)\begin{Bmatrix} -1 \\ \frac{1}{2} + a \\ -\frac{1}{2\pi}T_{12} \end{Bmatrix}\left\{0 \quad 1 \quad \frac{1}{\pi}T_{10}\right\}$$

and

$$\hat{C} = \bar{C} - 2\kappa(a_0 + a_1)\begin{Bmatrix} -1 \\ \frac{1}{2} + a \\ -\frac{1}{2\pi}T_{12} \end{Bmatrix}\left\{1 \quad \left(\frac{1}{2} - a\right) \quad \frac{1}{2\pi}T_{11}\right\}$$

The aeroelastic state-space coefficient matrices are thus the following:

$$A_e = \begin{pmatrix} 0 & I & 0 \\ -\bar{M}^{-1}\hat{K} & -\bar{M}^{-1}\hat{C} & -\bar{M}^{-1}N_a \\ \Gamma_a & & -b_1 \end{pmatrix}$$

and

$$B_e = (0_{1\times 5}, 1, 0)^T.$$

3.8 Generalized Aeroelastic Plant

The typical-section model with incompressible, flat-plate aerodynamics of the previous section is a special case of the general aeroelastic model with a three-dimensional structural dynamics system, and a compressible (but linearized) aerodynamic representation, resulting in a finite-state plant. The concept of RFA can be extended to three-dimensional wing and tails by applying frequency-domain aerodynamic models of lifting surfaces. As in the typical airfoil section, the thickness and camber effects are neglected and the lifting surface is represented by its average chord plane taken to be the mean flat surface, on which the flow boundary conditions are applied. For details of lifting-surface aerodynamic models used in aeroelastic applications in various speed regimes, the reader may refer to Tewari (2015, Chapter 3). Here, we will assume that the frequency-domain (harmonic) aerodynamic data is available for curve-fitting in deriving an appropriate RFA model for the lifting surface. The structural model in such a case utilizes a finite-element approximation (Tewari 2015) for obtaining a finite-order model of the structure. Consider a linear aeroelastic system with the following governing equations of motion:

$$M\ddot{q} + Kq = Q, \tag{3.172}$$

where the structural damping is neglected for convenience, $q(t) \in \mathbb{R}^n$ are the generalized coordinates based on a finite number, n, of structural d.o.f.s, $M \in \mathbb{R}^{n\times n}$ is the generalized mass matrix, $C \in \mathbb{R}^{n\times n}$ is the generalized damping matrix, $K \in \mathbb{R}^{n\times n}$ is the generalized stiffness matrix, and $Q(t) \in \mathbb{R}^n$ is the vector of generalized aerodynamic forces. The unsteady aerodynamic forces depend upon structural motion coordinates through a linear relationship given by

$$\mathcal{D}(Q) = \mathcal{F}(q), \tag{3.173}$$

where $\mathcal{D}(.) : \mathbb{R}^n \to \mathbb{R}^n$ is a linear differential operator, and $\mathcal{F}(.) : \mathbb{R}^n \to \mathbb{R}^n$ a functional operator. A solution for the motion coordinates, $q(t)$, requires a simultaneous solution to the structural dynamics equations, Eq. (3.172), and unsteady aerodynamic field equations, Eq. (3.173), which could be linear partial differential equations. The coupled integration of such fluid-structure equations requires an iterative solution in the time domain, which is not very amenable to aeroelastic computations.

In order to simplify the aerodynamic operational relationship, a transfer matrix representation given by the following relationship is employed

$$Q(s) = G(s)q(s), \tag{3.174}$$

where $Q(s)$ is the Laplace transform of the unsteady aerodynamic generalized force vector, $G(s)$ is the unsteady aerodynamic transfer matrix and $q(s)$ is the Laplace transform of the generalized coordinates vector. The use of transfer matrix relationship allows one to employ linear systems theory for aeroelastic analysis. For the in vacuo structural response ($Q(t) = 0$), the structural modes $(\lambda_i, \bar{q}_i), i = 1, \ldots, n$, are identified from the solution of the following eigenvalue problem:

$$(K + \lambda_i^2 M)\bar{q}_i = 0 \quad (i = 1, \ldots, n) \tag{3.175}$$

and result in the following structural modal matrix:

$$\Phi = (\bar{q}_1, \bar{q}_2, \ldots, \bar{q}_n). \tag{3.176}$$

An RFA for the elements of $G(s)$ results in a linear, time-invariant (LTI) aeroelastic plant. There are various RFA methods available in the literature (Tewari 2015), which can be employed for a suitable unsteady aerodynamic model. The emphasis is on those models which give the smallest aeroelastic plant dimensions, for a given fit accuracy with the frequency-domain aerodynamic data. Some form of nonlinear optimization is invariably required for the aerodynamic poles in a practical RFA method. The technique devised by Sevart (1975), Roger *et al.* (1975) and Abel *et al.* (1978), which was pole-optimized by Tiffany and Adams (1987) and Eversman and Tewari (1991), is referred to as the least-squares RFA, and is the most straightforward method among the various RFA approaches. The least-squares RFA is of the following type:

$$G(s) = A_0 + A_1 s + A_2 s^2 + \sum_{j=1}^{N} A_{j+2} \frac{s}{s + b_j}, \tag{3.177}$$

where $b_j > 0, j = 1, \ldots, N$, are the aerodynamic poles. The numerator coefficient matrices,

$$A_0, A_1, A_2, \ldots, A_{N+2},$$

are determined by fitting $G(ik)$ to the data, $D(ik)$, at a discrete set of reduced frequencies, k, derived from a frequency-domain aerodynamic theory, such as the doublet-lattice method for a subsonic flow. The curve-fitting is carried out by a least-squares process, where the squared, normalized fit error, ε, averaged over m selected reduced frequencies is given by

$$\varepsilon^2 = \sum_{i=1}^{n} \sum_{j=1}^{n} \sum_{r=1}^{m} \left[\bar{g}_{ij}(ik_r) - \bar{d}_{ij}(ik_r)\right] [g_{ij}(ik_r) - d_{ij}(ik_r)], \tag{3.178}$$

where g_{ij} is the (i,j) element of G, and d_{ij} is the (i,j) element of D. The order of the aerodynamic transfer function, n, must be same as the number of structural d.o.f.s.

For deriving the frequency-domain aerodynamics data matrix, D, the following procedure is adopted in the lifting-surface theory, such as doublet-lattice, doublet-point or Mach box schemes (Tewari 2015). After discretizing the wing geometry into wing panels (or boxes) given by discrete corner points, the coordinates and sweep angles required for calculating the aerodynamic influence coefficients (AICs) by an appropriate theory are computed and stored. Then the AIC computation begins by enforcing the upwash boundary condition at the selected chordwise location, x_i (3/4-chord in doublet-lattice/subsonic doublet-point, and mid-chord in Mach box and supersonic doublet-point (Tewari 2015)), on the ith box, based on its vertical deflection, z_i, and slope, $(\mathrm{d}z/\mathrm{d}x)_i$, which in turn, are determined by the normalized, in vacuo structural mode shapes (including the control surfaces), $\bar{q}_i, i = 1, \ldots, n$:

$$w_i = \dot{z}_i + U(\mathrm{d}z/\mathrm{d}x)_i. \tag{3.179}$$

The resulting AICs matrix, A_{ic}, is inverted and pre- and post-multiplied by the structural modal matrix, $\Psi \in \mathbb{R}^{n\times n}$, to yield the generalized (and normalized) aerodynamic matrix:

$$D = \Psi^T A_{\mathrm{ic}}^{-1} \Psi. \tag{3.180}$$

If the non-dimensional form of the AIC matrix is available, it must be converted to the dimensional form by multiplication by the factor $1/2\rho U$.

For the curve-fitting process, the frequency-domain aerodynamic data, D, is collected in the following matrix, F:

$$F = \begin{pmatrix} 1 & ik & (ik)^2 & \frac{ik_1}{ik_1+b_1} & \frac{ik_1}{ik_1+b_2} & \cdots & \frac{ik_1}{ik_1+b_N} \\ 1 & ik & (ik)^2 & \frac{ik_2}{ik_2+b_1} & \frac{ik_2}{ik_2+b_2} & \cdots & \frac{ik_2}{ik_2+b_N} \\ \vdots & \vdots & \vdots & \vdots & & & \\ 1 & ik & (ik)^2 & \frac{ik_m}{ik_m+b_1} & \frac{ik_m}{ik_m+b_2} & \cdots & \frac{ik_m}{ik_m+b_N} \end{pmatrix}. \tag{3.181}$$

Define

$$\mathbf{A} = \left(A_0^T, A_1^T, A_2^T, \ldots, A_{N+2}^T\right)^T$$

as the $[(N+3) \times n \times n]$ array of the unknown numerator coefficients to be determined from the fitting process. The squared fit error matrix is then given by

$$E^2 = \left(\bar{D}^T - \mathbf{A}^T \bar{F}^T\right)(D - F\mathbf{A}), \tag{3.182}$$

where D is the $(m \times n \times n)$ array of the generalized aerodynamics data evaluated at the selected frequency points,

$$D = \left[D^T\left(ik_1\right), D^T\left(ik_2\right), \ldots, D^T\left(ik_m\right)\right]^T.$$

For minimizing the squared fit error with respect to the numerator coefficients, we must have

$$\frac{\partial E^2}{\partial \mathbf{A}} = -\bar{F}^T (D - F\mathbf{A}) - F^T\left(\bar{D}^T - \mathbf{A}^T \bar{F}^T\right)^T = 0, \tag{3.183}$$

which yields

$$\mathbf{A} = (\bar{F}^T F + F^T \bar{F})^{-1} (\bar{F}^T D + F^T \bar{D}). \tag{3.184}$$

The order of the aeroelastic plant produced by the least-squares RFA is $n(2+N)$, where N is the number of lag parameters (poles) in the aerodynamic transfer matrix. As discussed later, additional aerodynamic states are introduced if $q(t)$ does not include control-surface d.o.f.s. If a suitable optimization can be performed to reduce the total number of aerodynamic (lag) states, then the overall size of the aeroelastic plant can be significantly reduced. This is the objective of the nonlinear optimization methods (Edwards 1979, Karpel 1981, Tiffany and Adams 1987). The tendency of some of the RFA poles to coalesce together indicates the need of a multiple-pole RFA (Eversman and Tewari 1991), which is consistent with the optimization process. Otherwise, using simple poles that have nearly identical values results in the ill-conditioned (nearly singular) state-space representation. For further discussion on RFA methods, the reader is referred to Tewari (2015).

Substituting the RFA into the Laplace transform of structural dynamics equations with zero initial conditions,

$$(Ms^2 + K)q(s) = G(s)q(s) + Q_c(s), \tag{3.185}$$

yields the following state equations of the aeroelastic plant:

$$\begin{Bmatrix} \dot{q} \\ \ddot{q} \end{Bmatrix} = \begin{pmatrix} 0 & I \\ -\bar{M}^{-1}\bar{K} & -\bar{M}^{-1}\bar{C} \end{pmatrix} \begin{Bmatrix} q \\ \dot{q} \end{Bmatrix} + \begin{pmatrix} 0 \\ \bar{M}^{-1}N_a \end{pmatrix} x_a + \begin{pmatrix} 0 \\ \bar{M}^{-1}N_g \end{pmatrix} x_g \begin{pmatrix} 0 \\ \bar{M}^{-1}I \end{pmatrix} Q_c, \tag{3.186}$$

and

$$\dot{x}_a = A_a x_a + \Gamma_a \begin{Bmatrix} q \\ \dot{q} \end{Bmatrix}, \tag{3.187}$$

$$\dot{x}_g = A_g x_g + \Gamma_g \begin{Bmatrix} q \\ \dot{q} \end{Bmatrix}. \tag{3.188}$$

Here, the generalized control-surface forces, $Q_c(t)$, are separated from the generalized aerodynamic forces, $Q(t)$, purely driven by the structural motion, $q(t)$. This is done because $Q_c(t)$ depend upon the control laws driving the control-surface actuators of the overall ASE system. The generalized matrices, $\bar{M}, \bar{K}, \bar{C}$ are derived by clubbing the relevant terms of the RFA transfer matrix with the corresponding structural matrices, M, K, C, respectively. The number n_a of the additional aerodynamic state variables, x_a, and the dimensions of their state-space coefficient matrices, A_a, N_a, Γ_a, depend upon the type of the RFA employed. The gust influence on the unsteady aerodynamic forces is modelled by a total number n_g of gust states, x_g, with corresponding gust coefficient matrices, A_g, N_g, Γ_g.

When the least-squares RFA of Eq. (3.177) is introduced (without considering any gust states and control-surface states for the moment), we have the following expression for the generalized aerodynamic forces in the time domain:

$$Q(t) = A_0 q + A_1 \dot{q} + A_2 \ddot{q} + N_a x_a, \tag{3.189}$$

where

$$N_a = (A_3,\ A_4,\ A_5,\ \ldots,\ A_{N+2}) \in \mathbb{R}^{n \times nN}, \tag{3.190}$$

is the numerator coefficients matrix and $x_a \in \mathbb{R}^{nN}$ is the aerodynamic state vector satisfying the following state equation:

$$\dot{x}_a = A_a x_a + \Gamma_a \dot{q}, \tag{3.191}$$

with

$$A_a = \begin{pmatrix} -b_1 I_n & 0 & 0 & \dots & 0 \\ 0 & -b_2 I_n & 0 & \dots & 0 \\ \vdots & \vdots & \vdots & \vdots & \vdots \\ 0 & 0 & 0 & \dots & -b_N I_n \end{pmatrix}, \tag{3.192}$$

$$\Gamma_a = \begin{pmatrix} I_n \\ I_n \\ \vdots \\ I_n \end{pmatrix}, \tag{3.193}$$

where I_n denotes the identity matrix of size n, $b_j > 0, j = 1, \dots, N$, are the lag parameters (or aerodynamic poles), $A_a \in \mathbb{R}^{nN \times nN}$ and $\Gamma_a \in \mathbb{R}^{nN \times n}$. This results in the following state-space coefficients without gusts and control-surface states:

$$A = \begin{pmatrix} 0_n & I_n & 0_{n \times nN} \\ -\bar{M}^{-1}\bar{K} & -\bar{M}^{-1}\bar{C} & \bar{M}^{-1}N_a \\ 0_{nN \times n} & \Gamma_a & A_a \end{pmatrix}, \tag{3.194}$$

$$B = \begin{pmatrix} 0_{n \times m} \\ \bar{M}^{-1} I_n \\ 0_{nN \times m} \end{pmatrix}. \tag{3.195}$$

where $\bar{M} = M - A_2$, $\bar{C} = -A_1$, $\bar{K} = K - A_0$ and the single subscript n indicates a square matrix.

Additional state variables result from the control-surface d.o.f.s, generating the generalized aerodynamic control force vector, $Q_c(t)$. The most accurate model of a control surface would include the bending deformation of the hinge line, a rigid rotation of the surface about the hinge-line, as well as spanwise bending and twisting of the surface. However, owing to its generally small size in comparison to the wing, the bending and twisting deformations can be neglected, resulting in the approximation of a rigid rotation angle, δ, about the hinge line. This allows a significant reduction in the d.o.f.s of the overall structural system, as well as an ease of control-law development. The control input for each surface can either be regarded as the hinge moment applied by the actuator or the control-surface deflection, which are treated as separate generalized coordinates. The latter approach of treating control-surface deflections as control inputs amounts to a quasi-steady approximation, in which the unsteady aerodynamic inertia and lag effect of the wake are neglected. Here, full account is given of the aerodynamic non-circulatory and circulatory effects of control surfaces by using an RFA for each control surface. Thus a control surface can be regarded as a separate lifting surface with its own d.o.f.s and aerodynamics.

As in the typical-section model, we consider a control surface with the following actuator dynamics:

$$I_\delta \ddot{\delta} + c_\delta \dot{\delta} + k_\delta \delta = u + H, \tag{3.196}$$

where $\delta(t)$ is the control-surface deflection about the hinge line, $I_\delta, c_\delta, k_\delta$ are the moment of inertia, damping constant and rotational stiffness, respectively, $H(t)$ is the unsteady aerodynamic hinge moment acting on the control surface and $u(t)$ is the hinge moment control input applied by the actuator. The aerodynamic hinge moment and the generalized aerodynamic force vector created on the wing by the control-surface deflection, $Q_c(t)$, are assumed to be linearly related to $\delta(t)(t)$, $\dot{\delta}(t)(t)$ and $\ddot{\delta}(t)(t)$, as well as to certain additional state variables collected into the control-surface aerodynamic lag state vector, $x_c(t)$. This relationship can be represented by the following the least-squares RFA:

$$Q_c(t) = B_0\delta + B_1\dot{\delta} + B_2\ddot{\delta} + N_c\xi_c, \tag{3.197}$$

where

$$N_c = (B_3,\ B_4,\ B_5,\ \dots,\ B_{\ell+2}) \in \mathbb{R}^{n\times\ell}, \tag{3.198}$$

is the numerators coefficient matrix and $\xi_c \in \mathbb{R}^\ell$ is the control aerodynamic state vector satisfying the following state equation:

$$\dot{\xi}_c = \Lambda_c\xi_c + \Gamma_c\dot{\delta}, \tag{3.199}$$

with

$$\Lambda_c = \begin{pmatrix} -b_1 & 0 & 0 & \dots & 0 \\ 0 & -b_2 & 0 & \dots & 0 \\ \vdots & \vdots & \vdots & \vdots & \vdots \\ 0 & 0 & 0 & \dots & -b_\ell \end{pmatrix}, \tag{3.200}$$

$$\Gamma_c = \begin{pmatrix} 1 \\ 1 \\ \vdots \\ 1 \end{pmatrix}, \tag{3.201}$$

Similarly, the aerodynamic hinge moment can be expressed by

$$H(t) = a_0\delta + a_1\dot{\delta} + a_2\ddot{\delta} + N_h^T\xi_c, \tag{3.202}$$

where

$$N_h^T = (a_3,\ a_4,\ a_5,\ \dots,\ a_{\ell+2}). \tag{3.203}$$

The coefficients $(B_0, B_1, B_2, B_3, \dots, B_\ell)$ and $(a_0, a_1, a_2, a_3, \dots, a_\ell)$ are the numerator coefficients of the RFAs for the unsteady aerodynamic generalized forces and the hinge moment, respectively, with ℓ lag parameters determined by a least-squares curve fit with the harmonic aerodynamic data.

Defining the control-surface state vector by

$$x_c = \left(\delta, \dot{\delta}, \xi_c^T\right)^T \tag{3.204}$$

enables the following state-space representation for the control-surface dynamics:

$$\dot{x}_c = A_c x_c + B_c u, \tag{3.205}$$

$$Q_c = C_c x_c + D_c u, \tag{3.206}$$

where

$$A_c = \begin{pmatrix} 0 & 1 & 0_{1\times\ell} \\ \frac{a_0-k_\delta}{I_\delta-a_2} & \frac{a_1-c_\delta}{I_\delta-a_2} & \frac{1}{I_\delta-a_2}N_h^T \\ 0 & \Gamma_c & \Lambda_c \end{pmatrix}, \tag{3.207}$$

$$B_c = \begin{pmatrix} 0 \\ \frac{1}{I_\delta-a_2} \\ 0_{\ell\times 1} \end{pmatrix}, \tag{3.208}$$

$$C_c = \left(B_0 + \frac{(a_0 - k_\delta)}{(I_\delta - a_2)}B_2,\ B_1 + \frac{(a_1 - c_\delta)}{(I_\delta - a_2)}B_2,\ \frac{1}{(I_\delta - a_2)}N_c \right), \tag{3.209}$$

and

$$D_c = \frac{B_2}{(I_\delta - a_2)}. \tag{3.210}$$

An example of the aeroelastic model for the modified DAST-ARW1 wing with a trailing-edge control surface is given in Appendix C. If a wing has several control surfaces, then the overall control-surface state vector, x_c, is augmented by the state variables of the additional surfaces, and each surface adds a corresponding row and column to the coefficient matrices, A_c, B_c, C_c, D_c.

The state equations of the overall aeroelastic plant, including the control-surface and gust states, can be expressed as follows:

$$\dot{x} = Ax + Bu, \tag{3.211}$$

where

$$x = \left(q^T, \dot{q}^T, x_a^T, x_g^T, x_c^T\right)^T,$$

$$A = \begin{pmatrix} 0 & I & 0 & 0 & 0 \\ -\bar{M}^{-1}\bar{K} & -\bar{M}^{-1}\bar{C} & \bar{M}^{-1}N_a & \bar{M}^{-1}N_g & \bar{M}^{-1}C_c \\ & \Gamma_a & A_a & 0 & 0 \\ & \Gamma_g & 0 & A_g & 0 \\ 0 & 0 & 0 & 0 & A_c \end{pmatrix}, \tag{3.212}$$

and

$$B = \begin{pmatrix} 0 \\ \bar{M}^{-1}D_c \\ 0 \\ \bar{M}^{-1}B_c \end{pmatrix}. \tag{3.213}$$

When sensors are added to pick up aeroelastic motion, output variables, $y \in \mathbb{R}^p$, become available for use in observer-based output-feedback and adaptive-feedback designs. An output equation is thus necessary for the aeroelastic plant before any control can be applied to it, and is given by

$$y = Cx + Du. \tag{3.214}$$

The output variables, $y(t)$, can consist of a set of normal accelerations measured by accelerometers at selected locations and laser-optically sensed vertical deflections. The sensor locations must be selected such that the resulting plant is observable (see Chapter 2) with the given coefficient matrices, (A, C). The most common outputs for an aeroelastic wing are the normal accelerations, $y_i = \ddot{q}_i, i = 1, \dots, n_s$, measured at selected points by accelerometers. Let the coordinates of the sensor locations correspond to the grid points, $i_s \in \mathbb{I}^{n_s}$, of the discretized wing geometry. The output vector picked up by the sensors is then given by

$$\begin{aligned} y = &-\left(\bar{M}^{-1}K\right)_{i_s j} q - \left(\bar{M}^{-1}C\right)_{i_s j} \dot{q} \\ &+ \left(\bar{M}^{-1}N_a\right)_{i_s j} + \left(\bar{M}^{-1}N_g\right)_{i_s j} + \bar{M}^{-1}_{i_s j} Q_c, j = 1, \dots, n, \end{aligned} \tag{3.215}$$

where $A_{i_s j}, j = 1, \dots, n$, represents the submatrix constructed out of i_s rows of the original matrix A. Hence, the output coefficient matrices are given by

$$\begin{aligned} C &= \left[-\left(\bar{M}^{-1}K\right)_{i_s j}, -\left(\bar{M}^{-1}C\right)_{i_s j}, \left(\bar{M}^{-1}N_a\right)_{i_s j}, \left(\bar{M}^{-1}N_g\right)_{i_s j}\right] \\ D &= \bar{M}^{-1}_{i_s, j}, j = 1, \dots, n. \end{aligned} \tag{3.216}$$

References

Abel I, Perry B, and Murrow HN 1978 Two synthesis techniques applied to active flutter suppression on a flight research wing. *J. Guid. Control* **1**, 340–346.

Albano E and Rodden WP 1969 A doublet-lattice method for calculating lift distributions on oscillating surfaces in subsonic flows. *AIAA J.* **7**, 279–285.

Ballhaus WF and Steger JL 1975 Implicit approximate-factorization schemes for the low-frequency transonic equation. *NASA Technical Memorandum* **TM X-73082**.

Birnbaum W 1924 Das ebene problem des schlagenden flügels. *Z. Angew Math. Mech.* **4**, 277–292.

Bisplinghoff RL, Ashley H, and Halfman RL 1955 *Aeroelasticity*. Addison-Wesley, Cambridge.

Dietze F 1947 I. The air forces of the harmonically vibrating wing in a compressible medium at subsonic velocity. II. Numerical tables and curves. *Air Materiel Command, U.S. Air Force* **F-TS-506-RE and F-TS-948-RE**.

Edwards JW 1979 Applications of Laplace transform methods to airfoil motion and stability calculations. *AIAA Paper* **79-0772**.

Eversman W and Tewari A 1991 Consistent rational function approximations for unsteady aerodynamics. *J. Aircr.* **28**, 545–552.

Fettis HE 1952 An approximate method for the calculation of nonstationary air forces at subsonic speeds. *Wright Air Develop. Center, U.S. Air Force* **Tech. Rept. 52-56**.

Fung YC 1955 *An Introduction to the Theory of Aeroelasticity*. John Wiley & Sons, Inc., New York.

Garrick IE 1938 On some reciprocal relations in the theory of non-stationary flow. *NACA* **Rept. 629**.

Garrick IE 1957 Nonsteady wing characteristics. In *Section F, Vol. VII of High Speed Aerodynamics and Jet Propulsion*, Princeton University Press, Princeton, NJ.

Garrick IE and Rubinow SI 1946 Flutter and oscillating air force calculations for an airfoil in a two-dimensional supersonic flow. *NACA* **Rept. 846**.

Hafez M, South J, and Murman E 1979 Artificial compressibility methods for numerical solutions of transonic full potential equation. *AIAA J.* **17**, 838–844.

Hayes WD and Probstein RF 1959 *Hypersonic Flow Theory*. Academic Press, New York.

Holst TL and Ballhaus WF 1979 Fast conservative schemes for the full potential equation applied to transonic flows. *AIAA J.* **17**, 145–152.

Karpel M 1981 Design of active and passive flutter suppression and gust alleviation. *NASA Contractor Report* **CR-3482**.

Küssner HG 1954 A general method for solving problems of the unsteady lifting surface theory in the subsonic range. *J. Aeronaut. Sci.* **21**, 17–27.

Küssner HG and Schwarz L 1940 Der schwingende flügel mit aerodynamisch ausgeglichenem ruder. *Luftfahrtforschung* **17**, 377–384.

Landahl MT 1961 *Unsteady Transonic Flow*. Pergamon, Oxford.

Leishman JG 1994 Unsteady lift of a flapped airfoil by indicial concepts. *J. Aircr.* **31**, 288–297.

Lin J and Iliff KW 2000 Aerodynamic lift and moment calculations using a closed-form solution of the Possio equation. *NASA Technical Memorandum* **TM-2000-209019**.

Murman EM 1973 Analysis of embedded shock waves calculated by relaxation methods. In *Proceedings of AIAA CFD Conference*, pp. 27–40.

Osher S, Hafez M, and Whitlow W Jr. 1985 Entropy conditions satisfying approximations for the full-potential equation of transonic flow. *Math. Comput.* **44**, 1–29.

Possio C 1937 Aerodynamic forces on an oscillating profile at supersonic speeds. *Pontif. Acad. Sci., Acta I* **11**, 93–106.

Possio C 1938 Aerodynamic forces on an oscillating profile in a compressible fluid at subsonic speeds. *Aerotecnica* **18**, 441–458.

Reissner E 1951 On the application of Mathieu functions in the theory of subsonic compressible flow past oscillating airfoils. *NACA* **Tech. Note 2363**.

Roger KL, Hodges GE, and Felt L 1975 Active flutter suppression–A flight test demonstration. *J. Aircr.* **12**, 551–556.

Runyan HL 1952 Single-degree-of-freedom flutter calculations for a wing in subsonic potential flow and a comparison with an experiment. *NACA* **Rept. 1089**.

Scanlan RH and Rosenbaum R 1951 *Introduction to the Theory of Aircraft Vibration and Flutter*. Macmillan USA, New York.

Schade T 1944 Beitrag zu zahlentafeln zur luftkraftberechnung der schwingenden tragfläche in ebener unterschallströmung. *ZWB* **UM 3211**.

Sevart FD 1975 Development of active flutter suppression wind tunnel testing technology. *U.S. Air Force Tech. Report* **AFFDL-TR-74-126**.

Smilg B and Wasserman LS 1942 Application of three-dimensional flutter theory to aircraft structures. *Air Materiel Command, U.S. Air Force* **Tech. Rept. 4798**.

Söhngen H 1940 Bestimmung der auftriebsverteilung für beliebige instationäre bewegungen (ebenes problem). *Luftfahrtforschung* **17**, 401–419.

Steger JL and Caradonna FX 1980 A conservative implicit finite difference algorithm for the unsteady transonic full potential equation. *AIAA Paper* **80-1368**.

Stewartson K 1950 On the linearized potential theory of unsteady supersonic motion. *Q. J. Mech. Appl. Math.* **III**, 182–199.

Tannehill JC, Anderson DA, and Pletcher RH 1997 *Computational Fluid Mechanics and Heat Transfer*, 2nd ed., Taylor and Francis, Philadelphia, PA.

Tewari A 1994 Doublet-point method for supersonic unsteady aerodynamics of nonplanar lifting surfaces. *J. Aircr.* **31**, 745–752.

Tewari A 2015 *Aeroservoelasticity: Modeling and Control*. Birkhäuser, Boston, MA.

Theodorsen T 1935 General theory of aerodynamic instability and the mechanism of flutter. *NACA* **Rept. 496**.

Theodorsen T and Garrick IE 1942 Non-stationary flow about a wing-aileron-tab combination including aerodynamic balance. *NACA* **Rept. 736**.

Tiffany SH and Adams WM 1987 Nonlinear programming extensions to rational function approximations of unsteady aerodynamics. *Proceedings AIAA Symposium on Structural Dynamics and Aeroelasticity* **Paper 87-0854**, 406–420.

Timman R and van de Vooren AI 1949 Theory of the oscillating wing with aerodynamically balanced control surface in a two-dimensional subsonic compressible flow. *National Aeronautics Research Institute, Amsterdam* **NLL Rept. F54**.

Timman R, van de Vooren AI, and Greidanus, JH, 1951 Aerodynamic coefficients of an oscillating airfoil in two-dimensional subsonic flow. *J. Aeronaut. Sci.* **18**, 717–802. (Also 1954 Tables of aerodynamic coefficients for an oscillating wing-flap system in subsonic compressible flow. *Natl. Aeronaut. Research Inst., Amsterdam* **NLL Rept. F151**.)

Turner MJ and Rabinowitz S 1950 Aerodynamic coefficients for an oscillating airfoil with hinged flap, with tables for a Mach number of 0.7. *NACA* **Tech. Note 2213**.

Ueda T and Dowell EH 1982 A new solution method for lifting surfaces in subsonic flow. *AIAA J.* **20**, 348–355.

Wagner H 1925 Über die entstehung des dynamischen auftriebs von tragflügeln. *Z. Angew Math. Mech.* **5**, 17–35.

4

Active Flutter Suppression

Flutter is the catastrophic dynamic instability encountered at a particular flight condition (ρ, U) when a pair (or more) of the complex eigenvalues of the aeroelastic dynamics matrix A crosses the imaginary axis into the right-half Laplace domain. This results in an exponentially growing structural motion, ultimately leading to wing (or tail) failure. The classical flutter mechanism is a dynamic interaction between the pitch (primary torsion) and plunge (primary bending) modes, and is a characteristic of high-aspect-ratio wings of transport type, subsonic aircraft. In contrast, the small-aspect-ratio wings of fighter-type aircraft have a much higher torsional stiffness, and thus do not encounter the classical bending-torsion flutter in their operating envelope. However, interaction with a control-surface mode can excite flutter in combination with bending and/or torsion modes. Wing-mounted external stores (bombs, missiles and drop tanks) can cause a reduction in the flutter speed of fighter/bomber-type aircraft, when compared with that of a clean wing. This is due to a dynamic interaction between a bending mode and the primary torsion mode in the changed mass distribution of the wing. Sometimes, merely dropping a particular external store can cause flutter. Similarly, partially empty, internal wing fuel tanks have the potential for causing flutter, especially in larger transport-type aircraft. Consequently, every new design has to be carefully tested and analysed for flutter in every possible flight condition and configuration.

Modern high-performance aircraft have automatic flight control systems for stabilizing and actively damping the rigid-body modes, which require a rapid actuation of control surfaces. Such a control system opens up the possibility of interaction among all the three types of structural modes (bending, torsion and control-surface rotation), leading to flutter-type instability. Unstable aeroservoelastic (ASE) interactions can arise well below the open-loop flutter speed of the aircraft, hence a careful ASE analysis is required throughout the flight envelope.

Preventing and suppressing flutter has been a primary interest of aircraft designers. The most common technique is passive stabilization by structural modifications, wherein the mass and stiffness distributions are changed such that aeroelastic interaction is avoided at any point inside the flight envelope. Historically, most low-speed aircraft have successfully avoided flutter simply by having ballast weights placed at appropriate structural locations, and in extreme cases, by redesigning the wing and tail spars in order to increase torsional stiffness. However, passive flutter prevention can be a quite cumbersome process for a high-performance aircraft with a large range of operating speeds, altitudes and loading configurations, wherein

Adaptive Aeroservoelastic Control, First Edition. Ashish Tewari.

many operating conditions can have conflicting flutter mechanisms. Avoiding flutter at one condition may lead to increased flutter susceptibility at another. In order to avoid the costs overruns and delays in a new design due to an extended flutter clearance programme, many aircraft manufacturers began studying active flutter suppression in the late 1960s and early 1970s. Several new fighter and airliner prototypes required flutter tests in that era and active flutter suppression appeared to be a practical alternative. Thus papers and technical reports on mathematical modelling, design and practical implementation of flutter suppression systems began appearing in the 1970s. There were enthusiastic developments in ASE in the 1970s and 1980s in all aeronautical establishments and laboratories, culminating in wind-tunnel tests of aeroelastically scaled models and flight testing of drones especially constructed for studying active flutter suppression systems. It was quickly recognized that robustness of the controller is a special requirement for active flutter suppression, due to the uncertainties in the aeroelastic models. The robust control techniques of LQG/LTR, H_2/H_∞ and μ-synthesis, which were newly minted in the 1980s, were readily applied to flutter suppression systems resulting in a plethora of research literature on the topic. These are sampled in the review paper of Mukhopadhyay (2003). Subsequent ASE research has built upon the landmark developments of that era, and has resulted in many novel ideas (such as the active flexible wing, morphing wing and flapping wing flight) that are currently being explored for the design of unmanned aerial vehicles.

Despite active research spanning several decades, there is as yet no aircraft in production that takes advantage of active flutter suppression. The reason for this discrepancy can be summarized as follows. Most of the modern aircraft cruise and manoeuvre at high-subsonic and transonic speeds; this is due to the fact that both the best cruising range and the largest turning rate of jet-powered aircraft nearly always occur at such speeds. Since their airframes are designed for efficient high-subsonic flight, such aircraft are expected to have their open-loop flutter speeds in the transonic regime. Even the fighter-type aircraft designed for brief supersonic flight encounter flutter in the transonic regime due to the transonic flutter-dip phenomena, which will be discussed in Chapter 11. Thus transonic aerodynamic modelling becomes an integral part of a practical flutter suppression system. However, the uncertainties of transonic unsteady aerodynamics – especially those due to rapid control-surface deflections – are notorious because of the presence of shock waves. Transonic unsteady flows are still not understood sufficiently well to be modelled accurately and simply for a practical control-system design. Hence, a flutter suppression system that can be certified to be safe for normal operation must wait until either the transonic aerodynamic models improve in their accuracy or the control systems can be made robust to uncertainties in transonic aeroelastic modelling. For rendering a flutter suppression system robust to even the large modelling uncertainties of the transonic regime, adaptive control laws must be explored. This is the thrust of the current research in ASE and also the topic of this chapter.

Trying to design feedback controllers for poorly understood plants is fraught with danger, because control signals could be easily driven to very large values, thereby destabilizing the system (see Chapter 2). A practical way of avoiding high loop gains is the use of structural damping to decouple the frequencies of the natural modes responsible for causing flutter. This approach has been successfully applied to semi-actively suppress the flutter caused by the wing-mounted external stores in fighter/attack-type aircraft. An active spring-dashpot-type plunge mechanism, combined with pitching of the external store pylon about a pivot, can make the open-loop torsional frequency sufficiently large so that it does not interact with the wing's bending mode. Semi-active structural stabilization of flutter modes can be regarded as an alternative to the typical flutter suppression concept, namely the use of aerodynamic coupling provided by the control surfaces to suppress flutter. It also offers a promise for the transonic regime, because the control-law design does not require an accurate unsteady aerodynamic model. However, the structural decoupling concept is very much like an active shock absorber for an automobile, and must have dampers and actuators that are sufficiently powerful to absorb and dissipate the energy of flutter. Clearly, this concept requires a new structural design philosophy before a general application is possible.

4.1 Single Degree-of-Freedom Flutter

Since quasi-steady aerodynamics of motion in each degree of freedom (d.o.f.) provides viscous damping, it is not expected that any one of the three d.o.f.s taken alone will produce a self-sustained oscillation forced by quasi-steady flow effects. However, when unsteady aerodynamics of wake-coupled motion is accounted for, there is a possibility of single d.o.f. flutter. In order to investigate such a possibility, consider first a wind-tunnel model of an airfoil of mass m mounted on a linear spring of stiffness K_h, and constrained to plunge along a slide with viscous damping constant c_h. This is representative of pure bending motion with an infinitely stiff (rigid) torsional spring. The airfoil is driven by an actuating mechanism that generates a plunge force, $P_e(t)$. If the actuation is simple harmonic with amplitude, P_0, and excitation frequency, ω, we write

$$P_e = P_0 e^{i\omega t} \tag{4.1}$$

with the understanding that only the real part of the term on the right-hand side is to be taken. The resulting plunge displacement, $h(t)$, is given by

$$h = \bar{h} e^{i\omega t}, \tag{4.2}$$

where $\bar{h}$ is the complex plunge amplitude. The instantaneous upward force experienced by the airfoil is the driving force, $P_e(t)$, subtracted from the aerodynamic lift, $L(t)$, created by the plunging motion. Note that an aerodynamic pitching moment, M_y, is also produced because of the plunging motion, but is opposed by the restraining apparatus (considered to be rigid) such that a pitch displacement (rotation) does not take place.

For an airfoil of chord $2b$ and flow density ρ, undergoing a pure plunging motion, the non-circulatory lift per unit span is given by (see Chapter 3):

$$L_{nc}(t) = \pi b^2 \rho \ddot{h}, \tag{4.3}$$

and acts at the airfoil's mid-chord point. The structural motion in a pure plunge is described by the following equation:

$$m\ddot{h} + c_h \dot{h} + K_h h = P_e - L. \tag{4.4}$$

The non-circulatory term is clubbed with the airfoil mass in the revised structural equation of motion, leaving only the circulatory lift as the aerodynamic forcing term:

$$(m + \pi b^2 \rho)\ddot{h} + c_h \dot{h} + K_h h = P_e - L_c, \tag{4.5}$$

Assuming zero initial condition, $h(0-) = \dot{h}(0-) = 0$, the structural transfer function, $G_s(s)$, is the following:

$$G_s(s) = \frac{h(s)}{L_c(s)} = -\frac{1}{(m + \pi b^2 \rho)s^2 + cs + k}. \tag{4.6}$$

The aerodynamic system is in a feedback loop with the structural system (see Fig. 3.4), hence the overall aeroelastic system can be represented by the following closed-loop transfer function, $G_c(s)$:

$$G_c(s) = \frac{\alpha(s)}{P(s)} = \frac{G_s(s)}{1 + sG_s(s)G_a(s).} \tag{4.7}$$

For the pure plunging motion, the change in the angle of attack is the same at all points on the chord, and is given by

$$\alpha = \tan^{-1} \frac{\dot{h}}{U},$$

which in the linear case (small $| \alpha |$) is approximated by

$$\alpha \simeq \frac{\dot{h}}{U}.$$

Thus we have

$$L_c(t) = \frac{1}{U} \int_0^\infty \dot{h}(\tau) g(t - \tau) \mathrm{d}\tau \tag{4.8}$$

or

$$L_c(t) = \frac{1}{U} \left[\alpha(t)\dot{h}(0) + \int_0^t \ddot{h}(\tau)\alpha(t - \tau)\mathrm{d}\tau \right]. \tag{4.9}$$

This becomes the following in terms of Wagner's function:

$$L_c(\hat{t}) = 2\pi\rho U^2 \left[\Phi(\hat{t})h'(0) + \int_0^{\hat{t}} h''(\tau)\Phi(\hat{t} - \tau)\mathrm{d}\tau \right]. \tag{4.10}$$

Here $'$ denotes taking the derivative with respect to non-dimensional time, $\hat{t}$, thereby implying

$$h' = \dot{h}\left(\frac{b}{U}\right)$$
$$h'' = \ddot{h}\left(\frac{b}{U}\right)^2. \tag{4.11}$$

The circulatory lift induced by the plunging oscillation can also be expressed as follows:

$$L_c = \bar{L}e^{i\omega t} = \bar{L}e^{ik\hat{t}}, \tag{4.12}$$

where $\bar{L} = L_0 e^{i\phi}$ is the complex lift amplitude. The complex amplitude ratios, $\bar{h}/P_0$, $\bar{L}/\bar{h}$, give rise to phase differences between the various signals, and depend upon the forcing frequency, ω. The linear aerodynamic dependence of the circulatory lift on the plunge rate, $\mathrm{d}h/\mathrm{d}t$, is described by the following transfer function:

$$G_a(s) = \frac{L_c(s)}{sh(s)},$$

wherein the magnitude ratio and phase angle, ϕ, between circulatory lift and plunge are given by

$$\left|\frac{\bar{L}}{\bar{h}(\omega)}\right| = i\omega \mid G_a(i\omega) \mid, \quad \phi(\omega) = \tan^{-1}\frac{\text{Im}[G_a(i\omega)]}{\text{Re}[G_a(i\omega)]}. \tag{4.13}$$

After a steady harmonic motion is established, the phase between the external excitation, $P_e(t)$, and the plunge displacement is unimportant. Thus we can regard the aeroelastic plunge response taking place with real amplitude, h_0, as follows:

$$h = h_0 e^{i\omega t}.$$

When integrated over a complete plunge cycle, the non-circulatory lift does not perform any net work:

$$\begin{aligned} W_{nc} &= -\int_0^{2\pi/\omega} \dot{h}(t) L_{nc}(t)\text{d}t \\ &= \frac{1}{2}\pi b^2 \rho h_0^2 \omega^2 [\cos(4\pi) - 1] = 0. \end{aligned} \tag{4.14}$$

Therefore, the only net work done on the airfoil per cycle due to aerodynamic lift is by its circulatory part, calculated as follows:

$$\begin{aligned} W_c &= -\int_0^{2\pi/\omega} \dot{h}(t) L_c(t)\text{d}t \\ &= -\rho U \omega h_0 L_0 \int_0^{2\pi/\omega} \sin(\omega t)\cos(\omega t + \phi)\text{d}t \\ &= -\rho U \omega h_0 L_0 \int_0^{2\pi/\omega} \sin(\omega t)\sin(\omega t + \phi)\text{d}t \\ &= -\pi \rho U h_0 L_0 \cos\phi. \end{aligned} \tag{4.15}$$

The phase angle ϕ can be calculated by Eq. (4.13) by putting $\bar{\alpha} = ikh_0/b$ for the amplitude of the non-dimensional angle of attack:

$$\phi = \tan^{-1}\frac{\text{Im}[ikC(k)]}{\text{Re}[ikC(k)]} = \tan^{-1}\frac{F(k)}{-G(k)}, \tag{4.16}$$

which is always in the range $-\pi/2 \leq \phi \leq \pi/2$, as shown in Fig. 4.1. Hence net aerodynamic work done per cycle, W_c, is always negative in the case of pure plunging oscillation. Therefore, energy is always extracted by the airflow from the system, and flutter in pure plunge (bending) is ruled out. While this analysis is for the ideal (inviscid, incompressible) flow, the rate of energy extraction from the system increases (becomes even more negative) if the effects of viscosity and compressibility are included. This is due to viscous flow dissipation and the energy lost in compressing the airflow. Thus it can be said that pure bending (plunging) oscillation can never lead to a self-sustained oscillation like flutter.

Next, the possibility of flutter by pure pitching oscillation is investigated. The wind-tunnel model of the airfoil is now mounted on a torsional spring of stiffness K_θ and restrained from plunging by a rigid support in the vertical direction. The moment of inertia of the airfoil about

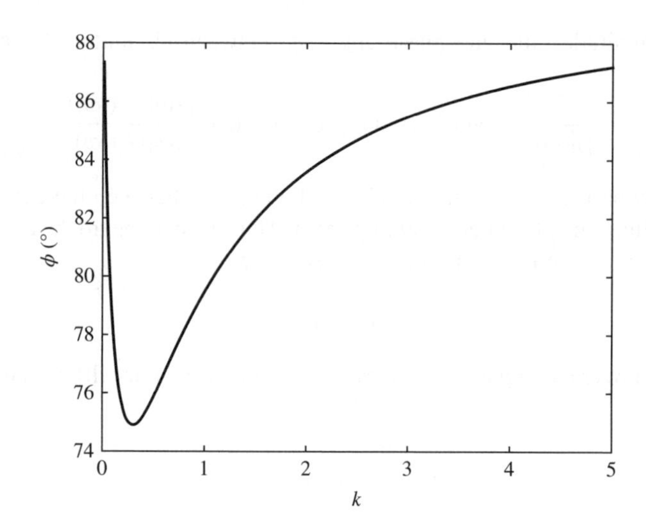

Figure 4.1 Phase lead angle, ϕ, of the circulatory lift with respect to the plunge displacement in a pure plunging oscillation

the pitch axis is I_θ and the viscous damping constant in pitch is c_θ. The airfoil is driven by an actuating mechanism that applies a torque about the pitch axis, such that a steady pitching oscillation is established:

$$\theta = \theta_0 e^{i\omega t} = \theta_0 e^{ikUt/b}. \tag{4.17}$$

Since plunge displacement is absent, the angle of attack is equal to the pitch displacement ($\alpha = \theta$). From Chapter 3, the following expressions are derived for the aerodynamic lift, L, per unit span,

$$L = -\pi\rho b^3 a\ddot{\theta} + \pi\rho b^2 U\dot{\theta} + 2\pi\rho b U \left[U\theta + b\left(\frac{1}{2} - a\right)\dot{\theta}\right] C(k) \tag{4.18}$$

and the pitching moment per unit span, M_y, about the pitch (elastic) axis located at $x = ab$ behind the mid-chord point:

$$\begin{aligned} M_y = &-\left(a^2 + \frac{1}{8}\right)\pi\rho b^4\ddot{\theta} - \pi\rho b^3 U\left(\frac{1}{2} - a\right)\dot{\theta} \\ &+ 2\pi\rho b^2 U\left(\frac{1}{2} + a\right)\left[U\theta + b\left(\frac{1}{2} - a\right)\dot{\theta}\right] C(k). \end{aligned} \tag{4.19}$$

The aerodynamic lift is opposed by the rigid vertical support, hence no plunge displacement takes place ($z(t) = 0$). However, the aerodynamic pitching moment contributes to the pitching oscillation. After establishing the constant amplitude oscillation, the actuating torque is removed ($M_{ye} = 0$), leading to a free aeroelastic motion. The non-circulatory inertia term is added to the pitch moment of inertia, and the non-circulatory damping term contributes to the structural viscous dissipation, while the circulatory moment due to the wake acts as the forcing term on the right-hand side of the following aeroelastic equation of motion:

$$\begin{aligned} &\left[I_\theta + \left(a^2 + \frac{1}{8}\right)\pi\rho b^4\right]\ddot{\theta} + \left[c_\theta + \pi\rho b^3 U\left(\frac{1}{2} - a\right)\right]\dot{\theta} + K_\theta\theta \\ &\qquad = 2\pi\rho b^2 U\left(\frac{1}{2} + a\right)\left[U\theta + b\left(\frac{1}{2} - a\right)\dot{\theta}\right] C(k). \end{aligned} \tag{4.20}$$

It is clear in Eq. (4.20) that the non-circulatory moments have no possibility of exciting a self-sustained oscillation like flutter, because they merely add to the structural inertia and damping terms. Furthermore, if the pitch axis $x = ab$ is moved to the 3/4-chord location ($a = 1/2$), the circulatory damping term vanishes, and only the circulatory stiffness term remains. This implies that even the slightest structural damping is sufficient to drive the oscillation to zero in the steady state for pitching about the 3/4-chord point. For a pitch axis located between the mid-chord and the 3/4-chord points ($0 < a < 1/2$), the circulatory damping term is positive, which effectively damps out the pitching oscillation. Hence, for the possibility of flutter, we must look for the cases where the pitch axis is located forward of the mid-chord point $a < 0$, for which the circulatory damping term becomes sufficiently negative to overcome structural damping in order to produce self-sustained pitching oscillation. The classical study conducted by Greidanus (as reported in Fung (1955)) indicated a single d.o.f. flutter condition in pitch for $a < -1/2$ (but not too far forward of the leading edge, $a = -1$) and $k < 0.0435$. Bisplinghoff and Ashley (1962) gave a physical explanation of the low-frequency single d.o.f. flutter. They required that the aerodynamic pitching-moment derivative with respect to pitch rate must change in sign from negative to positive at the flutter condition. This implies that the imaginary part of the moment derivative with respect to pitch amplitude should change in sign from the positive to the negative at the flutter frequency. The exact flutter condition is therefore obtained at the reduced frequency at which the imaginary part of the circulatory moment derivative with respect to pitch amplitude, given by

$$\frac{\partial \bar{M}_c / \partial \theta_0}{\frac{1}{2}\rho U^2 b^2} = 4\pi \left(\frac{1}{2} + a\right) \left[1 + ik\left(\frac{1}{2} - a\right)\right] C(k), \tag{4.21}$$

changes from positive to negative values, thereby producing negative aerodynamic damping. At reduced frequencies below the flutter point, the oscillation is unstable, while for those above the flutter point, we have stable pitching oscillation. A plot of the imaginary part of the non-dimensional circulatory moment derivative with respect to pitching amplitude versus the reduced frequency is shown in Fig. 4.2 for various values of the pitch axis location forward of the 1/4-chord point ($a < -1/2$). It is seen that as the pitch axis moves forward, the critical reduced frequency decreases, thereby diminishing the range of frequencies for which flutter instability can occur. For $a = -0.55$, the critical reduced frequency is found to be $k = 0.076$, while that for $a = -1.75$ is $k = 0.03$ (Fig. 4.2). The analytical result of Greidanus (Fung 1955) is obtained for $a = -1.29$, at which the reduced frequency range for single d.o.f. flutter is $0 < k < 0.0435$.

A similar analysis reveals that the single d.o.f. flutter in control-surface rotation can also occur in a specific range of reduced frequencies, for specific locations of the hinge line forward of the 1/4-chord point of the control surface. An aerodynamically balanced flap can also lead to single d.o.f. flutter (Runyan 1952). However, such a phenomenon can only be driven by an unsteady wake. If a quasi-steady aerodynamic assumption is used in the analysis, completely mistaken results can be produced, as shown by Bisplinghoff and Ashley (1962). Fortunately, quasi-steady analysis can be applied to explain the multi-degree-of-freedom flutter mechanism, as given below.

Runyan (1952) applied subsonic and supersonic compressibility corrections in his single d.o.f. flutter calculations using series expansion for the unsteady aerodynamic moment in terms of the compressible reduced frequency, $k/(M_\infty^2 - 1)$. It was shown that the compressibility effect can have a pronounced reduction in flutter stability margin. For single d.o.f.

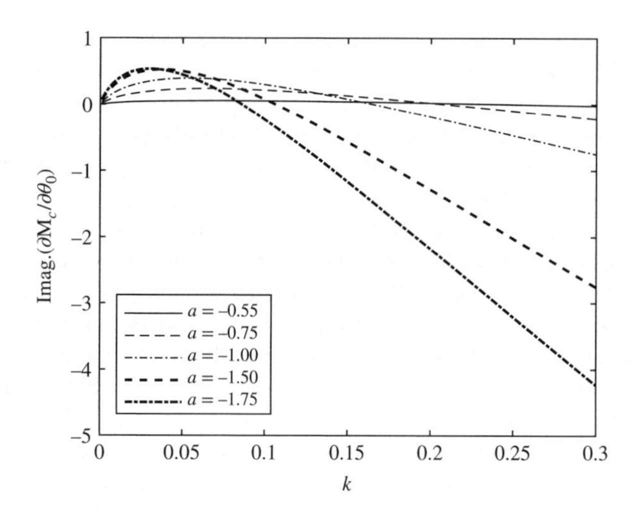

Figure 4.2 Circulatory pitching-moment derivative with respect to pitch amplitude versus reduced frequency in a pure pitching oscillation for various pitch axis locations, showing the low-frequency flutter condition

control-surface flutter, Runyan's approximate, linear analysis indicated a critical speed in the transonic or low-supersonic regimes. However, these are the very regimes where the effects of unsteady shock waves are equally (or perhaps more) important than those of an unsteady wake. When the nonlinear aerodynamic effects of shock-induced flow separation (SITES) are involved (to be considered in Chapter 11), single d.o.f. flutter in either pitch, or control-surface rotation becomes more likely. The SITES phenomenon usually lead to a self-sustained, constant-amplitude limit-cycle oscillation (LCO) in pitch, where the energy extracted from the freestream by the single d.o.f. flutter mechanism is dissipated in viscous flow separation caused by nearly normal shock waves.

Subsonic single d.o.f. flutter involving attached flows – due to its second-order dynamics – can be prevented either by the passive means of increasing the torsional stiffness of the wing and/or the control surface, or by placing high-gain, linear filters in a feedback control loop (to be discussed later in this chapter). Suppression of SITES LCO may require a nonlinear feedback strategy, to be covered in Chapter 11. Owing to the inherent uncertainty in the aerodynamic parameters, an adaptive control loop must be added to suppress both linear and nonlinear s.d.o.f. flutter.

4.2 Bending-Torsion Flutter

When two (or more) d.o.f.s are excited simultaneously, they can interact in such a way as to cause flutter, even though the structural stiffness in each may be large enough to preclude single d.o.f. flutter. When only two d.o.f.s are involved, the classical flutter mechanism is termed binary flutter, whereas that involving all the three d.o.f.s is called ternary flutter. Such a flutter mechanism is much more difficult to prevent and suppress owing to the complex aeroelastic interaction taking place among the various d.o.f.s.

The most common binary flutter is that involving the plunging and pitching motions (called bending-torsion flutter). Since the physical mechanism behind bending-torsion flutter involves a coupling of the pitch and plunge d.o.f.s, its explanation is not very intuitive, but requires analytical methods. Here two such explanations are presented. The first is based on the approximate quasi-steady approximation of Pines (Bisplinghoff and Ashley 1962), which offers a surprisingly accurate insight into the actual phenomenon. The second analysis is by the energy approach applied to the unsteady aerodynamic effects caused by the coupled pitch-plunge oscillation.

Quasi-steady aerodynamics of typical section can be expressed by the following coupled equations of motion:

$$m\ddot{h} + K_h h + S_\theta \ddot{\theta} = -qSC_{L_\alpha}\left(\theta + \frac{\dot{h}}{U}\right)$$

$$I_\theta \ddot{\theta} + K_\theta \theta + S_\theta \ddot{h} = qSeC_{L_\alpha}\left(\theta + \frac{\dot{h}}{U}\right), \tag{4.22}$$

where C_{L_α} is the sectional lift-curve slope, and e is the distance of the pitch (elastic) axis behind the aerodynamic centre. For subsonic flow, we have $e = (a + 1/2)b$, whereas for supersonic flow, $e = ab$.

This can be explained by considering a pitch oscillation that has a phase difference, ϕ, with the plunging motion. The two responses are expressed as follows:

$$h = h_0 e^{i\omega t}$$

$$\theta = \theta_0 e^{i\omega t - \phi}. \tag{4.23}$$

Such a motion can arise through a harmonic plunge force, $P_e(t) = P_0 e^{i\omega t}$, and then removing the excitation when the constant harmonic amplitudes (h_0, θ_0) are reached. Assuming simple harmonic motion, the aerodynamic lift and pitching moment are expressed as follows in terms of the Theodorsen function (Appendix A):

$$L = \pi\rho b^2[\ddot{h} + U\dot{\theta} - ab\ddot{\theta}] + 2\pi\rho bU\left[\dot{h} + U\theta + b\left(\frac{1}{2} - a\right)\dot{\theta}\right]C(k), \tag{4.24}$$

$$M_y = \pi\rho b^2\left[ab\ddot{h} - bU\left(\frac{1}{2} - a\right)\dot{\theta} - b^2\left(a^2 + \frac{1}{8}\right)\ddot{\theta}\right]$$
$$+ 2\pi\rho b^2 U\left(\frac{1}{2} + a\right)\left[\dot{h} + U\theta + b\left(\frac{1}{2} - a\right)\dot{\theta}\right]C(k). \tag{4.25}$$

From the above discussion, it becomes clear that the pitch (torsion) mode has the ability to cause flutter, either by itself or in combination with a plunge (bending) mode. The same can be said of the control-surface mode, which can cause flutter by itself, or in combination with the pitch and/or plunge modes. Conversely, the control-surface mode, if properly controlled, can be employed to suppress flutter caused by the other two d.o.f.s. This is the basis of active flutter suppression systems, which will be the focus of the remaining chapter.

4.3 Active Suppression of Single Degree-of-Freedom Flutter

A linear aeroelastic system with pitching motion, $\theta(t)$, coupled with control-surface motion, $\beta(t)$, is described by the following vector differential equation:

$$M\ddot{q} + Kq = Q + (0, 1)^T u, \tag{4.26}$$

where $q = (\theta, \beta)^T$ is the generalized coordinates vector, $Q = (M_y, H)^T$ is the generalized air loads vector, and M, K are the following generalized mass and stiffness matrices, respectively, of the structure:

$$M = \begin{pmatrix} I_\theta & m_c x_c x_\beta \\ m_c x_c x_\beta & I_\beta \end{pmatrix} \tag{4.27}$$

$$K = \begin{pmatrix} k_\theta & 0 \\ 0 & k_\beta \end{pmatrix}. \tag{4.28}$$

The structural damping is neglected for a conservative flutter suppression design. However, if the damping is estimated by an experiment, it can be added to the simulation model of the ASE system. The generalized aerodynamic loads vector consists of the pitching moment, $M_y(t)$, and the control-surface hinge moment, $H(t)$. For simplicity, it is assumed that the aerodynamic moments can be modelled primarily as first-order lag (or circulatory) effects of the unsteady wake shed by the airfoil, as well as the non-circulatory contributions of the aerodynamics to inertia, damping and stiffness. Such an RFA model of the unsteady aerodynamics with a single lag parameter is given in the Laplace domain by

$$Q(s) = \frac{1}{2}\rho U \left(a_1 + a_2 s + \frac{a_3 s}{s + b_1} \right) \begin{Bmatrix} A_1 \\ A_2 \end{Bmatrix} w(s), \tag{4.29}$$

where $w(s)$ is the following upwash at a specific location on the airfoil:

$$w(s) = (C_1, C_2, C_3, C_4) \begin{Bmatrix} q(s) \\ sq(s) \end{Bmatrix}, \tag{4.30}$$

and $A_1, A_2, a_1, \cdots, a_3, b_1, C_1, \cdots, C_4$ are constant aerodynamic parameters (in addition to ρ and U). Equations (4.29) and (4.30) result in the following relationship for the generalized unsteady aerodynamic loads:

$$Q(s) = G(s) \begin{Bmatrix} q(s) \\ sq(s) \end{Bmatrix}, \tag{4.31}$$

where the aerodynamic transfer matrix is given by

$$G(s) = \frac{1}{2}\rho U \left(a_1 + a_2 s + \frac{a_3 s}{s + b_1} \right) \begin{Bmatrix} A_1 \\ A_2 \end{Bmatrix} (C_1, C_2, C_3, C_4), \tag{4.32}$$

This leads to the following state-space representation

$$\dot{x} = Ax + Bu \tag{4.33}$$

$$y = \theta = Ex + Du, \tag{4.34}$$

where $x = (q^T, \dot{q}^T, x_a)^T$ is the state vector,

$$A = \begin{bmatrix} 0 & I & 0 \\ -\bar{M}^{-1}\bar{K} & \bar{M}^{-1}\bar{C} & -\frac{1}{2}\rho U b_1 a_3 \bar{M}^{-1} \begin{Bmatrix} A_1 \\ A_2 \end{Bmatrix} \\ (C_1, C_2) & (C_3, C_4) & -b_1 \end{bmatrix}, \tag{4.35}$$

$$B = \begin{bmatrix} 0 \\ \bar{M}^{-1}(0,1)^T \\ 0 \end{bmatrix}, \tag{4.36}$$

$$E = (1,0,0,0,0)\,; \qquad D = 0, \tag{4.37}$$

$$\bar{M} = M - \frac{1}{2}\rho U a_2 \begin{Bmatrix} A_1 \\ A_2 \end{Bmatrix} (C_3, C_4), \tag{4.38}$$

$$\bar{C} = \frac{1}{2}\rho U \begin{Bmatrix} A_1 \\ A_2 \end{Bmatrix} \left[a_2 \left(C_1, C_2\right) + \left(a_1 + a_3\right)\left(C_3, C_4\right) \right], \tag{4.39}$$

$$\bar{K} = K - \frac{1}{2}\rho U \left(a_1 + a_3\right) \begin{Bmatrix} A_1 \\ A_2 \end{Bmatrix} \left(C_1, C_2\right). \tag{4.40}$$

The matrices $\bar{K}, \bar{C}, \bar{M}$ are the generalized stiffness, damping and mass matrices of the aeroelastic system, which reduce to $K, 0, M$, respectively, for the in vacuo case ($\rho = 0$). Note that the order of the aeroelastic system is increased by one due to the aerodynamic state, $x_a(t)$, arising out of the single lag term, $a_3 s/(s + b_1)$, which augments the dynamics matrix A. The given aeroelastic plant is controllable with the motor torque input, which can be verified from the rank of the controllability test matrix for the pair (A, B). The aeroelastic stability is determined from the eigenvalues of A:

$$\det(sI - A) = 0$$

at a given flight condition, (ρ, U). Considering the flow to be incompressible, a Theodorsen type (Chapter 3) model can be used to derive $w(s)$, hence $G(s)$.

Example 4.3.1 *Consider an airfoil of semi-chord $b = 1$ m equipped with a trailing-edge control surface mounted on a frictionless torsional spring of stiffness $k_\theta = 40$ N m/rad. The control-surface hinge is also frictionless, and has a rotational spring of stiffness $k_\beta = 9$ N m/rad. The hinge line of the control surface is located at a distance of $c = 0.3$ m behind the pitch axis. The mass of the control surface is $m_c = 0.1$ kg, and the distance of the control surface's centre of mass behind its own hinge line is $x_\beta = 0.1$ m. The moment of inertia of the setup about the pitch axis is $I_\theta = 10$ kg m^2, while that of the control surface about its hinge is $I_\beta = 1$ kg m^2. The setup is placed in a uniform, incompressible flow of speed U and density $\rho = 1.225$ kg/m^3. The control surface is equipped with a DC motor that can apply a torque, u(t), about the control-surface hinge line, relative to the airfoil. The relevant aerodynamic parameters calculated by Theodorsen's model (Chapter 3) are the following:*

$$A_1 = \frac{1}{2}\,; \quad A_2 = -\frac{T_{12}}{(2\pi)}$$

$$C_1 = 1\,; \quad C_2 = \frac{T_{10}}{\pi}\,; \quad C_3 = \frac{1}{2}\,; \quad C_4 = \frac{T_{12}}{(2\pi)},$$

where the coefficients $T_i, i = 10, 11, 12$ are the following (Chapter 3):

$$\begin{aligned} T_{10} &= \sqrt{1 - c^2} + \cos^{-1} c \\ T_{11} &= (2 - c)\sqrt{1 - c^2} + (1 - 2c)\cos^{-1} c \\ T_{12} &= (2 + c)\sqrt{1 - c^2} - (2c + 1)\cos^{-1} c \end{aligned}$$

A simple-pole RFA of the least-squares type is applied to approximate the Theodorsen function, $C(k)$, with one lag parameter, b_1, given by

$$C(s) = a_1 + a_2 s + \frac{a_3 s}{s + b_1},$$

and has the following coefficients:

$$a_1 = 1.0\ ,\ a_2 = 0\ ,\ a_3 = -2.0\ ,\ b_1 = 0.05.$$

It can be shown that the aeroelastic plant is unstable at any speed, $U > 0$, due to the circulatory aerodynamic characteristics associated with the wake. This is the single d.o.f. (pitch) flutter, which needs to be suppressed by controlling the deflection of the control surface via a feedback loop.

In order to stabilize the plant, the following linear, state-feedback control law is explored:

$$u = -\left(\theta_1, \theta_2, \theta_3, \theta_4, \theta_5\right) \begin{Bmatrix} \theta \\ \beta \\ \dot{\theta} \\ \dot{\beta} \\ x_a \end{Bmatrix},$$

which is abbreviated as $u = -\Theta^T x$, with $x = (\theta, \beta, \dot{\theta}, \dot{\beta}, x_a)^T$ being the state vector, and $\Theta = (\theta_1, \theta_2, \theta_3, \theta_4)^T$ the controller parameter vector. When substituted into the state equation, Eq. (4.33), the control law produces the following regulated ASE system:

$$\dot{x} = (A - B\Theta^T)x.$$

The characteristic polynomial of the regulated system, $\det(sI - A + B\Theta^T)$, is chosen to have all the roots in the left-half plane at the following locations:

$$s_{1,2} = -1 \pm i, \quad s_{3,4} = -2 \pm 2i, \quad s_5 = -0.05.$$

This results in a particular value of the controller parameter vector, Θ, at each flight condition (ρ, U), for making the regulated ASE system asymptotically stable.

Unfortunately, a practical implementation of the state-feedback law is impossible in the present case, because it requires the direct measurement of all the state variables. While the angles, θ, β, and rates, $\dot{\theta}, \dot{\beta}$, can be picked up by angle encoders and rate sensors mounted on the airfoil's pitch axis and the control-surface hinge line, there is no way of measuring the aerodynamic lag state, x_a, which is not even a physical quantity. A design possibility is to use only some of the state variables as outputs, $y(t)$, and then estimate the entire state vector through an observer, which is an electrical subsystem of the controller. The derivation of the control law in that case will first require the design of an observer for estimating the state variables from a finite record of the input, $u(t)$, and the output vector, $y(t)$. The estimated states, $\hat{x}(t)$, are then electrical signals to be fed into a multi-channel amplifier (regulator) for driving the control-surface motor. Alternatively, a single output, $y(t)$, can be selected, which is already a physical combination of all the state variables. Such an output is the normal acceleration measured by an accelerometer placed at a point on the control surface. The controller is then a single-channel amplifier producing the control input, $u(t)$, by a transfer-function relationship *of the output, $u(s) = F(s)y(s)$, and can be designed by classical frequency-domain methods. Of course, the state-space and transfer-function representations are mathematically equivalent (an observer can be represented by its own transfer matrix), but the discussion here is about practical implementation.*

Let us consider the first (observer-based) approach, by selecting the output vector to be $y = (\theta, \beta)^T$, which implies the following output coefficient matrices:

$$E = \begin{pmatrix} 1 & 0 & 0 & 0 & 0 \\ 0 & 1 & 0 & 0 & 0 \end{pmatrix}, \quad D = \begin{pmatrix} 0 \\ 0 \end{pmatrix}.$$

The observer is represented by the following state equation:

$$\dot{\hat{x}} = (A - LE)\hat{x} + (B - LD)u + Ly,$$

where the observer gain matrix, L, is to be selected for placing the observer poles (i.e. eigenvalues of $A - LE$) at desired locations. The observer poles are chosen as follows to be deeper into the left-half plane than the regulator for a faster state estimation dynamics:

$$s_{1,2} = -3 \pm 3i, \quad s_{3,4} = -4 \pm 4i, \quad s_5 = -0.2.$$

If one selects a suitable reference flight condition for the controller (regulator and observer) design, it may not be necessary to change the controller gains, Θ, L, with a changing flight condition in order to have a stable closed-loop system. Suppose the operating speed range for this aircraft is $10 \le U \le 50$ m/s at standard sea level. If the design is based on $U = 50$ m/s and standard sea level, there would be no possibility of flutter occurring in the flight envelope. A similar argument applies at higher altitudes, where the flight speeds could be larger (but within the incompressible range), but the limiting dynamic pressure, $1/2\rho U^2 \le q_m$, is essentially the same as that at sea level. A plot of the closed-loop poles varying with the flight speed at standard sea level ($\rho = 1.225$ kg/m^3) is shown in Fig. 4.3 for the following fixed chosen values of Θ, L derived for sea-level, $U = 50$ m/s, condition:

$$\Theta^T = (27.1407, -0.5769, 19.0384, 6.4921, -0.6453)$$

$$L = \begin{pmatrix} 8.2562 & 1.7684 \\ -1.0813 & 6.3802 \\ 27.4451 & 5.6388 \\ -1.1005 & 11.2779 \\ 23.2311 & -4.1531 \end{pmatrix}.$$

The closed-loop system is therefore stable in the range of flight speeds $10 \le U \le 60$ m/s, which is adequate. However, if a process noise input is present, the stability can be eroded. The stability margin of the system is analysed by the largest singular value of its sensitivity matrix shown in Fig. 4.4 for the extreme condition of $U = 60$ m/s, and sea level. The gain margin is seen to be 5.44 dB with a crossover frequency of 0.23 rad/s, and high-frequency roll-off is −40 dB per decade. The simulated response in the same flight condition to an initial pitch perturbation in the presence of a normally distributed random process noise, $p(t)$, of standard deviation 0.05 units (which is quite large) in the input channel is plotted in Fig. 4.5, showing a stable system. Hence the designed stabilization system for flutter suppression is quite robust, even at a speed higher than the design range.

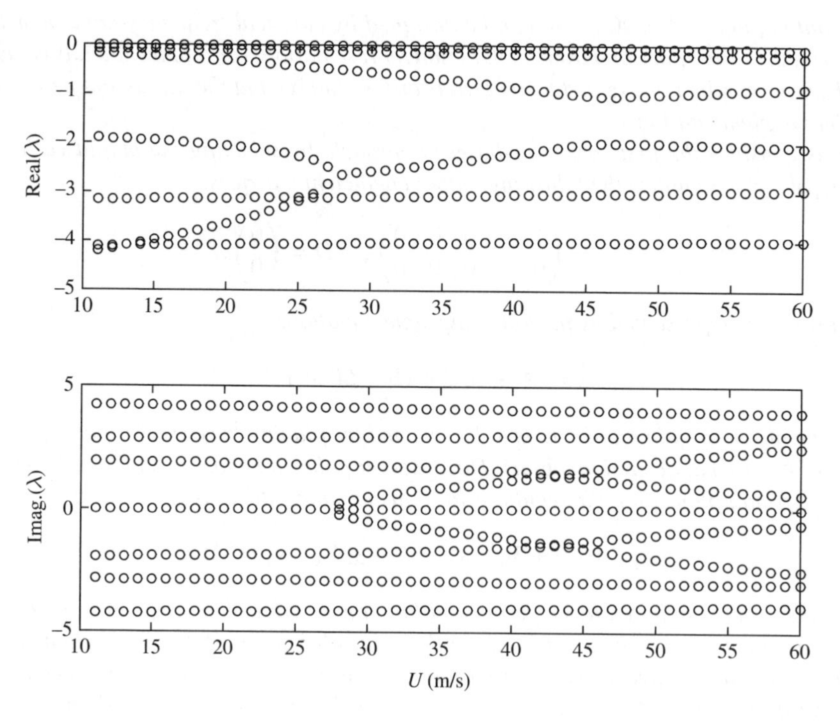

Figure 4.3 Variation of the closed-loop poles with flight speed at standard sea level

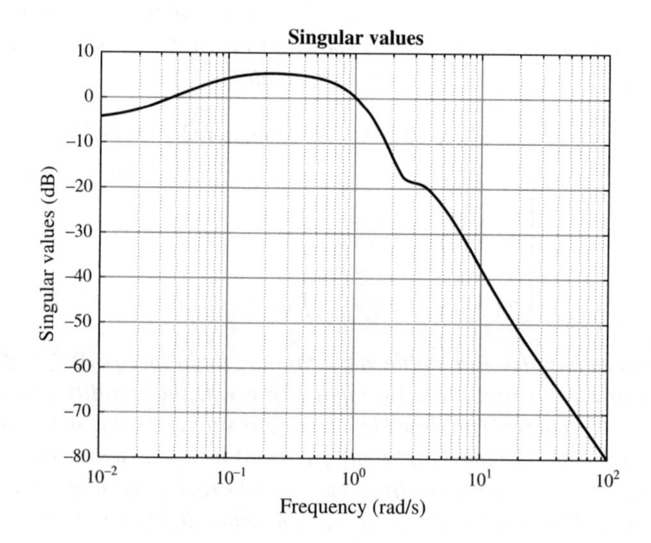

Figure 4.4 Largest singular value spectrum of the closed-loop sensitivity matrix

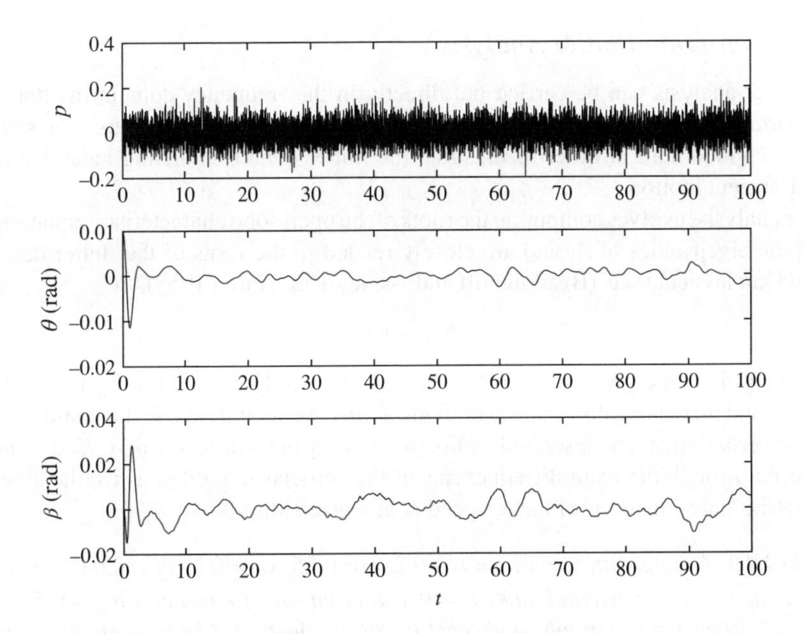

Figure 4.5 Closed-loop response for an initial pitch perturbation at $U = 60$ m/s, standard sea-level condition, in the presence of normally distributed random process noise, $p(t)$, of standard deviation 0.05 units

4.4 Active Flutter Suppression of Typical Section

Consider a typical wing section model derived above with an augmented state vector,

$$x = \begin{Bmatrix} q \\ \dot{q} \\ x_a \end{Bmatrix}, \tag{4.41}$$

and the following state equation:

$$\dot{x} = Ax + Bu, \tag{4.42}$$

where

$$A = \begin{pmatrix} 0 & I & 0 \\ -\bar{M}^{-1}\bar{K} & -\bar{M}^{-1}\bar{C} & -\bar{M}^{-1}N_a \\ \Gamma_a & & F_a \end{pmatrix} \tag{4.43}$$

and

$$B = \begin{pmatrix} 0 \\ I \\ 0 \end{pmatrix}. \tag{4.44}$$

Here $\bar{M} = M - M_a$, $\bar{C} = C - C_a$, $\bar{K} = K - K_a$ are the generalized mass, damping and stiffness matrices, respectively, of the aeroelastic system.

4.4.1 Open-Loop Flutter Analysis

While flutter analysis can be carried out directly in the frequency domain (as flutter is the simple harmonic motion at the boundary of stable and unstable conditions), the state-space method is applied here in order to illustrate the flutter condition arising naturally from the general transient motion.

Flutter analysis involves computing the roots of the open-loop characteristic equation, which are also the eigenvalues of A, and are closely related to the roots of the flutter determinant employed in classical texts (Bisplinghoff and Ashley 1962, Fung 1955),

$$\det(sI - A) = 0, \tag{4.45}$$

for a varying dynamic pressure $1/2\rho U^2$. This is practically carried out by fixing the flight altitude (ρ) and increasing the airspeed until instability is encountered. In the non-dimensional, open-loop aeroelastic plant described by Eq. (4.42), only the stiffness matrix, $\hat{K}$, depends upon the airspeed through the natural frequencies of the aeroelastic modes. Thus, the eigenvalues of the matrix A are functions of the airspeed U at a given altitude.

Example 4.4.1 *A wing with $b = 1$ m, $a = -0.2$, $c = 0.35$, $x_\theta = 0.2$, $x_\beta = 0.04$, $r_\theta^2 = 0.25$ and $r_\beta^2 = 0.02$, has non-dimensional mass $\kappa = 0.1$, and natural frequencies $\omega_h = 0.5$, $\omega_\theta = 1.5$ and $\omega_\beta = 2$ when flying straight and level at an airspeed of 30 m/s at standard sea level ($\rho = 1.225$ kg/m^3). It is required to carry out the open-loop flutter analysis of the wing.*

Consider the following rational function approximation (RFA) of the Theodorsen function (Eversman and Tewari 1991):

$$C(s) = 0.9962 - \frac{0.1667s}{s + 0.0553} - \frac{0.3119s}{s + 0.2861}.$$

This nonlinear-optimized approximation (the second entry on the right-hand side of Table 1 of (Eversman and Tewari 1991)) results in a net curve-fit error of only 0.000561 over a range of reduced frequencies $0 \leq k \leq 1$. Figure 4.6 compares the exact $C(ik)$ with the selected rational approximation for $s = ik$, $0 \leq k \leq 1$. Clearly, in the given range of frequencies (where flutter is expected to occur), the approximation of the Theodorsen function is quite good.

From the given data, the in vacuo, dimensional natural frequencies are computed to be the following:

$$\sqrt{\frac{k_h}{m}} = \omega_h \frac{U}{b} = 15 \text{ rad/s},$$

$$\sqrt{\frac{k_\theta}{I_\theta}} = \omega_\theta \frac{U}{b} = 45 \text{ rad/s},$$

$$\sqrt{\frac{k_\beta}{I_\beta}} = \omega_\beta \frac{U}{b} = 60 \text{ rad/s}.$$

Since these frequencies are far apart, there is little possibility of modal interaction in vacuum. However, unsteady aerodynamic interaction among the modes is possible, leading to the flutter condition.

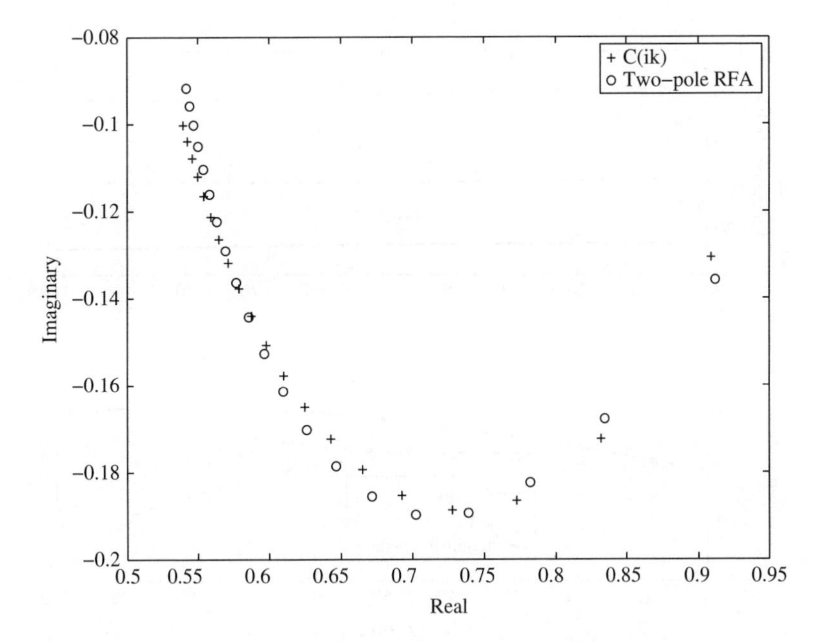

Figure 4.6 Two-pole rational function approximation of the Theodorsen function

Figure 4.7 is a plot of the natural frequencies and damping ratios versus airspeed corresponding to the plunge (h/b), pitch (θ), and control-surface (β) modes. All the natural frequencies decrease slightly with increasing U. While the natural frequencies of the plunge and pitch are only changed slightly from their in vacuo values, the change in the natural frequency of the control-surface mode due to aeroelasticity is much greater (an increase of about 33%). The damping ratios of the pitch and plunge modes are seen to decrease in slope with increase in airspeed (Fig. 4.7), while the control-surface mode – with the largest natural frequency – has the opposite trend. The pitch and plunge modes are predominant because of their relative proximity to the imaginary axis, when compared with the control-surface mode. The pitch mode is seen to become unstable at $U = 63.35$ *m/s with a corresponding flutter frequency of 41.18 rad/s. Such an interaction between the plunge and pitch modes is primarily responsible for the classical bending-torsion flutter condition, which is a characteristic of high-aspect-ratio wings.*

In order to complete the flutter analysis, the effect of atmospheric density on the aeroelastic modes is examined. It is expected that the flutter speed would increase with altitude due to a decreasing value of the non-dimensional mass κ. Consider the wing flying at a constant airspeed of 70 m/s at various altitudes in the density range of $0.255 \le \rho \le 1.225$ kg$/m^3$. *This is a supercritical speed at standard sea level, hence the flutter boundary would occur at a larger altitude. Figure 4.8 shows the variation of natural frequencies and damping ratios of the aeroelastic modes with density. As the atmospheric density is decreased from* 0.255 kg$/m^3$, *the flutter condition is first reached at the altitude where the density is* 0.7947 kg$/m^3$. *This corresponds to a standard altitude of 4283.3 m. Below this critical altitude, the aeroelastic modes are unstable for* $U = 70$ *m/s, while at higher altitudes they are stable. As investigated above, the instability is due to the pitch mode and corresponds to a critical frequency of 47.535 rad/s for the given airspeed and altitude.*

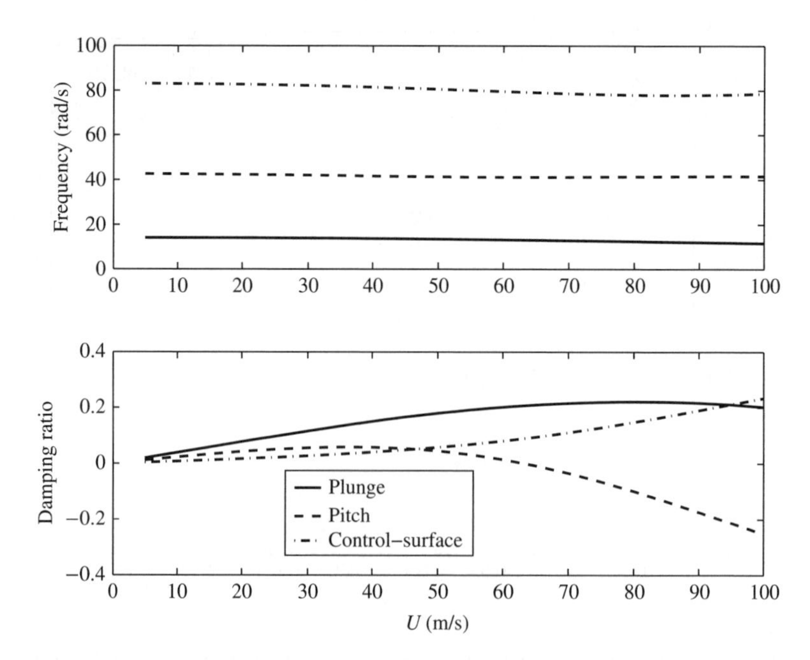

Figure 4.7 Variation of natural frequencies and damping ratios with airspeed at standard sea level, indicating flutter condition

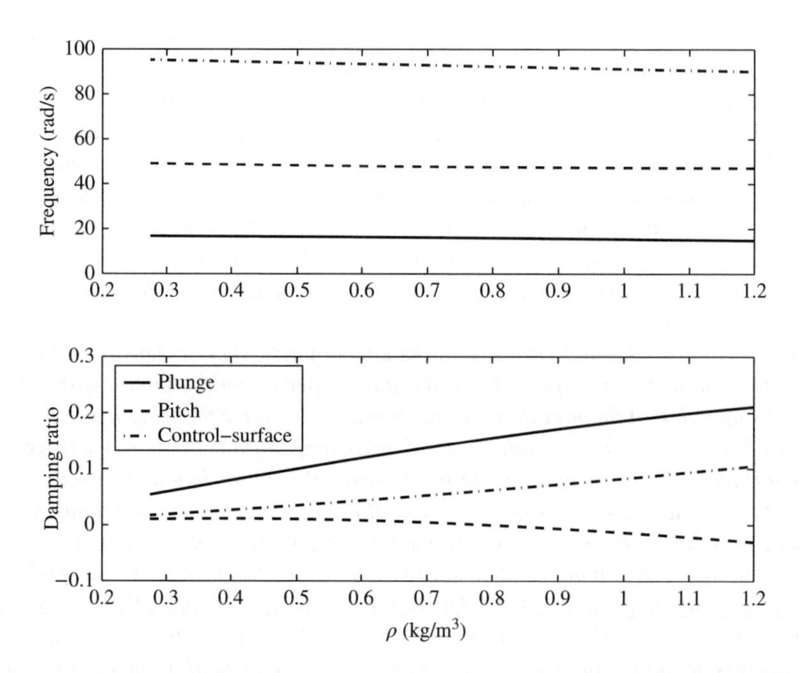

Figure 4.8 Variation of natural frequencies and damping ratios with atmospheric density for a constant airspeed of 70 m/s

4.5 Linear Feedback Stabilization

The ideal aerodynamics plant derived in the previous section is linear, time-invariant and unconditionally controllable. Hence, a linear feedback controller can be designed in the ideal case using the traditional methods for a given flight condition ρ, U. For the LTI plant given by

$$\dot{x} = Ax + Bu, \tag{4.46}$$

a linear controller is based on a state feedback regulator with the following control law:

$$u = -Kx, \tag{4.47}$$

where K is the regulator gain matrix considered to be constant in the nominal case. The task of the regulator is to stabilize the closed-loop system by feeding back the state vector of the plant. Two alternative design methods can be applied for linear, state-feedback control: pole placement and linear optimal control.

4.5.1 Pole-Placement Regulator Design

Design of a linear, state feedback regulator for the LTI plant of Eq. (4.46) with the control law Eq. (4.47) can be carried out by assigning a structure for the eigenvalues and eigenvectors of the closed-loop dynamics matrix, $\mathbf{A} - BK$. In case of single-input plants, this process reduces to selecting locations for the closed-loop poles (called the pole-placement method) by following Ackermann's formula, which yields the desired closed-loop characteristics (Chapter 2):

$$K = (a_d - a)(PP')^{-1}, \tag{4.48}$$

where a is the row vector formed by the coefficients, a_i, of the plant's characteristic polynomial in *descending order* $[a = (a_n, a_{n-1}, \cdots, a_2, a_1)]$:

$$\det(sI - A) = s^n + a_n s^{n-1} + a_{n-1} s^{n-2} + \cdots + a_2 s + a_1 \quad \text{and} \tag{4.49}$$

a_d is the row vector formed by the characteristic coefficients of the closed-loop system in descending order $[a_d = (a_{dn}, a_{d(n-1)}, \cdots, a_{d2}, a_{d1})]$:

$$\det(sI - A + BK) = s^n + a_{dn} s^{n-1} + a_{d(n-1)} s^{n-2} + \cdots + a_{d2} s + a_{d1}. \tag{4.50}$$

Here it may be recalled from Chapter 2 that P is the controllability test matrix of the plant, and P' is the following upper triangular matrix:

$$P' = \begin{pmatrix} 1 & a_n & a_{n-1} & \cdots & a_3 & a_2 \\ 0 & 1 & a_n & \cdots & a_4 & a_3 \\ 0 & 0 & 1 & \cdots & a_5 & a_4 \\ \cdots & \cdots & \cdots & \cdots & \cdots & \cdots \\ 0 & 0 & 0 & \cdots & 1 & a_n \\ 0 & 0 & 0 & \cdots & 0 & 1 \end{pmatrix}. \tag{4.51}$$

The matrix inversion involved in Ackermann's formula requires that the plant must be controllable, $\det(P) \neq 0$.

Example 4.5.1 *For the typical wing section considered in Example 4.4.1, consider a pole-placement regulator for keeping the closed-loop ASE poles at constant stable locations of*

$$s_{1,2} = -0.1 \pm 0.1i, \quad s_{3,4} = -0.2 \pm 0.2i, \quad s_{5,6} = -0.3 \pm 0.3i,$$

while having the two aerodynamic poles unchanged at $s_7 = -0.0553, s_8 = -0.2861$. *The resulting plots of the closed-loop natural frequencies are shown in Fig. 4.9, where all frequencies are seen to increase nearly linearly with the flight speed. The required feedback gains,* $K = (K_1, \cdots, K_8)$*, are plotted in Fig. 4.10 against the flight speed for the supercritical range (above the open-loop flutter velocity). While* K_1, K_6, K_8 *remain small throughout the speed range considered here, the other gains vary significantly, but remain bounded in magnitude. The variation of the natural frequencies and regulator gains with the atmospheric density is evident in Figs. 4.11 and 4.12. Such behaviour of the regulator gains indicates that it is possible to apply a gain-scheduling system for adjusting the controller parameters with the flight speed and atmospheric density, while maintaining a constant stability margin.*

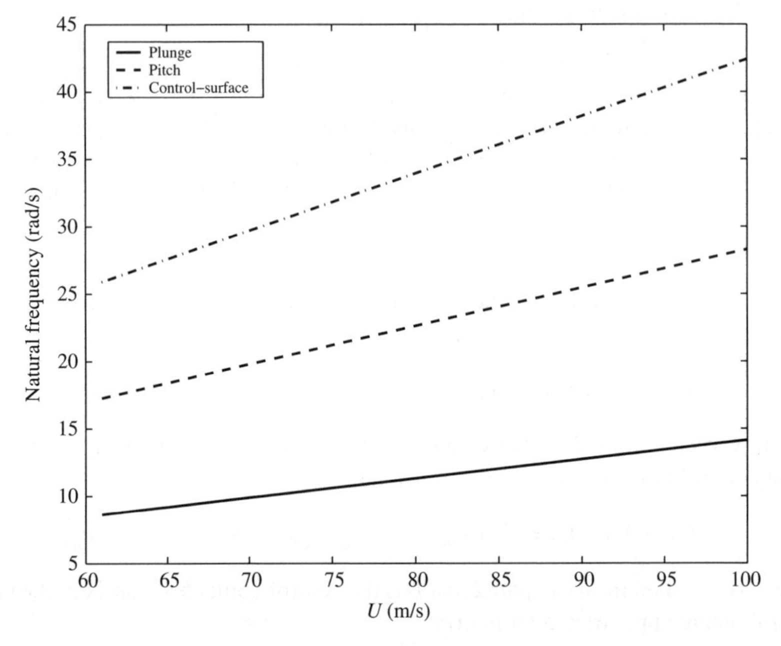

Figure 4.9 Variation of the closed-loop natural frequencies with airspeed at standard sea level for the supercritical (above open-loop flutter velocity) condition

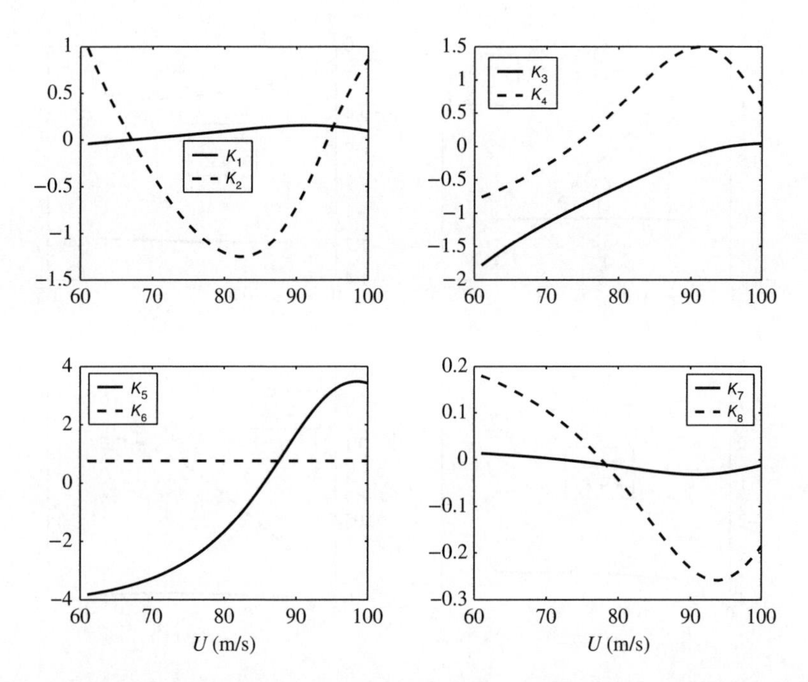

Figure 4.10 Variation of the regulator gains with airspeed at standard sea level for the supercritical (above open-loop flutter velocity) condition

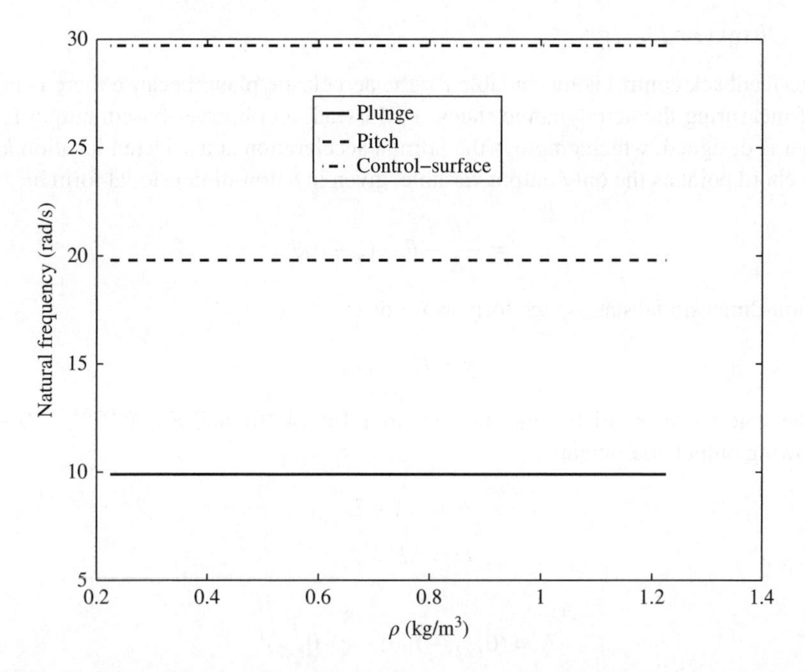

Figure 4.11 Variation of the closed-loop natural frequencies with atmospheric density at the supercritical airspeed of 70 m/s

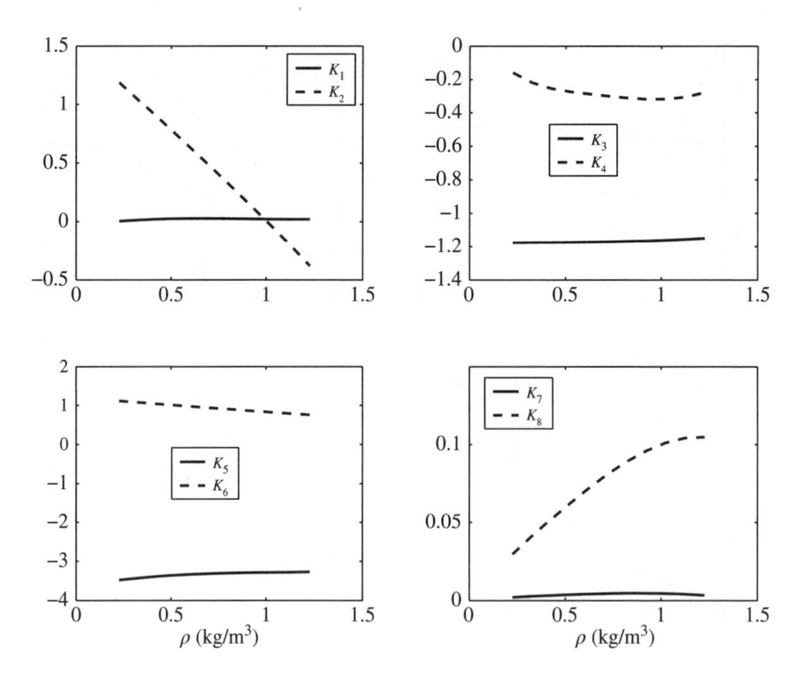

Figure 4.12 Variation of the regulator gains with atmospheric density at the supercritical airspeed of 70 m/s

4.5.2 *Observer Design*

The state-feedback control is unavailable for the aeroelastic plant, because there is no possibility of measuring the aerodynamic states, x_a. Instead, an observer-based, output feedback controller is designed, which employs the normal acceleration at a selected location $b\xi$ aft of the mid-chord point as the only output variable, given in a non-dimensional form by

$$y = -\frac{\ddot{h}}{b} - \dot{\theta} - (\xi - a)\ddot{\theta}, \tag{4.52}$$

or in a non-dimensional state-space form as follows:

$$y = Ex + Du, \tag{4.53}$$

where the state vector x satisfies the state equation Eq. (4.46), and $E \in \mathbb{R}^{1\times(6+\ell)}$, $D \in \mathbb{R}$ are the following output coefficients:

$$E = \Lambda A + \Xi,$$
$$D = \Lambda B, \tag{4.54}$$

where

$$\Lambda = (0_{1\times3}, -1, (a - \xi), 0_{1\times2})^T \tag{4.55}$$

and

$$\Xi = (0_{1\times4}, -1, 0_{1\times2})^T. \tag{4.56}$$

The observability of the plant in the open-loop case can be verified by the rank of the following observability test matrix (Chapter 2):

$$N = (E^T,\ A^T E^T,\ (A^T)^2 E^T,\ (A^T)^3 E^T,\ (A^T)^4 E^T,\ (A^T)^5 E^T,\ (A^T)^6 E^T). \tag{4.57}$$

Since E is a row vector, N is a square matrix. The determinant $\mid N \mid$ is non-zero provided $\xi \neq a$. Thus, the plant is observable with the normal acceleration output measured at a point not collocated with the pitch axis. The output equation Eq. (4.53) is used in the design of a full-order observer with the following state equation:

$$\dot{\hat{x}} = (A - LE)\hat{x} + (B - LD)u + Ly, \tag{4.58}$$

where $\hat{x}(t)$ is the estimated state and L the observer gain matrix, provided the plant, (A, E), is observable. The observer gain matrix, L, is selected by pole placement in a manner similar to the regulator gain, K, for the observer dynamics matrix, $A - LE$, where A is replaced by A^T, and B by E^T.

Example 4.5.2 *For the typical wing section considered in Example 4.5.1, let us design a full-order observer for placing the poles of the observer dynamics matrix, A – LE, at the following Hurwitz locations:*

$$s_{1,2} = -0.5 \pm 0.5i,\ \ s_{3,4} = -0.8 \pm 0.8i,\ \ s_{5,6} = -1.0 \pm 1.0i,\ \ s_7 = -0.1,\ \ s_8 = -0.3,$$

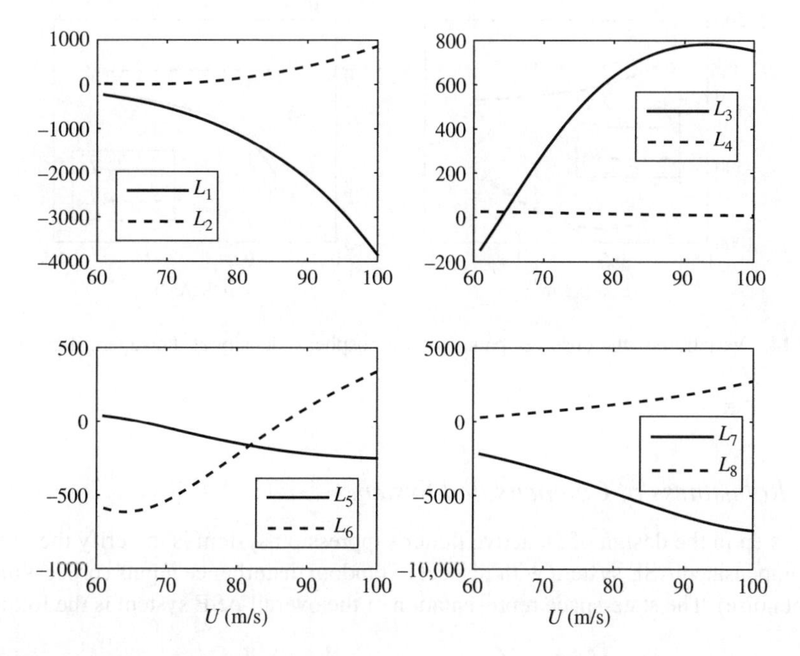

Figure 4.13 Variation of the observer gains with airspeed at standard sea level for the supercritical (above open-loop flutter velocity) condition

These poles are selected to be deeper in the left-half plane than those of the state-feedback regulator, as the estimated state must settle much faster than the plant state. However, too deep a location must be avoided owing to noise feedback considerations (discussed in Chapter 2). The measured output, $y = Ex + Du$, is the normal acceleration sensed at a location 1.0% semi-chord forward of the elastic axis, with

$$E = (0, \quad 0, \quad 0, \quad 0.1750, \quad -0.0098, \quad 0, \quad 0, \quad 0), \quad D = 0.$$

Thus the aeroelastic plant is strictly proper. The observer gain matrix, L, is easily derived by comparing the characteristic polynomials, $\det(sI - A)$ *and* $\det[sI - (A - LE)]$*, and is plotted in Figs. 4.13 and 4.14 against flight speed, U, and atmospheric density, ρ, respectively, for the supercritical range. Since the variation of the gains is smooth, a gain-scheduling system can be devised for the observer as well.*

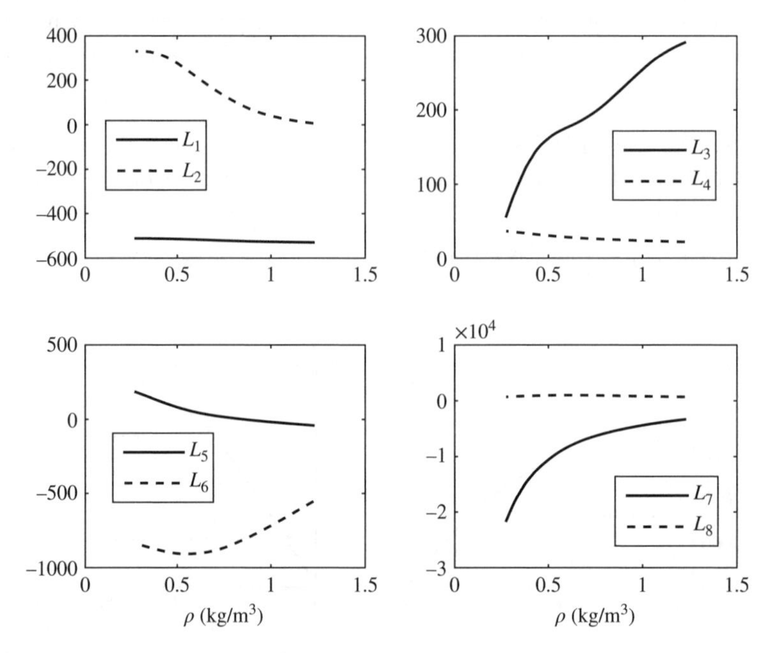

Figure 4.14 Variation of the observer gains with atmospheric density at the supercritical airspeed of 70 m/s

4.5.3 Robustness of Compensated System

The final step in the design of an active flutter suppression system is to verify the robustness of the compensated ASE system with respect to random disturbance inputs (process and measurement noise). The state-space representation of the overall ASE system is the following:

$$\begin{Bmatrix} \dot{x} \\ \dot{\hat{x}} \end{Bmatrix} = \begin{pmatrix} A & -BK \\ LE & A - BK - LE \end{pmatrix} \begin{Bmatrix} x \\ \hat{x} \end{Bmatrix} \tag{4.59}$$

$$y = Ex - DK\hat{x} = (E, -DK)\begin{Bmatrix} x \\ \hat{x} \end{Bmatrix}. \tag{4.60}$$

The Bode plot of the closed-loop ASE system for $U = 70$ m/s and standard sea level is shown in Fig. 4.15, indicating an infinite phase margin and a gain margin of 9.9 dB with a crossover

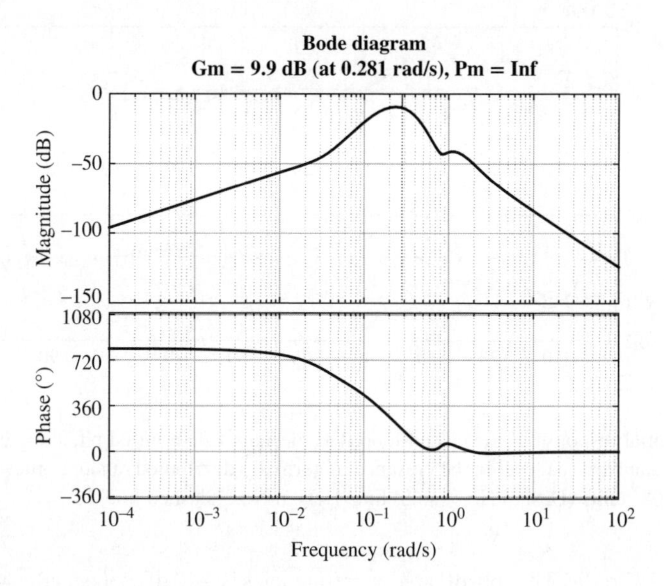

Figure 4.15 Bode plot of the closed-loop ASE system for $U = 70$ m/s and standard sea level with observer Design No.1

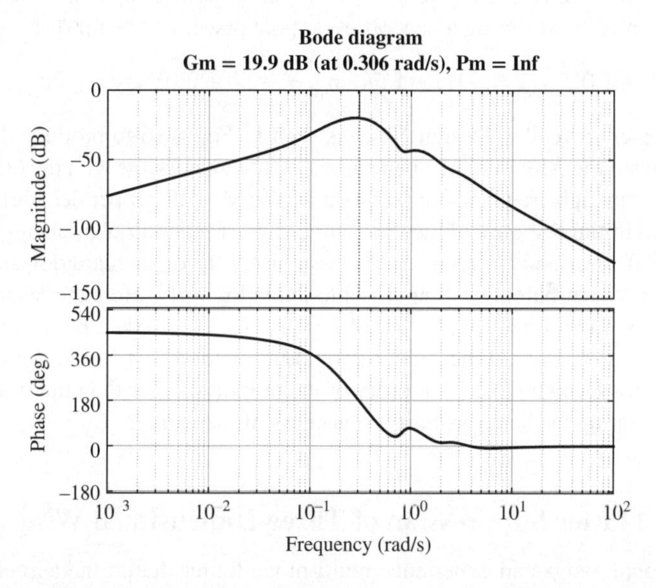

Figure 4.16 Bode plot of the closed-loop ASE system for $U = 70$ m/s and standard sea level with observer Design No.2 showing an increased stability robustness with respect to disturbance inputs

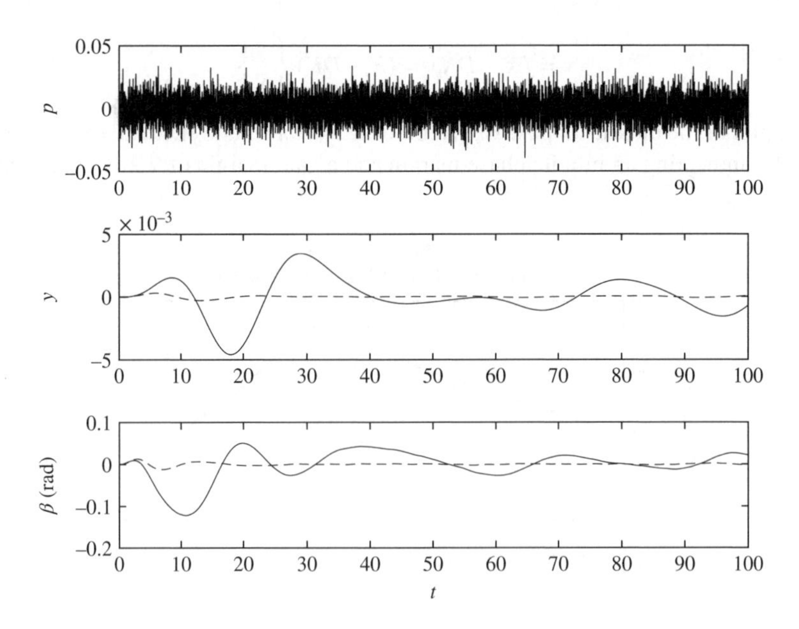

Figure 4.17 Initial response of the closed-loop ASE designs for an initial pitch angle perturbation at $U = 70$ m/s and standard sea level in the presence of normally distributed random process noise of standard deviation 10^{-2} units (Design No.1: solid line; Design No.2: dashed line)

frequency of 0.281 rad/s. The roll-off at high frequencies is −40 dB per decade, which indicates a good noise rejection property. Hence, it appears to be a sufficiently robust design, although the gain margin can be increased by moving the observer poles deeper into the left-half plane. This is demonstrated by selecting a new observer pole placement as follows:

$$s_{1,2} = -1.0 \pm 1.0i, \quad s_{3,4} = -1.5 \pm 1.5i, \quad s_{5,6} = -2.0 \pm 2.0i, \quad s_7 = -0.5, \quad s_8 = -0.8,$$

The changed observer (called Design No.2) is seen in Fig. 4.16 to produce double the gain margin (19.9 dB) of the ASE system when compared to that possible with the original observer (Design No.1). The high-frequency roll-off is unchanged (−40 dB per decade).

The simulated initial response of the two ASE designs for an initial pitch angle perturbation in the presence of a normally distributed process noise, $p(t)$, of standard deviation 10^{-2} units, representing a severe turbulent gust input, is plotted in Fig. 4.17. Here the response of Design No.1 is shown by a solid line, while that of Design No.2 is marked by a dashed line. Note that the response of the non-dimensional normal acceleration, $y(t)$, and the control-surface deflection, $\beta(t)$, for Design No.2 is an order of magnitude smaller than that for Design No.1, thereby demonstrating the better stability robustness of the former.

4.6 Active Flutter Suppression of Three-Dimensional Wings

Three-dimensional wings can experience multi-mode flutter due to the aeroelastic coupling between several structural modes and unsteady aerodynamic loads. The modelling of such structures is based on the finite-element approach and the unsteady aerodynamic behaviour is estimated by linearized, frequency-domain lifting surface theories extended to transient motion by RFAs. Chapter 3 briefly highlights these steps, and for greater details about them, the reader is referred to the companion book on aeroservoelasticity (Tewari 2015).

Assume that a finite-element (or assumed modes)-based structural analysis has been carried out, and the generalized mass, stiffness and damping matrices of the structure are available with respect to the generalized coordinates, $q(t)$. Furthermore, let us also assume that an RFA of the unsteady aerodynamic transfer matrix is also available, whose non-circulatory terms are clubbed with the corresponding structural coefficients, resulting in the overall generalized mass, $\bar{M}$, stiffness, $\bar{K}$, and damping, $\bar{C}$, matrices. Then the aeroelastic system is represented by the following state equations (Chapter 3):

$$\dot{x} = Ax + BQ_c, \tag{4.61}$$

where

$$x = (q^T, \dot{q}^T, x_a^T, x_g^T)^T$$

is the state vector, $x_a(t)$ is the aerodynamic state vector representing wake-induced circulatory dynamics, $x_g(t)$ the gust state vector and

$$A = \begin{pmatrix} 0 & I & 0 & 0 \\ -\bar{M}^{-1}\bar{K} & -\bar{M}^{-1}\bar{C} & \bar{M}^{-1}N_a & {}^{-1}N_g \\ & \Gamma_a & A_a & 0 \\ & \Gamma_g & 0 & A_g \end{pmatrix}, \quad B = \begin{pmatrix} 0 \\ \bar{M}^{-1}I \\ 0 \end{pmatrix}, \tag{4.62}$$

with 0 and I being the null and identity matrices, respectively, and the other coefficient matrices having their usual definitions given in Chapter 3. The output equation is expressed as follows:

$$y = Ex + DQ_c. \tag{4.63}$$

Here y is the output vector (usually the normal accelerations picked up by sensors at selected locations). Since the generalized coordinates, $q(t)$, do not include the control-surface deflections, these latter must be handled separately. The state equations of m control-surface actuators with the deflections vector, $\delta = (\delta_1, \delta_2, \cdots, \delta_m)^T$, and control-torques input vector, $\vec{u} = (u_1, u_2, \cdots, u_m)^T$, are added as follows:

$$\dot{x}_c = A_c x_c + B_c u, \tag{4.64}$$

where

$$x_c = \left(\delta, \dot{\delta}, \xi_c^T\right)^T$$

is the actuator state vector with $\xi(t)$ being the aerodynamic state vector of the actuator subsystem and

$$Q_c = C_c x_c + D_c u$$

is the generalized unsteady aerodynamics force vector produced by control-surface motion. Thus the actuators system and the unforced aeroelastic system are two subsystems of the open-loop ASE plant, with the overall state vector given by

$$X = \left(x^T, x_c^T\right)^T$$

and the overall open-loop ASE state equation is the following:

$$\dot{X} = \bar{A}X + \bar{B}u, \tag{4.65}$$

with the output equation,

$$y = \bar{E}X + \bar{D}u, \tag{4.66}$$

and

$$\bar{A} = \begin{pmatrix} A & BC_c \\ 0 & A_c \end{pmatrix}, \; \bar{B} = \begin{pmatrix} BD_c \\ B_c \end{pmatrix}, \tag{4.67}$$

$$\bar{E} = \begin{pmatrix} E & DC_c \end{pmatrix}, \; \bar{D} = DD_c. \tag{4.68}$$

The active flutter suppression problem is to derive a linear feedback control law,

$$u = -K\hat{X}, \tag{4.69}$$

such that the closed-loop ASE system is asymptotically stable. Here K denotes the regulator gain matrix to be selected such that the matrix $A - BK$ has its eigenvalues at the desired locations, and $\hat{X}$ is the estimated state vector generated by a linear observer with the following state equation:

$$\dot{\hat{X}} = (\bar{A} - L\bar{E})\hat{X} + (\bar{B} - L\bar{D})\,u + Ly, \tag{4.70}$$

based on the measured output vector, y. Alternatively, a reduced-order observer can be designed for a reduction in the overall order of the ASE system. The regulator and the observer are designed separately by the separation principle (Chapter 2) and then combined to yield the closed-loop ASE system. However, different control laws for the servo-actuator and the aeroelastic subsystems are not being considered here. Instead, the regulator and observer gains of the actuator and aeroelastic subsystems can be extracted from the respective gain matrices for a practical implementation. Since the flutter suppression problem is a regulator problem, the desired states (hence the outputs) are taken to be zeros. Hence, the state equations of the regulated ASE system are the following:

$$\begin{Bmatrix} \dot{X} \\ \dot{\hat{X}} \end{Bmatrix} = \begin{pmatrix} \bar{A} & -\bar{B}K \\ L\bar{E} & \bar{A} - \bar{B}K - L\bar{E} \end{pmatrix} \begin{Bmatrix} X \\ \hat{X} \end{Bmatrix}. \tag{4.71}$$

Example 4.6.1 *Consider a modification of the DAST-ARW1 wing, which was especially designed by NASA-Langley for conducting aeroelastic flight tests in a drone aircraft. Detailed aeroelastic characteristics and the open-loop flutter analysis of the modified DAST-ARW1 wing are based on experimental data and can be found in the companion reference (Tewari 2015). The relevant characteristics of the modified DAST-ARW1 wing are briefly listed in Appendix C of the present book. An open-loop flutter condition for this wing is seen at a speed of 284.7 m/s, corresponding to the Mach number 0.9192 at standard altitude 7.6 km. The wing is equipped with a trailing-edge control surface driven by a fourth-order actuator with scalar torque input, u. An outboard accelerometer provides the single output y, contributed by up to six aeroelastic modes. The resulting aeroelastic subsystem (based on two aerodynamic lag parameters) is of order 24, hence the overall ASE plant is of order 28.*

A regulator and observer designed by the LQG/LTR procedure are added to this plant, in order to yield the best combination of maximum control-torque magnitude and robustness.

The selected design parameters (see Chapter 2) in terms of Gaussian white, process and measurement noise, p(t) and m(t), with a zero mean (ZMGWN), are as follows:

$$S_{pm} = 0,\ S_p = 10^{-12}\bar{B}\bar{B}^T,\ S_m = 1,\ F = I.$$

These imply an equal process noise perturbation in all the channels of the plant state, and no correlation between the process and measurement noise. The LQR parameters are taken to be the following:

$$S = 0,\ Q = 5000\bar{C}^T\bar{C},\ R = 1.$$

The resulting gain matrices of the Kalman filter and the regulator are listed as follows:

$$\begin{aligned} K = 10^6\big(&3.9766, && -70.627, && -16.821, && -82.738,\\ &-126.08, && -90.589, && -4240.5, && -4121,\\ &-2864.1, && -6318, && 2117.6, && 1204.6,\\ &-0.00985, && -0.05264, && -0.09416, && -0.1139,\\ &-0.1487, && -0.3899, && 791,250, && 4,071,400,\\ &7,801,200, && 11,572,000, && 17,344,000, && 30,728,000,\\ &17,930, && -6.1962, && -3148.3, && 494.87\big) \end{aligned}$$

$$\begin{aligned} L^T = 10^{-5}\big(&-9.6977, && 44.932, && -57.742, && -33.756,\\ &3.2843, && -80.993, && 0.0272, && -0.0298,\\ &-0.2454, && 0.1233, && -0.03616, && -0.102,\\ &-0.0123, && 0.006, && -0.246, && 0.0774,\\ &-0.0738, && -0.102, && -0.00002, && 0.00011,\\ &-0.00014, && -0.00008, && 0.000008, && -0.000195,\\ &0.0268, && 0.57031, && 0.0268, && 0.000001\big) \end{aligned}$$

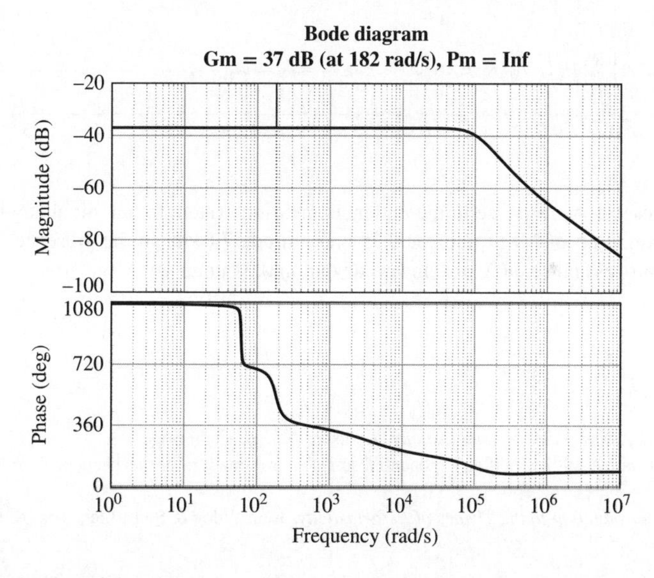

Figure 4.18 Bode plot of the closed-loop flutter suppression system at the supercritical condition of Mach number 0.95 and altitude 7.6 km

A Bode plot of the closed-loop ASE system is shown in Fig. 4.18, indicating a gain margin of 37 dB and an infinite phase margin. Therefore, the design is quite robust with respect to the process noise in the ASE plant. Good robustness to high-frequency measurement noise is also evident in the gain roll-off of 20 dB per decade. The closed-loop response to an initial tip displacement in the presence of normally distributed process noise, p(t), of standard deviation 0.01 in the input channel is shown in Fig. 4.19 for a supercritical condition of Mach number 0.95 at altitude 7.6 km. The noise is a realistic representation of atmospheric turbulence of high-to-severe intensity. Note that the non-dimensional acceleration response settles in less than 1 s at the given supercritical condition, while the control-surface deflection is limited to ± 0.002 *rad., but its oscillation persists for a longer time (about 3 s), indicating increased control activity due to the noise. The system's robustness is therefore demonstrated. We shall return to this example in a later chapter when carrying out adaptive control design.*

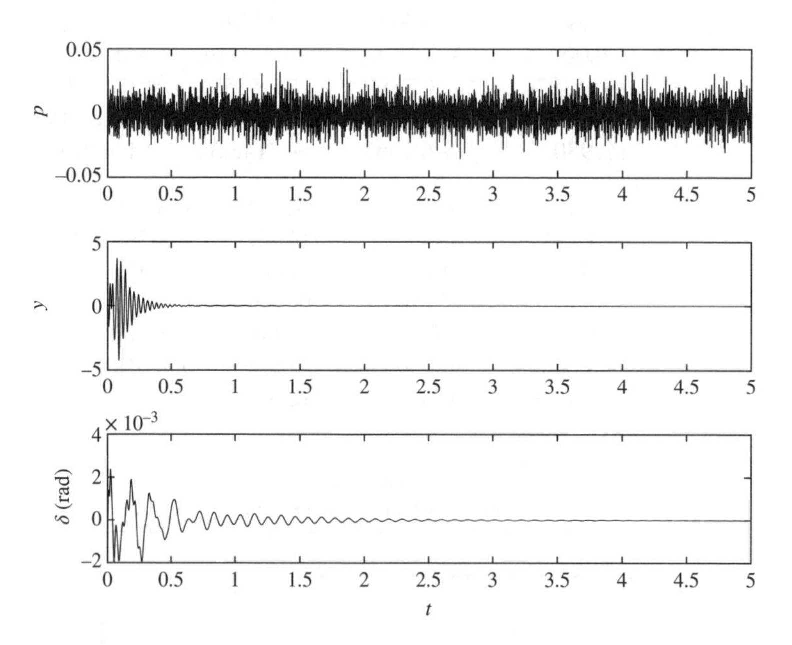

Figure 4.19 Closed-loop response of the flutter suppression system for an initial unit tip displacement at supercritical condition of Mach number 0.95 and altitude 7.6 km, in the presence of normally distributed random process noise, $p(t)$, of standard deviation 0.01 units

References

Bisplinghoff RL and Ashley H 1962 *Aeroelasticity*. Addison-Wesley, Cambridge.

Eversman W and Tewari A 1991 Modified exponential series approximation for the Theodorsen function. *J. Aircr.* **28**, 553–557.

Fung YC 1955 *An Introduction to the Theory of Aeroelasticity*. John Wiley & Sons, Inc., New York.

Mukhopadhyay V 2003 Historical perspective on analysis and control of aeroelastic responses. *J. Guid. Control Dyn.* **26**, 673–684.

Runyan HL 1952 Single-degree-of-freedom flutter calculations for a wing in subsonic potential flow and a comparison with an experiment. *NACA* **Rept. 1089**.

Tewari A 2015 *Aeroservoelasticity: Modeling and Control.* Birkhäuser, Boston.

5

Self-Tuning Regulation

5.1 Introduction

Self-tuning regulators (STRs) offer an attractive adaptive control option for uncertain aeroservoelastic (ASE) systems, because they are based upon online identification of the actual plant characteristics that can change in a theoretically unpredictable manner with flight conditions (speed, altitude and loading). An STR estimates the plant's uncertain parameters from a record of its input–output relationship, which is continuously updated in time. On the basis of the identified plant parameters, the underlying control design problem is solved at every time instant, resulting in an adaptation (or continuous update) of the controller parameters. Owing to the separate online identification and controller design modules, the self-tuning approach is considered an indirect (or modular) adaptation scheme. While our focus in this book is on the continuous-time (rather than discrete-time) adaptive control systems (because its nonlinear behaviour is best analysed in continuous time), a recursive identification algorithm is actually implemented in discrete time. By taking a sufficiently small sampling interval, the digital controller implementation can be approximated to be time continuous. For linear systems, an equivalence exists between continuous-time and discrete-time systems, and hence a direct conversion is possible from the one to the other. Since most system identification techniques are linear (an example of which is the least-squares rational function approximation of Chapter 3), they can be alternatively expressed in either discrete- or continuous-time format.

For the controller design module, both pole-placement (eigenstructure assignment) and robust optimal control (LQG/LTR, H_∞, μ-synthesis) methods are possible options. The latter have the advantage of resulting in a robust control system in the presence of process and measurement noise. When the plant parameters are known with some confidence, these disturbance inputs are of small magnitude and a linear controller can be designed to guarantee system stability in their presence. However, for larger perturbations caused by modelling errors, the stability margin offered by linear feedback control is exceeded. Then the controller parameters (observer and regulator gains) must be revised at every time instant by iteratively solving the algebraic Riccati equations with the identified parameters. Such an iterative scheme – although quite systematic – can have problems of convergence. It may also require inordinately large computational resources for an online implementation, due to the large size of the ASE plant.

Adaptive Aeroservoelastic Control, First Edition. Ashish Tewari.

Hence, simpler (possibly non-iterative) algorithms can prove beneficial in a STR design for an ASE system.

The STR has an additional (and often more serious) difficulty of converging to spurious (or unrealistic) conditions. Since the STR method works without an external reference signal, there is no way for the system to 'know' a priori if a stable set-point is also a physical equilibrium of the process. In some situations with large parametric perturbations, it is possible to have the STR driving the plant to unrealistic equilibria. This could be considered a failure of the scheme, and in ASE applications, such an unforeseen failure is often disastrous.

5.2 Online Plant Identification

Consider a linear plant with p unknown plant parameters vectors, $\theta \in \mathbb{R}^p$, which are assumed to be linearly related to the plant's output vector, $y(t) : \mathbb{R} \to \mathbb{R}^n$ by the following dynamic equation:

$$y(t) = \Phi(t)\theta, \tag{5.1}$$

where $\Phi(t) : \mathbb{R} \to \mathbb{R}^{n \times p}$ is the matrix representing the plant dynamics with known applied inputs. The problem of identification is to determine an estimate of the p unknown parameters, θ, from a measurement of $y(t)$ at various times. Since $n \neq p$, a unique estimate cannot be obtained at any given time. Thus an online identification algorithm must be devised for continuously updating the estimate θ, as new values of $y(t)$ become available for a given $\Phi(t)$. Equation (5.1) is called a regression model and the matrix $\Phi(t)$ is termed the regressor matrix of the model. Since there is rarely a system that can be exactly modelled by Eq. (5.1), there will always be an estimation error (called the residual), given by

$$e(t) = y(t) - \Phi(t)\theta, \tag{5.2}$$

which must be reduced to small values in a successful identification process.

5.2.1 *Least-Squares Parameter Estimation*

The earliest and the most popular systematic parameter estimation procedure is the least-squares method of Gauss, who successfully applied it for the orbital determination of the asteroid *Ceres* in 1801. The method is based on minimizing the sum of the squares of the residuals, $e(t)$, evaluated at different time instants. In a continuous-time formulation, the least-squares method consists of minimizing the following cost function with respect to θ, expressed as the time-integral of the square (quadratic form) of the residual:

$$\begin{aligned} V(\theta, t) &= \frac{1}{2}\int_0^t e^T(\tau)e(\tau)\mathrm{d}\tau \\ &= \frac{1}{2}\int_0^t [y(\tau) - \Phi(\tau)\theta]^T[y(\tau) - \Phi(\tau)\theta]\mathrm{d}\tau. \end{aligned} \tag{5.3}$$

The estimate $\hat{\theta}$ is the value of θ that minimizes the quadratic cost function at a given time t.

Theorem 5.2.1 *The parameter estimate $\hat{\theta}$ satisfying Eq. (5.1) minimizes the cost function of Eq. (5.3) with respect to θ at a given time t, only if the following identity is satisfied:*

$$\left[\int_0^t \Phi^T(\tau)\Phi(\tau)\mathrm{d}\tau\right]\hat{\theta} = \int_0^t \Phi^T(\tau)y(\tau)\mathrm{d}\tau. \tag{5.4}$$

Proof. The proof is obtained from the following necessary condition of $V(\theta, t)$ having a minimum value $V(\hat{\theta}, t)$:

$$\left.\frac{\partial V}{\partial \theta}\right|_{\theta=\hat{\theta}} = 0. \tag{5.5}$$

Equation (5.4) can be utilized for deriving a recursive estimate $\hat{\theta}$ as follows. If we define an estimator gain (or the estimation covariance) matrix, $P(t)$, such that

$$P^{-1}(t) = \int_0^t \Phi^T(\tau)\Phi(\tau)\mathrm{d}\tau, \tag{5.6}$$

then the following differential equation is satisfied:

$$\frac{\mathrm{d}}{\mathrm{d}t}P^{-1}(t) = \Phi^T(t)\Phi(t), \tag{5.7}$$

which is substituted in the following identity:

$$\frac{\mathrm{d}}{\mathrm{d}t}PP^{-1} = \dot{P}P^{-1} + P\frac{\mathrm{d}}{\mathrm{d}t}P^{-1}, \tag{5.8}$$

to obtain

$$\dot{P} = -P\Phi^T\Phi P. \tag{5.9}$$

The relationship between $\hat{\theta}$ and P is derived to be the following by differentiating Eq. (5.4) with time:

$$\dot{\hat{\theta}} = P(t)\Phi^T(t)\hat{e}(t), \tag{5.10}$$

where

$$\hat{e}(t) = y(t) - \Phi(t)\hat{\theta} \tag{5.11}$$

is the minimal estimation error. Thus, the parameter estimate $\hat{\theta}$ is derived from a recursive solution of Eqs. (5.9)–(5.11), beginning from an initial condition, $\hat{\theta}(0)$, $P(0)$, and utilizing the current output, $y(t)$. The convergence property of such a recursive estimate is well established (Slotine 1995), provided the initial covariance matrix, $P(0)$, is chosen to be sufficiently large. The equivalence between the least-squares method and the Kalman filter state estimation for the process

$$\dot{x} = x(t), \quad y(t) = \Phi(t)x(t) + e(t)$$

is to be noted.

An equivalent discrete-time representation of the least-squares method is expressed by the following recursive scheme:

$$\begin{aligned}\hat{\theta}(k) &= \hat{\theta}(k-1) + P(k)\Phi^T(k)[y(k) - \Phi(k)\hat{\theta}(k-1)] \\ P(k)\Phi^T(k) &= P(k-1)\Phi^T(k)[I + \Phi(k)P(k-1)\Phi^T(k)]^{-1} \\ P(k) &= [I - P(k)\Phi^T(k)\Phi(k)]P(k-1),\end{aligned} \tag{5.12}$$

where k is the discrete time index, and $\hat{\theta}(k-1)$ denotes the estimate based on $(k-1)$ measurements. The covariance inverse in discrete time given by

$$P^{-1}(k) = \sum_{i=1}^{k} \Phi^T(i)\Phi(i) = W(k)W(k)^T, \tag{5.13}$$

where

$$W(k)^T = [\Phi^T(1), \Phi^T(2), \ldots, \Phi^T(k)], \tag{5.14}$$

is a square matrix. The requirement for the recursive scheme to work successfully is that $W(k)W(k)^T$ must be non-singular for any given instant k.

5.2.2 *Least-Squares Method with Exponential Forgetting*

When the plant parameters are time-varying, it is more efficient to replace the standard least-squares method by one that discounts the past data exponentially with time. In this manner, recent estimation error is given much more importance than those of the past in the minimization process. The cost function of Eq. (5.3) is changed as follows by applying an exponential-time weighting factor:

$$V(\theta, t) = \frac{1}{2}\int_0^t e^{-\int_\tau^t \lambda(r)\mathrm{d}r[y(\tau)-\Phi(\tau)\theta]^T[y(\tau)-\Phi(\tau)\theta]\mathrm{d}\tau}, \tag{5.15}$$

where $\lambda(t) \geq 0$ is the time-forgetting factor.

An equivalent discrete-time representation of the least-squares method with exponential forgetting is the following:

$$\begin{aligned}
\hat{\theta}(k) &= \hat{\theta}(k-1) + P(k)\Phi^T(k)[y(k) - \Phi(k)\hat{\theta}(k-1)] \\
P(k)\Phi^T(k) &= P(k-1)\Phi^T(k)[\lambda_k I + \Phi(k)P(k-1)\Phi^T(k)]^{-1} \\
P(k) &= \frac{[I - P(k)\Phi^T(k)\Phi(k)]P(k-1)}{\lambda_k},
\end{aligned} \tag{5.16}$$

where $0 \leq \lambda_k < 1$.

5.2.3 *Projection Algorithm*

A least-squares type estimation method requires two sets of parameters, $\hat{\theta}, P$, to be stored at every time instant, which can be quite cumbersome for large-order plants (such as ASE systems). A simplification in the recursive method is possible by treating the measurement $y(k)$ to be a projection of the true parameter vector θ on the regressor vector, $\Phi(k)$, space. Every new estimate $\hat{\theta}(k)$ is therefore forced to satisfy the constraint $y(k) = \Phi(k)\hat{\theta}(k)$, while minimizing the error, $\mid \hat{\theta}(k) - \hat{\theta}(k-1) \mid$. This is carried out in practice by introducing the constraint via a Lagrange multiplier, λ, in the minimization of the quadratic loss function as follows (Aström and Wittenmark 1995):

$$V(\hat{\theta}, k) = \frac{1}{2}\left[\hat{\theta}(k) - \hat{\theta}(k-1)\right]^T \left[\hat{\theta}(k) - \hat{\theta}(k-1)\right] + \lambda\left[y(k) - \Phi(k)\hat{\theta}(k)\right]. \tag{5.17}$$

The minimization of the loss function with respect to $\hat{\theta}$ and λ leads to the following projection algorithm, which is implemented directly to obtain an update, $\hat{\theta}(k)$:

$$\hat{\theta}(k) = \hat{\theta}(k-1) + \frac{\Phi^T(k)}{\Phi(k)\Phi^T(k)}[y(k) - \Phi(k)\hat{\theta}(k)]. \tag{5.18}$$

The following modification of Eq. (5.18) is actually more robust in terms of convergence, and also avoids the singularity at $\Phi(k)\Phi^T(k) = 0$:

$$\hat{\theta}(k) = \hat{\theta}(k-1) + \frac{\gamma\Phi^T(k)}{\alpha + \Phi(k)\Phi^T(k)}[y(k) - \Phi(k)\hat{\theta}(k)], \tag{5.19}$$

where $\alpha \geq 0$ and $0 < \gamma < 2$ are adaptation constants.

5.2.4 *Autoregressive Identification*

While the methods discussed previously can be applied to either static or dynamic processes, a modification of the recursive least-squares method is especially suited to identification of dynamic systems (such as ASE plants). In a dynamic system, the response depends not only on the values of the inputs at previous time instants but also on its own past record. If a transfer function of such a system were to be identified, it would require an estimation of both the numerator and denominator polynomial coefficients, which in turn requires a regression model including the output's time history. Hence, it is reasonable to introduce the following modification into the least-squares method, illustrated here for a single-input, single-output system:

$$y(k) = \Phi(k-1)\theta, \tag{5.20}$$

whose parameter estimate is given by

$$\theta(k)^T = (a_1, a_2, \ldots, a_n, b_1, b_2, \ldots, b_m), \tag{5.21}$$

with the following time-delayed regressor vector including the outputs:

$$\begin{aligned}\Phi(k-1) = [&-y(k-1), -y(k-2), \ldots, -y(k-n), u(k+m+n-1),\\ &\ldots, u(k-n)]^T,\end{aligned} \tag{5.22}$$

with $a_1, \ldots, a_n$ being the denominator coefficients and $b_1, \ldots, b_m$ the numerator coefficients of the transfer function. The least-squares recursive scheme for the autoregressive model is the following:

$$\begin{aligned}\hat{\theta}(k) &= \hat{\theta}(k-1) + P(k)\Phi^T(k-1)[y(k) - \Phi(k-1)\hat{\theta}(k-1)]\\ P(k)\Phi^T(k-1) &= P(k-1)\Phi^T(k-1)[I + \Phi(k-1)P(k-1)\Phi^T(k-1)]^{-1}\\ P(k) &= [I - P(k)\Phi^T(k-1)\Phi(k-1)]P(k-1).\end{aligned} \tag{5.23}$$

In an autoregressive model applied to nonlinear systems, or to systems driven by random noise inputs, a stochastic regression model of the following type must be used:

$$y(k) = \Phi(k-1)\theta + \epsilon(k), \tag{5.24}$$

where $\epsilon(k)$ is an unknown disturbance. A recursive method based on such a model is the extended least-squares (ELSs) method, which has a direct equivalence with the extended Kalman filter (EKF) of the process, $x(k+1) = x(k)$, $y(k) = \Phi(k)x(k) + e(k)$.

5.3 Design Methods for Stochastic Self-Tuning Regulators

When some statistical information is available about the noise inputs (such as their mean and variance), it is possible to use a more accurate plant identification procedure than the one given above, which in turn leads to a better performing STR. Here we briefly discuss the ideas behind such designs. For details, the reader is referred to a textbook on identification-based adaptive control (Aström and Wittenmark 1995).

Since regulation is our objective, we shall only focus on the control design methods that guarantee asymptotic stability in the presence of noise inputs, which are modelled as zero-mean Gaussian white noise (ZMGWN) (Chapter 2). An appropriate objective function to be minimized is the expected value of the quadratic control and state cost, expressed by

$$J = E\left\{\int_0^\infty (x^T Q x + u^T R u)\mathrm{d}t\right\} \tag{5.25}$$

where Q, R are the selected cost coefficient matrices. This is the standard linear quadratic Gaussian (LQG) problem solved by taking $u = -K\hat{x}$, where $\hat{x}(t)$ is the estimated state of a Kalman filter. The solution to the LQG problem involves a pair of algebraic Riccati equations (Chapter 2). If $R = 0$, the controller derivation problem is termed the minimum-variance problem, and its solution can be obtained much more easily, because control inputs need not be minimized. When the autoregressive least-squares method is applied to the minimum-variance problem, the result is a control law whose output is a moving-average signal, and the controller is termed an autoregressive moving average (ARMA) controller. For a single-input, single-output system, it is easy to understand how a moving-average process can arise out of a finite impulse response (FIR) model driven by a convolution of random signals. Since such a controller has pole-zero cancellations of the plant, the degree of the denominator polynomial is less than that of the plant. The ARMA design process is therefore simply a pole-placement method applied to noise-driven plants.

As opposed to an ARMA controller, an LQG design is much more robust as it takes into account the minimization of the feedback signals, while the ARMA controller does not. When combined with an observer (Kalman filter) whose gains are adjusted for a better recovery of the return-ratio at the plant's input (see Chapter 2), the resulting LQG/LTR controller offers a suitably high stability margin for practical implementation in a noisy situation. This is our recommendation for an STR design for ASE applications, which are notorious for their uncertain and unmodelled dynamics acting as process noise disturbances.

5.4 Aeroservoelastic Applications

A STR requires that the controller parameters be modified on the basis of an online estimation of the aeroelastic plant's parameters, A, B. While the structural properties are well known from ground-based experiments and do not vary in flight, it is only the unsteady aerodynamics parameters that are capable of changing with flight conditions (Mach number, Reynolds number, etc.). In the subsonic and supersonic regimes (Chapter 3), the linearity of the unsteady aerodynamics may enable one to use gain scheduling with Mach number and dynamic pressure (or altitude) to adapt the regulator gains for flutter suppression, if the flutter point falls in either the subsonic or the supersonic regime. However, gain scheduling is an unsystematic procedure and requires an extensive database to be first constructed from either flight tests or wind-tunnel experiments. It may also fail crucially in the transonic regime where the aerodynamic behaviour is highly uncertain. Therefore, feedback adaptation is the only practical alternative for flutter suppression.

The greatest source of parametric uncertainty in an aeroelastic model is the unsteady aerodynamic loading, which must be identified accurately in an STR. Consider an aeroelastic plant represented by the following equations of motion:

$$M\ddot{\xi} + K\xi = qF + Tu, \tag{5.26}$$

where $\xi(t) : \mathbb{R} \to \mathbb{R}^n$ is the generalized coordinates vector corresponding to the n degrees of structural freedom (including the m control-surface degrees of freedom), $u(t) : \mathbb{R} \to \mathbb{R}^m$ is the vector of control torque inputs applied by m servo-actuators, and $M \in \mathbb{R}^{n\times n}$, $K \in \mathbb{R}^{n\times n}$ and $T \in \mathbb{R}^{n\times m}$ are the generalized mass, stiffness and control transmission matrices, respectively, assumed to be known. Here the generalized aerodynamic force vector has been expressed as $Q(t) = qF(t)$, where q is the dynamic pressure and $F(t) : \mathbb{R} \to \mathbb{R}^n$ is an unknown vector, which must be identified from a finite record of the plant's input–output behaviour. Before this can be carried out, it is necessary to have a structure for the functional dependence of $Q(t)$ on the structural motion coordinates, $\xi(t)$. For open-loop flutter analysis, it is customary to assume a simple harmonic motion in the critical flutter condition, resulting in the following flutter determinant to be solved for flutter frequency, ω_f, and flutter dynamic pressure, q_f (see Chapter 4):

$$\det[-\omega_f^2 M + K - q_f F(i\omega_f)] = 0, \tag{5.27}$$

where $F(i\omega)$ is the generalized air force (GAF) data derived from a frequency-domain aerodynamic theory (Chapter 3). For the general transient motion, the GAF data is used to derive a rational function approximation (RFA) in the Laplace domain by analytic continuation:

$$F(s) = G(s)\xi(s), \tag{5.28}$$

where $G(s)$ is the unsteady aerodynamic transfer matrix approximated by

$$G(s) = A_0 + A_1 s + A_2 s^2 + \Gamma s(sI - R)^{-1}E, \tag{5.29}$$

with $A_0, A_1, A_2, \Gamma, E, R$ being the constant coefficient matrices to be determined by a curve-fit with the harmonic GAF data, $F(i\omega)$.

As in the case of the RFA used to determine an unsteady transfer matrix from the frequency-domain data, it is logical to assume that $F(t)$ is linearly dependent upon $\xi(t)$ and its time derivatives. It is therefore appropriate to use an ARMA model for the identification of $F(t)$ from $\xi(t)$, expressed in discrete time as follows:

$$\begin{aligned} F(k) = &-C_1F(k-1) - C_2F(k-2) - \cdots - C_rF(k-r) + D_0\xi(k) + D_1\xi(k-1) \\ &+D_2\xi(k-2) + \cdots + D_p\xi(k-p), \end{aligned} \tag{5.30}$$

where $C_i, i = 1, \ldots, r$ and $C_j, j = 0, \ldots, p$ are the coefficient matrices to be estimated by an identification method. Since Eq. (5.30) is in a regressive form, it can be expressed as the following estimation problem:

$$\hat{F}(k) = \Theta^T \Phi(k), \tag{5.31}$$

where $\hat{F}$ is the predicted GAF vector,

$$\Theta^T = (C_1, C_2, \ldots, C_r, D_0, D_1, \ldots, D_p) \tag{5.32}$$

is the parameters matrix to be estimated and

$$\Phi(k) = [-F^T(k-1), -F^T(k-2), \ldots, -F^T(k-r), \xi^T(k), \xi^T(k-1), \ldots, \xi^T(k-p)]^T \tag{5.33}$$

is the regressor matrix to be supplied by measurements (or an online CFD method). The size of the regression model is specified by r, p, which determine the number of time steps at which the GAFs and structural coordinates must be stored, respectively. The parameter estimate, $\hat{\Theta}$, can be then derived by the least-squares method as the set that minimizes the square of the prediction error, $\tilde{F} = F - \hat{F} = F - \hat{\Theta}^T \Phi(k)$, over a sample length, N:

$$\begin{aligned} J(N) &= \frac{1}{N} \sum_{k=1}^{N} \tilde{F}^T(k) \tilde{F}(k) \\ &= \frac{1}{N} \sum_{k=1}^{N} [F - \hat{\Theta}^T \Phi(k)]^T [F - \hat{\Theta}^T \Phi(k)], \end{aligned} \tag{5.34}$$

that is, we must have

$$\frac{\partial J(N)}{\partial \hat{\Theta}} = 0 \tag{5.35}$$

or

$$\hat{\Theta}^T = (\Phi \Phi^T)^{-1} F^T \Phi^T. \tag{5.36}$$

Once an estimate, $\hat{\Theta}$, is available, the following discrete-time aeroelastic model also becomes available:

$$x_d(n+1) = \hat{A}_d x_d(n) + \hat{B}_D u_d(n), \tag{5.37}$$

which is used to find the equivalent continuous-time model:

$$\dot{x} = \hat{A}x + \hat{B}u, \tag{5.38}$$

where

$$\hat{A}_d = e^{\hat{A}T}, \quad \hat{B}_d = \int_0^T e^{\hat{A}\tau} \hat{B} d\tau, \tag{5.39}$$

T being the sampling interval. Now any linear, time-invariant (LTI) control strategy (such as eigenstructure assignment, LQG/LTR, H_∞) can be applied to determine the regulator and observer gains based on the identified plant, $\hat{A}, \hat{B}$. Control surface actuators are easily added to the aeroelastic model by augmenting the generalized coordinates, $\xi(t)$, with control

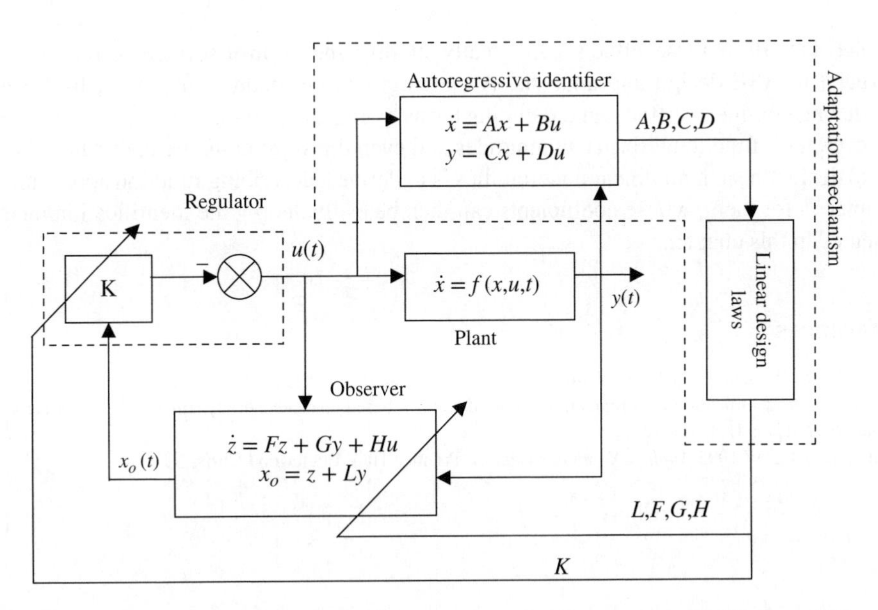

Figure 5.1 Schematic block-diagram representation of a self-tuning regulator (STR) for an ASE system based on an autoregressive identification of the plant parameters

surface deflections, $\delta(t)$. The overall STR method for a typical ASE system is depicted by the block-diagram of Fig. 5.1.

An advantage of the STR method is that there is no need to employ a large number of state variables (including the fictitious ones required for the lag states of RFAs) for an accurate estimation of the ASE plant. For example, reduced-order aeroelastic models have been derived (Raveh 2004) from CFD computations by the ARMA identification of aerodynamic forces, and used for flutter calculations. This approach offers great promise for the determination of controller parameters in the adaptive ASE loop, especially if the latter is based on iterative solutions to algebraic Riccati equations (see Chapter 2).

In the typical section model of the linear plant, it is easy to identify the aerodynamic parameters that can vary with Mach number and Reynolds number of real flows (although they are constants in Theodorsen's model). These are

(a) the non-circulatory lift, pitching-moment and hinge-moment coefficient parameters;
(b) the circulatory transfer function, $G(s)$, approximated by the numerator coefficients, $a_0, a_1, \ldots, a_\ell$, and lag parameters, $b_1, \ldots, b_\ell$, for an RFA. Note $G(ik) = C(k)$ for Theodorsen's model, but could vary with Mach number and Reynolds number in a real system;
(c) additional nonlinear forcing functions, which are not present in the original model of the plant.

The last of the effects enumerated here, namely, the unmodelled nonlinear forcing behaviour, have the greatest possibility of destabilizing the ASE system. Examples of nonlinear excitation include the unsteady flow separation near the trailing-edge due to large angle-of-attack or the separation caused by the shock waves in the transonic regime (see

Chapter 11). Both these effects can greatly modify the control-surface response, which is crucial in ASE design and analysis. An attempt can be made to model such effects by introducing nonlinear stiffness and damping terms by regressor models into the original plant. However, their functional forms, magnitudes and even the signs could be uncertain. The best way to deal with such nonlinear uncertainties is to derive a describing function approximation (Chapter 7) for them, whose coefficients can then be estimated by the identification methods presented in this chapter.

References

Aström KJ and Wittenmark B 1995 *Adaptive Control*, 2nd ed., Addison-Wesley, New York.

Raveh DE 2004 Identification of computational-fluid-dynamics based unsteady aerodynamics models for aeroelastic analysis. *J. Aircr.* **41**, 620–632.

Slotine JJE and Li W 1995 *Applied Nonlinear Control*. Prentice-Hall, Englewood Cliffs, NJ.

6

Nonlinear Systems Analysis and Design

6.1 Introduction

Control systems can generally be designed to perform well in a given set of conditions. If the domain of operating conditions is quite small, then many simplifying assumptions can be made about the system's behaviour, often resulting in linearized dynamics with well-known coefficients for which a plethora of control design tools are readily available. However, in many cases, such an approach would fail if the plant's behaviour is either essentially nonlinear or highly uncertain over the given range of operating conditions. In such cases, the controller must necessarily adapt itself to changing operating conditions, as well as to uncertain plant parameters, which generally implies that the control system must be governed by nonlinear ordinary differential equations in the time domain. It is the aim of this chapter to introduce nonlinear dynamical systems, the existence and uniqueness of their solutions to specified initial conditions and their stability analysis. Established techniques of nonlinear stability analysis can be classified into input–output transfer operators and state-space-based Lyapunov theorems. Our focus is on aeroservoelastic (ASE) design by adaptive techniques, therefore rather than detailing the nonlinear analysis techniques, we only consider concepts that are relevant to our purpose.

Even if the aeroelastic plant is linear, the use of an adaptive feedback control loop renders the overall ASE system nonlinear. Signals of nonlinear systems do not enjoy the basic properties of additivity, invariance under scalar multiplication and frequency fidelity. These are the very properties that enable systematic and straightforward analysis and design of linear systems. Because of these deficiencies, analysing and designing nonlinear control systems require special and mathematically much more rigorous techniques than those employed for linear systems. In the present chapter, we consider the basic concepts of nonlinear stability analysis and Lyapunov-based control design. However, as ASE systems are autonomous (i.e. they have time-independent characteristics), the treatment is limited to autonomous systems. Furthermore, only those concepts are covered that are necessary in adaptive control design. For other details, the reader is referred to specialized textbooks on nonlinear systems, such as Haddad and Chellaboina (2008), Khalil (2002) and Slotine and Li (1995).

Adaptive Aeroservoelastic Control, First Edition. Ashish Tewari.

6.2 Preliminaries

A finite-dimensional dynamical system of order n is completely described by a set of time-dependent internal *state variables*, $\xi(t) : \mathbb{R} \to \mathbb{R}^n$, and a set of external *input variables*, $\eta(t) : \mathbb{R} \to \mathbb{R}^m$. The system's behaviour is governed by a set of n scalar, first-order, ordinary differential equations in time, collectively referred to as the *state equation*, expressed as follows:

$$\dot{\xi} = F(\xi, \eta, t), \tag{6.1}$$

where $F : \mathbb{R}^n \times \mathbb{R}^m \times \mathbb{R} \to \mathbb{R}^n$ represents the state functional mapping. If the time t does not appear explicitly in $F(.)$, then the system is said to be *time-invariant* (or *autonomous*); otherwise, it is called a *time-varying* (*non-autonomous*) system. In addition to the state and input variables, the system might produce certain other variables as its *outputs*, described by the set $r(t) : \mathbb{R} \to \mathbb{R}^p$ and governed by the following set of algebraic equations, collectively called the *output equation*:

$$r = H(\xi, \eta, t), \tag{6.2}$$

where $H : \mathbb{R}^n \times \mathbb{R}^m \times \mathbb{R} \to \mathbb{R}^p$ is the output mapping functional. While the state vector, ξ, is wholly internal to the system, the output vector, r, is accessible to external measurements, and can be used to determine the system's characteristics. On the other hand, the input vector, η, is generated by processes external to the system and can be used to influence the state and the output vectors. For this reason, η is also called the *control input* vector. The complete input–output description of a system is provided by the state and output equations, thus Eqs. (6.1) and (6.2) are said to comprise a *state-space representation* of the system. Of course, the choice of the state vector is not unique and one can select an appropriate set of state variables, ξ, for a convenient state-space representation.

A special class of systems with a linear control contribution is often of interest in practical applications, such as in ASE. Such a system is called a *control affine* system and has the following state-space representation:

$$\dot{\xi} = a(\xi, t) + b(\xi, t)\eta, \tag{6.3}$$

$$r = c(\xi, t) + g(\xi, t)\eta, \tag{6.4}$$

where $a : \mathbb{R}^n \times \times \mathbb{R} \to \mathbb{R}^n$, $b : \mathbb{R}^n \times \times \mathbb{R} \to \mathbb{R}^{n\times m}$, $c : \mathbb{R}^n \times \times \mathbb{R} \to \mathbb{R}^p$ and $g : \mathbb{R}^n \times \times \mathbb{R} \to \mathbb{R}^{p\times m}$ are functional maps. Owing to the linear contribution of the input, the design of control systems for a control affine plant is much simpler than that for a fully nonlinear plant. Of course, a *linear system* is a member of the class of control-affine systems and is represented by

$$\dot{\xi} = A(t)\xi + B(t)\eta, \tag{6.5}$$

$$r = C(t)\xi + D(t)\eta, \tag{6.6}$$

where $A(t) : \mathbb{R} \to \mathbb{R}^{n\times n}$, $B(t) : \mathbb{R} \to \mathbb{R}^{n\times m}$, $C(t) : \mathbb{R} \to \mathbb{R}^{p\times n}$ and $D(t) : \mathbb{R} \to \mathbb{R}^{p\times m}$ are the *coefficient matrices* of the state-space representation. The simplest class of systems to analyse is that of *linear, time-invariant* (LTI) systems, which is defined by constant coefficient matrices, A, B, C, D, and is considered in Chapter 2.

6.2.1 *Existence and Uniqueness of Solution*

Consider a system described by the state equation Eq. (6.1). In order to understand the basic characteristics of such a system, one must have a solution to the state equation for a given control history and a set of initial conditions. A solution to the state equation, Eq. (6.1), for a prescribed input history, $\eta(t)$, and beginning from a given initial state, $\xi(t_0) = \xi_0$, is called a *trajectory*. The existence and uniqueness of such a solution may not always be guaranteed and crucially depends upon the characteristics of the system described by Eq. (6.1), as well as on the nature of the applied inputs. However, the system's characteristics (such as stability) are independent of the applied inputs and thus can be understood from the solution in the *unforced* case, that is, $\eta(t) = 0$, resulting in the homogeneous state equation,

$$\dot{\xi} = F(\xi, t), \tag{6.7}$$

where $F : \mathbb{R}^n \times \mathbb{R} \to \mathbb{R}^n$. To determine a solution to the unforced system, subject to initial condition, $\xi(t_0) = \xi_0$, is called the initial value problem (IVP), and is crucial in nonlinear systems analysis. The following theorem establishes a sufficient condition for the existence of a unique solution to the IVP.

6.2.1.1 Lipschitz Condition

Theorem 6.2.1 *Let $F(\xi, t)$ be piecewise continuous in t and satisfy the Lipschitz condition*

$$|| \; F(\xi, t) - F(x, t) \; || \leq L \; || \; \xi - x \; ||$$

for $t_0 \leq t \leq T$, for all $(\xi, x) \in \mathbb{R}^n \times \mathbb{R}^n$, where $L > 0$ and $|| \, . \, ||$ denotes a vector norm. Then the state equation Eq. (6.7), with $\xi(t_0) = \xi_0$, has a unique solution in the interval $t_0 \leq t \leq T$.

Proof. The proof is obtained by integrating Eq. (6.1) to write

$$\xi(t) = \xi_0 + \int_{t_0}^{t} F(\xi(\tau), \tau) \mathrm{d}\tau$$

and applying the contraction mapping theorem in a Banach space (Khalil 2002).

If the Lipschitz condition is not satisfied for all real (ξ, x), but only for those in a finite neighbourhood of the initial state ξ_0, that is, within a ball of radius R,

$$\beta = \{\xi \in \mathbb{R}^n : || \; \xi - \xi_0 \; || \leq R\}$$

then the global sufficiency condition of Theorem 6.2.1 reduces to that of local existence and uniqueness. This implies that there exists a $\delta > 0$ such that Eq. (6.7), with $\xi(t_0) = \xi_0$, has a unique solution only in the interval $t_0 \leq t \leq t_0 + \delta$.

It can be shown (Khalil 2002) that the Lipschitz condition will be satisfied in a domain $\mathbb{D} \subset \mathbb{R}^n$ for $a \leq t \leq b$, if and only if both $F(\xi, t)$ and $\partial F / \partial \xi(\xi, t)$ are continuous in $\xi \in \mathbb{D}$ for $a \leq t \leq b$. Furthermore, if $\partial F / \partial \xi(\xi, t)$ is uniformly bounded for all $\xi \in \mathbb{R}^n$ and $a \leq t \leq b$, then it satisfies the global Lipschitz condition and thus has a globally unique solution.

6.2.2 Expanded Solution

Let a reference state trajectory, $\xi_r(t)$, and a corresponding reference input history, $\eta_r(t)$, satisfy Eq. (6.1),

$$\dot{\xi}_r = F(\xi_r, \eta_r, t), \tag{6.8}$$

subject to the initial condition,

$$\xi_r(0) = \xi_{r0}. \tag{6.9}$$

It is further assumed that the functional maps, $F(.), H(.)$, possesses continuous derivatives of any given order with respect to state and control variables at the reference solution, (ξ_r, η_r). Let $x(t)$ and $u(t)$ be the state and control deviations, respectively, from the reference solution, such that the perturbed solution is given by

$$\begin{aligned} \xi(t) &= \xi_r(t) + x(t) \\ \eta(t) &= \eta_r(t) + u(t), \end{aligned} \tag{6.10}$$

subject to initial conditions given by Eq. (6.9) and $x(0) = x_0$. Let $y(t)$ measure the deviation of the plant's output vector from the reference output. Then Eqs. (6.1) and (6.2) can be expanded in Taylor series about the reference solution as follows:

$$\dot{x} = F(\xi_r + x, \eta_r + u, t) - F(\xi_r, \eta_r, t) \tag{6.11}$$

and

$$y = H(\xi_r + x, \eta_r + u, t) - H(\xi_r, \eta_r, t), \tag{6.12}$$

where

$$\begin{aligned} F(\xi_r + x, \eta_r + u, t) &= F(\xi_r, \eta_r, t) + \frac{\partial F}{\partial \xi}(\xi_r, \eta_r, t)x \\ &\quad + \frac{\partial F}{\partial \eta}(\xi_r, \eta_r, t)u + f(x, u, t), \end{aligned} \tag{6.13}$$

$$\begin{aligned} H(\xi_r + x, \eta_r + u, t) &= H(\xi_r, \eta_r, t) + \frac{\partial H}{\partial \xi}(\xi_r, \eta_r, t)x \\ &\quad + \frac{\partial H}{\partial \eta}(\xi_r, \eta_r, t)u + h(x, u, t). \end{aligned} \tag{6.14}$$

Here $f(.), h(.)$ are nonlinear functionals involving second- and higher-order terms of the state and input variables. Substitution of Eq. (6.13) into Eqs. (6.11) and (6.12) yields the following:

$$\dot{x} = A(t)x + B(t)u + f(x, u, t), \tag{6.15}$$

$$y = C(t)x + D(t)u + h(x, u, t), \tag{6.16}$$

where $A(t), B(t), C(t), D(t)$ are the following Jacobian matrices:

$$\begin{aligned} A(t) &= \frac{\partial F}{\partial \xi}(\xi_r, \eta_r, t), \quad B(t) = \frac{\partial F}{\partial \eta}(\xi_r, \eta_r, t) \\ C(t) &= \frac{\partial H}{\partial \xi}(\xi_r, \eta_r, t), \quad D(t) = \frac{\partial H}{\partial \eta}(\xi_r, \eta_r, t). \end{aligned} \tag{6.17}$$

Most ASE applications involve time-invariant (autonomous) systems, which must be maintained in equilibrium near a constant operating condition. This control application is referred to as *set-point regulation*. In such a case, explicit time dependence is dropped from the state and output equations, resulting in the following description of the plant:

$$\dot{\xi} = F(\xi, \eta)$$
$$r = H(\xi, \eta). \quad (6.18)$$

Let the system be at rest at a point, ξ_e, when not forced by inputs ($\eta = 0$):

$$\dot{\xi} = F(\xi_e, 0) = 0. \quad (6.19)$$

Then the solution ξ_e is called an *equilibrium point* of the system. A system can have many equilibrium points (e.g. a simple pendulum system has two) or it may have none. An equilibrium point of the autonomous system is thus a solution of $F(\xi_0, 0) = 0$. The state and input deviations from the equilibrium point are denoted by $x(t) = \xi(t) - \xi_e$ and $u(t) = \eta(t)$, respectively. Without any loss of generality, we can assume that an autonomous system has an equilibrium point at $x = 0, u = 0$. The state-space description in the expanded form about the equilibrium point can be expressed as follows:

$$\dot{x} = f(x, u) = Ax + Bu + g(x, u)$$
$$y = Cx + Du + h(x, u), \quad (6.20)$$

where A, B, C, D are the time-invariant coefficient matrices evaluated at $(\xi_0, 0)$ and $g(.), h(.)$ are nonlinear functional maps.

If the operating condition changes with time, the error dynamics of an autonomous nonlinear system is governed by a non-autonomous (time-varying) system, and thus requires a different approach for stabilization from the one presented here. Trying to follow a time-dependent nominal solution is referred to as the *tracking problem*. This is the problem associated with minimizing the deviation of the plant's state from the reference state in a model-reference adaptation system (MRAS), the topic of Chapter 8. Stability of error dynamics in an MRAS therefore requires the treatment of the system as if it were non-autonomous. In this chapter, only the stability concepts of autonomous systems are presented, which are extended in Chapter 8 to cover non-autonomous systems.

6.3 Stability in the Sense of Lyapunov

Stability of an autonomous system is broadly defined as a tendency to remain close to a given equilibrium point in an unforced state, despite small initial disturbances. Consider an unforced, autonomous system with state vector, $x(t)$, governed by the state equation,

$$\dot{x} = f(x), \quad (6.21)$$

with an equilibrium point at the origin. If an arbitrarily small initial condition exists,

$$x(0) = x_0, \quad (6.22)$$

then the perturbed solution, $x(t), t \geq 0$, may (or may not) remain arbitrarily close to the equilibrium point. Thus stability is concerned with the boundedness of the initial response, $\mid x(t) \mid; t \geq 0$, if $\mid x_0 \mid < \delta$, where δ is an arbitrarily small positive number. If there are several equilibrium points of a system, the system may have a different stability in the neighbourhood of each of them.

In order to analyse stability, it is assumed that the functional, $f(x)$, is Lipschitz, continuous and possesses continuous derivatives with respect to the state variables. The widely accepted mathematical definition of stability applied to mechanical systems is the *stability in the sense of Lyapunov*, which addresses the Euclidean norm of the state deviation from the zero equilibrium point, $\mid x(t) \mid$.

Definition 6.3.1 *A system described by Eq. (6.21) is said to be stable about the equilibrium point, $x_e = 0$ in the sense of Lyapunov if for each real and positive number, ϵ, however small, there exists another real and positive number, $\delta(\epsilon)$, such that*

$$\mid x(0) \mid < \delta \tag{6.23}$$

implies that

$$\mid x(t) \mid < \epsilon, \quad t \geq 0. \tag{6.24}$$

Stability in the sense of Lyapunov thus requires that a solution starting in a neighbourhood of the equilibrium point at the origin should always remain in a prescribed small neighbourhood of the origin. In this book, we will regard a stable equilibrium point to be the one that satisfies the stability criterion in the sense of Lyapunov. It is quite possible to have an unbounded solution to a large (but bounded) initial perturbation, but the same system may remain in the neighbourhood of the origin for small perturbations, and hence be stable in the sense of Lyapunov. Furthermore, it is possible for a system to have a departure from the equilibrium for even small perturbations from the origin (instability in the sense of Lyapunov), but the solution could still be norm bounded, $\mid x(t) \mid < a$. However, a is not arbitrarily small. Thus boundedness of an unforced system's response to small initial perturbation (called *stability in the sense of Lagrange*) is insufficient to guarantee stability in the sense of Lyapunov. Many nonlinear systems display a bounded oscillatory response to a small initial perturbation – called a limit-cycle oscillation (LCO) – which is considered to be unstable. An example of such a system is the van der Pol oscillator described by the following state equations:

$$\dot{x}_1 = x_2, \quad \dot{x}_2 = -x_1 + (1 - x_1^2)x_2.$$

There are special theorems that analytically predict the existence of LCOs of second-order systems about the equilibrium, thereby precluding stability in the sense of Lyapunov. These require the construction of phase portraits, which are the plots of one state variable, $x_2(t)$, against the other, $x_1(t)$, for $t \geq 0$. For phase plane analysis and the basic theorems for the existence of limit cycles (Poincare, Bendixson and Poincare–Bendixson theorems), the reader is referred to nonlinear systems textbooks (Slotine 1995). Here it suffices to say that the existence of a limit cycle in a system with energy dissipation (damping) indicates the presence of a positive energy source, hence instability. For example a transonic LCO occurs because of the energy fed by unsteady shock waves cancelling the energy lost because of viscous flow separation, resulting in a constant amplitude oscillation.

Definition 6.3.2 *A system described by Eq. (6.21) is said to be* asymptotically stable *about the origin, $x_e = 0$, if it is stable in the sense of Lyapunov, and if for each real and positive number, ϵ, however small, there exist real and positive numbers, δ and τ, such that*

$$| x(0) | < \delta \tag{6.25}$$

implies that

$$| x(t) | < \epsilon, \quad t > \tau. \tag{6.26}$$

Asymptotic stability is thus possessed by a special class of Lyapunov stable systems, whose slightly perturbed solutions approach the origin asymptotically for large times (i.e. in the limit $t \to \infty$). Thus, equilibrium is eventually restored. However, it is possible that an asymptotically stable system may need infinite time in regaining equilibrium.

Definition 6.3.3 *A system described by Eq. (6.21) is said to be* exponentially stable *about the origin, $x_e = 0$, if there exist two positive numbers, ϵ and λ, such that for all $t > 0$,*

$$| x(0) | < \delta \tag{6.27}$$

implies that

$$| x(t) | \leq \epsilon \, | x(0) | \, e^{-\lambda t}. \tag{6.28}$$

Exponential stability thus implies that the state of the system converges to equilibrium at a rate faster than an exponentially decaying function. Exponential stability implies asymptotic stability, but not vice versa.

Definition 6.3.4 *A system described by Eq. (6.21) is said to be* globally asymptotically stable *about the origin, $x_e = 0$, if it is stable in the sense of Lyapunov, and if for each real and positive pair, (δ, ϵ), there exists a real and positive number, τ, such that*

$$| x(0) | < \delta \tag{6.29}$$

implies that

$$| x(t) | < \epsilon, \quad t > \tau. \tag{6.30}$$

Global asymptotic stability thus refers to asymptotic stability with respect to *all* possible initial conditions, $| x(0) | < \delta$, and not only a few, specific ones. Global asymptotic stability is alternatively referred to as the asymptotic stability in the large. A similar definition can be given for global exponential stability.

6.3.1 Local Linearization about Equilibrium Point

Stability of an equilibrium point can be investigated by locally linearizing the unforced system in its neighbourhood. For an unforced, time-invariant system with the state equation given by Eq. (6.21), the linearization is carried by the following Taylor series expansion about the equilibrium point at the origin:

$$\dot{x} = Ax + \Delta f(x), \tag{6.31}$$

where A is the following Jacobian matrix:

$$A = \left(\frac{\partial f}{\partial x}\right)_{x=0}, \tag{6.32}$$

and $\Delta f(x)$ represents higher-order terms of the series, which are neglected. The stability of the system about the given equilibrium point is then examined from the eigenvalues of the Jacobian matrix, A, in a manner similar to that of the linear system, $\dot{x} = Ax$, covered in Chapter 2. However, if $f(x)$ is not differentiable at the equilibrium point, the local linearization cannot be carried out by the Taylor series expansion. The Lyapunov stability criteria in terms of the eigenvalues of A are stated as follows:

(a) If all the eigenvalues of the Jacobian, A, are in the left-half plane, then A is said to be a Hurwitz matrix, and the system given by Eq. (6.21) is asymptotically stable about the equilibrium point, $x_e = 0$.
(b) If at least one eigenvalue of the Jacobian, A, is in the right-half plane, then the system given by Eq. (6.21) is unstable about the equilibrium point, $x_e = 0$.
(c) If none of the eigenvalues of the Jacobian, A, is in the right-half plane, but at least one eigenvalue is on the imaginary axis, then nothing can be determined regarding the stability of the system given by Eq. (6.21) about the equilibrium point, $x_e = 0$.

6.3.1.1 Centre Manifold Theorem

As seen above, local linearization about an equilibrium point may not provide any information about stability in the sense of Lyapunov, if the Jacobian evaluated at the equilibrium point does not have any eigenvalues in the right-half plane, but some of its eigenvalues have zero real parts (i.e. they lie on the imaginary axis). An alternative approach is thus required to study stability when local linearization fails. Systems with multiple eigenvalues on the imaginary axis are typical in undamped structural dynamics (hence aeroservoelasticity), consequently their stability analysis is enabled by the following centre manifold theorem (Haddad and Chellaboina 2008).

If an unforced autonomous system,

$$\dot{x} = f(x), \tag{6.33}$$

where $f(.) : \mathbb{D} \rightarrow \mathbb{R}^n$ is a twice continuously differentiable map with $f(0) = 0$ and $\mathbb{D} \subset \mathbb{R}^n$ containing the origin, has k eigenvalues of the Jacobian

$$A = \frac{\partial f}{\partial x}(0), \tag{6.34}$$

on the imaginary axis, and the remaining $m = n - k$ eigenvalues in the left-half plane, then it can be linearly transformed into the following form:

$$\begin{aligned} \dot{y} &= A_1 y + g_1(y, z) \\ \dot{z} &= A_2 y + g_2(y, z), \end{aligned} \tag{6.35}$$

where

$$\begin{pmatrix} y \\ z \end{pmatrix} = Tx, \quad y \in \mathbb{R}^k, \quad z \in \mathbb{R}^m \tag{6.36}$$

with

$$TAT^{-1} = \begin{pmatrix} A_1 & 0 \\ 0 & A_2 \end{pmatrix} \tag{6.37}$$

then the functional maps g_1, g_2 are also twice continuously differentiable with the property $g_i(0,0) = 0, \partial g_i/\partial y(0,0) = 0, \partial g_i/\partial z(0,0) = 0$ for $i = 1,2$.

Definition 6.3.5 *A k-dimensional manifold in $\mathbb{R}^n$ for $1 \leq k < n$ is given by the solution of $g(x) = 0$, where $g(.) : \mathbb{R}^n \to \mathbb{R}^{n-k}$ is sufficiently many times continuously differentiable.*

Definition 6.3.6 *A manifold $g(x) = 0$ of the system given by Eq. (6.33) is said to be an invariant manifold, if $g(x(0)) = 0$ implies that $g(x(t)) = 0$ for all $t \in [0, \tau) \subset \mathbb{R}$ is a time interval over which the solution is defined.*

Definition 6.3.7 *For the system given by Eq. (6.35), if $z = h(y)$ an invariant manifold, h is smooth, and $h(0) = 0, \partial h/\partial y(0) = 0$, then it is called a centre manifold of the system.*

Theorem 6.3.8 *For a system described by Eq. (6.35), if the functional maps g_1, g_2 are twice continuously differentiable with the property $g_i(0,0) = 0, \partial g_i/\partial y(0,0) = 0, \partial g_i/\partial z(0,0) = 0$ for $i = 1,2$, all eigenvalues of matrix A_1 have zero real parts, and all eigenvalues of matrix A_2 have negative real parts, then there exists a real positive number δ and a continuously differentiable function $h(y)$ defined for all $| y | < \delta$, such that $z = h(y)$ is a centre manifold for Eq. (6.35).*

If the conditions of Theorem 6.3.8 are satisfied then the system can be reduced to the following *k*th-order equation:

$$\dot{y} = A_1 y + g_1(y, h(y)). \tag{6.38}$$

Theorem 6.3.9 *For a system described by Eq. (6.35) and satisfying the conditions of Theorem 6.3.8, if the origin $y = 0$ of the reduced system given by Eq. (6.36) is asymptotically stable, then the origin of the original system, $x = 0$, is also asymptotically stable.*

Theorem 6.3.10 *For a system described by Eq. (6.35) and satisfying the conditions of Theorem 6.3.8, if the origin $y = 0$ of the reduced system given by Eq. (6.36) is unstable, then the origin of the original system, $x = 0$, is also unstable.*

6.3.2 Lyapunov Stability Theorem

Definition 6.3.11 *If a scalar function, $V(x) : \mathbb{R}^n \to \mathbb{R}$, satisfies the following conditions:*

$$V(0) = 0, \quad V(x) > 0, x \in U, \quad \text{for all } x \neq 0 \tag{6.39}$$

where $U \subset \mathbb{R}^n$ containing the origin $x = 0$, then $V(x)$ is said to be a positive-definite function of x in the region U.

If the condition in the definition is replaced by $V(x) \geq 0$, then $V(x)$ is said to be a positive semi-definite function of x in the region U. If the region U covers the entire space $\mathbb{R}^n$, then $V(x)$ is said to be a positive-definite (or semi-definite) function of x.

Theorem 6.3.12 *Let $V(x) : \mathbb{R}^n \to \mathbb{R}$ be a continuously differentiable, positive-definite, scalar function of the state variables of a system described by Eq. (6.21), whose equilibrium point is $x_e = 0$.*

(a) If the following conditions are satisfied:

$$V(0) = 0, \quad V(x) > 0, \quad \frac{dV}{dt}(x) = -W(x) < 0; \qquad \text{for all } x \neq 0, \tag{6.40}$$

then the equilibrium point, $x_e = 0$, is asymptotically stable.

(b) If in addition to the conditions given in (a), the following condition is also satisfied:

$$|\, x \,| \to \infty \text{ implies } V(x) \to \infty, \tag{6.41}$$

then the equilibrium point, $x_e = 0$, is globally asymptotically stable.

Proof of the Lyapunov stability theorem (Slotine 1995) is obtained from the unbounded, positive-definite nature of $V(x)$ and negative definite nature of $\dot{V}(x)$, implying that for any initial perturbation from the origin, $x(0) \neq 0$, the resulting solution satisfies $V[x(t)] \leq V[x(0)], t > 0$ (i.e. remains in a bounded neighbourhood of the origin). Furthermore, the same also implies that $V[x(t_2)] \leq V[x(t_1)], t_2 > t_1$, which means a convergence of every solution to the origin. The existence of a positive-definite scalar function, $W(x) : \mathbb{R}^n \to \mathbb{R}$, in Eq. (6.40) confirms the negative definite nature of $\dot{V}(x)$. The property of a scalar function, $V(x)$, given by Eq. (6.41), is termed *radial unboundedness* and is used to prove global asymptotic stability. A continuously differentiable, positive-definite, scalar function of the state variables, $V(x)$, satisfying the conditions of Lyapunov theorem is called a Lyapunov function.

The Lyapunov theorem gives a test of asymptotic stability without the necessity of solving the system's state equations, and is thus a powerful tool in nonlinear control design. It merely requires finding a suitable Lyapunov function of the state variables, $V(x)$. However, the Lyapunov theorem gives the sufficient (but not necessary) condition for stability. It is thus possible to have an asymptotically stable system for which none of the conditions given in the theorem are satisfied. A misunderstanding of this point has sometimes led to an unnecessary confusion in the applied control literature.

Example 6.3.13 *Consider an autonomous system described by*

$$\begin{aligned} \dot{x}_1 &= -x_1 - x_2 \\ \dot{x}_2 &= x_1 - x_2^3. \end{aligned} \tag{6.42}$$

Clearly, the system, $\dot{x} = f(x)$, with $x = (x_1, x_2)^T$, has an equilibrium point at the origin, $x = 0$. Let us select the following Lyapunov function:

$$V(x) = x_1^2 + x_2^2. \tag{6.43}$$

The function $V(x)$ is radially unbounded, as it satisfies Eq. (6.41):

$$|\, x \,| = \sqrt{x_1^2 + x_2^2} \to \infty \text{ implies } V(x) = x_1^2 + x_2^2 \to \infty. \tag{6.44}$$

Furthermore, we have

$$V(x) > 0, \quad \text{for all } x \neq 0 \tag{6.45}$$

and

$$\begin{aligned}\frac{\mathrm{d}V}{\mathrm{d}t} &= \dot{V}(x) = \frac{\partial V}{\partial x} f(x) \\ &= 2\left(x_1, x_2\right)\left(-x_1 - x_2, x_1 - x_2^3\right)^T \\ &= -2\left(x_1^2 + x_2^4\right) \\ &< 0, \quad \text{for all } x \neq 0.\end{aligned} \tag{6.46}$$

Therefore, all the sufficient conditions of the Lyapunov theorem are satisfied by $V(x)$ and thus the origin is a globally asymptotically stable equilibrium point of the system.

Definition 6.3.14 *If all the eigenvalues of a square matrix $P \in \mathbb{R}^{n \times n}$ are real, positive numbers, then P is said to be a positive-definite matrix. A positive-definite matrix P can be expressed in the following positive-definite quadratic form:*

$$x^T P x > 0, \quad \text{for all } x \in \mathbb{R}^n \neq 0. \tag{6.47}$$

If some of the eigenvalues of P are zeros, while the rest are real and positive, then P is said to be a positive semi-definite matrix. In that case, it has the following property:

$$x^T P x \geq 0, \quad \text{for all } x \in \mathbb{R}^n \neq 0. \tag{6.48}$$

A direct method of constructing a valid Lyapunov function for autonomous systems is offered by the following theorem (Slotine 1995).

6.3.2.1 Krasovski Theorem

Theorem 6.3.15 *Let $A(x)$ be the Jacobian matrix,*

$$A = \frac{\partial f}{\partial x}, \tag{6.49}$$

of an autonomous system described by

$$\dot{x} = f(x),$$

whose equilibrium point is $x_e = 0$. If there exist two symmetric and positive-definite matrices, (P, Q), such that for all $x \neq 0$, the matrix

$$A^T P + PA + Q \tag{6.50}$$

is globally negative semi-definite, and that the Lyapunov function of the system,

$$V(x) = f^T P f, \tag{6.51}$$

is radially unbounded,

$$\mid x \mid \rightarrow \infty \text{ implies } V(x) \rightarrow \infty, \tag{6.52}$$

then the origin is globally asymptotically stable.

6.3.3 LaSalle Invariance Theorem

Lyapunov stability requires the convergence of the state solution to an arbitrarily small region in the neighbourhood of the equilibrium point. Such a region is termed an invariant set.

Definition 6.3.16 *A set M is called an invariant set of the autonomous system,*

$$\dot{x} = f(x),$$

if any solution $x(t_0) \in M$ *implies that* $x(t) \in M$ *for all* $t \neq t_0$.

A positively invariant set M_+ is the set for which invariance holds only for all future times, $t > t_0$.

A powerful method of applying the Lyapunov stability theorem is by testing whether the solution to Eq. (6.21) converges to a desired invariant set. This is carried out by the following theorem, called the *LaSalle local invariance* theorem.

Theorem 6.3.17 *Let* Ω *be a positively invariant set of a system described by Eq. (6.21) with* $f(x)$ *a continuous function of x. Let* $V(x) : \Omega \rightarrow \mathbb{R}$ *be a positive semi-definite, continuously differentiable function such that*

$$\frac{dV}{dt}(x) \leq 0, \quad \text{for all } x \in \Omega. \tag{6.53}$$

Let $E = \{x \in \Omega | \dot{V}(x) = 0\}$ *and let M be the largest invariant set contained in E. Then every bounded solution* $x(t)$ *starting in* Ω *at some time* t_0, $x(t_0) \in \Omega$, *converges to M in the limit* $t \rightarrow \infty$.

If the domain Ω covers the entire state space, $\mathbb{R}^n$, and the Lyapunov function $V(x)$ is radially unbounded, then all solutions globally and asymptotically converge to M in the limit $t \rightarrow \infty$, where M is the largest invariant set in $\mathbb{R}^n$. This is the LaSalle global invariance theorem.

The Lyapunov stability theorem is easily applied to linear systems. For LTI systems, Lyapunov stability requirements lead to the well-known stability criteria in terms of the eigenvalues (characteristic roots) of the state dynamics matrix (Chapter 2). Similarly, control of LTI systems is carried out by selecting quadratic state and controls cost functions as Lyapunov functions, which is the basis of linear, optimal control theory (Tewari 2011).

6.4 Input–Output Stability

Alternative to the concept of stability of an equilibrium point in the sense of Lyapunov is the concept of input–output stability, also referred to as bounded input, bounded output (BIBO) stability. Input–output stability was introduced in Chapter 2 as the property of a causal system possessing a finite gain, which was defined as the supremum function of the ratio of $\mathcal{H}_2$ norm of the output to that of the input. A system comprising two causal subsystems connected in a feedback loop is input–output stable by the small gain theorem, if the product of their gains is less than unity. The existence of the $\mathcal{H}_2$ norms of the input and output vectors require them to be well-behaved (square-integrable) signals, whereas a finite gain requires the response to finite inputs to be finite. One such property of a square-integrable vector signal, $y(t)$, defined for $t \geq 0$ is given by Parseval's theorem:

$$\int_0^\infty y^T(t)y(t)\mathrm{d}t = \frac{1}{2\pi}\int_{-\infty}^{\infty} Y^H(i\omega)Y(i\omega)\mathrm{d}\omega, \tag{6.54}$$

where $Y(i\omega)$ is the Fourier transform of $y(t)$. Square-integrable signals are thus finite energy signals, and if a system produces such a signal as its output vector when driven by a finite input energy, then the system has input–output stability. However, input–output stability does not generally require the stability of an equilibrium point of the system, and can therefore be termed a more basic stability requirement than Lyapunov stability.

Input–output stability of an LTI system is simply investigated from the gain of its transfer matrix.

Theorem 6.4.1 *For the LTI system with following state-space representation*

$$\dot{x} = Ax + Bu$$
$$y = Cx + Du, \tag{6.55}$$

with A a Hurwitz matrix, and $G(s) = C(sI - A)^{-1}B + D$ *a transfer matrix realization, the gain is given by the induced* H_2 *norm (also the largest singular value) of the frequency-response matrix evaluated over all frequencies,* ω*:*

$$\sup_{\omega\in\mathbb{R}} \| G(i\omega) \|_2 = \bar{\sigma}(G(i\omega)). \tag{6.56}$$

6.4.1 Hamilton–Jacobi Inequality

While input–output stability can be investigated for any nonlinear system, the focus in this book is on control-affine, proper systems described by

$$\dot{x} = f(x) + G(x)u, \quad x(0) = x_0$$
$$y = h(x), \tag{6.57}$$

where $f(.) : \mathbb{R}^n \rightarrow \mathbb{R}^n$ is locally Lipschitz with property $f(0) = 0$, and $G(.) : \mathbb{R}^n \rightarrow \mathbb{R}(n \times m)$, $h(.) : \mathbb{R}^n \rightarrow \mathbb{R}^p$ are continuous over $\mathbb{R}^n$ with the property $h(0) = 0$.

Theorem 6.4.2 *Let there exist a positive real number,* γ*, and a positive semi-definite function,* $V(x)$*, satisfying the following Hamilton–Jacobi inequality for the control-affine, autonomous system represented by Eq. (6.57):*

$$H(V, f, G, h, \gamma) \doteq \frac{\partial V}{\partial x}f(x) + \frac{1}{2\gamma^2}\frac{\partial V}{\partial x}G(x)G^T(x)\left(\frac{\partial V}{\partial x}\right)^T + \frac{1}{2}h^T(x)h(x) \leq 0, \tag{6.58}$$

for all $x \in \mathbb{R}^n$*. Then for every initial state,* $x_0 \in \mathbb{R}^n$*, the system represented by Eq. (6.57) is input–output stable, and its gain does not exceed* γ*.*

Proof. The proof is obtained by completing the squares of $\mathcal{H}_2$ norms of the input and output vectors and showing them to be lower-bounded by the use of the triangle inequality (Khalil 2002).

The Hamilton–Jacobi inequality given by Eq. (6.58) gives a search condition for the stabilizing Lyapunov function, $V(x)$, which is to be found by solving the partial differential equation (Hamilton–Jacobi equation) derived by replacing the inequality by equality. This is a formidable problem requiring a numerical iterative solution, and is closely related to the condition derived by the optimal control formulation (Tewari 2011). The relationship between Lyapunov stability and input–output stability is explored by the following lemmas.

Lemma 6.4.3 *Suppose a system given by Eq. (6.57) satisfies the conditions of Theorem 6.4.2 on a domain $D_R$opfn containing the origin, $x = 0$, with $f(.) : \mathbb{D} \to \mathbb{R}^n$ being a continuously differentiablemap such that $f(0) = 0$ and the origin, $x = 0$, an asymptotically stable equilibrium point of $x = f(x)$.*

Lemma 6.4.4 *Suppose a system given by Eq. (6.57) satisfies the conditions of Theorem 6.4.2 on a domain $\mathbb{D} \subset \mathbb{R}^n$ containing the origin, $x = 0$, with $f(.) : \mathbb{D} \to \mathbb{R}^n$ being a continuously differentiable map such that no solution of $\dot{x} = f(x)$, except the trivial solution, $x(t) = 0$, can stay identically in the subdomain $\mathbb{S} : \{x \in \mathbb{D} | h(x) = 0\}$. Then the system, $\dot{x} = f(x)$, is asymptotically stable and there exists a real and positive number k_1, such that for each $| x_0 | \leq k_1$, the system is input–output stable, with its gain not exceeding γ.*

6.4.2 *Input-State Stability*

A useful stability requirement related to input–output stability is input-state stability, which requires that the state response due to a non-zero initial condition and a bounded input must be bounded by a scalar function of the supremum norm of the input taken over all times, and a contribution from the norm of the initial condition exponentially decaying with time.

Definition 6.4.5 *The autonomous system represented by*

$$\dot{x} = f(x, u) \tag{6.59}$$

and possessing the equilibrium point $x = 0, u = 0$, is said to be input-state stable (ISS) if for any given initial state, $x(0)$, and for any continuous and bounded input, $u(t)$, for $0 \leq t < \infty$, the state solution, $x(t)$, exists for all $t \geq 0$ and satisfies the following condition:

$$| x(t) | \leq \beta (| x(0) |, t) + \gamma \left(\sup_{0 \leq \gamma \leq t} | u(\tau) |\right), \tag{6.60}$$

where $\beta(\alpha, t)$ and $\gamma(\alpha)$ are strictly increasing functions of α (a positive real number), with the property $\beta(0, t) = \gamma(0) = 0$, and β is decrescent with time, with the property $\lim_{t\to\infty} \beta(\alpha, t) = 0$.

6.5 Passivity

Passivity of a nonlinear system is a physical property where external energy must be continuously supplied to drive the system's response. In other words, a passive system does not have an active energy source that can sustain the system's output when the external energy sources are removed. In aeroelastic applications, passivity comes into the picture when we examine whether self-sustained oscillations such as flutter and transonic limit cycles are possible. If the system can extract energy from the airstream to sustain its oscillations, then it is not passive (i.e. it is active) and thus unstable in the Lyapunov sense. While passivity theory is generally used for dealing with non-autonomous systems, it can be applied to the design of stabilizing controllers for autonomous systems. A stronger stabilizing property than passivity is dissipativity (or strict passivity), which requires a decrease of a system's stored energy with time. The concept of dissipativity can be used to design nonlinear feedback control systems with appropriate Lyapunov functions, which are now considered to be positive-definite, energy storage functions. Thus dissipativity is a useful tool not only in stability analysis but also in deriving Lyapunov-based nonlinear regulators for ASE systems.

Consider an autonomous system described as follows:

$$\begin{aligned} \dot{x} &= f(x, u) \\ y &= h(x, u), \end{aligned} \tag{6.61}$$

where $f(.) : \mathbb{R}^n \times \mathbb{R}^m \to \mathbb{R}^n$ is locally Lipschitz, with property $f(0, 0) = 0$ and $h(.) : \mathbb{R}^n \times \mathbb{R}^m \to \mathbb{R}^p$ is continuous and has the property $h(0, 0) = 0$.

Definition 6.5.1 *The autonomous system represented by Eq. (6.61) is said to be passive with respect to a supply rate function, $r(u, y) : \mathbb{R}^m \times \mathbb{R}^p \to \mathbb{R}$, with the property $r(0, 0) = 0$, if there exists a continuously differentiable storage function, $V(x)$, such that*

$$r(u, y) \geq \dot{V} = \frac{\partial V}{\partial x} f(x, u), \quad \text{for all } (x, u) \in \mathbb{R}^n \times \mathbb{R}^m. \tag{6.62}$$

If the inequality in Eq. (6.62) is replaced by an equality, then the system is said to be *loss-less*. If the inequality is changed to the following with $R(u)$ a scalar function:

$$r(u, y) \geq \dot{V} + R(u), \quad R(u) > 0 \quad \text{for all } u \neq 0 \in \mathbb{R}^m, \tag{6.63}$$

then the system is said to be *input strictly passive*. If the following inequality is satisfied for $Q(y)$ a scalar function:

$$r(u, y) \geq \dot{V} + Q(y), \quad Q(y) > 0 \quad \text{for all } y \neq 0 \in \mathbb{R}^p, \tag{6.64}$$

then the system is said to be *output strictly passive*. Finally, if the inequality satisfied by $r(u, y)$ $V(x)$, and a positive-definite scalar function, $\psi(x)$, is the following:

$$r(u, y) \geq \dot{V} + \psi(x), \tag{6.65}$$

then the system is said to be *strictly passive* (or dissipative).

The supply rate, $r(u, y)$ is the rate of energy (power) supplied to the system, while the time derivative of $V(x)$ represents the rate of change of energy stored in the system. The following is an alternative expression of Eq. (6.62):

$$\dot{V} = r(u, y) - g(t), \tag{6.66}$$

where $g(t) : \mathbb{R} \to \mathbb{R}$ is the power dissipation function. A passive system must have $g(t) \geq 0$, provided $V(x) > 0$ (i.e. there is a positive energy stored initially). Furthermore, if the net power input is non-zero, that is

$$\int_0^\infty r[u(t), y(t)]\mathrm{d}t \neq 0, \tag{6.67}$$

and the net power dissipation is positive,

$$\int_0^\infty g(t)\mathrm{d}t > 0, \tag{6.68}$$

then the passive system is dissipative.

For a square system ($m = p$), the scalar product $r(u, y) = u^T y$ is an example of the supply rate function, and much of passivity literature is devoted to such systems. The aeroelastic plant derived by Lagrange's equations (Chapter 3) with u being the generalized forces vector and y the generalized coordinates vector is an example of such a system. It can be shown that a structural dynamics system with a non-zero viscous damping matrix is a dissipative system. However, when the feedback loop between structural dynamics and unsteady aerodynamics subsystems is closed, the resulting aeroelastic system could become active (thus unstable) leading to a phenomenon like flutter (see Chapter 4). It can be shown (Slotine 1995) that a system formed by two subsystems, (V_1, g_1) and (V_2, g_2), in either a negative feedback configuration or in a parallel connection, is passive, provided that the sum of the respective dissipation functions is positive, $g_1(t) + g_2(t) > 0$, and the sum of their storage functions is lower bounded, $V_1 + V_2 > 0$. This includes the possibility of one of the subsystems being active, that is, $g_1 < 0$, hence a formal procedure can be applied to design a nonlinear feedback (or feedforward) controller with a dissipative function, $g_2(t) > 0$, in order to stabilize the overall system. This approach can be extended to several subsystems in either a feedback or a parallel connection.

6.5.1 Positive Real Transfer Matrix

Definition 6.5.2 *The transfer matrix, $G(s) : \mathbb{C} \to \mathbb{R}^{m\times m}$, of a square, strictly proper, LTI system is called positive real if the following conditions are satisfied:*

(a) The poles of all its elements have non-negative real parts.
(b) For all frequencies $\omega \in \mathbb{R}$ for which $s = i\omega$ is not a pole of $G(s)$, the matrix $G(i\omega) + G^T(-i\omega)$ is positive semi-definite.
(c) Any pole on the imaginary axis, $s = i\omega$, is simple and the following residue matrix is positive semi-definite Hermitian:

$$\lim_{s\to i\omega} (s - i\omega)G(s).$$

If $G(s - \epsilon)$ is positive real for some $\epsilon > 0$, then $G(s)$ is said to be strictly positive real (SPR). A transfer matrix must satisfy the following necessary conditions for being SPR:

(a) The poles of all its elements have positive real parts.
(b) For all frequencies $\omega \in \mathbb{R}$, $\mathrm{Re}[G(i\omega)] > 0$.

These conditions imply that the Nyquist plot of $G(i\omega)$ must lie entirely in the right-half plane, hence the phase shift of the system's response to sinusoidal inputs is always less than 90°. This property can be used to demonstrate the impossibility of flutter in a simple plunging oscillation (Chapter 4).

6.5.1.1 Positive Real Lemma

Lemma 6.5.3 *A transfer matrix realization, $G(s) = C(sI - A)^{-1}B + D : \mathbb{C} \to \mathbb{R}^{m\times m}$, of a square, strictly proper, LTI system, with (A, B) controllable and (A, C) observable, is positive real if and only if there exists a symmetric, positive-definite matrix P, and matrices L and W, such that*

$$\begin{aligned} A^T P + PA &= -L^T L \\ PB &= C^T - L^T W \\ W^T W &= D + D^T. \end{aligned} \tag{6.69}$$

Proof. The proof is obtained (Khalil 2002) by substituting the state-transition matrix of the LTI system into the conditions of Definition 6.5.2 of a positive real transfer matrix.

6.5.1.2 Kalman–Yakubovich–Popov Lemma

Lemma 6.5.4 *A transfer matrix realization, $G(s) = C(sI - A)^{-1}B + D : \mathbb{C} \to \mathbb{R}^{m\times m}$, of a square, strictly proper, LTI system, with (A, B) controllable and (A, C) observable, is strictly positive real if and only if there exists a symmetric, positive-definite matrix P, matrices L and W, and a positive real number ϵ, such that*

$$\begin{aligned} A^T P + PA &= -L^T L - \epsilon P \\ PB &= C^T - L^T W \\ W^T W &= D + D^T. \end{aligned} \tag{6.70}$$

Proof. The proof is obtained by showing that the transfer matrix, $G(s - \epsilon/2)$, is positive real.

For the single-input, single-output case, the following variant of Lemma 6.5.4 is available, called the Meyer–Kalman–Yakubovich lemma.

Lemma 6.5.5 *Given a real number $\gamma \geq 0$, a constant Hurwitz matrix, A, constant vectors, b, c, and a constant, symmetric positive-definite matrix, L, if the transfer function realization,*

$$G(s) = \frac{\gamma}{2} + c(sI - A)^{-1}b,$$

is strictly positive real, then there exists a positive real number, ϵ, a vector q and a symmetric, positive-definite matrix P, such that

$$\begin{aligned} A^T P + PA &= -qq^T - \epsilon L \\ Pb &= c^T + q\sqrt{\gamma}. \end{aligned} \tag{6.71}$$

This lemma requires the system to be only stabilizable, and has the following output equation:

$$y = cx + \frac{\gamma}{2}u. \tag{6.72}$$

We now see how an SPR transfer matrix implies a strictly passive LTI system by the following lemma.

Lemma 6.5.6 *The LTI system with a minimal realization given by state-space coefficients (A, B, C, D) is strictly passive if the transfer matrix,*

$$G(s) = C(sI - A)^{-1}B + D,$$

is strictly positive real.

In Lemma 6.5.6, if $G(s)$ is only positive real (not SPR), then the system is passive (and not strictly passive).

6.5.2 *Stability of Passive Systems*

The main utility of passivity is in the stability analysis of nonlinear, passive systems given by Eq. (6.61). In this book, we are concerned only with Hamiltonian systems (such as the aeroelastic plant) whose equations can be derived by Lagrange's formulation to be the following:

$$\frac{\mathrm{d}}{\mathrm{d}t}\left[\frac{\partial \mathcal{L}}{\partial \dot{q}}(q, \dot{q})\right]^T - \left[\frac{\partial \mathcal{L}}{\partial q}(q, \dot{q})\right]^T + \left[\frac{\partial \mathcal{D}}{\partial \dot{q}}(\dot{q})\right]^T = Q, \tag{6.73}$$

where $q(t) \in \mathbb{R}^n$ are the generalized coordinates, $Q(t) \in \mathbb{R}^n$ the generalized aerodynamics and control forces, $\mathcal{L}(q, \dot{q}) : \mathbb{R}^n \times \mathbb{R}^n \to \mathbb{R}$ is the Lagrangian, and $\mathcal{D}(\dot{q}) : \mathbb{R}^n \to \mathbb{R}$ is a viscous dissipation (Rayleigh damping) function. If the external inputs are treated to be the generalized aerodynamics and control forces vector, $u = Q$, while the outputs are generalized coordinates, $y = q$, then such a system is square and its supply rate can be given by $r = u^T y$. If $\mathcal{D}(.) = 0$, the system is passive, but not strictly passive (dissipative). The following lemmas analyse the stability of such systems.

Lemma 6.5.7 *If the system given by Eq. (6.61) is output strictly passive with $u^T y \geq \dot{V} + \delta y^T y$ for some real, positive number, δ, then the system is input–output stable, and its gain is less than or equal to $1/\delta$.*

Proof. The proof is produced by completing the squares of $\dot{V}$, integrating and applying the triangle inequality (Khalil 2002).

Lemma 6.5.8 *If the system given by Eq. (6.61) is passive with a positive-definite storage function, $V(x)$, then the origin of the unforced system, $\dot{x} = f(x, 0)$, is stable in the sense of Lyapunov.*

Proof. The proof is obtained by taking $V(x)$ to be the candidate Lyapunov function for the unforced system ($u = 0$) and showing that $\dot{V} \leq u^T y = 0$.

Definition 6.5.9 *The system given by Eq. (6.61) is said to be zero-state observable if no solution of the unforced system, $\dot{x} = f(x, 0)$, can identically stay in the region $S = \{x \in \mathbb{R}^n | h(x, 0) = 0\}$, other than the trivial solution, $x(t) = 0$.*

Lemma 6.5.10 *If the system given by Eq. (6.61) is either strictly passive, or it is output strictly passive and zero-state observable, then the origin of the unforced system, $\dot{x} = f(x, 0)$, is asymptotically stable in the sense of Lyapunov. If, in addition, the storage function, $V(x)$, is radially unbounded, then the system is globally asymptotically stable.*

Proof. The proof is obtained by taking $V(x)$ to be the candidate Lyapunov function for the unforced system ($u = 0$), integrating the inequality involving the supply rate for $u = 0$ and showing that either $\dot{V} \leq -\psi(x)$ or $\dot{V} \leq 0$, depending upon whether the system is strictly passive, or output strictly passive and zero-state observable. Radial unboundedness of $V(x)$ guarantees global asymptotic stability by Theorem 6.2.1.

Definition 6.5.11 *The system given by Eq. (6.61) is said to be completely reachable if for all $x_0 \in \mathbb{R}^n$, there exists a finite time, $\tau < t_0$, and a square-integrable input, $u(t)$, defined in the interval $\tau \leq t \leq t_0$, such that the state $x(t), t \geq \tau$ can be driven from $x(\tau) = 0$ to $x(t_0) = x_0$.*

6.5.2.1 Extended Kalman–Yakubovich–Popov Lemma

The Kalman–Yakubovich–Popov lemma, which was presented originally for LTI systems, can be extended to nonlinear, control-affine systems as follows (Haddad and Chellaboina 2008).

Lemma 6.5.12 *Let $Q \in \mathbb{R}^p$ and $R \in \mathbb{R}^m$ be constant, symmetric matrices, $S \in \mathbb{R}^{p \times m}$ be a constant matrix, and the following system be zero-state observable and completely reachable:*

$$\dot{x} = f(x) + G(x)u, \quad x(0) = x_0$$
$$y = h(x) + J(x)u, \tag{6.74}$$

where $f(.) : \mathbb{R}^n \to \mathbb{R}^n$ is locally Lipschitz and continuously differentiable with property $f(0) = 0$, and $G(.) : \mathbb{R}^n \to \mathbb{R}(n \times m)$, $h(.) : \mathbb{R}^n \to \mathbb{R}^p$ and $J(.) : \mathbb{R}^n \to \mathbb{R}^{p \times m}$ are continuously differentiable over $\mathbb{R}^n$ with the property $h(0) = 0$. Then the system is passive with respect to a quadratic supply rate, $r(u, y) = y^T Q y + 2y^T S u + u^T R u$, if and only if there exists a continuously differentiable, positive-definite scalar function, $V(x)$, with property $V(0) = 0$, *a vector function,* $\ell(x) : \mathbb{R}^n \to \mathbb{R}^\nu$*, and a matrix function,* $W(x) : \mathbb{R}^n \to \mathbb{R}^{\nu \times m}$*, such that for all* $x \in \mathbb{R}^n$*,*

$$\begin{aligned} V'(x)f(x) - h^T(x)Qh(x) + \ell^T(x)\ell(x) &= 0 \\ \frac{1}{2}V'(x)G(x) - h^T(x)[QJ(x) + S] + \ell^T(x)W(x) &= 0 \\ R + S^T J(x) + J^T(x)S + J^T(x)QJ(x) - W^T(x)W(x) &= 0. \end{aligned} \tag{6.75}$$

Alternatively, if we have $N(x) = R + S^T J(x) + J^T(x)S + J^T(x)QJ(x) > 0$ *for all* $x \in \mathbb{R}^n$*, then the system is passive with respect to a quadratic supply rate,* $r(u, y) = y^T Qy + 2y^T Su + u^T Ru$*, if and only if there exists a continuously differentiable, positive-definite scalar function,* $V(x)$*, with property* $V(0) = 0$*, such that for all* $x \in \mathbb{R}^n$*,*

$$\begin{aligned} 0 \geq\ & V'(x)f(x) - h^T(x)Qh(x) + \ell^T(x)\ell(x) \\ & + \left[\frac{1}{2}V'(x)G(x) - h^T(x)\{QJ(x) + S\}\right] \qquad (6.76) \\ & \times N^{-1}(x)\left[\frac{1}{2}V'(x)G(x) - h^T(x)\{QJ(x) + S\}\right]^T. \qquad (6.77) \end{aligned}$$

Proof. See Haddad and Chellaboina (Haddad and Chellaboina 2008), pp. 347–348.

If the system satisfies the necessary and sufficient conditions for passivity given by Lemma 6.5.12, then the origin, $x = 0, u = 0$, is stable in the sense of Lyapunov.

6.5.3 Feedback Design for Passive Systems

Consider the following autonomous, square system described as follows:

$$\begin{aligned} \dot{x} &= f(x, u) \\ y &= h(x), \end{aligned} \tag{6.78}$$

where $f(.) : \mathbb{R}^n \times \mathbb{R}^m \to \mathbb{R}^n$ is locally Lipschitz and continuous, with property $f(0, 0) = 0$, such that $x = 0, u = 0$ is an equilibrium point, and $h(.) : \mathbb{R}^n \to \mathbb{R}^m$ is continuous, with the property $h(0) = 0$. Such a system is passive with respect to the supply function $u^T y$ (Def. 6.5.1), if there exists a continuously differentiable storage function, $V(x)$, such that

$$u^T y \geq \dot{V} = \frac{\partial V}{\partial x} f(x, u), \quad \text{for all } (x, u) \in \mathbb{R}^n \times \mathbb{R}^m.$$

Furthermore, if the system is zero-state observable, it can be rendered asymptotically stable by selecting a suitable output feedback law, provided a positive-definite, radially unbounded storage function, $V(x)$, can be found. These facts are summarized in the following theorem.

Theorem 6.5.13 *If the system given by Eq. (6.78) is passive with a positive-definite, radially unbounded storage function, and is also zero-state observable, then the origin,* $x = 0, u = 0$*,* *can be globally asymptotically stabilized by an output feedback law,* $u = -g(y)$*, where* $g(.)$ *is any locally Lipschitz function, such that* $g(0) = 0$ *and* $y^T g(y) > 0$ *for all* $y \neq 0$*.*

Proof. The proof can be obtained by selecting the storage function, $V(x)$, as the Lyapunov function for the closed-loop system, and showing the negative-definite nature of $\dot{V}$.

The physical principle behind Theorem 6.5.13 is evident when the storage function is treated as the energy stored in the system. A passive system can remain in the arbitrarily small neighbourhood of the equilibrium point at the origin by itself. However, if damping is added by closing the loop by $u = -g(y)$, the system becomes asymptotically stable (dissipative), and thus tends to the origin in the steady state. The choice of the function $g(y)$ hinges upon the physical constraints on the magnitude (and possibly the rate) of the inputs, $u(t)$. If a control-affine system with state equation

$$\dot{x} = f(x) + G(x)u.$$

is not passive (active) with some outputs, it can be rendered passive merely by selecting different outputs, such that

$$y = h(x) = G^T(x)\frac{\partial V}{\partial x}^T.$$

Then a feedback law, $u = -g(y)$, is selected to render the system globally asymptotically stable. Such a design approach is very useful in dealing with ASE systems, wherein passivity can be achieved by merely changing the location of the sensors such that a suitable output equation is derived.

References

Haddad WM and Chellaboina V 2008 *Nonlinear Dynamical Systems and Control.* Princeton University Press, Princeton, NJ.

Khalil HS 2002 *Nonlinear Systems*, 3rd ed. Prentice-Hall, Upper Saddle River, NJ.

Slotine JJE and Li W 1995 *Applied Nonlinear Control.* Prentice-Hall, Englewood Cliffs, NJ.

Tewari A 2011 *Advanced Control of Aircraft, Spacecraft, and Rockets.* John Wiley & Sons, Ltd., Chichester.

7

Nonlinear Oscillatory Systems and Describing Functions

7.1 Introduction

Unstable nonlinear systems often display an oscillatory behaviour that can be analysed in an approximate manner using some of the frequency-domain techniques employed in linear systems analysis. Deriving a describing function model for a nonlinear aeroelastic plant is one such method. This approach involves an approximation of all the signals in the unforced control system by only the first (fundamental) harmonic component of a Fourier series expansion. Such an approximation is seen to be valid when the control system can be represented by a linear subsystem in a feedback loop with a time-independent (static) nonlinearity. In such a case, the linear subsystem acts as a low-pass filter, and all higher harmonic components except the fundamental one are attenuated. The nonlinear part of the system usually provides an energy dissipation mechanism, such that the response of the unstable linear subsystem does not tend to infinite magnitudes, but instead remains in a bounded, constant amplitude oscillation called a limit cycle. The describing function analysis therefore consists of a solution for the frequency and amplitude of the limit-cycle oscillation (LCO).

A typical aeroservoelastic (ASE) system can exhibit nonlinear oscillatory behaviour, either due to inertial properties and restoring mechanisms inherently present in the aeroelastic plant or due to deliberately designed feedback controller elements for achieving desired closed-loop stability objectives. In the former case, the plant's nonlinear behaviour is seldom dynamic (time-dependent) in nature, and hence can be represented by the following functional relationship between input, $e(t)$, and output, $z(t)$:

$$z = f(e), \tag{7.1}$$

where $f(.)$ is a nonlinear operator. Such a nonlinearity is called a static (also memoryless or time-invariant) nonlinearity, because it is independent of the history of the input operated by it. In a feedback loop depicted in Fig. 7.1, a static nonlinearity is seen to act upon the signal generated by a linear subsystem, which can be modelled by a transfer function, $G(s)$, and understood to be the product of the transfer functions of linear subsystems of the plant and

Adaptive Aeroservoelastic Control, First Edition. Ashish Tewari.

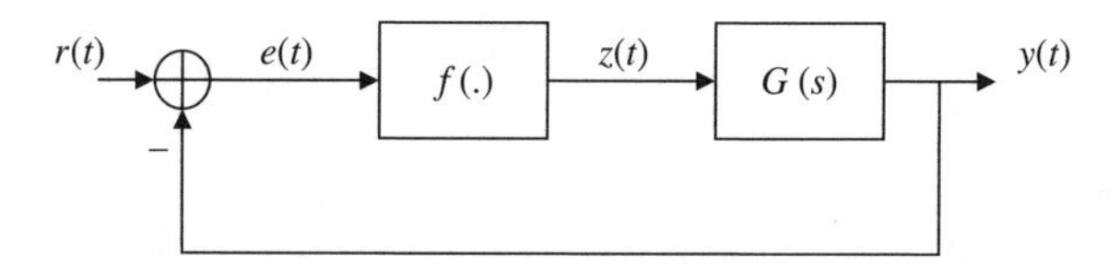

Figure 7.1 Basic control system with a static nonlinearity

the controller.[1] The presence of the nonlinearity can cause a change in the overall control system's characteristics, such as stability and response, to known inputs and initial conditions. For example, for a given set of initial conditions and bounded inputs, a nonlinear control system can produce a bounded response, an unbounded response or an LCO. Apart from modifying the stability characteristics, the nonlinearity in a typical aeroelastic system can change the oscillatory behaviour displayed by the output waveform (frequency, amplitude and phase). This is illustrated by the system shown in Fig. 7.2, where the aeroelastic plant is assumed to be an undamped, single degree-of-freedom, second-order, linear system of transfer function,

$$\frac{Y(s)}{Z(s)} = \frac{9}{s^2 + 9},$$

in series with an aerodynamic nonlinearity due to either unsteady flow separation or oscillating shock waves, represented by a relay with hysteresis. For $0 \leq e(t) \leq 1.01$, the relay output, $z(t)$, is not uniquely defined. If the error signal supplied to the nonlinear relay block, $e(t) = r(t) - u(t)$, is greater than 1.01, then the relay output remains constant at $z = 1$. When $e(t)$ drops below 0, the output is constant at $z = 0.3$. The negative feedback loop is closed by a proportional-derivative (PD) controller of the transfer function

$$\frac{U(s)}{Y(s)} = 0.2 + 0.1s.$$

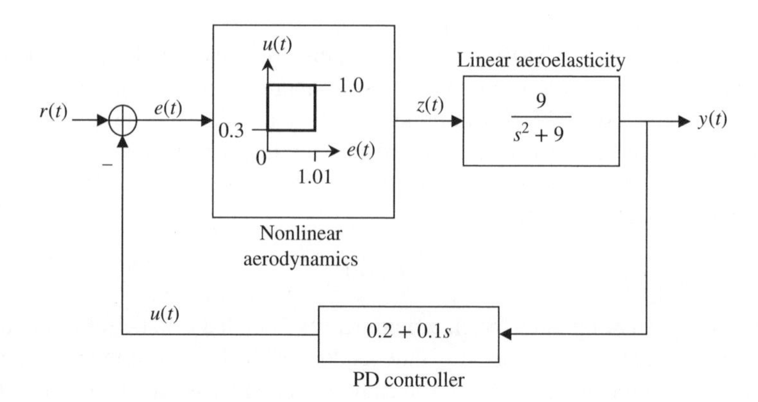

Figure 7.2 A simple nonlinear ASE system with hysteresis model for unsteady aerodynamics

[1] Irrespective of the type of the closed-loop connection, the linear subsystems of the plant, $H(s)$, and controller, $K(s)$, can be combined into a resultant linear block, $G(s) = H(s)K(s)$, through successive block operations.

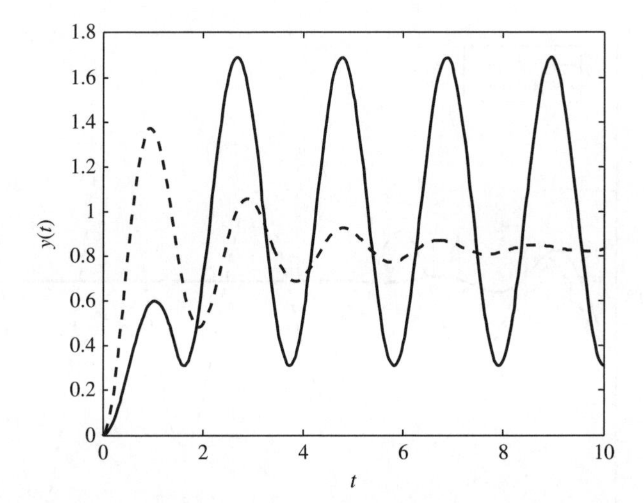

Figure 7.3 Step response of the nonlinear ASE system (solid line) of Fig. 7.2 compared with that (dashed line) without the nonlinearity

The closed-loop response to a unit-step reference input, $R(s) = 1/s$, with zero initial condition is compared in Fig. 7.3 with what it would have been in the absence of the hysteresis nonlinearity. The drastic change in the system's stability as well as in the waveform of the output signal is clearly illustrated. While the linear closed-loop system is stable and is seen to reach a constant steady state, the nonlinear system has a LCO with a frequency smaller than the natural frequency of the linear system (3 rad/s), and with a phase difference.

If the hysteresis block in Fig. 7.2 is replaced by the nonlinearity $f(u) = u^3$, then the result is a stable step response with a faster convergence to the steady state than that of the basic linear system, as shown by Fig. 7.4. However, with $f(u) = e^{-u}$, the step response becomes unbounded (Fig. 7.4).

7.2 Absolute Stability

The concept of absolute stability of passive feedback systems with a static nonlinearity is very useful in the design of stabilizing controllers. It is based on the application of Nyquist-like stability criteria to nonlinear feedback systems. Consider a single-input, single-output (SISO) nonlinear ASE system represented by a closed-loop connection of a linear subsystem of a strictly proper transfer function, $G(s) = c(sI - A)^{-1}b$, and a static nonlinearity, $f(.)$ (Fig. 7.1). For the active stabilization problem, we are only concerned with the regulation of the system at a given set point (an equilibrium point) for which $r(t) = 0$. Thus $r(t)$ is taken to be 0 in further analysis. The state equations of the system can be expressed as follows:

$$\dot{x} = Ax - bf(y)$$

$$y = cx. \tag{7.2}$$

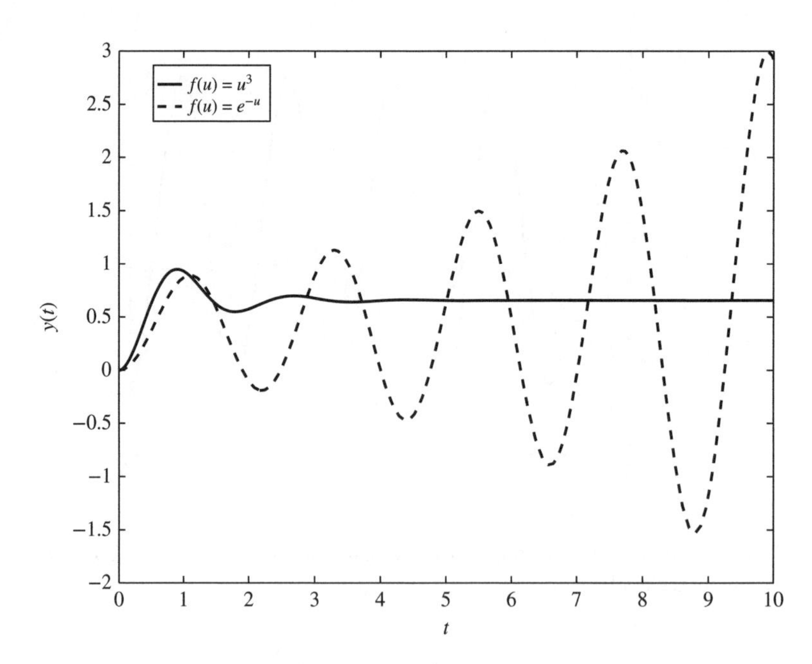

Figure 7.4 Step responses of the nonlinear ASE system with the nonlinear block in Fig. 7.2 replaced by the nonlinearity $f(u) = u^3$ (solid line) and $f(u) = e^{-u}$ (dashed line) showing, respectively, the stable and unbounded behaviour

If the nonlinearity were replaced by a linear feedback controller, $f(y) = ky$, with a constant gain $k > 0$, then the closed-loop stability would depend upon the eigenvalues of the matrix, $A - bkc$. This concept is extended to a general nonlinear function, $f(y)$, by the following Aizerman conjecture.

Definition 7.2.1 *If $f(y)$ can be represented by*

$$k_1 \leq \frac{f(y)}{y} \leq k_2, \quad \text{for all } y \neq 0, \tag{7.3}$$

then the nonlinearity $f(y)$ is said to belong to sector $[k_1, k_2]$.

The Aizerman conjecture postulates that if $f(y)$ belongs to the sector $[k_1, k_2]$, then the stability of the closed-loop system given by Eq. (7.2) can be analysed from the eigenvalues of the matrix, $A - bkc$, where $k_1 \leq k \leq k_2$.

If the Aizerman conjecture were true, then the stability of a nonlinear system could be easily analysed by treating it as if it was a linear system. Unfortunately, this is a false conjecture, because there are several examples to show its invalidity. However, if additional conditions to be satisfied by $f(y)$ and $G(s)$ were imposed, it might be possible to extend the linear systems analysis to nonlinear systems. The Popov stability criterion provides such a condition to be satisfied by $G(s)$.

7.2.1 Popov Stability Criteria

Theorem 7.2.2 *If the closed-loop system given by Eq. (7.2) satisfies the following conditions:*

(1) The linear dynamics matrix, A, is Hurwitz and the pair (A, b) is controllable.
(2) The nonlinearity, $f(y)$, belongs to the sector, $[0, k]$.
(3) There exist two positive real numbers a and ϵ, such that

$$\mathrm{Re}\{(1 + ia\omega)G(i\omega)\} \geq \epsilon - \frac{1}{k}, \quad \text{for all } \omega \geq 0, \tag{7.4}$$

then the equilibrium point at origin, $x = 0$, is globally asymptotically stable.

The proof of the Popov criterion can be established by the Kalman–Yakubovich lemma (Chapter 6) through a suitable candidate Lyapunov function. Since this stability criterion promises stability for all possible nonlinearities $f(y)$ belonging to a given sector, it is also termed as the criterion for absolute stability. However, as in the case of Lyapunov stability theorems, Popov criterion only gives a sufficient condition for stability. Popov's inequality (Eq. (7.4)) can be graphically represented by a condition where the polar plot of the modified frequency response function

$$H(i\omega) = \mathrm{Re}[G(i\omega)] + i\omega\mathrm{Im}[G(i\omega)] = X + iY,$$

(called the Popov plot) always lies below the straight line $X - aY + 1/k = 0$. Thus the Popov criterion is quite similar to the Nyquist stability criterion for linear SISO systems (Chapter 2), where the stability is examined by analysing the distance of the frequency response, $G(i\omega)$, from the point $(-1, 0)$.

7.2.2 Circle Criterion

Nyquist-like analysis for nonlinear oscillatory systems is enabled by the following theorem, called the circle criterion (Slotine 1995).

Theorem 7.2.3 *If the closed-loop system given by Eq. (7.2) satisfies the following conditions:*

(1) The linear dynamics matrix, A, has no eigenvalues on the imaginary axis and has N eigenvalues in the right-half plane.
(2) The nonlinearity, $f(y)$, belongs to the sector, $[k_1, k_2]$.
(3) Any one of the following is true:
 (a) $0 < k_1 \leq k_2$, and the Nyquist plot of $G(i\omega)$ circles the disk, $D(k_1, k_2)$ (Fig. 7.5) exactly N times in a counterclockwise direction, without entering it.
 (b) $0 = k_1 < k_2$, and the Nyquist plot of $G(i\omega)$ stays strictly in the half plane defined by $\mathrm{Re}(s) > -1/k_2$.
 (c) $k_1 < 0 < k_2$, and the Nyquist plot of $G(i\omega)$ stays in the interior of the disk, $D(k_1, k_2)$ (Fig. 7.5).

(d) $k_1 < k_2 < 0$, *and the Nyquist plot of* $-G(i\omega)$ *circles the disk,* $D(-k_1, -k_2)$ *(Fig. 7.5) exactly N times in a counterclockwise direction, without entering it.*

then the equilibrium point at origin, $x = 0$, *is globally asymptotically stable.*

The circle criterion replaces the point $-1/k$ in the Nyquist stability criterion by a circle of Fig. 7.5. For $k_2 \to k_1$, the circle shrinks to the point $-1/k_1$. It is also an absolute stability criterion, because it is not limited to a particular nonlinearity, $f(y)$, but to all possible nonlinearities belonging to a given sector, $[k_1, k_2]$.

In order to understand the derivation of the circle criterion, let us consider the block diagram of Fig. 7.6 showing the closed-loop connection between linear transfer function, $G(s)$, and a static nonlinearity, $f(.)$, driven by reference inputs r_1, and r_2. The system is in the form (Fig. 2.3) used for the application of the small gain theorem in Chapter 2 to derive input–output stability of closed-loop, causal systems. Since the small gain theorem is applied to the gain of signals, it can be applied even if a subsystem is nonlinear. If the nonlinearity $f(y)$ belongs to sector $[k_1, k_2]$, then the small gain theorem would require the following condition to be satisfied

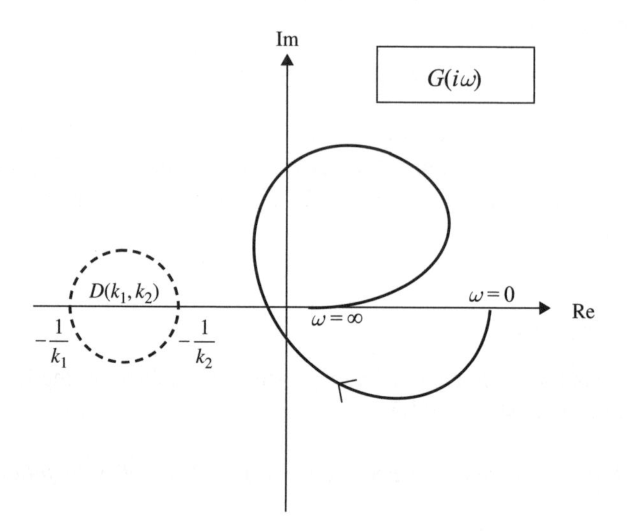

Figure 7.5 Nyquist plot of $G(i\omega)$ for applying the circle criterion

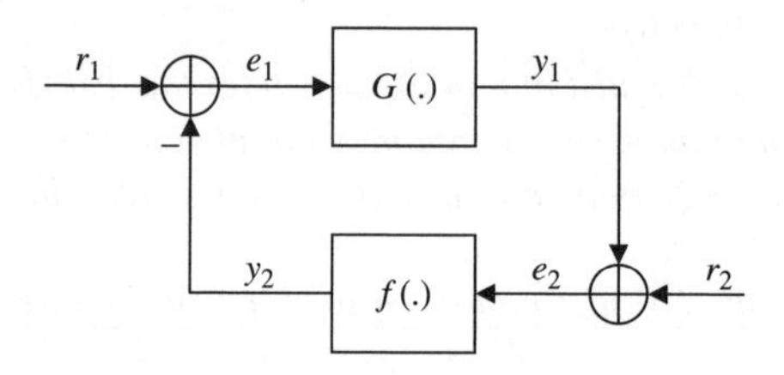

Figure 7.6 Block-diagram representation of a system with a static nonlinearity for the application of small gain theorem

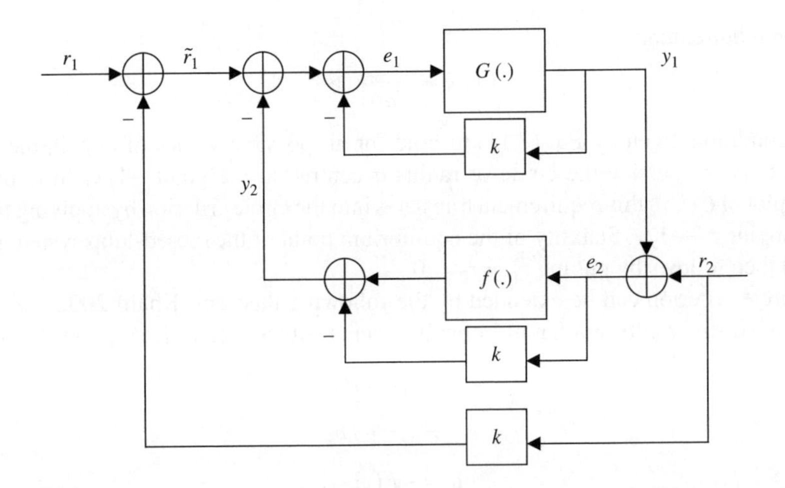

Figure 7.7 Transformation of the system of Fig. 7.6 by adding negative-feedback connections with a constant gain $k = (k_1 + k_2)/2$

by $G(s)$ for input–output stability:

$$k_2 \sup_{\omega} \mid G(i\omega) \mid < 1, \tag{7.5}$$

which is overly restrictive, because a large loop gain may be necessary at some frequencies in order to reduce the effects of disturbances (robustness). Instead, we define an average controller gain by

$$k = \frac{k_1 + k_2}{2}, \tag{7.6}$$

and transform the original system by adding negative feedback connections to $G(s), f(.)$, and between the reference signals, r_1, r_2, as shown in Fig. 7.7. The small gain theorem is now applied to the effective closed-loop subsystems represented by the linear transfer function,

$$\tilde{G}(s) = \frac{G(s)}{1 + kG(s)}, \tag{7.7}$$

and the nonlinearity,

$$\tilde{f}(y) = f(y) - ky, \tag{7.8}$$

driven by the reference inputs $\tilde{r}_1 = r_1 - kr_2$ and r_2. Since the nonlinearity, $\tilde{f}(y)$, is seen to satisfy the following inequality:

$$\frac{\tilde{f}(y)}{y} \leq \sigma, \quad \text{for all } y \neq 0, \tag{7.9}$$

where $\sigma = (k_2 - k_1)/2$, the small gain theorem results in the following criterion for stability:

$$\sigma \mid \tilde{G}(i\omega) \mid < 1, \tag{7.10}$$

in addition to the requirement that $\tilde{G}(s)$ must not have any of its poles in the right-half s-plane. Combination of the stability requirements given by Eqs (7.5) and (7.10) yields the following requirement:

$$\left|k + \frac{1}{G(i\omega)}\right| > \sigma. \tag{7.11}$$

For the condition given by Eq. (7.11) to hold for all possible values of $\omega \geq 0$, the locus of $1/G(i\omega)$ must lie outside the circle of radius σ centred at the point $-1/k$. In terms of the Nyquist plot of $G(i\omega)$, this requirement translates into the circle criterion by applying the reciprocal mapping $z \rightarrow 1/z$. Stability of the equilibrium point of the closed-loop system given by Eq. (7.2) then follows by putting $r_1 = r_2 = 0$.

The circle criterion can be extended by the following theorem (Khalil 2002) for absolute stability of square, multi-input, multi-output systems with a state-space representation:

$$\dot{x} = Ax + Bu$$

$$y = Cx + Du$$

$$u = -F(y), \tag{7.12}$$

where $x \in \mathbb{R}^n$, $u \in \mathbb{R}^m$, $y \in \mathbb{R}^m$, $F(.) : \mathbb{R}^m \in \mathbb{R}^m$, (A, B) is controllable and (A, C) is observable.

Definition 7.2.4 *A square mapping function, $F(.) : \mathbb{R}^m \in \mathbb{R}^m$, is said to belong to the sector $[K_1, \infty]$, where K_1 is a symmetric, positive-definite matrix, if*

$$y^T[F(y) - K_1 y] \geq 0. \tag{7.13}$$

$F(y)$ is said to belong to the sector, $[K_1, K_2]$, where $K = K_2 - K_1$ is a symmetric, positive-definite matrix, if

$$[F(y) - K_1 y]^T[F(y) - K_2 y] \leq 0. \tag{7.14}$$

Theorem 7.2.5 *The system given by Eq. (7.12) is absolutely stable, if any of the following conditions are satisfied:*

(1) The nonlinearity $F(y)$ belongs to the sector $[K_1, \infty]$ and $G(s)[I + K_1 G(s)]^{-1}$ is strictly positive real.
(2) The nonlinearity $F(y)$ belongs to the sector $[K_1, K_2]$ and $[I + K_2 G(s)][I + K_1 G(s)]^{-1}$ is strictly positive real.

7.3 Describing Function Approximation

Suppose we are interested in stabilizing a SISO nonlinear ASE system represented by the block diagram of Fig. 7.1, without any external reference input ($r(t) = 0$). The basic assumptions involved in the describing function analysis are that the nonlinear part of the system, $f(e)$, is time-invariant (static) and an odd function of the input $e(t)$. Furthermore, it is assumed that when driven by a simple harmonic input signal, $e(t) = A \sin \omega t$, all the harmonics of the signals $z(t)$ and $y(t)$ except the fundamental mode are neglected, because of the linear subsystem acting as a low-pass filter. This implies that the second- and higher-order harmonics are very much more attenuated by $G(s)$ compared to the fundamental harmonic,

$$| G(i\omega) | >> | G(in\omega) |, \quad n = 2, 3, \ldots, \tag{7.15}$$

By expanding $z(t)$ in Fourier series, we have

$$z(t) = \sum_{n=1}^{\infty} a_n \cos(n\omega t) + b_n \sin(n\omega t), \tag{7.16}$$

where the coefficients a_n, b_n, are given by

$$a_n = \frac{1}{\pi} \int_{-\pi}^{\pi} z(t) a_n \cos(n\omega t) \mathrm{d}(\omega t) \tag{7.17}$$

and

$$b_n = \frac{1}{\pi} \int_{-\pi}^{\pi} z(t) a_n \sin(n\omega t) \mathrm{d}(\omega t). \tag{7.18}$$

When only the fundamental mode of the signal $z(t)$ is taken, we have the approximation

$$z(t) \simeq a_1 \cos(\omega t) + b_1 \sin(\omega t) = M \sin(\omega t + \phi), \tag{7.19}$$

or in the complex representation,

$$z(t) \simeq M e^{\omega t + \phi}. \tag{7.20}$$

The describing function, $N(A)$, for the static nonlinearity, $f(.)$, is defined as the ratio of the fundamental output signal with the driving input signal, given by

$$N(A) = \frac{M e^{\omega t + \phi}}{A e^{i\omega t}} = \frac{M(A)}{A} e^{i\phi(A)}. \tag{7.21}$$

Note that while a linear system has the frequency dependence of the forced harmonic output only, for a nonlinear system, this changes to include amplitude dependence as well. Since the nonlinearity, $f(.)$, is static, the describing function is independent of the frequency of excitation and depends only upon the excitation amplitude, A.

By equating the driving error signal, $e(t)$, with the feedback-return signal, we have the following condition to be satisfied by the system:

$$N(A, \omega) G(i\omega) = -1. \tag{7.22}$$

If a real pair, A, ω, exists satisfying Eqs. (7.21) and (7.22), then a self-sustained LCO exists. If the Nyquist plot of $G(i\omega)$ and the plot of $-1/N(A)$ versus A, both plotted on the same figure, have a point of intersection, then the LCO condition exists. In order to analyse the stability of the LCO, consider a small perturbation on the amplitude, A. If the new amplitude is less than A (for which the LCO exists), then the oscillation is unstable by Nyquist stability criterion. Conversely, if the perturbed amplitude is greater than A, the oscillation is stable.

Describing functions for some common static nonlinearities are listed in Table 7.1. Other nonlinearities, such as hysteresis, backlash and so on, can be found in classical nonlinear control textbooks (Gibson 1963).

Table 7.1 Describing functions for some basic nonlinearities

$f(e)$	$N(A)$	Nomenclature
e^3	$\frac{3}{4}A^2$	Cubic gain
$\begin{cases} 1, & e > e_0 \\ 0, & \lvert e \rvert \leq e_0 \\ -1, & e < -e_0 \end{cases}$	$\frac{4}{A\pi}\sqrt{1-\left(\frac{e_0}{A}\right)^2}, \ (A > e_0)$	Relay with dead zone
$\begin{cases} a\left(\frac{e}{e_0}-1\right), & e > e_0 \\ 0, & \lvert e \rvert \leq e_0 \\ a\left(1-\frac{e}{e_0}\right), & e < -e_0 \end{cases}$	$\begin{cases} \frac{a}{e_0}\left[1-\frac{2}{\pi}\left(\sin^{-1}\left(\frac{e_0}{A}\right)+\frac{e_0}{A}\sqrt{1-\left(\frac{e_0}{A}\right)^2}\right)\right], & A \geq e_0 \\ 0, & A < e_0 \end{cases}$	Free play
$\begin{cases} 1, & e > 1 \\ e, & \lvert e \rvert \leq 1 \\ -1, & e < -1 \end{cases}$	$\begin{cases} \frac{2}{\pi}\left[\sin^{-1}\left(\frac{1}{A}\right)+\frac{1}{A}\sqrt{1-\left(\frac{1}{A}\right)^2}\right], & A > 1 \\ 1, & A \leq 1 \end{cases}$	Saturation

7.4 Applications to Aeroservoelastic Systems

The circle criterion and describing functions can be applied to design-adaptive controllers for nonlinear ASE systems where the nonlinear subsystem is uncertain. This is true for most aeroelastic plants with structural and aerodynamic nonlinearities, which can be represented by standard models such as dead-zone, free play, backlash, nonlinear damping, hysteresis and saturation. Before one attempts to derive a controller for such systems, it is advisable to know the extent of the nonlinearity in terms of its sector. Sometimes, the bounds on the nonlinear operator, $f(y)$, are known a priori, and can be used to determine the controller gains by using the small gain theorem. For example, the limits on a nonlinearly hardening spring (either structural or aerodynamic in source) can be determined by an analytical model or experimental tests. Similarly, complicated phenomena such as shock-induced and separated-flow buffet usually have well-defined frequency spectra, for which an appropriate describing function can be derived and used to adjust the stabilizing controller gains according to the sensed oscillatory response. In this manner, an adaptive control loop is easily established without recourse to more sophisticated adaptation mechanisms.

Most of the nonlinear aeroelastic/ASE strategies found in the literature fall into the describing functions category. Such an approach takes advantage of the fact that the presence of a linear aeroelastic subsystem, $G(s)$, acts as a low-pass filter for the higher Fourier series harmonics arising out of the nonlinearity, $f(.)$, thereby enabling a describing function approximation for the latter. While derived for structural nonlinearities (Gordon *et al.* 2008), these methods can be easily extended to include nonlinear aerodynamic behaviour, because of the feedback interconnection between structural and aerodynamic subsystems (see Chapter 3). Ueda and Dowell (1984) presented one of the few articles on the application of describing functions to specifically represent the aerodynamic nonlinearity associated with transonic flutter. The simplest computational model for unsteady transonic aerodynamics is possible through the low-frequency, transonic small-disturbance equation (see Chapter 3), which can be used to computationally determine the describing functions for transonic flutter analysis. However, while a simple describing function gives a reasonable flutter prediction, its utility is limited to small amplitude motions (Ueda and Dowell 1984). Dowell and Tang (2002) and Dowell *et al.* (2003) provide a survey of nonlinear aeroelastic applications, where the describing function models are seen as practical tools for ASE design and analysis with nonlinear effects. In contrast, while many CFD-based aeroelastic models have appeared in the literature, they are still too time-consuming for use as ASE design tools and are likely to remain so in the foreseeable future. Describing functions can be derived either from analytical and computational results, or wind-tunnel test data. However, the online determination of describing functions by autoregressive schemes (Raveh 2004) and their incorporation via an adaptation loop into self-tuning regulators (Chapter 8) offers the greatest promise in nonlinear adaptive ASE applications.

Describing functions of static nonlinearities can be directly used to build robust controllers for the suppression of flutter and LCOs suppression through the frequency-domain, multivariable design methods employing structured singular-value (μ) synthesis (Maciejowski 1989). The describing function approach fits naturally with the frequency-domain design offered by the μ-synthesis. However, an ASE state-space representation must be derived via linear fractional transformations (LFT) for a plant containing both linear and nonlinear subsystems, which is followed by the derivation of a stabilizing feedback controller. Prominent methods based on μ-synthesis for a robust ASE analysis are those by Lind and Brenner (1999), Borglund (2003), and Baldelli *et al.* (2005). The last of these employs a describing function for structural free-play nonlinearity in the LFT model of the aeroelastic system and analyses its effect on the flutter point. Such a systematic framework can be extended to the presence of aerodynamic describing functions, if available. The nonlinear operators can be updated by an identification loop, which allows an adaptive control implementation of the system. However, before controllers can be designed, a suitable aeroelastic plant to be used in stability analysis is necessary. It has been indicated earlier (see Chapter 4) that the simple incompressible flow models of the Theodorsen type do not lead to accurate flutter-point estimates in the high-subsonic regime. If an aerodynamic describing function for transonic shock-induced behaviour (see Chapter 11) is introduced in a Theodorsen-type model, there is little possibility of it giving a reasonable ASE design. The minimum level of sophistication required in a nonlinear transonic flutter/LCO-describing function model has to be based on the linear aeroelastic subsystem rational function approximation (RFA) in compressible, time-linearized aerodynamics. The latter could be either a doublet-lattice-type model (Tewari 2015) for the three-dimensional wing or a Possio-type compressible integral equation model for the typical section (Appendix B). The low-frequency transonic small-disturbance (LFTSD) model could be the best choice of time-linearized aerodynamics underlying the superimposed shock-induced nonlinear behaviour. However, such a model requires iterative computations at present. If analytical results can be derived for the LFTSD equation in the future, it will be a big leap in the adaptive transonic ASE design process.

7.4.1 Nonlinear and Uncertain Aeroelastic Plant

Consider an aeroelastic system represented by the following equations of motion:

$$M\ddot{\xi} + C\dot{\xi} + K\xi + f(\zeta) = Q + Tu, \tag{7.23}$$

where $\xi(t) : \mathbb{R} \to \mathbb{R}^n$ is the generalized coordinates vector corresponding to the n degrees of structural freedom (including the m control-surface degrees of freedom), $u(t) : \mathbb{R} \to \mathbb{R}^m$ is the vector of control torque inputs applied by m servo-actuators, and $M \in \mathbb{R}^{n\times n}$, $C \in \mathbb{R}^{n\times n}$, $K \in \mathbb{R}^{n\times n}$ and $T \in \mathbb{R}^{n\times m}$ are the generalized mass, damping, stiffness and control transmission matrices, respectively. The generalized aerodynamic force vector, $Q(t) : \mathbb{R} \to \mathbb{R}^n$ is assumed to be linearly related to $\xi(t)$, $\dot{\xi}(t)$ and $\ddot{\xi}(t)$, as well as to certain additional state variables collected into the aerodynamic lag state vector, $x_a(t) : \mathbb{R} \to \mathbb{R}^\ell$, which is necessary for modelling the aerodynamic lag caused by a circulatory wake by a RFA. All structural and aerodynamic nonlinearities are clubbed into the static nonlinear operator, $f(\xi) : \mathbb{R}^n \to \mathbb{R}^n$. When a least-squares-type RFA is used (see Chapter 3), the generalized aerodynamic forces are related to the generalized coordinates by a generalized aerodynamics transfer matrix, $G_a(s)$, as follows:

$$Q(s) = q_\infty G_a(s)\xi(s), \tag{7.24}$$

where q_∞ is the freestream dynamic pressure, s is the non-dimensional Laplace variable, and

$$G_a(s) = A_0 + A_1 s + A_2 s^2 + \Gamma s(sI - R)^{-1}E \tag{7.25}$$

with $A_0, A_1, A_2, \Gamma, E, R$ being the constant coefficient matrices to be determined by a curve-fit with the harmonic generalized air force (GAF) data, $G_a(i\omega)$. In terms of the lag parameters, $b_i, i = 1, 2, \ldots, N$, and the corresponding least-squares lag coefficient matrices, $A_3, A_4, \ldots, A_{N+2}$, the matrices $\Gamma \in \mathbb{R}^{n\times\ell}$, $E \in \mathbb{R}^{\ell\times n}$ and $R \in \mathbb{R}^{\ell\times\ell}$, where $\ell = nN$, are given by

$$\Gamma = (I, I, \ldots, I), \quad E = \begin{pmatrix} A_3 \\ A_4 \\ \vdots \\ A_{N+2} \end{pmatrix}, \tag{7.26}$$

and

$$R = -\begin{pmatrix} b_1 I & 0 & 0 & \cdots & 0 \\ 0 & b_2 I & 0 & \cdots & 0 \\ 0 & 0 & b_3 I & \cdots & 0 \\ \vdots & \vdots & \vdots & \vdots & \vdots \\ 0 & 0 & 0 & \cdots & b_N I \end{pmatrix}. \tag{7.27}$$

The unsteady aerodynamic force vector in the time domain is the following:

$$Q(t) = q_\infty \left(A_2\ddot{\xi} + A_1\dot{\xi} + A_0\zeta + E^T x_a\right), \tag{7.28}$$

where the aerodynamic lag states, $x_a(t)$, are related to the generalized coordinates, $\xi(t)$, by the following state equation:

$$\dot{x}_a = Rx_a + \Gamma^T\dot{\xi}. \tag{7.29}$$

The ASE plant is therefore represented by the following state-space model:

$$\dot{x} = Ax + F(y) + Bu, \tag{7.30}$$

where $x = \left(\xi^T, \dot{\xi}^T, x_a^T\right)^T \in \mathbb{R}^{2n+\ell}$ is the aeroelastic state vector, $y = \xi$ is the aeroelastic system's response,

$$A = \begin{pmatrix} 0 & I & 0 \\ -\bar{M}^{-1}\bar{K} & -\bar{M}^{-1}\bar{C} & q_\infty \bar{M}^{-1}E^T \\ 0 & \Gamma^T & R \end{pmatrix} \tag{7.31}$$

with $\bar{M} = M - q_\infty A_2$, $\bar{K} = K - q_\infty A_0$, $\bar{C} = C - q_\infty A_1$

$$F(y) = \begin{pmatrix} 0 \\ -\bar{M}^{-1}f(\xi) \\ 0 \end{pmatrix}, \tag{7.32}$$

and

$$B = \begin{pmatrix} 0 \\ \bar{M}^{-1}T \\ 0 \end{pmatrix}. \tag{7.33}$$

The plant's coefficients A, B, and the nonlinear operator, $f(.)$, are uncertain. If they were precisely known, and the pair (A, B) were stabilizable, regulation could be achieved by dynamic inversion and feedback linearization, such as the one given by the control law

$$u = -kx - f(\xi). \tag{7.34}$$

However, as the nonlinear operator and the linear coefficients are uncertain, such a simple method is not guaranteed to work, and a structured uncertainty model is necessary.

The uncertain linear parameters as well as the nonlinear operator driving the aeroelastic system can be handled by using a structured uncertainty model through LFT (see Chapter 3). Consider both structural and aerodynamic subsystems to be represented by block-diagonal, multiplicative uncertainty models depicted in Fig. 7.8. Here $G_A(s)$ and $G_S(s)$ are the aerodynamic and structural dynamic transfer matrices, respectively, and N_A and N_S denote the describing functions for aerodynamic and structural nonlinearities, respectively. The multiplicative uncertainty blocks of aerodynamic and structural subsystems are $\Delta_A(s)$ and $\Delta_S(s)$, respectively. The system is excited by external reference inputs, $r(s)$, which are the pilot's command signals. In addition, the process, $p(s)$, and measurement, $m(s)$, noise signals are denoted as random inputs driving the system. The main source of $p(s)$ is atmospheric turbulence that affects the aerodynamic subsystem, while the measurement noise is picked up by the controller, $H(s)$, through feedback of the measured signals, $z(s)$. The task of ASE design is to derive a controller transfer matrix, $H(s)$, such that the closed-loop system is stable in the presence of random noise inputs, system nonlinearities and structured uncertainties. A stability analysis of the overall ASE system is possible by the extended Nyquist method given by Eq. (7.22), with N representing the aeroelastic nonlinear subsystem in a block-diagonal feedback loop with the linear ASE plant, $G(s)$. The system can be used as the baseline plant for deriving adaptive control systems. However, the inclusion of block-diagonal uncertainties in an iterative adaptation loop can be cumbersome for the reasons mentioned at the end of Chapter 2. Hence, an adaptive control system almost never includes the uncertainty model in the design loop, but instead uses fixed controller gains that are derived from a robust design

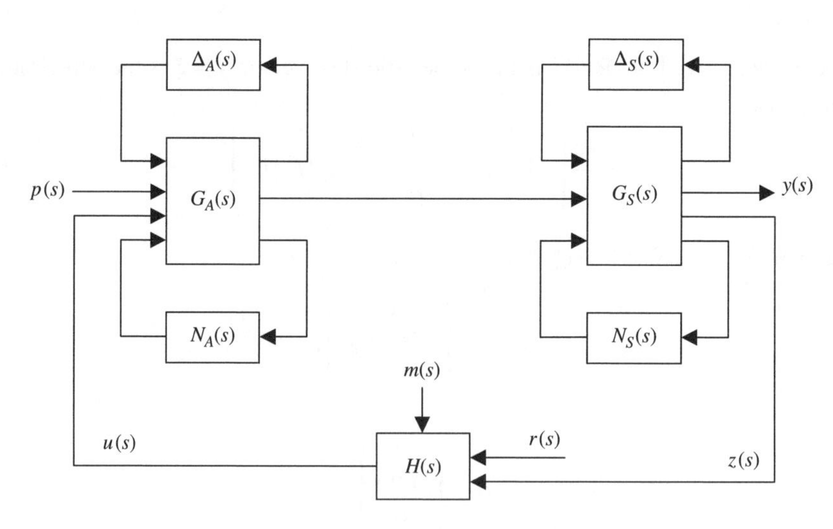

Figure 7.8 Linear fractional transformation of an ASE system with structured uncertainties, nonlinearities, reference signals and noise inputs

carried out by an uncertainty model (e.g., μ-synthesis). These gains are treated as the initial controller parameters to be evolved in time by a self-tuning identification loop (Chapter 5) or model-reference adaptation (Chapter 8).

References

Baldelli DH, Lind RC, and Brenner M 2005 Robust aeroelastic match-point solutions using describing function method. *J. Aircr.* **42**, 1597–1605.

Borglund D 2003 Robust aeroelastic stability analysis considering frequency-domain aerodynamic uncertainty. *J. Aircr.* **40**, 189–193.

Dowell EH, Edwards J, and Strganac T 2003 Nonlinear aeroelasticity. *J. Aircr.* **40**, 857–874.

Dowell EH and Tang D 2002 Nonlinear aeroelasticity and unsteady aerodynamics. *AIAA J.* **40**, 1697–1707.

Gibson JE 1963 *Nonlinear Automatic Control*, McGraw-Hill, New York.

Gordon JT, Meyer EE, and Minogue RL 2008 Nonlinear stability analysis of control surface flutter with free-play effects. *J. Aircr.* **45**, 1904–1916.

Khalil HS 2002 *Nonlinear Systems*, 3rd ed. Prentice-Hall, Upper Saddle River, NJ.

Lind R and Brenner M 1999 *Robust Aeroservoelastic Stability Analysis: Flight-Test Applications*. Springer-Verlag, London.

Maciejowski JM 1989 *Multivariable Feedback Design*. Addison-Wesley, Reading, MA.

Raveh DE 2004 Identification of computational-fluid-dynamics based unsteady aerodynamics models for aeroelastic analysis. *J. Aircr.* **41**, 620–632.

Slotine JJE and Li W 1995 *Applied Nonlinear Control*. Prentice-Hall, Englewood Cliffs, NJ.

Tewari A 2015 *Aeroservoelasticity: Modeling and Control*. Birkhäuser, Boston, MA.

Ueda T and Dowell EH 1984 Flutter analysis using nonlinear aerodynamic forces. *J. Aircr.* **21**, 101–109.

8

Model Reference Adaptation of Aeroservoelastic Systems

The model reference adaptation system (MRAS) approach offers the greatest promise of applications to aeroservoelastic (ASE) systems for the following reasons:

(a) It adapts controller parameters in the presence of modelling uncertainties and prescribed nonlinearities, without requiring an online estimation of uncertain plant parameters.
(b) On the basis of Lyapunov's direct method, it guarantees the uniform boundedness of the state error from the reference model by ensuring that the system's trajectories always remain inside a compact set.

Both these features are quite useful when applied to ASE systems, because the alternative method of the self-tuning regulator (STR), which requires an online parameter estimation, is not guaranteed to be stable in the presence of large, random perturbations, which are especially encountered in the transonic regime. Furthermore, the other classical method of dealing with prescribed nonlinearities, namely the describing functions method, can work only when the oscillatory response can be approximated by the first harmonic. Again, this assumption can be violated in the presence of large modelling uncertainties of an ASE design. Therefore, the MRAS appears to be well suited for an adaptive ASE design. However, being based on minimization of state error from a reference model, the success of MRAS hinges on selecting an appropriate reference model. Furthermore, as it has an essentially non-autonomous error dynamics, the MRAS requires projection-based adaptation laws for guaranteeing robustness in the presence of random signals.

The design process for model reference adaptation consists of the following steps:

(a) Derivation of a reference model
(b) Design of the basic feedback controller
(c) Design of the adaptation mechanism for the controller parameters.

The analysis in this chapter is limited to full-state-feedback control, which is later extended to output feedback systems in Chapter 10.

Adaptive Aeroservoelastic Control, First Edition. Ashish Tewari.

8.1 Lyapunov-Like Stability of Non-autonomous Systems

Consider an unforced, autonomous system with state vector, $x(t) \in \mathbb{R}^n$, governed by the state equation,

$$\dot{x} = f(x, t), \tag{8.1}$$

with an equilibrium point at the origin, $f(0, t) = 0$. If an arbitrarily small initial condition exists,

$$x(t_0) = x_0, \tag{8.2}$$

then the perturbed solution, $x(t), t \geq t_0 \geq 0$, may or may not remain arbitrarily close to the equilibrium point. Thus stability is concerned with the boundedness of the trajectory, $\mid x(t) \mid$; $t \geq t_0 \geq$, if $\mid x_0 \mid < \delta$, where δ is an arbitrarily small positive number.

Definition 8.1.1 *A system described by Eq. (8.1) is said to be stable about the equilibrium point, $x = 0$, in the sense of Lyapunov if for each real and positive number, ϵ, however small, and initial time, $t_0 \geq 0$, there exists another real and positive number, $\delta(\epsilon, t_0)$, such that for all initial conditions,*

$$\mid x(t_0) \mid < \delta \tag{8.3}$$

implies that

$$\mid x(t) \mid < \epsilon, \quad t \geq t_0 \geq 0. \tag{8.4}$$

1. *The equilibrium is said to be globally stable, if it is stable and* $\lim_{\epsilon \to \infty} \delta(\epsilon, t_0) = \infty$.
2. *The equilibrium is said to be uniformly stable, if it is stable and δ does not depend upon the initial time, t_0.*
3. *The equilibrium is said to be asymptotically stable, if it is stable and there exists a positive, real constant, $c(t_0)$, such that* $\lim_{t \to \infty} x(t) = 0$ *for every* $\mid x(t_0) \mid \leq c$.
4. *The equilibrium is said to be uniformly asymptotically stable, if it is uniformly stable and there exists a positive, real constant, c, independent of t_0, such that* $\lim_{t \to \infty} x(t) = 0$ *for every* $\mid x(t_0) \mid \leq c$. *Then the trajectory $x(t)$ is said to converge uniformly in t_0 to zero.*
5. *The equilibrium is said to be globally, uniformly asymptotically stable (GUAS), if it is uniformly asymptotically stable and* $\lim_{\epsilon \to \infty} \delta(\epsilon) = \infty$.

A geometric interpretation of stability is given by the ability to find an n-dimensional sphere of radius $\delta \leq \epsilon$ such that every trajectory starting in the smaller sphere at $t = t_0$ remains inside the larger sphere of radius ϵ at all times. If this happens irrespective of the value of t_0, then the equilibrium point is uniformly stable. Achieving uniform asymptotic stability in the presence of uncertain dynamics is the objective of model reference adaptation.

Further discussion assumes $f(x, t)$ to be piecewise continuous in t and locally Lipschitz in x, such that existence and uniqueness of solutions, $x(t)$, in a local neighbourhood of $x(t_0)$ are guaranteed. It can be shown that the Lipschitz condition will be satisfied in a domain containing the origin, $\mathbb{D} \subset \mathbb{R}^n$, for $a \leq t \leq b$, if and only if both $f(x, t)$ and $\partial f / \partial x$ are continuous in $x \in \mathbb{D}$ for $a \leq t \leq b$. In order to extend the concept of Lyapunov stability to the system of Eq. (8.1), assume that a positive-definite scalar function, $V(x) : \mathbb{D} \to \mathbb{R}$, exists whose time derivative along the system's trajectory is locally non-positive definite, given by

$$\dot{V} = \nabla V(x) f(x, t) \leq 0, \quad \text{for all } x \in \mathbb{D}, \quad \text{for all } t \geq 0. \tag{8.5}$$

Theorem 8.1.2 *If on a domain containing the origin, $\mathbb{D} \subset \mathbb{R}^n$, there exists a continuously differentiable, locally positive-definite scalar function, $V(x)$, such that the condition given by Eq. (8.5) is satisfied for the system given by Eq. (8.1), then the origin $x = 0$ is uniformly stable in the sense of Lyapunov. If the condition, Eq. (8.5), is satisfied by a strict inequality (i.e., $\dot{V}$ is locally negative definite) then the origin is locally uniformly asymptotically stable.*

Proof of Theorem 8.1.2 (called Lyapunov's direct method) can be found in textbooks on non-linear systems, (cf. (Haddad and Chellaboina 2008)). Thus we have a method of establishing uniform asymptotic stability by finding a suitable Lyapunov function. Geometric interpretation of Lyapunov's direct method is enabled by considering that the requirement of Eq. (8.5) dictates that at all times, the outward normal to a closed contour $V(x) = c$ given by the gradient vector, $\nabla V(x(t))$, makes an angle greater than $\pi/2$ with the system's dynamics vector, $f(x, t)$. This implies that a trajectory $x(t)$ cannot cross outside the boundary, $V(x) = c$, of a closed and bounded (compact) set containing the origin $x = 0$. The Lyapunov function $V(x)$ thus has energy like behaviour for the dynamical system of Eq. (8.1) and Lyapunov stability refers to energy dissipation along the system's trajectory.

Theorem 8.1.3 *If the Lyapunov function, $V(x) : \mathbb{R}^n \to \mathbb{R}$, satisfying the conditions of Theorem 8.1.2 is also radially unbounded, then the origin $x = 0$ is a globally uniformly, asymptotically stable (GUAS) equilibrium of the system given by Eq. (8.1).*

8.1.1 Uniform Ultimate Boundedness

Since a system is always under some uncertain perturbation, $w(t) \in \mathbb{R}^n$, its real dynamics is represented as follows:

$$\dot{x} = f(x, t) + w(t), \qquad x(t_0) = x_0, \tag{8.6}$$

and an equilibrium point is therefore undefined. Instead of stability of the equilibrium, it is more meaningful to consider the response of the system to a bounded disturbance, $| w(t) | \leq w_m$. However, Lyapunov's direct method can still be applied by finding a Lyapunov-like function to show that the system's trajectories remain bounded for all time.

Definition 8.1.4 *The solutions $x(t)$ to Eq. (8.6) are said to be uniformly ultimately bounded (UUB) with an ultimate bound b, if there exist positive, real numbers, b and c, independent of initial time $t_0 \geq 0$, and a time interval, $T = T(a, b)$ for every $0 < a < c$, also independent of t_0, such that*

$$| x(t_0) | \leq a, \tag{8.7}$$

implies that

$$| x(t) | \leq b, \tag{8.8}$$

for all times $t \geq t_0 + T$.

If the conditions given by Eqs. (8.7) and (8.8) hold for an arbitrarily large number a, then the solutions to Eq. (8.6) are said to be globally UUB.

Uniform ultimate boundedness is a weaker form of stability than stability in the sense of Lyapunov, because it does not ensure maintenance of the response in a small neighbourhood of equilibrium.

8.1.2 Barbalat's Lemma

Since it is difficult to find Lyapunov functions with a negative-definite time derivative for non-autonomous systems, the stability analysis is instead based on seeing whether a set of solutions, E, for which $\dot{V} = 0$ can be considered the set to which all solutions converge as $t \to \infty$. If E is a null set, then the origin is asymptotically stable, because all trajectories have $\dot{V} < 0$. In fact, the continuity of the time derivative of a function of time can be used to show whether a trajectory starting outside E (i.e. having $\dot{V} < 0$) will converge to E as $t \to \infty$. This is the extension of the Lyapunov stability analysis for non-autonomous systems.

Definition 8.1.5 *A scalar function $f(t)$ is said to be uniformly continuous, if for every $\epsilon > 0$, there exists a $\delta = \delta(\epsilon) > 0$, such that $\mid t_2 - t_1 \mid \leq \delta$ implies that $\mid f(t_2) - f(t_1) \mid \leq \epsilon$.*

Lemma 8.1.6 *If a scalar function $f(t)$ is uniformly continuous for $0 \leq t < \infty$, and the limit,*

$$\lim_{t\to\infty} \int_0^t f(\tau)\mathrm{d}\tau$$

exists and is finite, then $\lim_{t\to\infty} f(t) = 0$.

An equivalent form of Barbalat's lemma can be stated for a continuously differentiable scalar function, $f(t)$, with a finite limit as $t \to \infty$. If $\dot{f}(t)$ is uniformly continuous, then it converges to zero in the limit as $t \to \infty$.

8.1.3 LaSalle–Yoshizawa Theorem

Where non-autonomous systems are concerned, invariant sets cannot be defined. In such a case, Barbalat's lemma is applied to establish Lyapunov-like stability of an equilibrium point. The concept of invariant set is then extended to that of a bounded set to show uniform boundedness of non-autonomous trajectories by the LaSalle–Yoshizawa theorem.

Theorem 8.1.7 *If there exists a continuously differentiable, positive-definite, and radially unbounded scalar function, $V(x) : \mathbb{R}^n \to \mathbb{R}$, such that the following condition is satisfied for the system given by Eq. (8.1):*

$$\dot{V} = \nabla V(x) f(x,t) \leq -W(x) \leq 0, \quad \text{for all } x \in \mathbb{R}^n, \quad \text{for all } t \geq 0, \tag{8.9}$$

where $W(x) : \mathbb{R}^n \to \mathbb{R}$ is a continuous and non-negative function, then all solutions of Eq. (8.1) are UUB and satisfy

$$\lim_{t\to\infty} W(x(t)) = 0. \tag{8.10}$$

Furthermore, if $W(x)$ is positive definite, then the equilibrium point, $x = 0$, is GUAS.

The LaSalle–Yoshizawa theorem allows the construction of a Lyapunov function to establish convergence of non-autonomous trajectories to a set E where $W(x) = 0$. This is very useful in designing MRAS by Lyapunov-like methods.

Example 8.1.8 *In order to understand the basic aspects of Lyapunov-based control, consider a single degree of freedom mass–spring–damper system with displacement, $x(t)$, from the static equilibrium position. The system is linear with a constant mass m, but has unknown and randomly varying stiffness, $k(t)$, and damping coefficient, $c(t)$, with the equation of motion given by*

$$m\ddot{x} + c\dot{x} + kx = u, \tag{8.11}$$

where $u(t)$ is the force applied as the control input. The task of an automatic controller is simply to stabilize the system, such that any initial displacement or rate is quickly brought to zero without too many oscillations. If the bounds on the unknown parameters are known,

$$| k |\leq \bar{k}, \quad | c |\leq \bar{c}, \tag{8.12}$$

then the following linear feedback control law

$$u = -2(\bar{k}x + \bar{c}\dot{x}), \tag{8.13}$$

will stabilize the system, which can be proved by a Lyapunov stability analysis. Consider the following candidate Lyapunov function:

$$V = ax^2 + b\dot{x}^2 + 2abx\dot{x}, \tag{8.14}$$

which is positive definite and radially unbounded with the state vector $(x, \dot{x})$, if $a > 0, b > 0$. It is evident that the time derivative of the Lyapunov function along the system's trajectories is bounded by a non-positive function of the state variables:

$$\begin{aligned}
\dot{V} &= 2(ax\dot{x} + b\dot{x}\ddot{x} + abx\ddot{x} + ab\dot{x}^2) \\
&= 2ax\dot{x} + 2ab\dot{x}^2 + 2b(ax + \dot{x})\left(-\frac{c}{m}\dot{x} - \frac{k}{m}x + \frac{u}{m}\right) \\
&= 2ax\dot{x} + 2ab\dot{x}^2 - \frac{2b}{m}(ax + \dot{x})\left[(c + 2\bar{c})\,\dot{x} + (k + 2\bar{k})x\right] \\
&\leq 2ax\dot{x} + 2ab\dot{x}^2 - \frac{2b}{m}(ax + \dot{x})\left[\bar{c}\dot{x} + \bar{k}x\right] \\
&= -\frac{2b}{m}\left[a\bar{k}x^2 + (\bar{c} - am)\dot{x}^2 + \left(\bar{k} + a\bar{c} - \frac{a}{b}m\right)x\dot{x}\right] \\
&= -\frac{2b}{m}(x, \dot{x})\begin{pmatrix} a\bar{k} & \frac{1}{2}\left(\bar{k} + a\bar{c} - \frac{a}{b}m\right) \\ \frac{1}{2}\left(\bar{k} + a\bar{c} - \frac{a}{b}m\right) & \bar{c} - am \end{pmatrix}\begin{pmatrix} x \\ \dot{x} \end{pmatrix}.
\end{aligned} \tag{8.15}$$

There are many possible choices of the positive constants a, b for the candidate Lyapunov function. For example, by selecting

$$a = \frac{\bar{c}}{m} \qquad b = \frac{m\bar{c}}{m\bar{k} + \bar{c}^2},$$

it can be readily shown that the coefficient matrix in Eq. (8.15) is positive semi-definite. Thus the sufficient conditions for the LaSalle–Yoshizawa theorem are satisfied and globally uniform boundedness of all solutions starting from the equilibrium point at the origin is guaranteed. The asymptotic stability consistent with this choice of Lyapunov function is illustrated by the

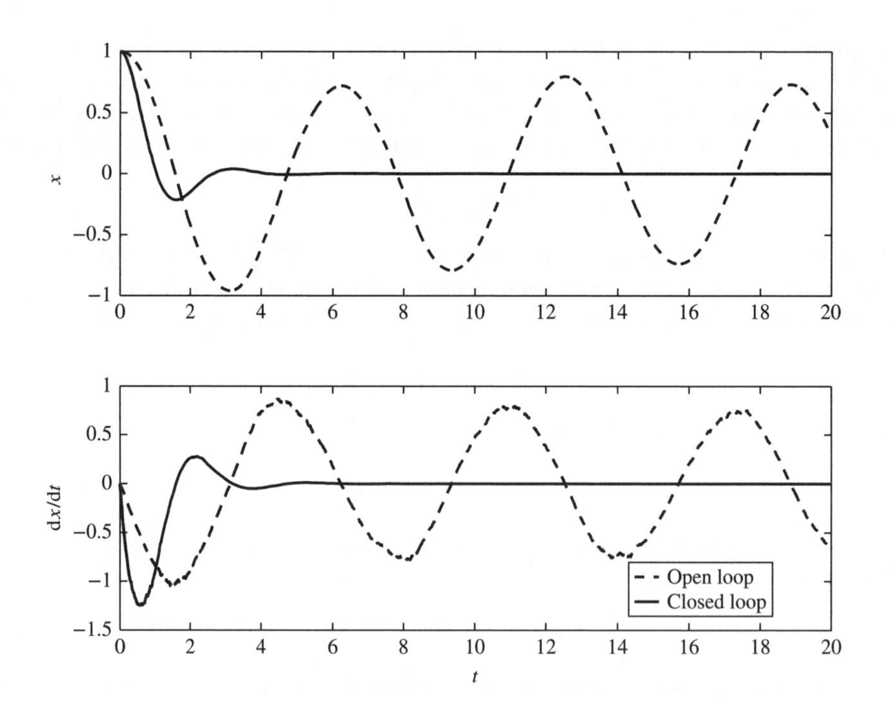

Figure 8.1 Simulation of a spring–mass–damper system with known bounds on randomly varying stiffness and damping ($m = 1, \bar{k} = 2, \bar{c} = 1$)

Runge–Kutta simulation plotted in Fig. 8.1 for $m = 1, \bar{k} = 2, \bar{c} = 1$ *and initial displacement* $x(0) = 1$ *(all in appropriate units), with the stiffness varying as a normal random distribution and the damping as a uniform random distribution with each time step. Another possibility of constants* a, b *that yields a positive-definite coefficient matrix in Eq. (8.15) (thus global asymptotic stability of the origin), is the following:*

$$a = \frac{\bar{c}}{2m} + \frac{1}{2m}\sqrt{\bar{c}^2 - \bar{k}m} \qquad b = \frac{m}{\bar{c}},$$

which is simulated with a much smaller stiffness bound, $m = 1, \bar{k} = 0.1, \bar{c} = 1$, *and is seen to have asymptotic stability in Fig. 8.2.*

While Example 8.1.8 devised a linear feedback controller with constant gains to stabilize a linear plant with known parameter bounds, such an approach would fail whenever such bounds are unknown. In such cases, adaptation is necessary, as explored next. In a stabilization problem (which is the focus in ASE design), the reference inputs are zeros, thus we have the following reference model:

$$\dot{x}_m = A_m x_m, \tag{8.16}$$

where $A_m \in \mathbb{R}^{n \times n}$ is a known dynamics coefficient matrix and $x_m(t) \in \mathbb{R}^n$ is the state vector of the reference model. Since the reference parameters, A_m, are different from those of the actual plant, there is always an error between the states of the two systems, which must be driven to small values by a model-reference adaptive scheme.

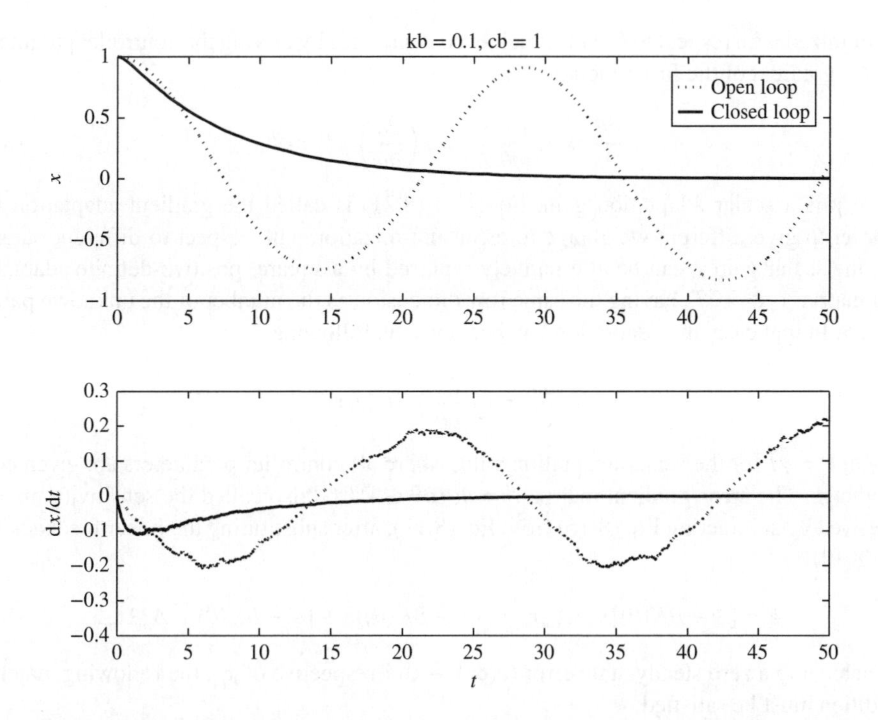

Figure 8.2 Simulation of a spring–mass–damper system with known bounds on randomly varying stiffness and damping ($m = 1, \bar{k} = 0.1, \bar{c} = 1$)

8.2 Gradient-Based Adaptation

A common MRAS adaptation mechanism is based on the minimization of a loss function of the error between the actual state and the state of the reference model. Consider a linear, time-invariant aeroelastic plant with state equation

$$\dot{x} = Ax + Bu, \tag{8.17}$$

where $A \in \mathbb{R}^{n\times n}$, $B \in \mathbb{R}^{n\times m}$ are uncertain (but constant) plant parameters matrices. The pair (A, B) is controllable. The stabilization reference model is described by Eq. (8.16), with A_m being a Hurwitz matrix (has all eigenvalues in the left-half plane). The state error is given by

$$e = x - x_m, \tag{8.18}$$

which is to be driven to zero in the steady state using a state-feedback law,

$$u = -K(\theta)x. \tag{8.19}$$

The state-feedback regulator gain matrix, $K \in \mathbb{R}^{m\times n}$, depends upon the controller parameters vector, $\theta \in \mathbb{R}^p$, $(p \leq mn)$, which must be adjusted such that a scalar, positive semi-definite loss function of the error vector,

$$V(\theta) = f(e), \tag{8.20}$$

is minimized with respect to θ. This is achieved in practice by driving the controller parameters along a gradient of the loss function,

$$\frac{d\theta}{dt} = \gamma\left(\frac{\partial V}{\partial \theta}\right)^T = \gamma\left(\frac{\partial e}{\partial \theta}\right)^T (f'(e))^T, \tag{8.21}$$

where γ is a scalar adaptation gain. Equation (8.21) is called the gradient adaptation law. In order to give different weightage to error minimization with respect to different parameters, the scalar gain γ can be alternatively replaced by a square, positive-definite adaptation gain matrix, $\Gamma \in \mathbb{R}^{p\times p}$, having the same row dimension as the number of the unknown parameters, θ. In that case, the adaptation law becomes the following:

$$\frac{d\theta}{dt} = \Gamma\left(\frac{\partial e}{\partial \theta}\right)^T (f'(e))^T. \tag{8.22}$$

Hence, $\Gamma = \gamma I$ for the scalar adaptation gain, where all controller parameters are given equal weightage. The error-gradient matrix, $S = \partial e/\partial \theta \in \mathbb{R}^{n\times p}$ (also called the sensitivity matrix), is derived by subtracting Eq. (8.16) from Eq. (8.17), after substituting the linear feedback law, Eq. (8.19):

$$\dot{e} = [A - BK(\theta)]x - A_m x_m = [A - BK(\theta)]e + [A - BK(\theta) - A_m]x_m. \tag{8.23}$$

For achieving a zero steady-state error ($e(\infty) \to 0$) irrespective of x_m, the following matching condition must be satisfied:

$$A - BK(\theta) = A_m. \tag{8.24}$$

However, in practice, the matching condition can be satisfied only at a specific time, say $t = 0$, for which $\theta(0) = \theta_0$ are known. If there exists such a parameters vector, θ_0, then we have

$$A - BK(\theta_0) = A_m, \tag{8.25}$$

and a residual error always remains, which should be driven to small values by an appropriate adaptation law. The deviation of the controller parameters from the initially known values, given by

$$\Delta\theta = \theta - \theta_0, \tag{8.26}$$

should be taken into account while deriving such an adaptation law. Suppose the error state equation can be expressed as follows in the presence of the parameter error, $\Delta\theta$:

$$\dot{e} = A_m e + \Psi\Delta\theta, \tag{8.27}$$

where $\Psi \in \mathbb{R}^{n\times p}$ is a known constant matrix. Then the following gradient law is shown to be stabilizing by Lyapunov's stability theory (to be seen later):

$$\dot{\theta} = -\gamma\Psi^T P e, \tag{8.28}$$

where $\gamma > 0$ is an adaptation gain and $P \in \mathbb{R}^{n\times n}$ is a symmetric, positive-definite real matrix.

8.2.1 Least-Squared Error Adaptation

A simple choice of the loss function is the squared-error function,

$$V(\theta) = \frac{1}{2}e^T e, \tag{8.29}$$

which leads to the adaptation law,

$$\dot{\theta} = \Gamma\left(\frac{\partial e}{\partial \theta}\right)^T e = \Gamma S^T e. \tag{8.30}$$

Substituting the exact matching condition Eq. (8.24) into Eq. (8.23) leads to the following identity:

$$\dot{e} = \frac{\partial e}{\partial \theta}\dot{\theta} = [A - BK(\theta)]e = A_m e, \tag{8.31}$$

which, on substitution of Eqs. (8.30) and (8.24), results in the following:

$$S\Gamma S^T = A_m. \tag{8.32}$$

Equation (8.32) must be solved for the sensitivity matrix, S, before the gradient adaptation law of Eq. (8.30) can be implemented. In such a solution procedure, θ is considered to be free, and subject to the initial condition, $\theta(0) = \theta_0$. The initial error, $e(0)$, must also be specified to complete the statement of the problem. It can be shown (see Chapter 6) that as A_m is Hurwitz, the adaptation law given by Eqs. (8.30) and (8.32) drive the loss function, $V(\theta)$, to a minimum. However, as Eq. (8.32) gives n equations to be solved for np elements of S, it does not have a unique solution unless $p = 1$ (single free parameter). Hence, $n(p-1)$ parameters of S must be selected from arbitrary conditions. Furthermore, nonlinear algebraic equations such as Eq. (8.32) require an iterative solution (except in some special cases). Thus, even when the exact matching condition is satisfied at all times, the solution for a sensitivity matrix for use in the adaptation law is problematic. The problem becomes more acute when the matching condition is satisfied only initially, that is, Eq. (8.24) is replaced by Eq. (8.25). Then there is no guarantee that a simple least-squares adaptation would lead to a stable system. As seen next, Lyapunov's direct method can be applied to yield sufficient conditions of stability with a proper choice of the Lyapunov function.

8.3 Lyapunov-Based Adaptation

In the previous section, it was found that choosing a gradient-based adaptation law with free controller parameters leads to a nonlinear algebraic equation to be solved for the sensitivity matrix. However, a simplification in the adaptation law occurs by assuming a controller structure. Such a structure can be based on any of the basic linear control strategies, such as pole placement (eigenstructure assignment) and linear optimal control. If the plant model (A, B) were known with certainty, a linear feedback control law based on the known plant,

$$u = -Kx, \tag{8.33}$$

could be designed with constant controller gains, K, such that the closed-loop system closely follows the reference model. Therefore, the matching condition,

$$A - BK = A_m, \tag{8.34}$$

would be satisfied. However, when the plant parameters are uncertain, the controller gain cannot precisely achieve the matching condition. In such a case, we have

$$u = -\hat{K}x, \tag{8.35}$$

$$\hat{K} = K + \delta K, \tag{8.36}$$

where $\delta K(t)$ is the variation (or error) in the controller gain matrix due to the plant's modelling uncertainty. Substitution of Eqs. (8.34)–(8.36) into the plant's state equation, Eq. (8.17), and subtracting the reference state equation, Eq. (8.16), leads to the following closed-loop error dynamics:

$$\dot{e} = A_m e - BK(e + x_m) = A_m e - B\delta Kx. \tag{8.37}$$

The term $B\delta Kx$ is the residual error term due to modelling uncertainty.

The controller design assumes that errors due to modelling uncertainty can be reduced by applying feedback as if the plant parameters were known with certainty. In other words, the feedback employs the estimated values of uncertain controller parameters as if they were certain. This is the certainty equivalence principle on which most adaptive control laws are based. Of course, this approach requires being able to compute the controller gains from either a knowledge or an estimate of the plant parameters, (A, B).

Before proceeding further, an uncertainty model for the controls coefficient matrix, B, is necessary. Assume that the uncertainty in B has a multiplicative structure given by

$$B = B_0\Lambda, \tag{8.38}$$

where $B_0 \in \mathbb{R}^{n\times m}$ is a known matrix and $\Lambda \in \mathbb{R}^{m\times m}$ is an unknown, constant, diagonal matrix with positive elements, $\lambda_{ii} > 0, i = 1, \ldots, m$. The system's state equation is thus the following:

$$\dot{x} = Ax + B_0\Lambda u = (A - B_0\Lambda\hat{K})x. \tag{8.39}$$

Define the following candidate Lyapunov function:

$$V = e^T Pe + \mathrm{tr}(\delta K\Gamma^{-1}\delta K^T\Lambda), \tag{8.40}$$

with P and Γ being symmetric and positive-definite cost matrices. The function V is radially unbounded in $(e, \delta K)$, and has the following time derivative:

$$\dot{V} = \dot{e}^T Pe + e^T P\dot{e} + 2\mathrm{tr}\left(\delta K\Gamma^{-1}\dot{\hat{K}}^T\Lambda\right), \tag{8.41}$$

or, by Eq. (8.37),

$$\dot{V} = e^T(A_m^T P + PA_m)e - 2e^T PB_0\Lambda\delta Kx + 2\mathrm{tr}\left(\delta K\Gamma^{-1}\dot{\hat{K}}^T\Lambda\right). \tag{8.42}$$

Since A_m is Hurwitz, the matrix P satisfies the following Lyapunov identity (Chapter 6):

$$A_m^T P + PA_m = -Q, \tag{8.43}$$

where Q is a positive-definite matrix. Furthermore, the following identity exists for the trace of a square matrix, ba^T, where a, b are two vectors of the same dimension:

$$a^T b = \mathrm{tr}(ba^T), \tag{8.44}$$

which can be applied as follows:

$$e^T P B_0 \Lambda \delta K x = \mathrm{tr}(\delta K x e^T P B_0 \Lambda). \tag{8.45}$$

By virtue of Eqs. (8.43) and (8.45), we have

$$\dot{V} = -e^T Q e + 2\mathrm{tr}\left\{ \delta K \left(\Gamma^{-1} \dot{\hat{K}}^T - x e^T P B_0 \right) \Lambda \right\}. \tag{8.46}$$

In order to make the non-quadratic term in Eq. (8.46) vanish, the following adaptation law is selected:

$$\dot{\hat{K}}^T = \Gamma x e^T P B_0, \tag{8.47}$$

or

$$\dot{\hat{K}} = B_0^T P e x^T \Gamma. \tag{8.48}$$

Hence, by substitution of Eqs. (8.43) and (8.47) into Eq. (8.42), we have the following result:

$$\dot{V} = -e^T Q e \leq 0, \tag{8.49}$$

which satisfies Lyapunov stability theorem for global asymptotic stability (Chapter 6). An implementation of the adaptation law of Eq. (8.47) is possible, because the matrix B_0 is known. By taking the second time derivative of the Lyapunov function, we obtain the result,

$$\ddot{V} = -2e^T Q \dot{e}, \tag{8.50}$$

which is bounded, because $\dot{e}$ is bounded, hence $\dot{V}$ is uniformly continuous. Additionally, by virtue of Eq. (8.49) and $V(e, \delta K)$ being lower bounded by $V(0, 0)$, $\dot{V}$ converges uniformly to zero in the limit $t \to \infty$ due to Barbalat's lemma. Thus the error $e(t)$ uniformly converges to zero in the limit $t \to \infty$, hence GUAS stability is guaranteed without the knowledge of the uncertain plant parameters (A, Λ).

The basic approach highlighted here is extended in Chapter 10 to handle the matched uncertainty in a plant's input variables due to uncertain nonlinearities and for a robust design in the output-feedback form.

Example 8.3.1 *Example 8.1.8 used a linear feedback controller with constant gains to stabilize a linear plant with known parameter bounds. This approach is inapplicable whenever such bounds are unknown. Furthermore, in many cases, only a bounded response – rather than asymptotic stability – can be achieved, thus the controller cannot be truly called stabilizing. In order to have model reference adaptation in such a case, let us define the following reference model:*

$$m_m \ddot{x}_m + c_m \dot{x}_m + k_m x_m = r, \tag{8.51}$$

where m_m, c_m, k_m are known constants and $x_m(t), r(t)$ are the displacement and reference input for the model. It is required to track the response of the reference model by minimizing the *tracking error,* $\delta x = x - x_m$, *and its time derivatives for a specified reference input,* $r(t)$. *For a regulation task, the error is to be minimized for* $r(t) = 0$. *In order to achieve this objective, the following feedback control law is proposed:*

$$u = -(k_1 \delta x + k_2 \delta \dot{x}), \tag{8.52}$$

where k_1, k_2 *are regulator constants. If* m, c, k *were known with certainty,* k_1, k_2 *could be chosen from the desired closed-loop characteristic polynomial,* $s^2 + a_2 s + a_1 = 0$, *where* a_1, a_2 *are specified constants. However, as* m, c, k *are unknown, we can only use the estimates of the regulator gains,* $\hat{k}_1, (\hat{k})_2$, *in the control law:*

$$u = -(\hat{k}_1 \delta x + \hat{k}_2 \delta \dot{x}), \tag{8.53}$$

with

$$\begin{aligned} \hat{k}_1 &= k_1 + \delta k_1 \\ \hat{k}_2 &= k_2 + \delta k_2, \end{aligned} \tag{8.54}$$

and determine the variation of the estimated controller parameter vector, $\hat{k}(t) = (\hat{k}_1(t), \hat{k}_2(t))^T$, *such that both the tracking error vector,* $e(t) = (\delta x(t), \delta \dot{x}(t))^T$, *and the parameter estimation error vector,* $\delta k = (\delta k_1, \delta k_2)^T$, *tend to zero in the steady-state limit,* $t \to \infty$. *The following adaptation law is known to be stabilizing (Eq. (8.48)):*

$$\dot{\hat{k}} = B_0^T P e x^T \Gamma. \tag{8.55}$$

where $B_0 = (0, 1)^T$ *and* P, Γ *are symmetric, positive-definite constant matrices weighting the following Lyapunov function:*

$$V = e^T P e + \text{tr}(\delta k \Gamma^{-1} \delta k^T \Lambda), \tag{8.56}$$

whose time derivative along the system's trajectories is given by Eq. (8.49) to be negative semi-definite, thereby guaranteeing GUAS stability. A simulated response of the adaptive closed-loop system is plotted in Figs. 8.3–8.5 for $k_m/m_m = 2, c_m/m_m = 1$, *initial displacement* $x(0) = 1, \dot{x}(0) = 0$ *(all in appropriate units), with the actual system's constants,* $k/m, c/m$, *varying as normal random distributions of mean 1.5 and 0.5, respectively, and a unit variance (which information is considered unknown, hence not utilized in the MRAS controller design). The largest error from reference plant parameters is thus 1.5 units in both* k/m *and* c/m, *which constitutes a variation of 75% and 150%, respectively. The weighting matrices in Eq. (8.55) are taken as* $P = 10I, \Gamma = I$. *Both the plant state (Fig. 8.3) and controller parameter (Fig. 8.4) vectors are seen to have a stable response, reaching converged, steady values in about* $t = 10$ *units. The simulated control input is plotted in Fig. 8.5, showing a zero steady-state value and a bounded magnitude,* $| u | < 0.4$ *units.*

8.3.1 Nonlinear Gain Evolution

A novel alternative adaptation law can be derived by using the following quadratic form of the Lyapunov function:

$$V = x^T \delta K^T \Gamma \delta K^T x, \tag{8.57}$$

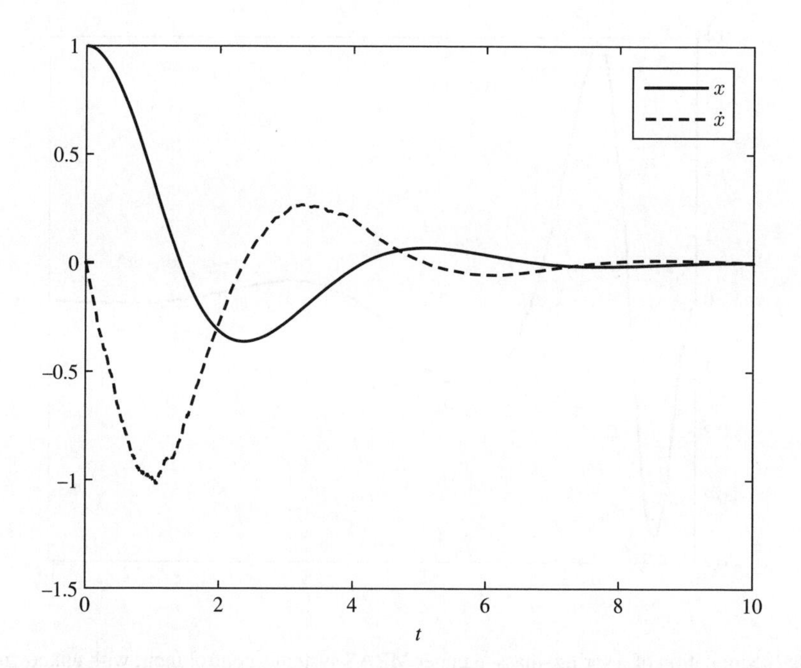

Figure 8.3 Simulation of a spring–mass–damper MRAS system's states with unknown bounds on randomly varying plant parameters

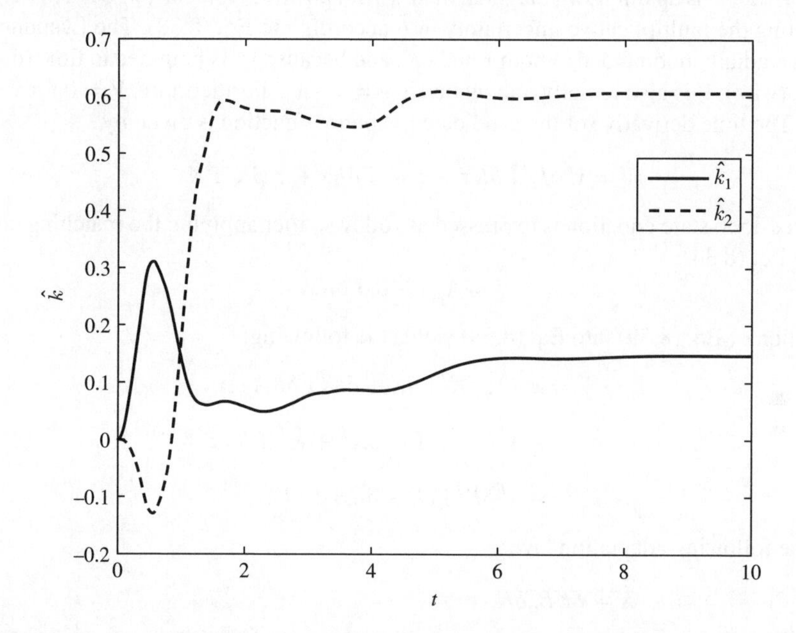

Figure 8.4 Simulation of a spring–mass–damper MRAS system's controller parameters with unknown bounds on randomly varying plant parameters

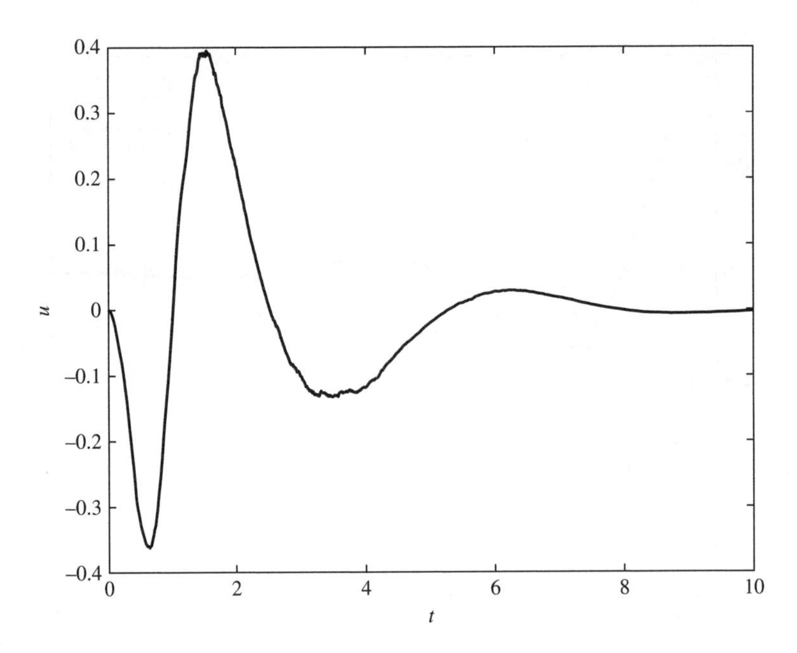

Figure 8.5 Simulation of a spring–mass–damper MRAS system's control input with unknown bounds on randomly varying plant parameters

where $\Gamma \in \mathbb{R}^{m\times m}$ is an unknown diagonal matrix with positive elements, $\lambda_{ii} > 0, i = 1, \dots, m$, representing the multiplicative uncertainty in B according to Eq. (8.38). The Lyapunov function, V, is radially unbounded in both x and δK, and because x_m is bounded in time (due to A_m being Hurwitz); V is also radially unbounded in $e = x - x_m$. Furthermore, $V = 0$ for $x = 0$ and $\delta K = 0$. The time derivative of the candidate Lyapunov function is given by

$$\dot{V} = \dot{x}^T \delta K^T \Gamma \delta K x + x^T \dot{\hat{K}}^T \Gamma \delta K x + x^T \delta K^T \Gamma \dot{\hat{K}} x. \tag{8.58}$$

The closed-loop state equation is expressed as follows, after applying the matching condition given by Eq. (8.34):

$$\dot{x} = A_m x - B_0 \Gamma \delta K x. \tag{8.59}$$

Substitution of Eq. (8.59) into Eq. (8.58) yields the following:

$$\begin{aligned}\dot{V} &= x^T (A_m^T \delta K^T \Gamma \delta K + \delta K^T \Gamma \delta K A_m) x \\ &\quad + x^T (-\delta K^T \Gamma B_0^T \delta K^T + \dot{\hat{K}}^T) \Gamma \delta K x \\ &\quad + x^T \delta K^T \Gamma (\dot{\hat{K}} - \delta K B_0 \Gamma \delta K) x.\end{aligned} \tag{8.60}$$

Select the following adaptation law:

$$\dot{\hat{K}} = \delta K B_0 \delta K. \tag{8.61}$$

$$\begin{aligned}\dot{V} &= x^T (A_m^T \delta K^T \Gamma \delta K + \delta K^T \Gamma \delta K A_m) x \\ &\quad + x^T (-\delta K^T \Gamma B_0^T \delta K^T + \delta K^T B_0^T \delta K^T) \Gamma \delta K x \\ &\quad + x^T \delta K^T \Gamma (\delta K B_0 \delta K - \delta K B_0 \Gamma \delta K) x.\end{aligned} \tag{8.62}$$

or

$$\dot{V} = x^T(A_m^T \delta K^T \Gamma \delta K + \delta K^T \Gamma \delta K A_m)x$$
$$+ x^T \delta K^T (-\Gamma + I) B_0^T \delta K^T \Gamma \delta K x$$
$$+ x^T \delta K^T \Gamma \delta K B_0 (I - \Gamma) \delta K x. \quad (8.63)$$

We note that if there is no uncertainty in B, we have $\Gamma = I$, hence $\dot{V} = -x^T Q x$ is easily established with $A_m^T P + P A_m = -Q$, and $P = \delta K^T \Gamma \delta K$, a symmetric, positive-definite matrix, thereby satisfying the conditions of the Kalman–Yakubovich lemma (Chapter 5). However, for the uncertain matrix, $B, \Gamma \neq I$, and we write

$$\dot{V} = x^T(A_m^T P + P A_m + R + R^T)x, \quad (8.64)$$

where

$$P = \delta K^T \Gamma \delta K, \quad (8.65)$$

and

$$R = \delta K^T \Gamma \delta K B_0 (I - \Gamma) \delta K = P B_0 (I - \Gamma) \delta K. \quad (8.66)$$

Thus we have

$$A_m^T P + P A_m + R + R^T = [A_m + B_0(I - \Gamma)\delta K]^T P + P[A_m + B_0(I - \Gamma)\delta K]. \quad (8.67)$$

By selecting the poles of A_m sufficiently deep inside the left-half plane, one can ensure that the matrix

$$A_m + B_0(I - \Gamma)\delta K$$

is always Hurwitz, therefore, a symmetric, positive-definite matrix Q can be found such that the following Lyapunov identity is satisfied:

$$A_m^T P + P A_m + R + R^T = -Q, \quad (8.68)$$

thereby implying that

$$\dot{V} = -x^T Q x \leq 0, \quad (8.69)$$

and the adaptive system is globally, asymptotically stable by the Kalman–Yakubovich lemma. This implies that in the limit $t \to \infty$, x (and hence $e = x - x_m$) converges to 0. The nonlinear gain evolution law, Eq. (8.61), does not require the feedback of x and e, and can therefore be regarded as an open-loop adaptation mechanism. However, it requires an initial condition for the controller gain error, $\delta K(0)$.

8.3.2 MRAS for Single-Input Systems

Model reference adaptation is readily applied to single-input plants, and forms the basis of many practical control systems. For example, an ASE system with a trailing-edge control surface driven by the actuating torque is a single-input system. Most of the theoretical developments in MRAS (Aström and Wittenmark, 1995, Lavretsky and Wise, 2013) are presented for such plants, although the scheme can be extended to multi-input systems. Consider an

nth-order aeroelastic system with the following state equations expressible in the controller companion form (Tewari 2002):

$$\begin{aligned}\dot{x}_1 &= x_2\\ \dot{x}_2 &= x_3\\ \dot{x}_{n-1} &= x_n\\ \dot{x}_n &= \frac{(u - a_1x_1 - a_2x_2 - \cdots - a_nx_n)}{a_{n+1}},\end{aligned} \tag{8.70}$$

where $a_i, i = 1, \ldots, n+1$ are unknown coefficients, but the sign of a_{n+1} is assumed to be known. The plant's parameter vector is given by $a^T = (a_{n+1}, a_n, \ldots, a_2, a_1)$.

The general control objective is to track the following asymptotically stable reference model, driven by a bounded reference signal, $r(t)$:

$$\begin{aligned}\dot{\xi}_1 &= \xi_2\\ \dot{\xi}_2 &= \xi_3\\ \dot{\xi}_{n-1} &= \xi_n\\ \dot{\xi}_n &= \frac{(r - \alpha_1\xi_1 - \alpha_2\xi_2 - \cdots - \alpha_n\xi_n)}{\alpha_{n+1}},\end{aligned} \tag{8.71}$$

where $\alpha_i > 0, i = 1, \ldots, n+1$ are known coefficients. However, for the regulation problem, we are only interested in achieving asymptotic stability at equilibrium, hence $r(t) = 0$. For closed-loop stability, we define the following signal:

$$z(t) = -\beta_1 e - \beta_2 \dot{e} - \cdots - \beta_n e^{(n-1)} + \dot{\xi}_n, \tag{8.72}$$

where, $e = x_1 - \xi_1$ and $\beta_i > 0, i = 1, \ldots, n$ are the coefficients of the following Hurwitz closed-loop polynomial:

$$\beta_1 + \beta_2 s + \cdots + \beta_n s^{n-1} + s^n. \tag{8.73}$$

Selecting the following state-feedback control law:

$$u = \hat{a}_{n+1}z + \hat{a}_n x_n + \cdots + \hat{a}_1 x_1 = v^T\hat{a}, \tag{8.74}$$

where

$$v = (z, x_n, \ldots, x_2, x_1)^T, \tag{8.75}$$

and

$$\hat{a} = (\hat{a}_{n+1}, \hat{a}_n, \ldots, \hat{a}_2, \hat{a}_1)^T, \tag{8.76}$$

the following closed-loop error dynamics is obtained:

$$a_{n+1}\left(e^{(n)} + \beta_n e^{(n-1)} + \cdots + \beta_1 e\right) = v^T\tilde{a}, \tag{8.77}$$

where

$$\tilde{a} = \hat{a} - a \tag{8.78}$$

is the parametric error vector. The closed-loop system is thus expressed in the following companion form:

$$\dot{x} = Ax + \frac{b}{a_{n+1}} v^T \tilde{a}$$
$$e = cx, \tag{8.79}$$

where

$$A = \begin{pmatrix} 0 & 1 & 0 & \dots & 0 & 0 \\ 0 & 0 & 1 & \dots & 0 & 0 \\ \vdots & \vdots & \vdots & \vdots & \vdots & \vdots \\ 0 & 0 & 0 & \dots & 0 & 1 \\ -\beta_1 & -\beta_2 & -\beta_3 & \dots & -\beta_{n-1} & -\beta_n \end{pmatrix}, \quad b = \begin{Bmatrix} 0 \\ 0 \\ \vdots \\ 0 \\ 1 \end{Bmatrix} \tag{8.80}$$

and $c = (1, 0, \dots, 0, 0)$. The following quadratic candidate Lyapunov function is now selected:

$$V = x^T P x + \tilde{a}^T \Gamma^{-1} \tilde{a}, \tag{8.81}$$

where P and Γ are symmetric and positive-definite matrices, with P satisfying the Lyapunov identity:

$$A^T P + PA = -Q, \tag{8.82}$$

where Q is a symmetric, positive-definite cost matrix. The time derivative of V is given by

$$\dot{V} = -x^T Q x + 2\tilde{a}^T v b^T P x + 2\tilde{a}^T \Gamma^{-1} \dot{\tilde{a}} \tag{8.83}$$

In order to ensure a globally, asymptotically stable closed-loop system, the following adaptation law is chosen:

$$\dot{\hat{a}} = -\Gamma v b^T P x, \tag{8.84}$$

resulting in

$$\dot{V} = -x^T Q x. \tag{8.85}$$

The convergence of $x(t)$ to zero in the limit $t \to \infty$ is proved by Lyapunov stability theorem (Chapter 5), as carried out previously. Thus the error $e(t)$ and its time derivatives of order up to $(n-1)$ are proved to vanish in the steady state.

8.4 Aeroservoelastic Applications

The ASE plant parameters are contained in the state-space coefficients A, B, C, D, and it could be tempting to treat all of them as being random and uncertain. However, a more practical approach is to assign uncertainty bounds to the plant parameters wherever it is possible to do so. For example, the structural coefficients are known with a much higher accuracy as compared to the aerodynamic parameters. In addition, as opposed to aerodynamic parameters, the structural coefficients normally do not depend on flight conditions[1] and hence can be regarded

[1] In very high-speed aircraft (such as the Lockheed SR-71, MiG-25/31 and futuristic hypersonic aeroplanes), structural stiffnesses can appreciably vary with the airspeed owing to thermal effects. Accounting for such a variation must be coupled with aerodynamics and falls into a special area called *aerothermoelasticity*.

as constants. It is rare to have an entirely unknown coefficient in the plant matrix and thus one can assign a range of expected values (or bounds) to each parameter, even to those that vary with flight conditions in an uncertain manner. A primary task of an adaptive controller of the self-tuner type (Chapter 7) is to identify the uncertain plant parameters by an adaptation loop based on a fixed framework of the plant equations. If a detailed knowledge of the various uncertainty bounds is available, the adaptation loop becomes unnecessary, and a simple linear feedback based on the known bounds can produce stability (see Example 8.1.8). However, when such bounds are unknown, a model reference approach is an attractive alternative to the STR, as it does not require online parameter estimation. A reference model based on linearized unsteady aerodynamics with a time-invariant representation is a reasonable one, because it captures most of the essential features of the aeroelastic plant, and can be employed as a basis of an adaptive control system for a more general, nonlinear and uncertain dynamics.

8.4.1 Reference Aeroelastic Model

Model reference adaptation requires a basic aeroelastic model with a well-known structure, having the same number of states and governing parameters as the actual, unknown plant. Such a model can be simply derived from the linear structural dynamics and inviscid, incompressible flow. The effects of compressibility and viscosity can then be regarded as parametric perturbations (uncertainties), to be taken care of by the adaptation mechanism. Let the reference model be governed by the linear ordinary differential equations derived in Chapter 3 and written in the following matrix form:

$$M\ddot{q} + C_d\dot{q} + Kq = Q_a + Q_c, \tag{8.86}$$

where $q(t) : \mathbb{R} \rightarrow \mathbb{R}^n$ is the generalized coordinates vector corresponding to the n degrees of structural freedom, and $M \in \mathbb{R}^{n\times n}$, $C_d \in \mathbb{R}^{n\times n}$ and $K \in \mathbb{R}^{n\times n}$ are the generalized mass, damping and stiffness matrices, respectively, of the structure. The generalized control force vector produced by control-surface deflections is given by $Q_c(t) : \mathbb{R} \rightarrow \mathbb{R}^n$. The generalized aerodynamic force vector, $Q_a(t) : \mathbb{R} \rightarrow \mathbb{R}^n$ is assumed to be linearly related to $q(t)$, $\dot{q}(t)$ and $\ddot{q}(t)$, as well as to certain additional state variables collected into the aerodynamic lag-state vector, $x_a(t) : \mathbb{R} \rightarrow \mathbb{R}^\ell$, which is necessary for modelling the aerodynamic lag caused by a circulatory wake by a rational-function approximation (see Chapter 3). Thus, we write

$$Q_a = M_a\ddot{q} + C_a\dot{q} + K_aq + N_ax_a, \tag{8.87}$$

where $M_a \in \mathbb{R}^{n\times n}$, $C_a \in \mathbb{R}^{n\times n}$ and $K_a \in \mathbb{R}^{n\times n}$ are the generalized aerodynamic inertia, aerodynamic damping and aerodynamic stiffness matrices, respectively, and $N_a \in \mathbb{R}^{n\times \ell}$ is the aerodynamic lag-coefficient matrix associated with the time lag due to a circulatory wake. Substitution of Eq. (8.87) into Eq. (8.86) yields

$$(M - M_a)\ddot{q} + (C_d - C_a)\dot{q} + (K - K_a)q = N_ax_a + Q_c. \tag{8.88}$$

The aerodynamic lag states are assumed to be governed by the following linear state equations (Chapter 2):

$$\dot{x}_a = F_ax_a + Z_a\begin{Bmatrix} q \\ \dot{q} \end{Bmatrix}, \tag{8.89}$$

where $F_a \in \mathbb{R}^{\ell\times\ell}$ and $Z_a \in \mathbb{R}^{\ell\times 2n}$ are the aerodynamic coefficient matrices corresponding to circulatory lag effects. By collecting the structural and aerodynamic state vectors into an *augmented state vector*, $x_r(t) : \mathbb{R} \to \mathbb{R}^{2n+\ell}$,

$$x_r = \begin{Bmatrix} q \\ \dot{q} \\ x_a \end{Bmatrix}, \tag{8.90}$$

and expressing the generalized control forces vector as the control input, $Q_c(t) : \mathbb{R} \to \mathbb{R}^m$, we have the following augmented state equations of the linear, reference plant:

$$\dot{x}_r = A_r x_r + B_r Q_c, \tag{8.91}$$

where

$$A_r = \begin{pmatrix} 0 & I & 0 \\ -\bar{M}^{-1}\bar{K} & -\bar{M}^{-1}\bar{C} & -\bar{M}^{-1}N_a \\ Z_a & & F_a \end{pmatrix} \tag{8.92}$$

and

$$B_r = \begin{pmatrix} 0 \\ I \\ 0 \end{pmatrix}, \tag{8.93}$$

where $\bar{M} = M - M_a$, $\bar{C} = C_d - C_a$, $\bar{K} = K - K_a$ are the generalized mass, damping and stiffness matrices, respectively, of the aeroelastic system, and 0 and I represent the null and identity matrices, respectively, of appropriate dimensions.

The reference model is assumed to be governed by a finite-order, linear, unsteady aerodynamic behaviour in the Laplace domain, which is expressed as follows:

$$Q_a(s) = G_a(s)q(s), \tag{8.94}$$

where $G_a(s) \in \mathbb{R}^{n\times n}$ is the unsteady aerodynamic transfer matrix, $Q_a(s) \in \mathbb{R}^n$ is the generalized aerodynamics force vector in the Laplace domain and $\vec{q}(s) \in \mathbb{R}^n$ is the vector of generalized motion coordinates based on a finite number, n, of structural degrees of freedom in the Laplace domain. The model state-space representation given by $A_r \in \mathbb{R}^{(2n+\ell)\times(2n+\ell)}$ and $B_r \in \mathbb{R}^{(2n+\ell)\times m}$ requires the approximation of the aerodynamic transfer matrix, $G_a(s)$, as a rational matrix function relating $Q_a(t)$ to the generalized coordinates vector $q(t)$ and the aerodynamic lag-state vector, $x_a(t)$ in the Laplace domain. This is carried out differently in various speed regimes (incompressible, subsonic or supersonic) and dimensions (typical-section or three-dimensional model), and the aerodynamic coefficients are functions of the flight Mach number. Considering the simple-pole, least-squares approximation (Chapter 3), we have the following rational function approximation:

$$G_a(s) = A_0 + A_1 s + A_2 s^2 + \sum_{j=1}^{N} A_{j+2}\frac{s}{s+b_j}, \tag{8.95}$$

where the numerator coefficient matrices,

$$A_0, A_1, A_2, \ldots, A_{N+2},$$

each of size $(n \times n)$, are determined by fitting $G_a(ik)$ to the frequency-domain aerodynamics data, $H(k)$, at a discrete set of reduced frequencies k. The lag parameters (or aerodynamic poles) $b_j > 0, j = 1, \ldots, N$, are evaluated by a nonlinear optimization process in order to minimize the total curve-fit error at the selected frequencies (Chapter 3). The state-space model of control-surface actuators is given by

$$\dot{x}_c = A_c x_c + B_c u$$
$$Q_c = C_c x_c + D_c u_m, \tag{8.96}$$

where $u_m(t) : \mathbb{R} \to \mathbb{R}^m$ is the vector of reference control torque inputs applied to m control surfaces.

The overall reference model is thus represented by

$$\dot{x}_m = A_m x_m + B_m u_m, \tag{8.97}$$

where $x_m = (x_r, x_c)$ and

$$A_m = \begin{pmatrix} A_r & B_r C_c \\ 0 & A_c \end{pmatrix}, B_m = \begin{pmatrix} B_r D_c \\ B_c \end{pmatrix}. \tag{8.98}$$

Note that A_m, B_m are known matrices, with A_m chosen to be Hurwitz at the reference flight condition. If the original model is not Hurwitz, it can be made so by an LQR-type state-feedback gain. If an output equation is added on the basis of the available sensors (accelerometers or optical sensors),

$$y_m = C_m x_m + D_m u_m, \tag{8.99}$$

then the matrices C_m, D_m are also known. The model is stabilizable with the pair (A_m, B_m), and detectable with (A_m, C_m). A transfer-matrix realization of the reference model can be given by

$$G_m(s) = C_m(sI - A_m)^{-1} B_m + D_m, \tag{8.100}$$

which is positive real, hence the reference model is input–output stable (see Chapter 6).

8.4.2 Adaptive Flutter Suppression of Typical Section

The MRAS scheme is directly applicable to the typical-section model with a trailing-edge control surface. The simplest reference model is for Theodorsen type incompressible flow aerodynamics (Chapter 3) and a simple-pole type, least-squares RFA, for which we have

$$\dot{x}_m = A_m x_m + B_m r, \tag{8.101}$$

where

$$x_m = (h/b, \theta, \beta, x_a, \dot{h}/b, \dot{\theta}, \dot{\beta}, \dot{x}_a)^T$$

is the state vector, with $x_a \in \mathbb{R}^\ell$ being the aerodynamic state vector,

$$A_m = \begin{pmatrix} 0_n & I_n & 0_{n\times\ell} \\ -\overline{M}^{-1}\hat{K} & -\overline{M}^{-1}\hat{C} & -\overline{M}^{-1}N_a \\ \Gamma_a & & F_a \end{pmatrix}, \tag{8.102}$$

$$B_m = (0_{1\times5}, 1, 0_{1\times\ell})^T, \tag{8.103}$$

where $\hat{C} = \bar{C} - C_\ell$, $\hat{K} = \bar{K} - K_\ell$, with $\bar{M}, \bar{C}, \bar{K}$ given in Chapter 3.

$$N_a = -2\kappa \begin{Bmatrix} -1 \\ \frac{1}{2} + a \\ -\frac{1}{2\pi} T_{12} \end{Bmatrix} \{ a_1 b_1 \quad a_2 b_2 \quad \dots \quad a_\ell b_\ell \}, \tag{8.104}$$

$$K_\ell = 2\kappa (a_0 + a_1 + \cdots + a_\ell) \begin{Bmatrix} -1 \\ \frac{1}{2} + a \\ -\frac{1}{2\pi} T_{12} \end{Bmatrix} \left\{ 0 \quad 1 \quad \frac{1}{\pi} T_{10} \right\}, \tag{8.105}$$

$$C_\ell = 2\kappa (a_0 + a_1 + \cdots + a_\ell) \begin{Bmatrix} -1 \\ \frac{1}{2} + a \\ -\frac{1}{2\pi} T_{12} \end{Bmatrix} \left\{ 1 \quad \left(\frac{1}{2} - a\right) \quad \frac{1}{2\pi} T_{11} \right\}, \tag{8.106}$$

$$\Gamma_a = \left\{ 0_{\ell\times1} \quad 1_{\ell\times1} \quad \frac{1}{\pi} T_{10} 1_{\ell\times1} \quad 1_{\ell\times1} \quad \left(\frac{1}{2} - a\right) 1_{\ell\times1} \quad \frac{1}{2\pi} T_{11} 1_{\ell\times1} \right\}, \tag{8.107}$$

with $0_{\ell\times1}$ and $1_{\ell\times1}$ being the arrays of zeros and ones, respectively, and

$$F_a = \begin{pmatrix} -b_1 & 0 & 0 & \dots & 0 \\ 0 & -b_2 & 0 & \dots & 0 \\ 0 & 0 & -b_3 & \dots & 0 \\ \vdots & \vdots & \vdots & \vdots & \vdots \\ 0 & 0 & 0 & \dots & -b_\ell \end{pmatrix}. \tag{8.108}$$

For the actual plant, we have a similar model given by

$$\dot{x} = Ax + B_0 \lambda u, \tag{8.109}$$

where A, λ are uncertain parameters, λ being a positive scalar, and $B_0 = B_m$ is assumed to be known. The objective is to stabilize the plant in the presence of uncertainties and with $r = 0$. The source of uncertainties in A can be the errors in the structural model modified by non-circulatory effects, $\bar{M}, \bar{C}, \bar{K}$, or in the aerodynamic circulatory model, $F_a, \Gamma_a, N_a, C_\ell, K_\ell$ or both. The modelling error in the torque produced by the control surface actuation mechanism is represented by the uncertain parameter, λ, which can have a random value in the range $-1 \leq \lambda \leq 1$. If A_m is chosen to be Hurwitz, then the modelling errors in A can be assumed to be random variations in the elements of A_m which are neither zero nor unity. The MRAS stabilization can be achieved by the following adaptation law (Eq. (8.48)) using the state-feedback approach:

$$\dot{\hat{K}} = B_m^T P e x^T R, \tag{8.110}$$

where P, R are symmetric, positive-definite weighting matrices and $e = x - x_m$ is the tracking error from the reference model.

Example 8.4.1 *The design for an active flutter-suppression system for a typical section was carried out in Chapter 4. Here, the same example is made adaptive by treating the matrices A, B to be uncertain and by applying the MRAS adaptation law for state-feedback. In order to* *derive the reference model, let us consider the following pole locations for a Hurwitz matrix,* A_m:

$$s_{1,2} = -0.1 \pm 0.1i, s_{3,4} = -0.2 \pm 0.2i, s_{5,6} = -0.3 \pm 0.3i, s_7 = -0.0553, s_8 = -0.2861.$$

These pole locations were also selected for the state-feedback active flutter suppression system designed for this model in Chapter 4. By mimicking the response of the asymptotically stable reference model, the MRAS system will remain stable at a higher than normal (supercritical) speed. Recall from Chapter 4 that the open-loop flutter speed for the original model is $U = 63.35$ *m/s, with a corresponding flutter frequency of 41.18 rad/s. An alternative method of deriving a stable, flutter-suppression reference model is to increase all the structural stiffness parameters,* k_h, k_θ, k_β, *of the original plant, such that the open-loop flutter occurs at a higher than normal dynamic pressure. However, such a method would yield unrealistic adaptation gains. The computed state dynamics coefficient matrix,* $A_m = A * -B_m\hat{k}(0)$, *for the reference model, with* $A*$ *being the open-loop matrix at the subcritical (below the flutter dynamic pressure) speed of 30 m/s and standard sea level, evaluated with two lag parameters, is the following:*

$$A_m = \begin{pmatrix} 0 & 0 & 0 & 1.0 & 0 & 0 & 0 & 0 \\ 0 & 0 & 0 & 0 & 1.0 & 0 & 0 & 0 \\ 0 & 0 & 0 & 0 & 0 & 1.0 & 0 & 0 \\ -0.2742 & 0.2989 & 0.0019 & -0.1358 & -0.1407 & -0.0369 & 0.0024 & 0.0234 \\ 0.1932 & -3.3639 & 0.9810 & 0.3111 & -0.0024 & -0.0208 & -0.0055 & -0.0536 \\ 0.7656 & -11.3171 & 3.3051 & 0.9493 & 3.7732 & -1.0618 & -0.0199 & -0.2008 \\ 0 & 1.0 & 0.6844 & 1.0 & 0.7 & 0.3039 & -0.0553 & 0 \\ 0 & 1.0 & 0.6844 & 1.0 & 0.7 & 0.3039 & 0 & -0.2861 \end{pmatrix}.$$

The initial value of the feedback gain matrix of MRAS is derived at the subcritical speed of $U = 30$ *m/s at sea level by pole placement to be the following:*

$$\hat{k}(0) = (-0.5749, 17.7842, -9.3744, -1.4245, -4.1330, 0.7578, 0.0283, 0.2827).$$

The measured output, $y = Cx + Du$, *is the normal acceleration sensed at a location 1.0% semi-chord forward of the elastic axis. The state variables as well as the input and output are rendered non-dimensional as in Theodorsen's model (Chapter 3).*

The adaptation law is applied with $P = R = I$ *and simulated using a normally distributed random perturbation per time step of up to* $\pm 150\%$ *variation in* λ, *as well as in all the elements of the actual plant matrix, A, except those which are either* 0 *or* 1. *Such a simulation allows for a large, time-dependent variation in the plant parameters due to such effects as structural modelling uncertainties, flow viscosity (including both leading-edge and trailing-edge flow separation of small magnitudes) and changes in the wake-induced circulatory forces and moments, all of which allow a linear, but random, time-varying dynamic description. However, as Lyapunov stability analysis assumes constant uncertain parameters,* A, λ, *it would be interesting to see whether a time-dependent random perturbation (considered process noise and covered in Chapter 10) can also be handled by the MRAS scheme. This would be a*

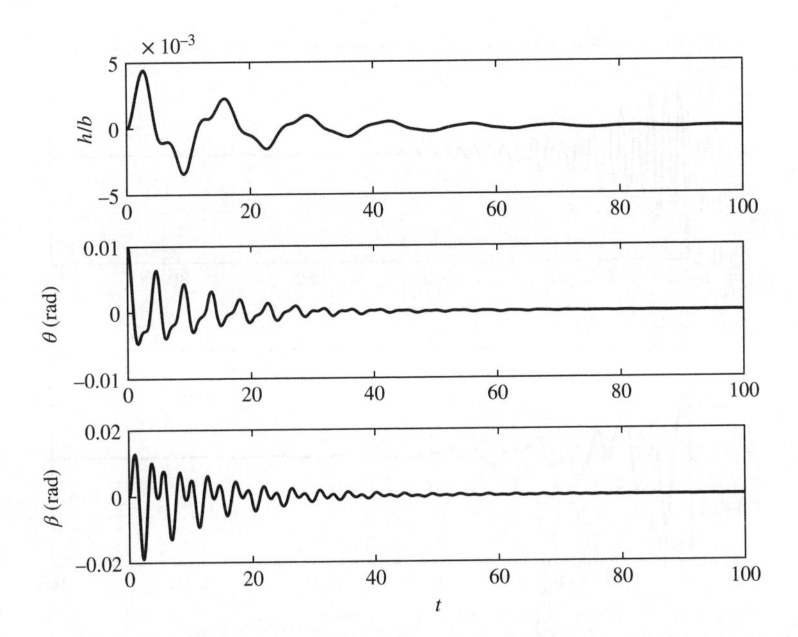

Figure 8.6 Simulated initial response of plunge displacement, h/b, pitch angle, θ, and control-surface deflection, β, of an adaptive flutter suppression system for randomly varying typical-section parameters at a subcritical speed and standard sea level

test of the method's robustness with respect to process noise in all the parameters of the plant. The simulated results for an initial pitch perturbation and a zero initial adaptation gain matrix ($\hat{k}(0) = 0$) at $U = 30$ m/s are plotted in Figs. 8.6–8.9, showing a stable subcritical behaviour of the plunge displacement, h/b, pitch angle, θ, control-surface deflection, β, input torque, u, and normal acceleration output, y, as well as the state-feedback adaptation gains, $\hat{k} = (\hat{k}_1, \hat{k}_2, \ldots, \hat{k}_8)$. While the system's response settles down in about $t = 100$, the adaptation gains are seen to remain bounded with small random variations due to the randomly varying A, λ. The MRAS for the subcritical case is therefore quite robust for the process noise, which is achieved without taking into account any knowledge of its bounds. Chapter 10 is devoted to the design of a robust MRAS, and we shall return to this example in that chapter.

Now the speed is increased to a supercritical value of $U = 65$ m/s at standard sea level. While the actual plant's parameters have now changed from what they were at the subcritical condition, the reference model remains the same as that given above, with $A_m = A * -B_m\hat{k}(0)$ having the aforementioned closed-loop pole locations. However, when the same constant gains are applied to the uncertain plant, $A, B_m\lambda$, there is no guarantee that the resulting closed-loop system will be stable. Therefore, the adaptation law of Eq. (8.110) is applied with $P = R = I$ and the following new values for the initial regulator gains intended to stabilize the reference model at the given supercritical condition:

$$\hat{k}(0) = (-0.0142, 0.3088, -1.4759, -0.5726, -3.6371, 0.7578, 0.0096, 0.1514).$$

We first simulate (as before) a normally distributed, random, time-dependent perturbation of up to $\pm 150\%$ variation in the actual plant parameters, A, λ, which is applied at each

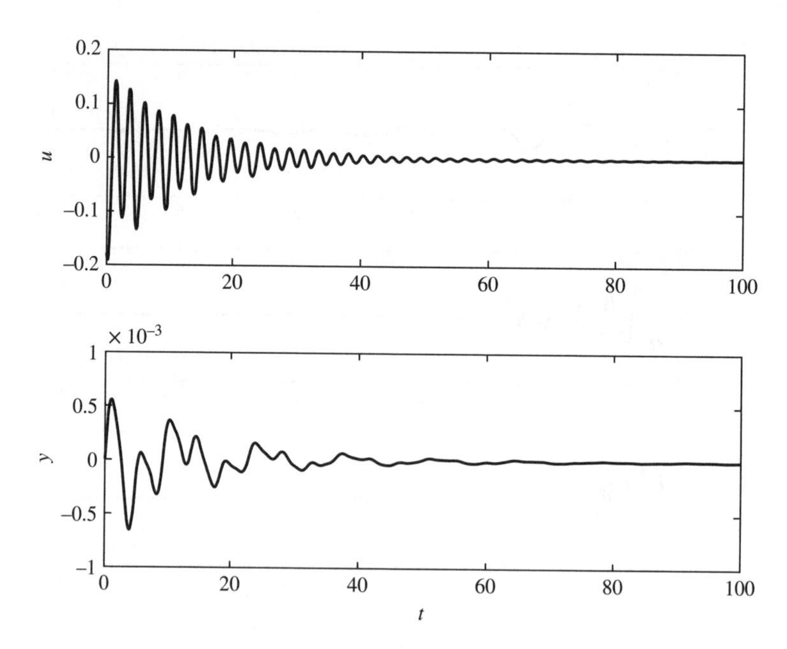

Figure 8.7 Simulated initial response of input torque, u, and normal acceleration output, y, of an adaptive flutter suppression system for randomly varying typical-section parameters at a subcritical speed and standard sea level

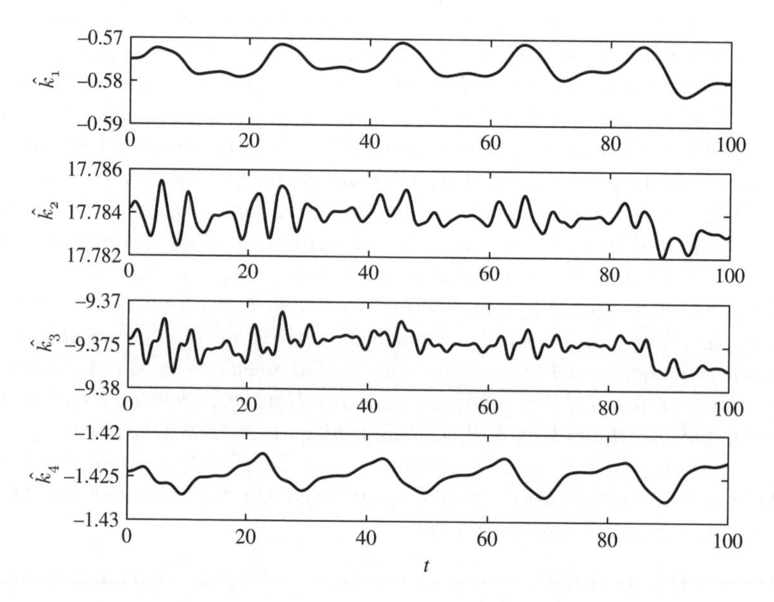

Figure 8.8 Simulated variation of state-feedback adaptation gains, $\hat{k}_1 - \hat{k}_4$, of an adaptive flutter suppression system for randomly varying typical-section parameters at a subcritical speed and standard sea level

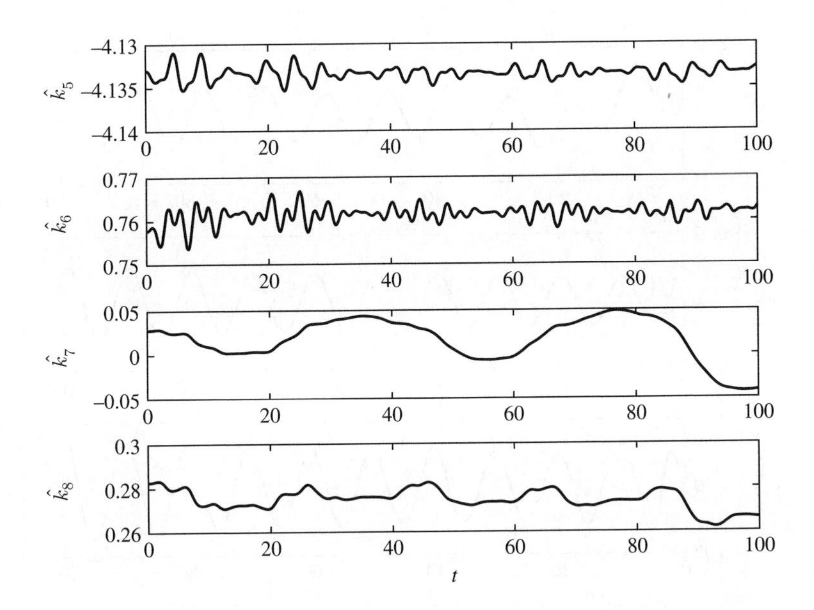

Figure 8.9 Simulated variation of state-feedback adaptation gains, $\hat{k}_5 - \hat{k}_8$, of an adaptive flutter suppression system for randomly varying typical-section parameters at a subcritical speed and standard sea level

time step in order to test robustness with respect to process noise. The response, plotted in Figs. 8.10–8.13, remains bounded, but does not converge to a steady state. In fact, the adaptation gains are seen to increase abruptly near t $= 90$, *causing the input torque to momentarily shoot up at that instant. Such a phenomenon is a peculiarity of the MRAS schemes called parameter drift, which happens when the controller parameters are allowed to slowly change even though the tracking error, e, has converged to a small magnitude. Parameter drift can ultimately destabilize the system, if allowed to persist for a sufficiently long time. Therefore, if a time-dependent process noise of unknown bounds is applied to the MRAS, the flutter suppression might not be successful. This points towards the need for a robust MRAS design, which can take into account the bounds on the process noise. Such a design is covered in Chapter 10.*

We end the example with a more realistic, constant, uniformly distributed random perturbation of up to $\pm 50\%$ *variation in A,* λ, *for which the responses at the supercritical condition,* $U = 65$ *m/s and standard sea level, are plotted in Figs. 8.14–8.17. As expected by the Lyapunov stability analysis, the state variables and the controller parameters are seen to be bounded and convergent with the UUB property. The flutter suppression is thus guaranteed with the MRAS, if the linear plant uncertainties are constant, even within a large random range (*$\pm 50\%$*).*

8.4.3 Adaptive Stabilization of Flexible Fighter Aircraft

Fighter-type aircraft can encounter ASE instabilities while manoeuvring at high-subsonic/transonic speeds. Usually, such aircraft are intentionally designed to be statically unstable for

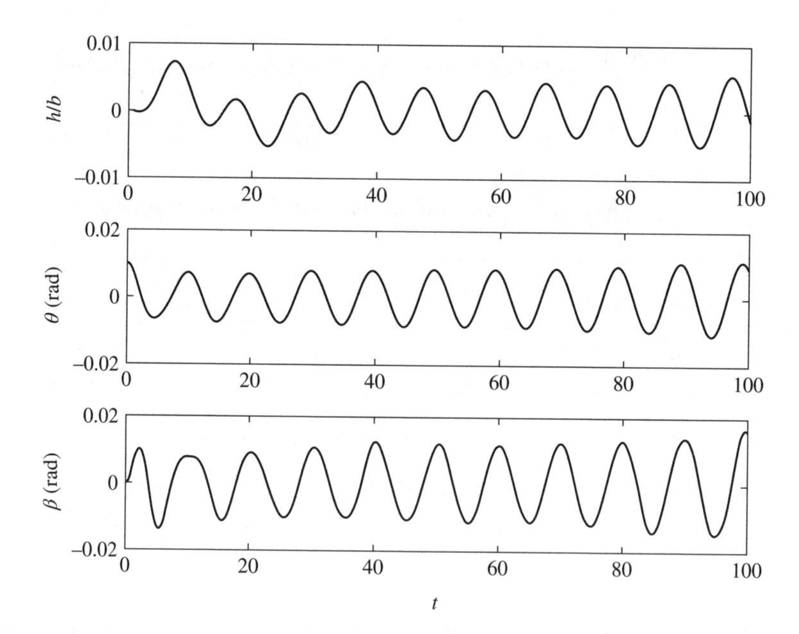

Figure 8.10 Simulated initial response of plunge displacement, h/b, pitch angle, θ, and control-surface deflection, β, of an adaptive flutter suppression system for randomly varying typical-section parameters at a supercritical speed and standard sea level

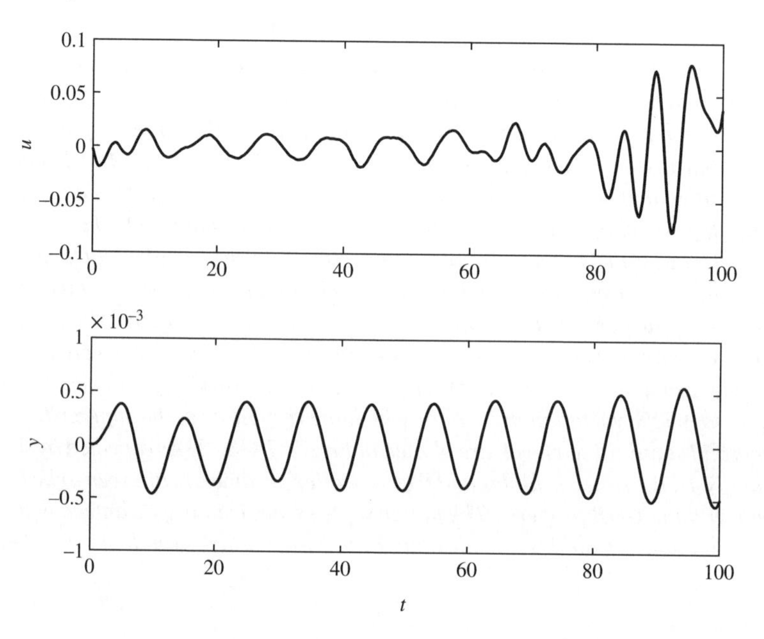

Figure 8.11 Simulated initial response of input torque, u, and normal acceleration output, y, of an adaptive flutter suppression system for randomly varying typical-section parameters at a supercritical speed and standard sea level

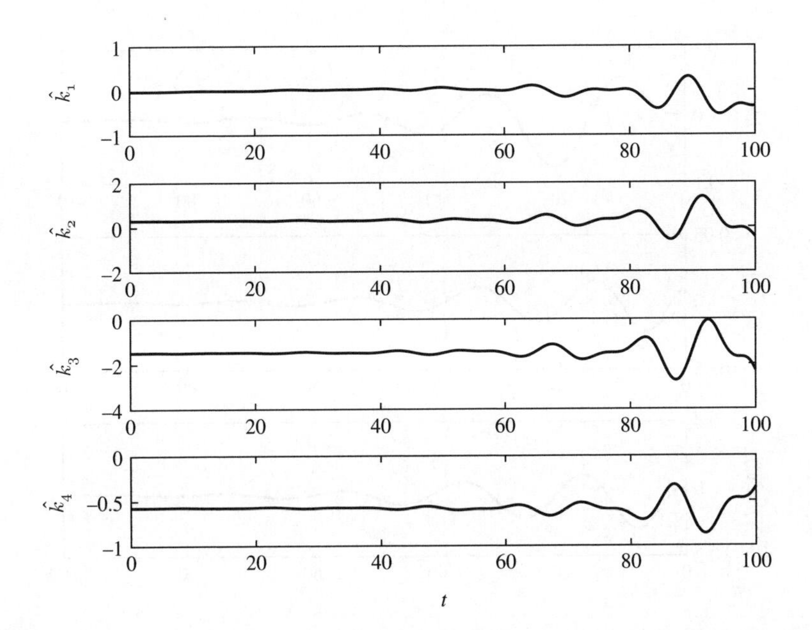

Figure 8.12 Simulated variation of state-feedback adaptation gains, $\hat{k}_1 - \hat{k}_4$, of an adaptive flutter suppression system for randomly varying typical-section parameters at a supercritical speed and standard sea level

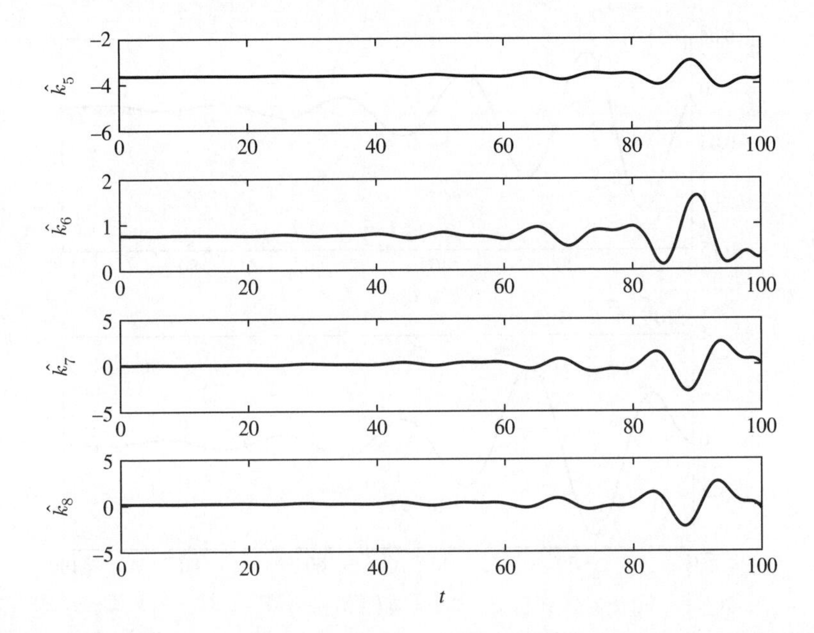

Figure 8.13 Simulated variation of state-feedback adaptation gains, $\hat{k}_5 - \hat{k}_8$, of an adaptive flutter suppression system for randomly varying typical-section parameters at a supercritical speed and standard sea level

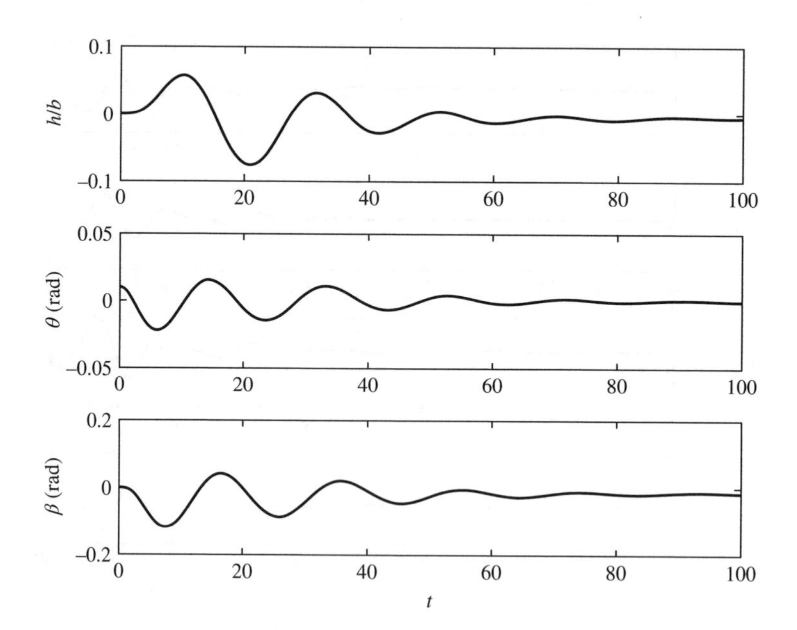

Figure 8.14 Simulated initial response of plunge displacement, h/b, pitch angle, θ, and control-surface deflection, β, of an adaptive flutter suppression system for constant random perturbation in typical-section parameters at a supercritical speed and standard sea level

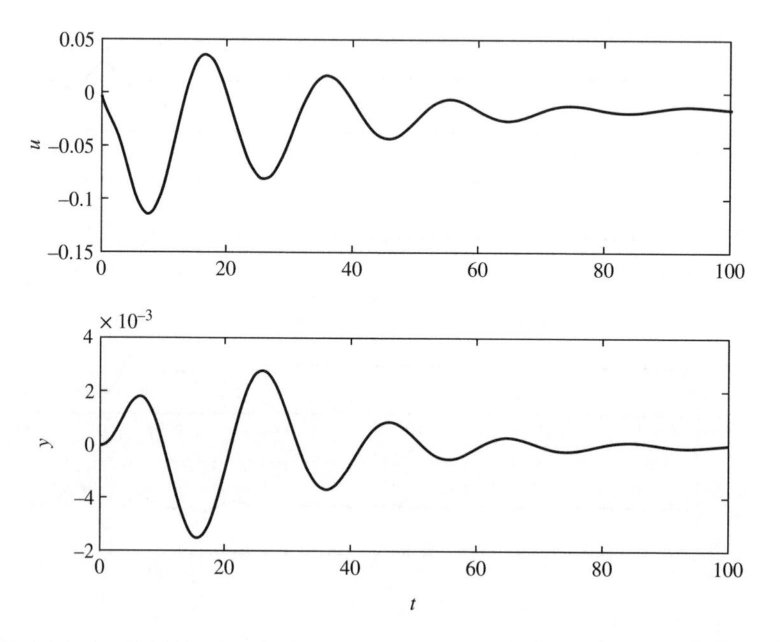

Figure 8.15 Simulated initial response of input torque, u, and normal acceleration output, y, of an adaptive flutter suppression system for constant random perturbation in typical-section parameters at a supercritical speed and standard sea level

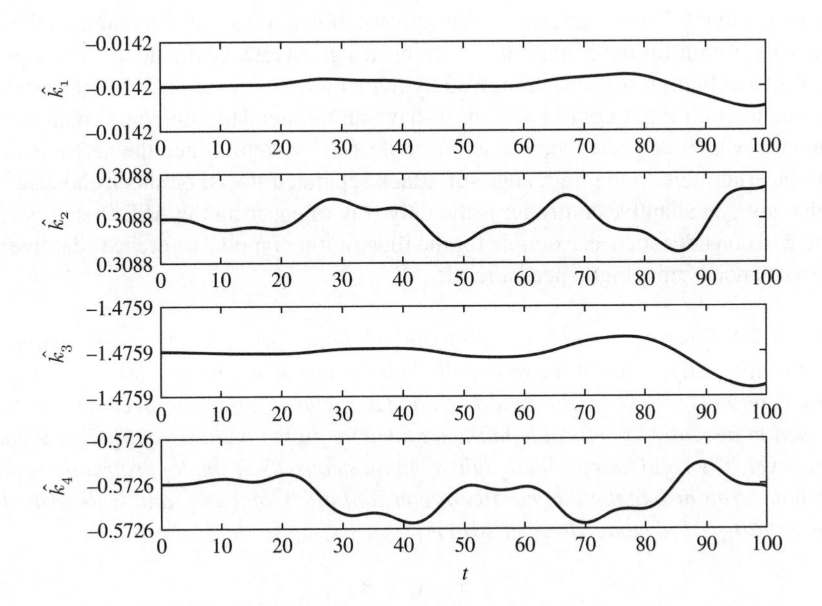

Figure 8.16 Simulated variation of state-feedback adaptation gains, $\hat{k}_1 - \hat{k}_4$, of an adaptive flutter suppression system for constant random perturbation in typical-section parameters at a supercritical speed and standard sea level

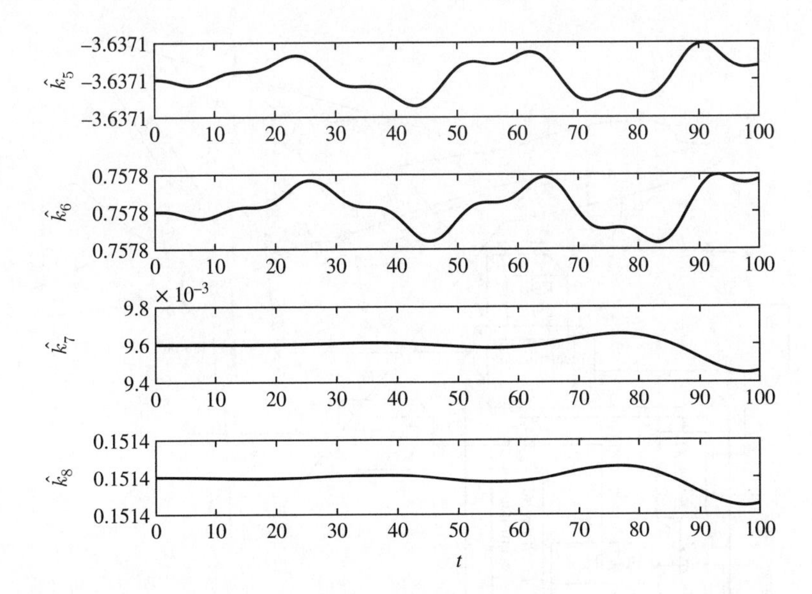

Figure 8.17 Simulated variation of state-feedback adaptation gains, $\hat{k}_5 - \hat{k}_8$, of an adaptive flutter suppression system for constant random perturbation in typical-section parameters at a supercritical speed and standard sea level

a better manoeuvrability, which requires an active stabilization system to maintain the vehicle either at a given equilibrium condition, or turning at a given rate. Owing to the low-aspect ratio wings of relatively high stiffness employed by fighter aircraft, there is a greater possibility of the high-bandwidth flight control system to have an unintended interaction with the aeroelastic modes, which can often lead to an unstable ASE system. Since the aeroelastic model is often uncertain due to the high angle-of-attack separated flows, or shock-induced oscillatory behaviour, an adaptive controller is the only way of ensuring the ASE system's stability. Here we will consider such an example for the illustration of model reference adaptive control applied to manoeuvring, high-speed aircraft.

Example 8.4.2 *Consider a tail-less, delta-winged fighter aircraft (Fig. 8.18) equipped with a pair of trailing-edge control surfaces called elevons (one on either side), a pitch rate gyro and an accelerometer to sense the rigid-body and aeroelastic modes, respectively. The aircraft is designed to be statically unstable in the manoeuvre flight condition of Mach 0.9 and standard sea level. The rigid longitudinal flight dynamics consists of small perturbations about an equilibrium, straight-line flight condition of constant pitch angle, θ_e, and is described by the following short-period mode (Tewari, 2011):*

$$\dot{x}_r = A_r x_r + B_r \delta_c, \tag{8.111}$$

where $x_r = (\alpha, \theta, q)^T$ is the rigid state vector consisting of angle-of-attack, α, pitch angle, θ, and pitch rate, q; δ_c is the commanded elevon angle and A_r, B_r are the following coefficient

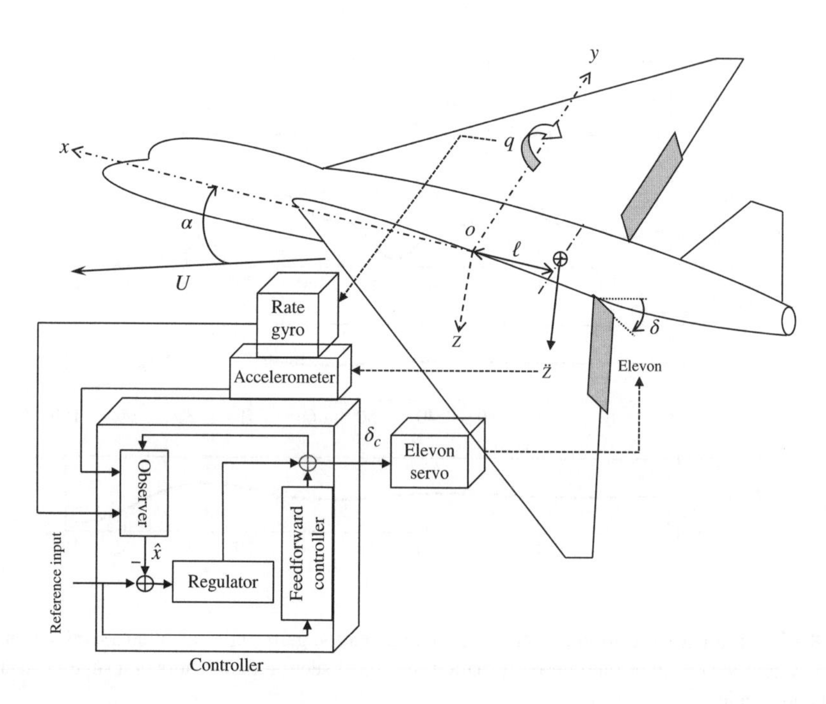

Figure 8.18 A tail-less delta-winged fighter aircraft equipped with a pair of elevons, a rate gyro and an accelerometer for and an automatic flight control system with programmable control laws

matrices:

$$A_r = \begin{pmatrix} \frac{Z_\alpha}{mU-Z_{\dot\alpha}} & -\frac{mg\sin\theta_e}{mU-Z_{\dot\alpha}} & \frac{mU+Z_q}{mU-Z_{\dot\alpha}} \\ 0 & 0 & 1 \\ \frac{M_\alpha}{J_{yy}} + \frac{M_{\dot\alpha}Z_\alpha}{J_{yy}(mU-Z_{\dot\alpha})} & -\frac{M_{\dot\alpha}(mg\sin\theta_e)}{J_{yy}(mU-Z_{\dot\alpha})} & \frac{M_q}{J_{yy}} + \frac{M_{\dot\alpha}(mU+Z_q)}{J_{yy}(mU-Z_{\dot\alpha})} \end{pmatrix}, \tag{8.112}$$

$$B_r = \begin{pmatrix} \frac{Z_\delta}{mU-Z_{\dot\alpha}} \\ 0 \\ \frac{M_\delta}{J_{yy}} + \frac{M_{\dot\alpha}Z_\delta}{J_{yy}(mU-Z_{\dot\alpha})} \end{pmatrix}. \tag{8.113}$$

Here, m denotes the aircraft's mass, g the acceleration due to gravity, J_{yy} the pitch moment of inertia, U the flight speed; $Z_\alpha, Z_{\dot\alpha}, Z_q, Z_\delta, M_\alpha, M_{\dot\alpha}, M_q, M_\delta$ are the aerodynamic stability derivatives, assumed to be constants at a given flight condition (either steady, or steadily turning). Static longitudinal stability requires the derivative, M_α, to be negative.

The actual elevon deflection, δ, which controls the aerodynamics, is governed by a second-order actuator called the elevon servo with the following transfer function:

$$\frac{\delta}{\delta_c} = \frac{\omega_a^2}{s^2 + 2\zeta_a\omega_a s + \omega_a^2}, \tag{8.114}$$

where ω_a and ζ_a are the natural frequency and damping ratio, respectively. The accelerometer is located on the fuselage at a distance ℓ aft of the centre of mass (Fig. 8.18), and senses the normal acceleration, $\ddot{z} = U(\dot\alpha - q) + \ell\dot{q}$, while the rate gyro separately measures the pitch rate, q. Hence, the rigid-body longitudinal dynamics has three vehicle states, x_r, two actuator states (resulting in a fifth-order system), a single input, δ_c, and two outputs, $\ddot{z}, q$.

The flexible structure of the aircraft is modelled by the structural dynamic state vector, x_s, governed by the following state equations:

$$\dot{x}_s = A_s x_s + B_s\delta_c. \tag{8.115}$$

These are combined with the rigid-body equations to yield the following state equations of the aeroelastic plant:

$$\dot{x} = Ax + B\delta_c, \tag{8.116}$$

where the augmented state-space model is given by

$$x = \begin{Bmatrix} x_r \\ x_s \\ \delta \\ \dot\delta \end{Bmatrix}, \tag{8.117}$$

$$A = \begin{pmatrix} A_r & 0 & B_r & 0 \\ 0 & A_s & B_s & 0 \\ 0 & 0 & 0 & 1 \\ 0 & 0 & -\omega_a^2 & -2\zeta_a\omega_a \end{pmatrix}, \tag{8.118}$$

$$B = \begin{pmatrix} 0 \\ 0 \\ 0 \\ \omega_a^2 \end{pmatrix}. \tag{8.119}$$

The common equilibrium condition is of straight and level flight ($\theta_e = 0$), for which the pitch angle, θ, is no longer regarded as a state variable, which simplifies the model considerably. The short-period dynamics of the tail-less fighter aircraft at Mach no. 0.9 ($U = 306\,m/s$) and standard sea level about a straight and level flight condition is the following:

$$A_r = \begin{pmatrix} -0.3078 & 1.0 \\ 11.3 & -3.17 \end{pmatrix}, B_r = \begin{pmatrix} -0.0114 \\ -8.25 \end{pmatrix}$$

$$C_r = \begin{pmatrix} 0 & 1.0 \\ -71.587 & -6.34 \end{pmatrix}, D_r = \begin{pmatrix} 0 \\ 20 \end{pmatrix}. \tag{8.120}$$

The rigid aircraft is thus unstable with eigenvalues $s = 1.9146, -5.3924$.

The aircraft has 20 in vacuo structural dynamic modes below 50 Hz natural frequency (Tewari, 2015). Two wing-bending modes, two combined wing-torsion/ fuselage-bending modes, and the actuator mode of $\omega_a = 9$ Hz, $\zeta = 0.41$, fall within the expected flight control system bandwidth of 10 Hz. When aerodynamic effects are added, the resulting aeroelastic model has much larger acceleration magnitudes (about 40 dB higher at the first bending mode), and a phase lag when compared to the rigid-body dynamics. The aeroelastic effect on the pitch-rate magnitude is about 10 dB more than the rigid response at the first bending mode. Hence, the aeroelastic modes contribute greatly to the overall outputs of the system, which are fed to the controller. For this reason, an attempt to stabilize the aircraft's rigid-body dynamics without taking into account the aeroelastic modes would be a disastrous failure, as demonstrated in the analysis in the companion text (Tewari, 2015). The reason for the failure is the destabilization of the aeroelastic modes by a high-gain feedback loop, resulting in an unstable ASE system.

The problem of unstable aeroelastic modes is addressed by reprogramming the flight control computer with the LQR and Kalman filter gains based on the overall rigid + aeroelastic (rather than only the rigid) plant. The Kalman filter is designed with the following parameters:

$$S_{\mathrm{pm}} = 0,\ S_p = 10^{-8} I,\ S_p = 1,\ F = I,$$

while the LQR parameters are taken to be

$$S = 0,\ Q = 10^{-2} C^T C,\ R = 1.$$

The dimensions of the aeroelastic plant are (38 × 38). *The regulator (linear, quadratic regulator (LQR)) and observer (Kalman filter) poles are in the left-half s-plane, which is seen by the pole-zero map of the ASE system in Fig. 8.19. This implies that the right-half plane pole and zeros of the open-loop plant, which indicated the unstable, non-minimum phase (Tewari, 2002) 'tail-wag-the-dog' behaviour, have been moved to the left-half s-plane in the closed loop. The stability robustness of the LQG design is demonstrated by the Nyquist diagram of the closed-loop transfer function $s^2 z(s)/\delta_c(s)$ shown in Fig. 8.20. The Nyquist locus does not circle the point $(-1, 0)$, showing an infinite stability margin. This is confirmed by the Bode plot of the same transfer function plotted in Fig. 8.21 showing an infinite gain margin and a* 90° *phase margin. This means that the originally negative gain margin of the open-loop system is increased to high positive values by raising the DC gain. The closed-loop response to an initial random perturbation in all the state variables, is plotted in Figs. 8.22 and 8.23. Note that while the open-loop response in $\ddot{z}, \alpha, q$ is unbounded, the closed-loop response of the same*

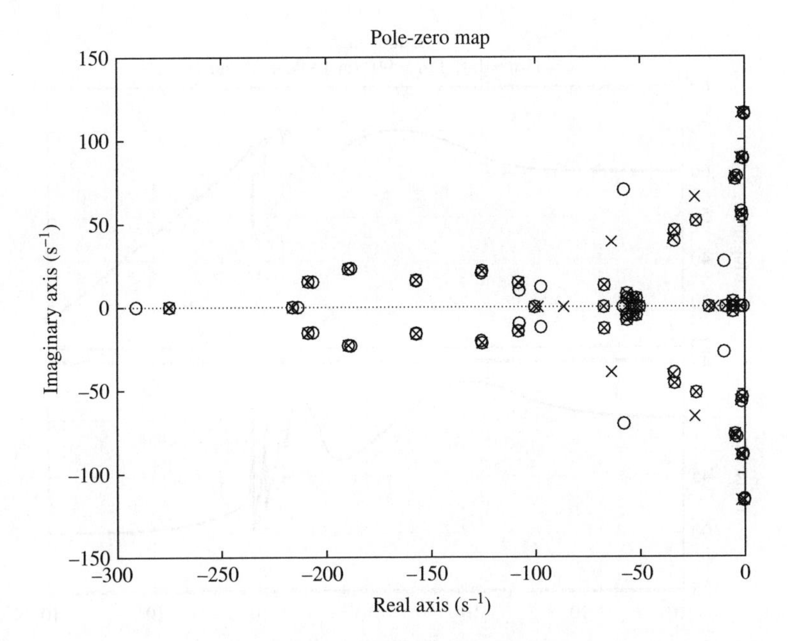

Figure 8.19 Pole-zero map of the aeroservoelastic system for the fighter aircraft for the transfer function $s^2z(s)/\delta_c(s)$ at $U = 306$ m/s and standard sea level (×: pole; o: zero)

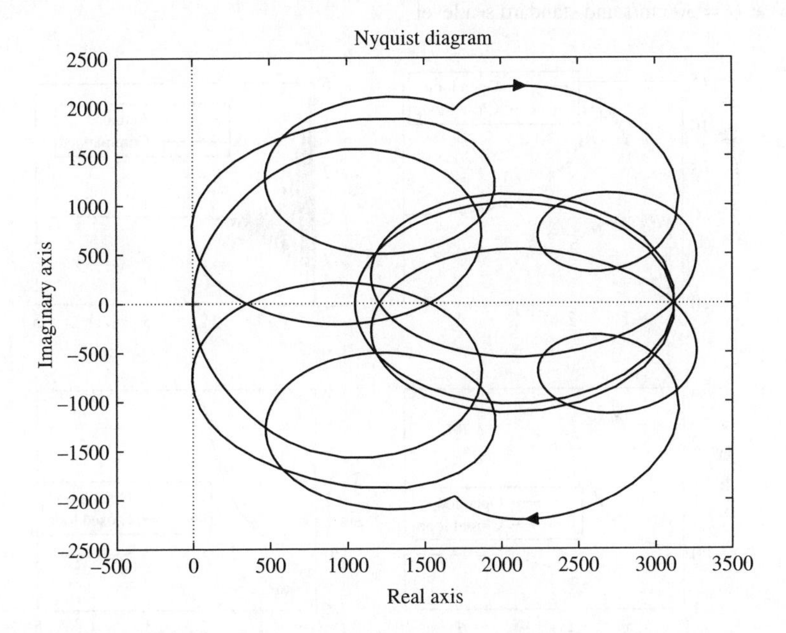

Figure 8.20 Nyquist plot of the aeroservoelastic system for the fighter aircraft for the transfer function $a_z(s)/\delta_c(s)$ at $U = 306$ m/s and standard sea level

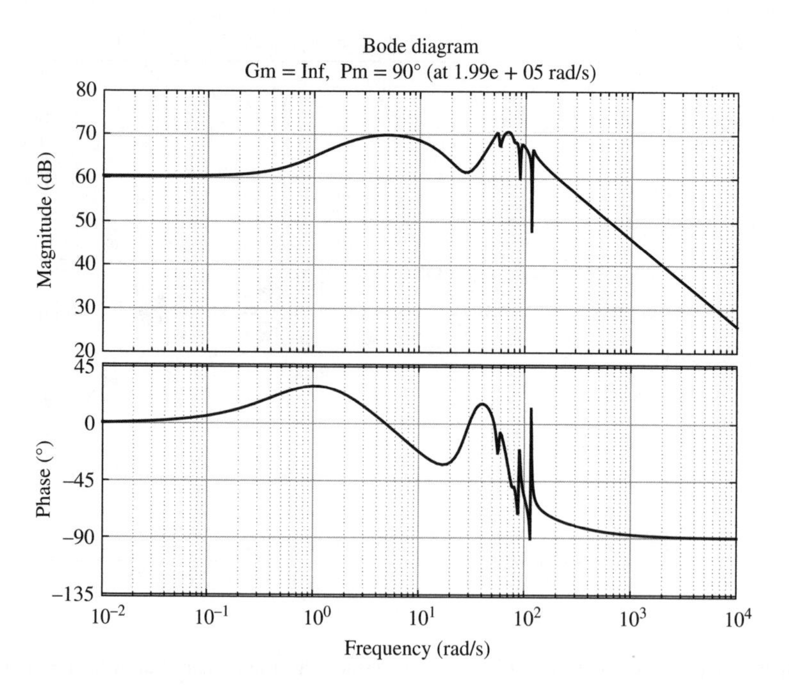

Figure 8.21 Bode plot of the aeroservoelastic system for the fighter aircraft for the transfer function $a_z(s)/\delta_c(s)$ at $U = 306$ m/s and standard sea level

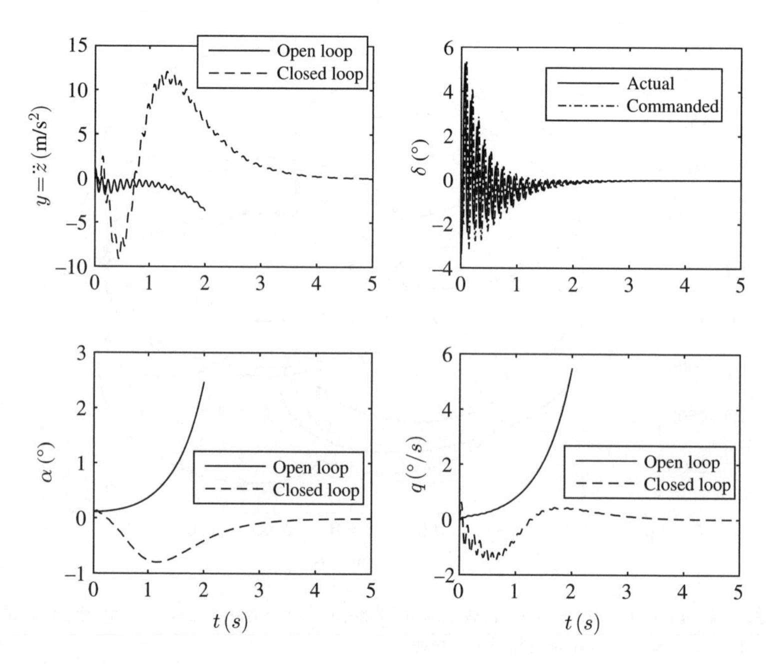

Figure 8.22 Closed-loop response for an initial random perturbation of the fighter aircraft with ASE stabilization at $U = 306$ m/s and standard sea level

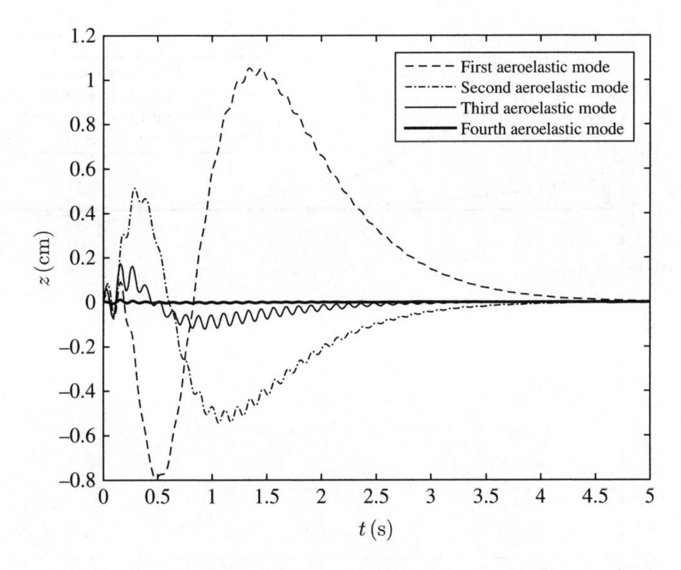

Figure 8.23 Closed-loop response of the first four aeroelastic modes for an initial random perturbation of the fighter aircraft with aeroelastic modes stabilization, confirming a stable and well-behaved aeroservoelastic system

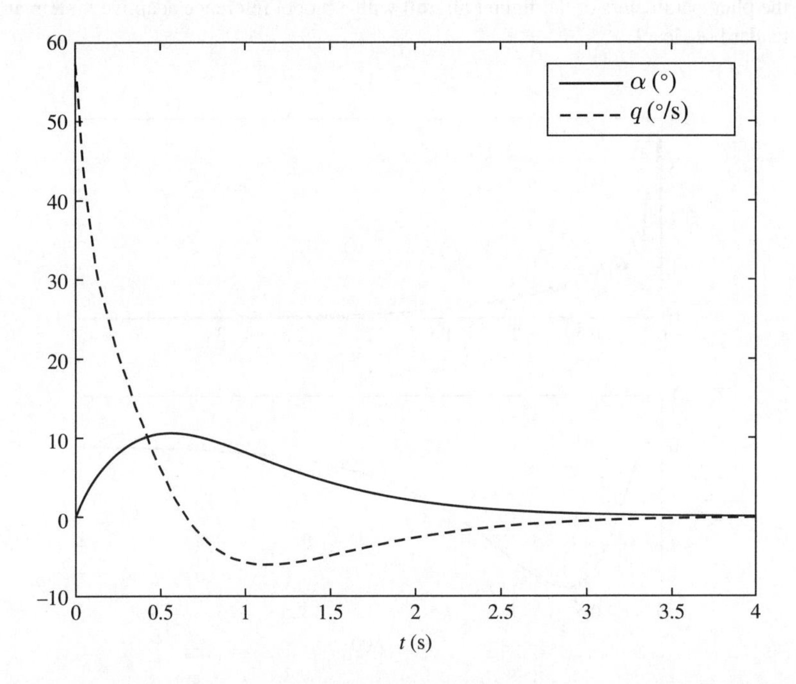

Figure 8.24 Closed-loop initial response of α and q for a random static perturbation in the plant parameters of the fighter aircraft with a model reference adaptive system at $U = 306$ m/s and standard sea level

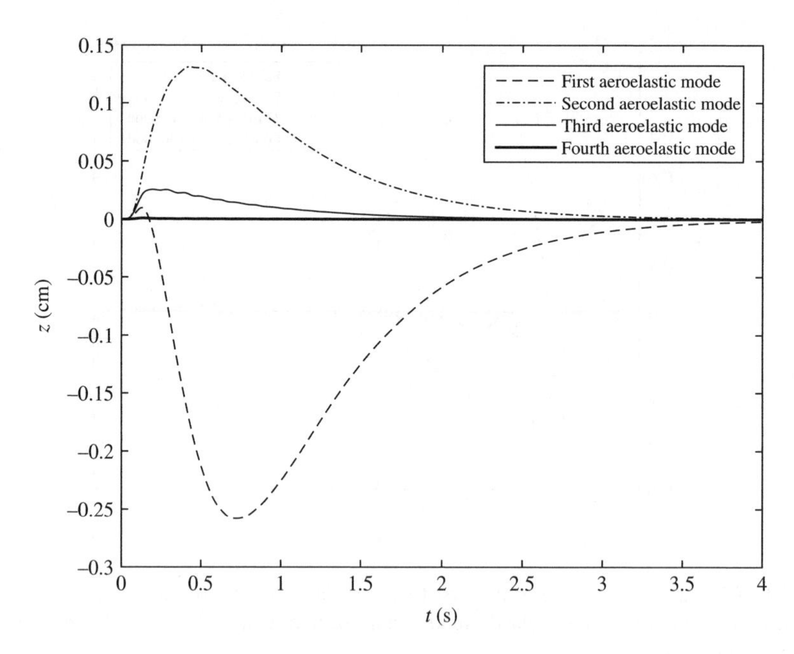

Figure 8.25 Closed-loop initial response of the first four aeroelastic modes for a random static perturbation in the plant parameters of the fighter aircraft with a model reference adaptive system at $U = 306$ m/s and standard sea level

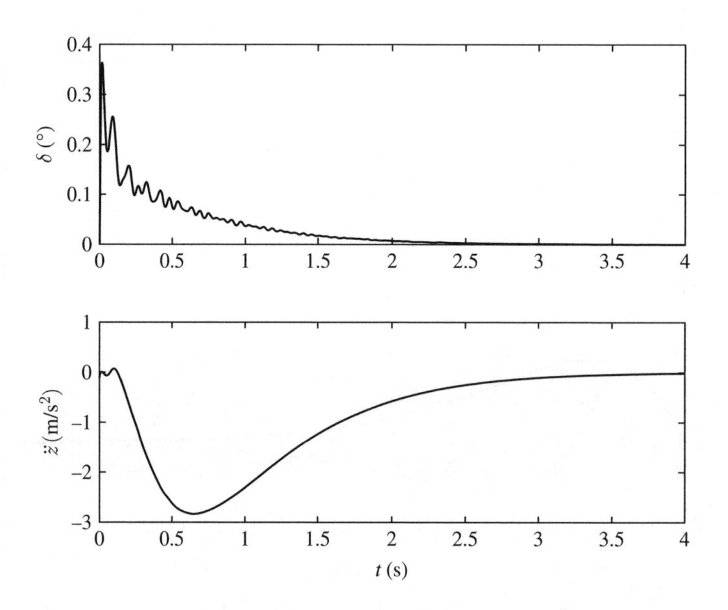

Figure 8.26 Closed-loop initial response of the elevon deflection, δ, and normal acceleration, $\ddot{z}$, for a random static perturbation in the plant parameters of the fighter aircraft with a model reference adaptive system at $U = 306$ m/s and standard sea level

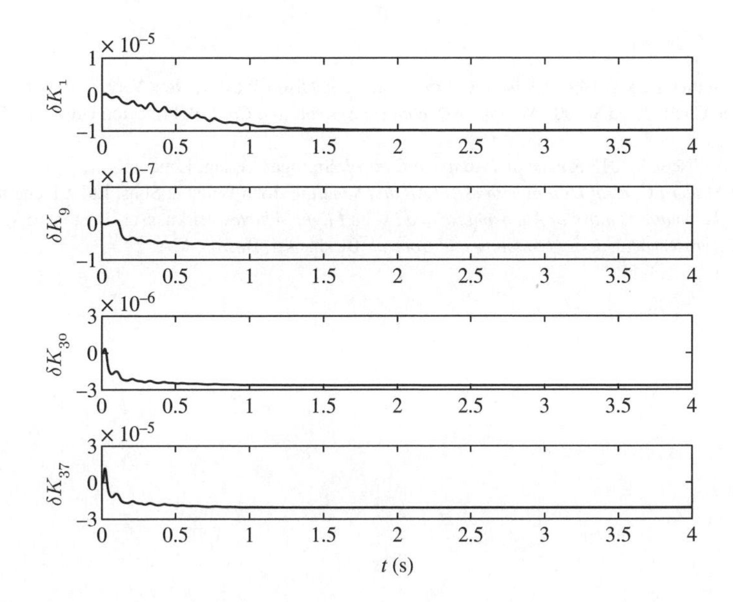

Figure 8.27 Variation of selected controller parameter errors, δK, for a random static perturbation in the plant parameters of the fighter aircraft with a model reference adaptive system at $U = 306$ m/s and standard sea level

variables converges to zero in about 5 s (Fig. 8.22). The elevon servo is also seen to behave well with only a slight difference between commanded and actual elevon deflections. Similarly, the first four aeroelastic modes are seen to decay to zero with some overshoots, in a similar duration.

To design a MRAS for this aircraft, we assume the reference model to be the closed-loop ASE system designed above with the given regulator and observer gains. The elevon servo gains are a part of the set of controller parameters, $\hat{K}$, and change in the adaptation process. It is possible to do so in this aircraft owing to its 'fly-by-wire' control system, where servos are driven by electrical signals. Most of the modern aircraft of the high-speed variety have such control systems. The normally distributed random static variations of up to 1% variation in the elements of A, Λ are handled by the following MRAS law:

$$\dot{\hat{K}} = B_0^T P e x^T R, \tag{8.121}$$

with $P = 10^{-6}I, R = I$, and the observer gains, L, are fixed at their reference values. The closed-loop simulation results are plotted in Figs. 8.24–8.27, showing an asymptotically stable adaptive control system in all the variables. With only the regulator gains varied, the overall adaptive ASE system is of the order 114. The present simulation required a memory of 1.5 GB. Owing to the large order of the system (even with the observer gains fixed), it is not possible to simulate the response for large random perturbations, which can exceed the memory resources of a personal computer.

References

Aström KJ and Wittenmark B 1995 *Adaptive Control*, 2nd ed. Addison-Wesley, New York.
Haddad WM and Chellaboina V 2008 *Nonlinear Dynamical Systems and Control*. Princeton University Press, Princeton, NJ.
Lavretsky E and Wise KA 2013 *Robust and Adaptive Control*. Springer-Verlag, London.
Tewari A 2002 *Modern Control Design with MATLAB and Simulink*. John Wiley & Sons, Ltd, Chichester.
Tewari A 2011 *Automatic Control of Atmospheric and Space Flight Vehicles*. Birkhäuser, Boston, MA.
Tewari A 2015 *Aeroservoelasticity: Modeling and Control*. Birkhäuser, Boston, MA.

9

Adaptive Backstepping Control

9.1 Introduction

Backstepping is a powerful adaptive control method when applied to uncertain nonlinear systems. It provides an alternative to the traditional model-reference adaptation system (MRAS) and self-tuning regulation (STR) methods covered in the previous chapters, which are based on the certainty equivalence principle. To recall, the certainty equivalence principle states that a controller can be designed to stabilize a plant by treating the uncertain plant parameters as if they are known with certainty. Parametric uncertainty is handled by supplying the parameter estimates to the controller design process and expecting the closed-loop system thus designed would be stable. The main advantage of the backstepping approach is the use of a recursive design procedure, which does not require that the parametric uncertainty should be matched by control inputs. The method is based on treating some state variables as virtual control and designing intermediate control laws for them, which in turn are substituted to determine those state variables that are integrals of the first virtual set. In this way, adaptive control is carried out recursively by applying control to the parametric uncertainty appearing one integrator previously. By not requiring a matching of the control with the uncertainty in the same state equation, the backstepping integrator allows a greater flexibility in handling parametric uncertainties. In fact, because of its departure from the certainty equivalence design, the backstepping integrator is said to cross the extended matching barrier that plagued the traditional Lyapunov-based adaptive schemes (MRAS and STR). Therefore, the resulting adaptive controller is considered to be more 'intelligent' than those designed by the certainty equivalence principle (Krstic *et al.* 1995). Another feature of the adaptive backstepping method is that it allows modular design, where any identifier can be combined with any stabilizing controller. Thus, the main disadvantage of the backstepping approach, that it requires many more controller parameters than necessary (over-parameterization), can be alleviated by using a modular approach. The treatment given here is based on the pioneering work by Krstic *et al.* (1995). As in the previous chapters, our focus will be on set-point regulation of aeroelastic systems with a single control surface, hence the analysis need not cover multi-input systems and tracking systems. For advanced concepts of the adaptive backstepping method, the reader is referred to the works by Krstic *et al.* (1995).

Adaptive Aeroservoelastic Control, First Edition. Ashish Tewari.

9.2 Integrator Backstepping

For an illustration of the backstepping method, consider a control affine system with state vector, $x \in \mathbb{R}^n$, and a scalar control input, $u \in \mathbb{R}$, represented by

$$\dot{x} = f(x) + g(x)u, \quad f(0) = 0. \tag{9.1}$$

Assume that there exists a feedback control law,

$$u = k(x), \quad k(0) = 0, \tag{9.2}$$

where $k(x) : \mathbb{R}^n \to \mathbb{R}$. If there exists a continuously differentiable, radially unbounded Lyapunov function, $V(x) : \mathbb{R}^n \to \mathbb{R} > 0$, such that

$$\dot{V} = \frac{\partial V}{\partial x}\dot{x} = \frac{\partial V}{\partial x}[f(x) + g(x)k(x)] \leq -W(x) \leq 0, \tag{9.3}$$

for all $x \in \mathbb{R}^n$, where $W(x) : \mathbb{R}^n \to \mathbb{R} \geq 0$, then by LaSalle's invariance theorem (Chapter 6), the state-feedback regulated system is globally stable and

$$\lim_{t\to\infty} W(x(t)) = 0.$$

However, if $W(x) > 0$, the state-feedback regulation achieves global asymptotic stability, and $x(t)$ converges to the origin.

Lemma 9.2.1 *Consider a control affine system augmented by a backstepping integrator as follows:*

$$\dot{x} = f(x) + g(x)\xi \tag{9.4}$$

$$\dot{\xi} = u, \tag{9.5}$$

with the state equation, Eq. (9.4), satisfying Eqs. (9.2) and (9.3) with ξ as the control input.

(a) If $W(x) > 0$ in Eq. (9.3), then there exists a control Lyapunov function given by

$$V_a(x, \xi) = V(x) + \frac{1}{2}[\xi - k(x)]^2, \tag{9.6}$$

and consequently, there exists a control law

$$u = k_a(x, \xi), \tag{9.7}$$

such that $x = 0, \xi = 0$ is a globally asymptotically stable equilibrium point of the system described by Eqs. (9.4) and (9.5). A possible control law is the following:

$$u = c[k(x) - \xi] + k'(x)[f(x) + g(x)\xi] - \frac{\partial V}{\partial x}g(x), \tag{9.8}$$

where c is a positive constant. Such a control law has an equivalence with the optimal control formulation, with $\dot{V}_a = (\partial V_a/\partial x)\dot{x}$ being the Lagrangian function.

(b) If $W(x) \geq 0$, then there exists a feedback law making $\dot{V}_a \leq -W_a(x, \xi) \leq 0$, such that $W_a(x, \xi) > 0$ when either $W(x) > 0$ or $\xi \neq k(x)$. Therefore, the trajectory $[x(t); \xi(t)] \in \mathbb{R}^{n+1}$ converges to the largest invariant set contained in E defined by the conditions $W(x) = 0, \xi = k(x)$, hence the system's trajectories originating in the neighbourhood of the equilibrium point, $x = 0, \xi = 0$, are globally bounded.

Proof. The proof can be found by differentiating the error variable, $e = \xi - k(x)$, with time along the system's trajectories and selecting a control law that satisfies LaSalle–Yoshizawa theorem (Chapter 8).

The lemma can be extended to a chain of backstepping integrators by the following corollary.

Corollary 9.2.2 *If the conditions given by Eqs. (9.1)–(9.3) are satisfied by a system, then the system augmented by the following chain of integrators:*

$$\begin{aligned}
\dot{x} &= f(x) + g(x)\xi_1 \\
\dot{\xi}_1 &= \xi_2 \\
&\vdots \\
\dot{\xi}_{k-1} &= \xi_k \\
\dot{\xi}_k &= u,
\end{aligned} \tag{9.9}$$

with the following control Lyapunov function,

$$V_a(x, \xi) = V(x) + \frac{1}{2} \sum_{i=1}^{k} [\xi_i - k_i(x, \xi_1, \dots, \xi_{i-1})^2, \tag{9.10}$$

where $\xi_1, \dots, \xi_k$ are virtual control inputs, has the same criteria of boundedness and global asymptotic stability as given by Lemma 9.2.1, with ξ replaced by the vector, $(\xi_1, \dots, \xi_k)$.

Proof. The proof is easily obtained by applying Lemma 9.2.1 repeatedly along the chain of integrators.

9.2.1 A Motivating Example

For illustration of the adaptive backstepping method, consider the following second-order, affine system with a time-invariant, unknown parameter vector, θ:

$$\begin{aligned}
\dot{x}_1 &= x_2 + \psi^T(x_1)\theta \\
\dot{x}_2 &= u + \phi^T(X)\theta,
\end{aligned} \tag{9.11}$$

where $X = (x_1, x_2)^T$ is the state vector and $\psi^T(x_1), \phi^T(X)$ are known continuous functions. The design task is to find a globally asymptotically stabilizing (GAS) control law to regulate the system without the knowledge of parameters, θ, such that the response, $X(t)$, from an initial condition, $x_1(0), x_2(0)$, is brought to equilibrium, $(0, 0)$, in the steady state. Select a virtual feedback from the displacement, x_1, to the virtual control, x_2, as a candidate control law, $x_2 = \alpha(x_1, \hat{\theta})$, based on the parameter estimate, $\hat{\theta}$, and define the following adaptation variables:

$$z_1 = x_1$$
$$z_2 = x_2 - \alpha(x_1, \hat{\theta}). \tag{9.12}$$

This easily gives the first scalar adaptation law as follows:

$$\dot{z}_1 = z_2 + \alpha(x_1, \hat{\theta}) + \psi^T(x_1)\theta. \tag{9.13}$$

However, the virtual control law, α, is yet to be designed. For this purpose, select the following positive-definite and radially unbounded candidate Lyapunov function:

$$V_1 = \frac{1}{2}z_1^2 + \frac{1}{2}\tilde{\theta}^T\Gamma^{-1}\tilde{\theta}, \tag{9.14}$$

where $\tilde{\theta} = \theta - \hat{\theta}$ is the parameter estimation error vector and Γ is a symmetric, positive-definite, invertible matrix. The time derivative of V_1 along the system's trajectories is given by

$$\begin{aligned}\dot{V}_1 &= z_1\dot{z}_1 + \tilde{\theta}^T\Gamma^{-1}\dot{\tilde{\theta}} \\ &= z_1[z_2 + \alpha(x_1, \hat{\theta}) + \psi^T(x_1)\theta] - \tilde{\theta}^T\Gamma^{-1}[\dot{\hat{\theta}} - \Gamma\psi(x_1)z_1].\end{aligned} \tag{9.15}$$

A control law of the form,

$$\alpha(x_1, \hat{\theta}) = -c_1z_1 - \psi^T(x_1)\theta, \tag{9.16}$$

with $c_1 > 0$ results in the following:

$$\dot{V}_1 = -c_1z_1^2 + z_1z_2 - \tilde{\theta}^T[\Gamma^{-1}\dot{\hat{\theta}} - \psi(x_1)z_1]. \tag{9.17}$$

A substitution of Eq. (9.16) into Eq. (9.13) yields the following:

$$\dot{z}_1 = -c_1z_1 + z_2 + \psi^T(x_1)\tilde{\theta}. \tag{9.18}$$

A second feedback law, $u = \beta(X, \hat{\theta})$, is selected to cancel the second term on the right-hand side of Eq. (9.17), as well as to render the overall adaptation system, (z_1, z_2), GAS. Substituting the new feedback law into Eqs. (9.11) and (9.12) and taking the time derivative along the closed-loop system's trajectories, we have

$$\dot{z}_2 = \beta(X, \hat{\theta}) + \phi^T(X)\theta - \frac{d\alpha}{dt} \tag{9.19}$$

or

$$\dot{z}_2 = \beta(X, \hat{\theta}) + w^T(X, \hat{\theta})\theta - \frac{\partial\alpha}{\partial x_1}x_2 - \frac{\partial\alpha}{\partial\hat{\theta}}\dot{\hat{\theta}}, \tag{9.20}$$

where the new regressor vector is defined by

$$w(X, \hat{\theta}) = \phi(X) - \frac{\partial\alpha}{\partial x_1}\psi(x_1). \tag{9.21}$$

In order to derive an expression for β, define a second candidate Lyapunov function as follows:

$$V_2 = V_1 + \frac{1}{2}z_2^2, \tag{9.22}$$

take its time derivative as follows along the closed-loop system's trajectories:

$$\dot{V}_2 = -c_1 z_1^2 + z_1 z_2 - \tilde{\Theta}^T[\Gamma^{-1}\dot{\hat{\theta}} - \psi(x_1)z_1] + z_2\dot{z}_2, \tag{9.23}$$

and substitute Eq. (9.20), which results in the following:

$$\dot{V}_2 = -c_1 z_1^2 + z_2\left[z_1 + \beta + w^T\hat{\theta} - \frac{\partial\alpha}{\partial x_1}x_2 - \frac{\partial\alpha}{\partial\hat{\theta}}\dot{\hat{\theta}}\right] + \tilde{\Theta}^T\left[\psi z_1 + wz_2 - \Gamma^{-1}\dot{\hat{\Theta}}\right]. \tag{9.24}$$

The following control law follows by inspection of Eq. (9.24), which makes $\dot{V}_2 = -c_1 z_1^2 - c_2 z_2^2 < 0, z_1 \neq 0, z_2 \neq 0$, thereby guaranteeing global asymptotic stability:

$$\beta(X, \hat{\theta}) = -z_1 - c_2 z_2 - w^T\hat{\theta} + \frac{\partial\alpha}{\partial x_1}x_2 + \frac{\partial\alpha}{\partial\hat{\theta}}\dot{\hat{\theta}}, \tag{9.25}$$

where $c_2 > 0$. A natural choice of the parameter update law follows from Eq. (9.24) as

$$\dot{\hat{\Theta}} = \Gamma[\psi z_1 + wz_2], \tag{9.26}$$

which results in the following control law,

$$\beta(X, \hat{\theta}) = -z_1 - c_2 z_2 - w^T\hat{\theta} + \frac{\partial\alpha}{\partial x_1}x_2 - \frac{\partial\alpha}{\partial\hat{\theta}}\Gamma[\psi z_1 + wz_2], \tag{9.27}$$

and the following backstepping integrator adaptation laws:

$$\begin{aligned} \dot{z}_1 &= -c_1 z_1 + z_2 + \psi^T\tilde{\theta} \\ \dot{z}_2 &= -z_1 - c_2 z_2 + w^T(X)\tilde{\theta}. \end{aligned} \tag{9.28}$$

However, in order to retain another degree of control in the choice of adaptation variables, z_1, z_2, Krstic *et al.* (1995) propose the following tuning function:

$$\tau(X, \hat{\theta}) = \psi z_1 + wz_2, \tag{9.29}$$

instead of using the parameter update law given by Eq. (9.26). This results in a negative definite $\dot{V}_2$ with a non-zero parameter error term, $\Gamma\tau - \dot{\hat{\theta}}$, and the following modified adaptation laws:

$$\begin{aligned} \dot{z}_1 &= -c_1 z_1 + z_2 + \psi^T\tilde{\theta} \\ \dot{z}_2 &= -z_1 - c_2 z_2 + \phi^T(X)\tilde{\theta} + \frac{\partial\alpha}{\partial\hat{\theta}}(\Gamma\tau - \dot{\hat{\theta}}). \end{aligned} \tag{9.30}$$

We will now consider a simple design example to illustrate the adaptive backstepping integrator method.

Example 9.2.3 *Let us revisit the single degree of freedom mass–spring–damper system of Chapter 8 with the equation of motion given by*

$$m\ddot{x} + c\dot{x} + kx = u, \tag{9.31}$$

where the constant parameters m, c, k *are unknown. We begin by writing Eq. (9.1) in the following state-space form of Eq. (9.31), with* $X = (x, \dot{x})^T = (x_1, x_2)^T$*, as the state vector. It is clear that* $\psi(x_1) = 0$, $w^T = \phi^T(X) = (-x_1, -x_2) = -X^T$, $\theta^T = \left(\frac{k}{m}, \frac{c}{m}\right)$ *and* $\alpha = -c_1 z_1 = -c_1 x_1$*. Then it follows that*

$$\frac{\partial \alpha}{\partial x_1} = -c_1, \quad \frac{\partial \alpha}{\partial \hat{\theta}} = 0, \tag{9.32}$$

thus we have,

$$\beta(X, \hat{\theta}) = -z_1 - c_2 z_2 - \phi^T(X)\hat{\theta} - c_1 x_2 \tag{9.33}$$

and

$$\begin{aligned} \dot{z}_1 &= -c_1 z_1 + z_2 \\ \dot{z}_2 &= -z_1 - c_2 z_2 + \phi^T(X)\tilde{\theta}, \end{aligned} \tag{9.34}$$

which results in the following GAS system:

$$\begin{aligned} \dot{x}_1 &= x_2 \\ \dot{x}_2 &= -c_1 x_2 - z_1 - c_2 z_2 + \phi^T(X)\tilde{\theta}. \end{aligned} \tag{9.35}$$

The closed-loop system thus requires a feedback of the variables, $x_1, x_2, \tilde{\theta}$*, the last of which requires a parameter update law,* $\dot{\hat{\Theta}} = \Gamma\phi(X)z_2$*, because the tuning function,* τ*, is inapplicable for the constant* c_1, c_2 *case.*

We simulate the response of the adaptive integrator feedback system for randomly time-varying parameters, $k/m, c/m$*, with normal distributions of mean 1.5 and 0.5, respectively, and a unit variance. The adaptation gain matrix is taken to be* $\Gamma = 0.001I$*, while the two controller constants are* $c_1 = 0.1, c_2 = 1.0$*. The initial condition is given by* $x(0) = 1, \dot{x}(0) = 0$, $z_1(0) = 0.1, z_2(0) = 0.01, \hat{\theta} = (0, 0)^T$*, and the simulated response is plotted in Figs. 9.1–9.4, confirming asymptotic stability for* $x, \dot{x}, z_1, z_2$*. However, the adaptation errors do not converge to a zero steady state due to the persistent random excitation (process noise). Hence, while the adaptive controller was designed assuming constant plant parameters, it is seen to be globally asymptotically stable even in the presence of a large random excitation (process noise). Therefore, the design is quite robust with respect to unmodelled disturbances. The backstepping integrator approach is seen to be equivalent to the MRAS scheme with state feedback (see Chapter 8) in terms of the total number of adaptation variables required. The regulator gains,* $\hat{K}$*, of the MRAS are the same in number as the adaptation variables,* z*, in the backstepping method, and the errors from the reference model,* e*, of the former are equivalent to the parameter estimation errors,* $\tilde{\theta}$*, of the latter. Hence, whenever an implementation for an ASE plant is possible by the MRAS approach, the same can be converted into a backstepping scheme.*

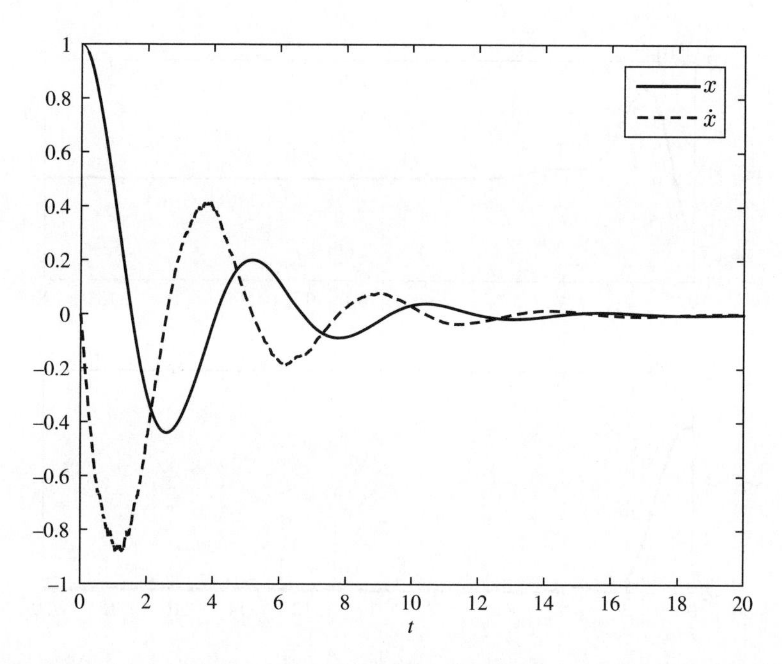

Figure 9.1 Simulation of the states of an integrator backstepping adaptive control scheme for a spring–mass–damper system with unknown, randomly time-varying parameters

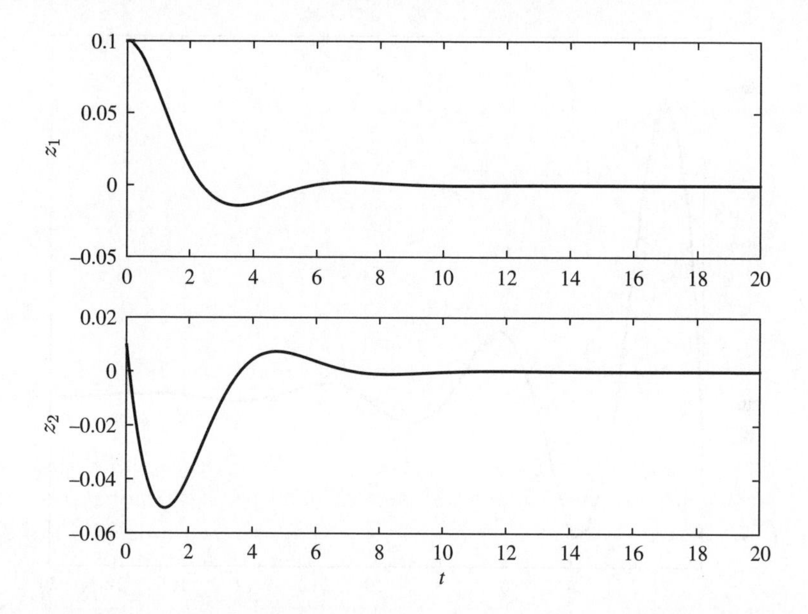

Figure 9.2 Simulation of the adaptation variables of an integrator backstepping adaptive control scheme for a spring–mass–damper system with unknown, randomly time-varying parameters

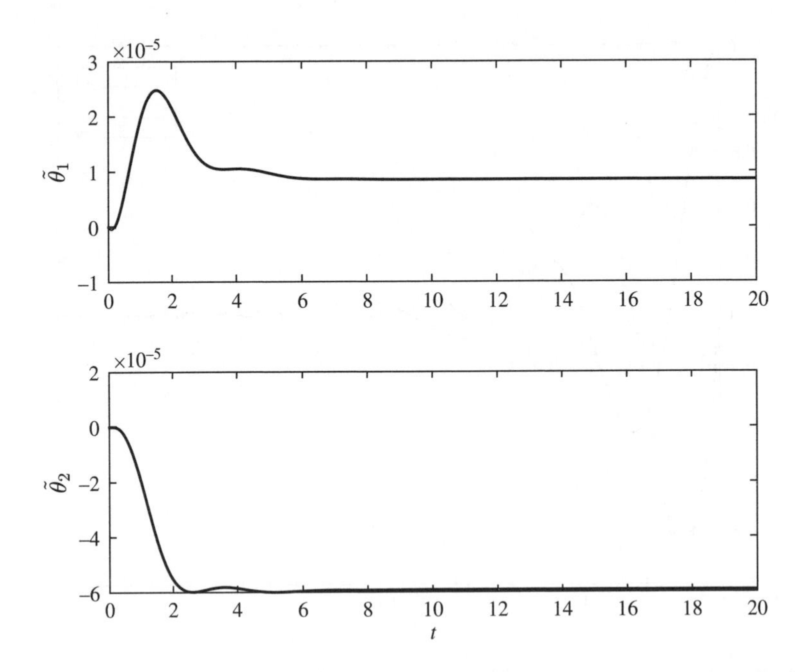

Figure 9.3 Simulation of the parameter estimation errors of an integrator backstepping adaptive control scheme for a spring–mass–damper system with unknown, randomly time-varying parameters

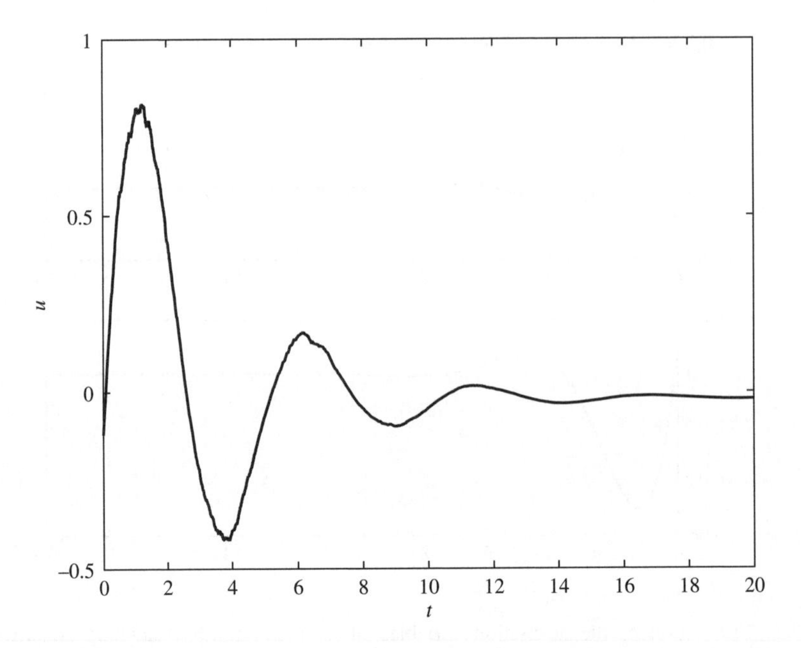

Figure 9.4 Simulation of the control input of an integrator backstepping adaptive control scheme for a spring–mass–damper system with unknown, randomly time-varying parameters

9.3 Aeroservoelastic Application

The integrator backstepping adaptation appears to be directly applicable to ASE plants, which are necessarily in a controllable, cascade structure. One can express the nominal ASE plant without an actuator in the following cascade form:

$$\begin{aligned}\dot{x}_1 &= x_3\\ \dot{x}_2 &= x_3 - \Gamma_a x_2\\ \dot{x}_3 &= -M^{-1}Kx_1 - M^{-1}Cx_3 + M^{-1}N_a x_2 + M^{-1}Q_c,\end{aligned} \tag{9.36}$$

where x_1 is the vector of generalized coordinates with x_3 being the vector of its time derivatives and x_2 is the aerodynamic state vector arising out a rational-function approximation (RFA) (see Chapter 3). Here M, K, C are the generalized structural matrices, N_a the numerator coefficient matrices of the RFA, Γ_a the diagonal matrix containing the aerodynamic poles (lag parameters) and Q_c the vector of generalized aerodynamic forces resulting from the motion of control surfaces. Each of the subsystems is controllable by Q_c, which only appears in the last subsystem. This is a classic configuration for the application of the backstepping approach, where x_3 acts as the virtual control for the first two subsystems. The scheme is likely to be successful in the presence of modelling uncertainties and nonlinearities, because they are matched by the control vector Q_c in the last subsystem.

Let us parameterize the ASE plant as follows:

$$\begin{aligned}\dot{\xi}_1 &= \xi_2 + \Theta^T\psi(\xi_1)\\ \dot{\xi}_2 &= u + \Theta^T\phi(\xi_1, \xi_2),\end{aligned} \tag{9.37}$$

where $\xi_1 = \left(x_1^T, x_2^T\right)^T$, $\xi_2 = x_3$, $u = M^{-1}Q_c$ and $\Theta^T = \left(M^{-1}K, M^{-1}N_a, M^{-1}C, \Gamma_a\right)$. This yields the following regressor matrices, each having a row dimension equal to the structural degrees of freedom:

$$\begin{aligned}\psi^T &= (0, 0, 0, -x_2)\\ \phi^T &= (-x_1, -x_2, -x_3, 0).\end{aligned} \tag{9.38}$$

By extension of the motivating example given above, we specify the following feedback control and adaptation laws:

$$u = \beta(X, \hat{\theta}) = -z_1 - C_2 z_2 - \Theta^T w + \frac{\partial\alpha}{\partial\xi_1}\xi_2 - \frac{\partial\alpha}{\partial\hat{\Theta}}\Gamma[\psi z_1 + w z_2], \tag{9.39}$$

where

$$\alpha(\xi_1, \hat{\theta}) = -C_1 z_1 - \Theta^T\psi(\xi_1), \tag{9.40}$$

$$w(X, \hat{\Theta}) = \phi(X) - \frac{\partial\alpha}{\partial\xi_1}\psi(\xi_1), \tag{9.41}$$

$$\begin{aligned}\dot{z}_1 &= -C_1 z_1 + z_2 + \tilde{\theta}^T\psi(\xi_1)\\ \dot{z}_2 &= -z_1 - C_2 z_2 + \tilde{\theta}^T w(X).\end{aligned} \tag{9.42}$$

Here, C_1, C_2, Γ are constant, symmetric positive-definite, square matrices of appropriate dimensions giving weightage to each parametric error. Owing the large size of a typical ASE plant, treating all the parameters as being uncertain would lead to a cumbersome design. Hence, selection of the weighting matrices by experience is an important task of an adaptive ASE designer.

Reference

Krstic M, Ioannis K, and Kokotovic P 1995 *Nonlinear and Adaptive Control Design*. John Wiley & Sons, Inc., New York.

10

Adaptive Control of Uncertain Nonlinear Systems

10.1 Introduction

A large class of engineering systems are control affine, control linear, autonomous, but have an uncertain nonlinear behaviour forced by complex dynamical processes that are difficult to model. Such nonlinearities include large and unknown changes in the process dynamics or external disturbances. An example is the aircraft plant for transonic flight, where the unsteady aerodynamic effects of shock-induced flow separation are nonlinear as well as notoriously difficult to predict (Lee 1990). The uncertain nonlinear motion in these cases is forced by randomly changing loads. If the forcing conditions are bounded, a stable linear system can display a limit-cycle oscillation. An aircraft wing subject to turbulent gusts has a similar behaviour at certain flight speeds. Other examples include mechanical and electrical systems with an unknown hysteresis-type nonlinearity superimposed over a basic linear, time-invariant, stable subsystem. In Chapter 7, the design and analysis of oscillatory nonlinear systems by approximate methods was presented. However, such an approach is unlikely to work for a more general case where the response is neither purely oscillatory nor limited in its amplitude. This chapter is mainly concerned with how a general nonlinear system can be controlled adaptively.

A vast literature exists on the adaptive control of uncertain nonlinear systems, with references listed in textbooks (Aström and Wittenmark 1995, Isidori 1989, Kokotovic 1991, Krstic *et al.* 1995, Slotine 1995) and survey articles (Kokotovic and Arcak 2001). Practical adaptive control techniques include autoregressive modelling (Aström and Wittenmark 1995), model-reference adaptation (Slotine 1995), adaptive feedback linearization (Kanellakopoulos *et al.* 1991), sliding-mode control (Isidori 1989) and integral backstepping and modular control (Krstic *et al.* 1995). Such methods have been successfully applied to a class of nonlinear systems, wherein the nonlinearities have a known structure (Chang 2001, Kojic and Annaswamy 2002). The nonlinear adaptive strategies are either Lyapunov based, or derived by parameter identification algorithms, and utilize nonlinear update laws Aström and Wittenmark (1995) or tuning functions (Krstic *et al.* 1995), whose stability with respect to changing operating conditions and unknown disturbances is often difficult to analyse. While robust adaptation for linear plants is well documented (Ioannou and Sun 1996), similar

Adaptive Aeroservoelastic Control, First Edition. Ashish Tewari.

effort for nonlinear systems is an area of active research, but currently limited to either scalar systems (Xu and Ioannou 2002) or to simple structural nonlinearities (Wang *et al.* 2004). The neural-network-based intelligent (learning) adaptation has been explored as an alternative technique applied to nonlinear uncertain systems (Chen and Liu 1994, Xu and Ioannou 2002, Yu and Annaswamy 1997).

The adaptive control problem is traditionally classified as being either direct or indirect, depending upon whether the controller parameters are directly manipulated or indirectly determined from the estimates of the process dynamics. The common assumption in the indirect adaptive control methods applied to linear plants with uncertain parameters is the certainty equivalence principle, wherein parameter estimates are treated as known values at each time instant when closing the primary feedback loop. While certainty equivalence is a workable approximation in designing self-tuning regulators (STRs) for most linear plants, it cannot be relied upon for a general nonlinear system. The stability and performance in adaptation of a nonlinear system can degrade significantly if the parameter estimates are highly inaccurate. In order to achieve acceptable performance, it is thus necessary to have a probing method for deriving uncertainty measures of the estimated parameters based on plant inputs and outputs, which can then be taken into account in updating the controller parameters. A rigorous method of carrying out such nonlinear-uncertainty-based corrections is via stochastic optimal control, wherein a dynamic objective function of the estimated plant states and the probability estimates of uncertain parameters is minimized with respect to the control variables. This results in the formidable Hamilton–Jacobi equation to be numerically solved for a large number of variables. While such solutions exist for isolated cases, they involve formidable online computations of partial differential equations, which are infeasible in a general application. Modular designs based on recursive, backstepping adaptation have been evolved (Krstic *et al.* 1995) as a practical alternative to the stochastic Hamilton–Jacobi solutions for a certain class of uncertain nonlinear systems.

Model reference adaptation systems (MRASs), STRs and backstepping integrators covered in the previous chapters are all inherently nonlinear control systems. Their robustness in the presence of process noise inputs is necessary for practical implementations. A part of this chapter considers the various robust adaptive methods, especially in possible application in aeroservoelastic design.

10.2 Integral Adaptation

While integral control lies at the heart of any adaptation method (e.g. model-reference control, integral backstepping control, autoregressive identification), the time integration involves a nonlinear function of both unknown parameters and input, output or state variables. The effectiveness of a simple linear integral control applied to nonlinear uncertain plants has not been explored in the literature. The present section explores how such a method can be simply but effectively applied to an uncertain systems with an unknown structural nonlinearity superimposed over a basic linearized system.

Consider an autonomous, causal nonlinear system represented by the following state equations:

$$\dot{x} = Ax + f(x) + Bu, \tag{10.1}$$

where $x(t) : \mathbf{R} \to \mathbf{R}^n$ is the state vector and $u(t) : \mathbf{R} \to \mathbf{R}^m$ is the control input vector, with $m \leq n$. The linear time-invariant part represented by the constant coefficients A, B is stabilizable, and $f(x) : \mathbf{R}^n \rightarrow \mathbf{R}^m$ is an unknown, Lipschitz continuous mapping, assumed to have a continuous derivative, $f'(x)$, with the property $f(0) = 0$. Therefore, $x = 0$ is an equilibrium point. Many practical systems are expressible in this form. Our control task is to regulate the plant about $x = 0$ in the absence of knowledge of $f(x)$, but without employing nonlinear feedback.

Consider the following linear feedback control law

$$u = -Kx - z, \tag{10.2}$$

where K is the regulator gain matrix and $z(t) : \mathbf{R} \rightarrow \mathbf{R}^m$ is an adaptation vector required for estimating the unknown function $f(x)$.

Theorem 10.2.1 *The control system comprising the state equation, Eq. (10.1), and control-law, Eq. (10.2), is globally, uniformly and asymptotically stable about the equilibrium $x = 0$ with the following adaptation law:*

$$B\dot{z} = x. \tag{10.3}$$

Proof. Consider the following control Lyapunov function:

$$V(x, z) = \frac{1}{2}x^T x + \frac{1}{2}[Bz - f(x)]^T[Bz - f(x)], \tag{10.4}$$

which is radially unbounded for $x \neq 0, z \neq 0$. Taking the time derivative, we have

$$\dot{V} = x^T\dot{x} + [Bz - f(x)]^T[B\dot{z} - f'(x)\dot{x}] \tag{10.5}$$

or

$$\dot{V} = x^T[(A - BK)x + f(x) - Bz] + [B\dot{z} - f'(x)\dot{x}]^T[Bz - f(x)]. \tag{10.6}$$

Substituting the adaptation law of Eq. (10.3) into Eq. (10.6) results in the following:

$$\dot{V} = x^T(A - BK)x + x^T[f(x) - Bz] + [x - f'(x)\dot{x}]^T[Bz - f(x)] \tag{10.7}$$

or

$$\dot{V} = x^T(A - BK)x - [f'(x)\dot{x}]^T\left[Bz - f(x)\right]. \tag{10.8}$$

The first term on the right-hand side of Eq. (10.8) is negative definite for a Hurwitz matrix $(A - BK)$, which is achieved either by a linear, quadratic regulator solution or eigenstructure assignment of k. A sufficient condition for global, uniform, asymptotic stability (GUAS) is for the adaptation error, $e = Bz - f(x)$, to vanish in the steady state irrespective of $f(x)$, which requires a Hurwitz error dynamics matrix, Q, such that

$$\dot{e} = Qe. \tag{10.9}$$

Therefore, the sufficient condition of GUAS translates into the following:

$$-\left[f'(x)\dot{x}\right]^T\left[Bz - f(x)\right] = \dot{e}^T e - x_2^T e \tag{10.10}$$

or

$$\dot{V} = -W(x, e) < 0, \tag{10.11}$$

where

$$-W(x,e) = (x^T, e^T)\begin{pmatrix} (A-BK) & -\frac{1}{2}I \\ -\frac{1}{2}I & Q \end{pmatrix}\begin{Bmatrix} x \\ e \end{Bmatrix}. \tag{10.12}$$

Therefore, the premise of the LaSalle–Yoshizawa theorem (see Chapter 8) specifying the sufficient conditions for GUAS are met by the control Lyapunov function. Hence, GUAS is established without the apriori knowledge of $f(x)$.

Equation (10.3) must be solved for the adaptation variable $z(t)$ before the feedback control, Eq. (10.2), can be implemented. Since the unknown adaptation variables, z, are no larger in number than the state variables, x, one can determine them uniquely by using a part of the state vector. For square plants with an invertible B matrix, there is no difficulty. Sometimes the plant Eq. (10.1) can be represented by N coupled subsystems, $x = (x_1^T, x_2^T, \dots, x_N^T)^T$, such that the controls coefficient matrix is accordingly partitioned into invertible square matrices, $B = (B_1, B_2, \dots, B_N)$. Then Eq. (10.3) results in the following

$$\dot{z}_i = B_i^{-1} x_i, \quad (i = 1, \dots, N). \tag{10.13}$$

However, in other cases the solution is not straightforward. For non-square plants, a 'squaring-up' scheme (Misra 1998) could be adopted, which adds fictitious inputs to the system such that $m = n$. A pseudo control coefficient matrix, B', thus becomes available to be used as follows:

$$\dot{z} = (B')^{-1} x. \tag{10.14}$$

While GUAS is guaranteed for the system, the appropriate feedback gain matrix, K, required to achieve it must be determined numerically, because $f(x)$ (hence E) is unknown. It is also noted that as the second subsystem is square and B is nonsingular, the adaptation variable z is the integral of $B^{-1}x$.

10.2.1 Extension to Observer-Based Feedback

The state-feedback control law of Eq. (10.2) is seldom employed in practice. Instead, the following observer-based feedback is applied:

$$u = -K\hat{x} - z, \tag{10.15}$$

with the estimated state $\hat{x}$ supplied by an observer. The observer is designed by either eigenstructure assignment or the Kalman filter method, based on an output vector $y = Cx + Du \in \mathbf{R}^p$ to provide the estimated state $\hat{x}$. The state equations of the adaptive system are the following:

$$\dot{x} = Ax + f(x) - BK\hat{x} - Bz \tag{10.16}$$

$$\dot{z} = B^{-1}\hat{x} \tag{10.17}$$

$$\dot{\hat{x}} = \hat{A}\hat{x} + \hat{B}u + Ly. \tag{10.18}$$

By constructing a modified Lyapunov function by augmenting the state vector as $\bar{x} = (x^T, \hat{x}^T)^T$, and replacing the matrix $(A - BK)$ with its compensator counterpart,

$$A_c = \begin{pmatrix} A & -BK \\ LC & A - BK - LC \end{pmatrix}, \tag{10.19}$$

which is made Hurwitz by a proper selection of the regulator and observer gain matrices, K, and L, respectively, the extension of Theorem 10.2.1 is easily established.

10.2.2 Modified Integral Adaptation with Observer

Instead of Eq. (10.16), which is difficult to implement for a non-square plant, a practical choice of the integral adaptation method is the following:

$$u = -z^T \hat{x}, \tag{10.20}$$

where $z \in \mathbb{R}^{n \times m}$ is evolved by the following adaptation law:

$$\dot{z} = \Gamma \Phi(\hat{x}), \tag{10.21}$$

with Γ being a symmetric and positive-definite weighting matrix and $\Phi(\hat{x})$ being a non-negative regressor matrix formed out of the elements of $\hat{x}$. The initial value of the adaptation matrix, z, can be taken as the transpose of the regulator gain matrix, K, which is known to be stabilizing with $f(x) = 0$. The modified adaptation law brings us closer to model reference adaptation for nonlinear systems, which is the topic of the next section. ultimate uniform boundedness (UUB) stability of the closed-loop system with the adaptation law of Eq. (10.21) can be established by the LaSalle–Yoshizawa theorem for an unknown (but bounded) nonlinearity, $f(x)$. The following example illustrates an important application of this approach for stabilizing an ASE plant with saturated control inputs and constrained state variables.

Example 10.2.2 *Consider the single d.o.f flutter-suppression system designed in Chapter 4. Suppose the torque produced by the control-surface actuator has a saturation limit described by the following relationship:*

$$u = \begin{cases} -K\hat{x}, & |u| < u_m \text{ or } |\beta| < \beta_m \\ u_m \text{sgn}(u), & |u| \geq u_m \text{ or } |\beta| \geq \beta_m \end{cases} \tag{10.22}$$

Furthermore, suppose that the control-surface deflection, β, is physically limited by another saturation limit, $\beta(t) \leq \beta_m \text{sgn}\{\beta(t)\}$. This is a state constraint on the system. Here $u_m > 0, \beta_m > 0$ are unknown parameters. If the flutter-suppression system designed in Chapter 4 were to be applied to the actual system, it would not be successful if saturation is reached at any point along the trajectory, $x(t)$. The simulation in Fig. 10.1, carried out for $U = 50$ m/ss, $\rho = 1.225$ kg/m^3, $u_m = 1, \beta_m = 28.65°$, demonstrates this fact.

Let us design an integral adaptation law without the knowledge of the saturation limits, u_m, β_m, by choosing

$$\Gamma = 10I, \quad \Phi(\hat{x}) = (\hat{x}^2), \tag{10.23}$$

where $(\hat{x}^2) \geq 0$ is the vector of the squares of the elements of $\hat{x}$. When this law is implemented with the previously determined values of the gains K, L, with $z(0) = K^T$, and treating the

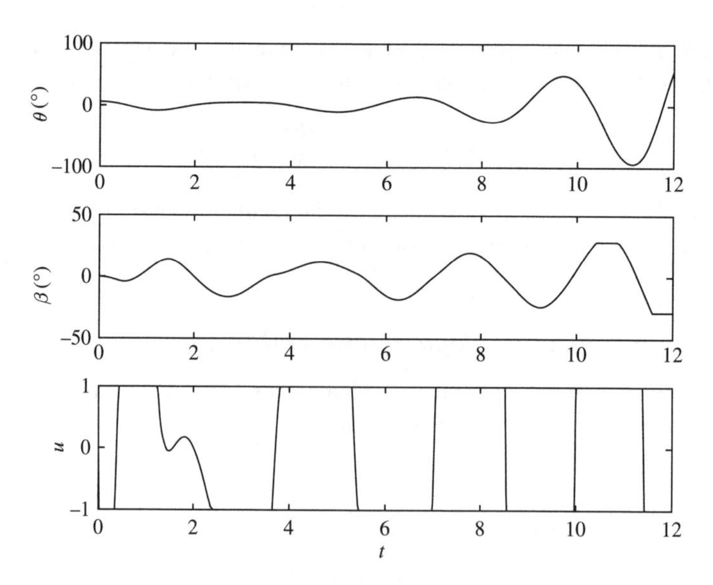

Figure 10.1 Closed-loop response for an initial pitch perturbation at $U = 60$ m/s, standard sea-level condition, with a saturated control-surface actuator of unmodelled characteristics, $u_m = 1, \beta_m = 28.65°$

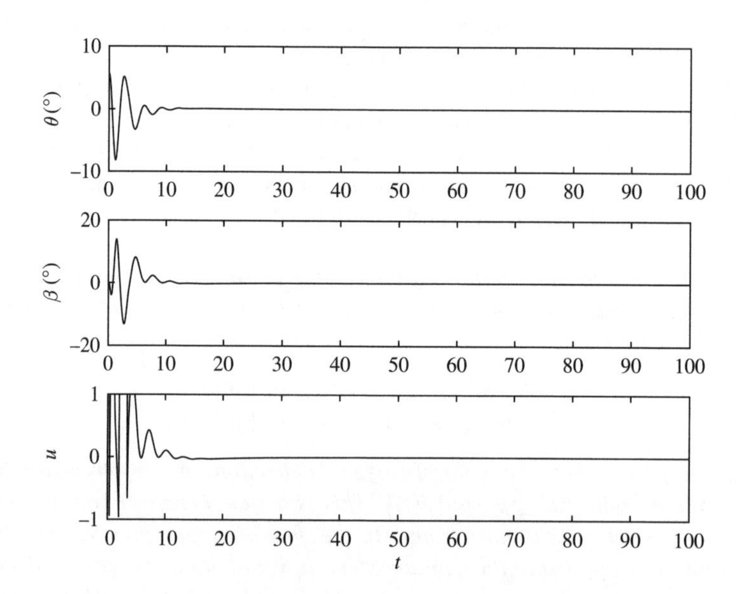

Figure 10.2 Closed-loop response of an integral adaptive control system for an initial pitch perturbation at $U = 60$ m/s, standard sea-level condition, with a saturated control-surface actuator of unmodelled characteristics, $u_m = 1, \beta_m = 28.65°$

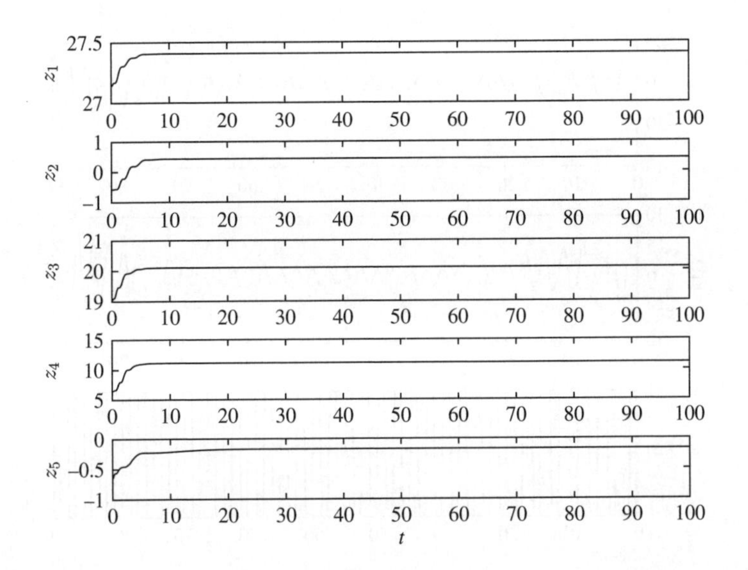

Figure 10.3 Variation of adaptation variables of an integral adaptive control system, for an initial pitch perturbation at $U = 60$ m/s, standard sea-level condition, with a saturated control-surface actuator of unmodelled characteristics, $u_m = 1, \beta_m = 28.65°$

observer gains, L, to be fixed, the simulated response is plotted in Figs. 10.2 and 10.3, with the same initial condition as considered in Fig. 10.1. The closed-loop response in all the variables, including the adaptation variables, $z = (z_1, \cdots, z_5)^T$*, is seen to be convergent to zero. However, as the nonlinearity in this example does not have a continuous derivative vector with respect to the state variables, global stability is not guaranteed. This is found to be true, as an increase in the initial pitch angle causes the response to be unstable. The system is asymptotically stable for all initial perturbations in either pitch angle or the control-surface angle, less than or equal to 5.73°. This range of stability can be increased by redesigning the regulator and observer.*

Let us consider an additional nonlinearity in an aeroservoelastic model, namely a hardening spring. This may arise because of either structural reasons when the torsional stiffness of the wing varies nonlinearly with twist angle, or aerodynamic effects such as strong normal shock waves (see Chapter 11), which increase the pitch stiffness by creating an opposing pitching moment in proportion to the square of the twist angle. If the aerodynamic coupling to structural motion were static, such an effect can lead to greater static stability. However, the dynamic aeroelastic coupling can have a destabilizing effect, which must be adaptively suppressed, as shown in the next example.

Example 10.2.3 *Reconsider the single d.o.f flutter-suppression system designed in Chapter 4. Suppose that in addition to the torque saturation of the control-surface actuator described in Example 10.2.2, the torque produced by the torsional spring about the pitch axis has a nonlinear hardening effect described by the following relationship:*

$$f(\theta) = \begin{cases} k_\theta \theta, & \theta < \bar{\theta} \\ k_\theta \theta + a\theta^2, & \theta \geq \bar{\theta} \end{cases} \tag{10.24}$$

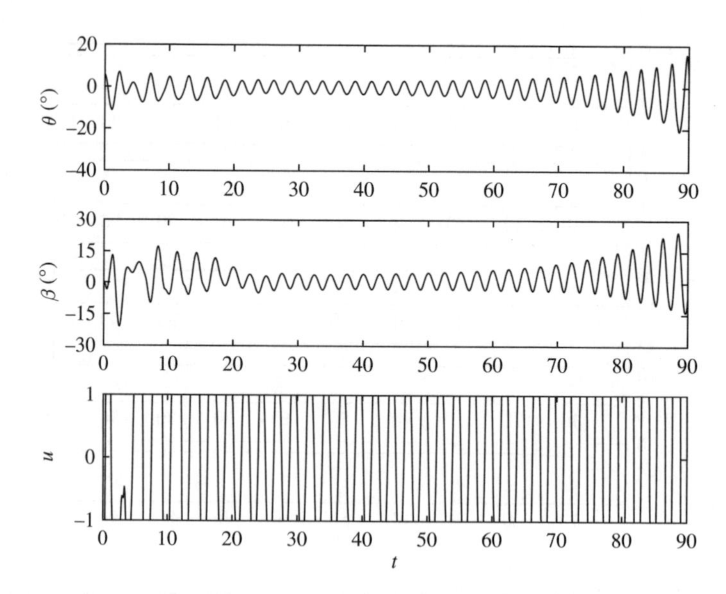

Figure 10.4 Closed-loop response of the integral adaptive controller with $\Gamma = 10I$, for an initial pitch perturbation at $U = 50$ m/s, standard sea-level condition, with a nonlinearly hardening pitch spring and a saturated control-surface actuator of unmodelled characteristics, $u_m = 1, \beta_m = 28.65°, a = 515, \bar{\theta} = 0.573°$

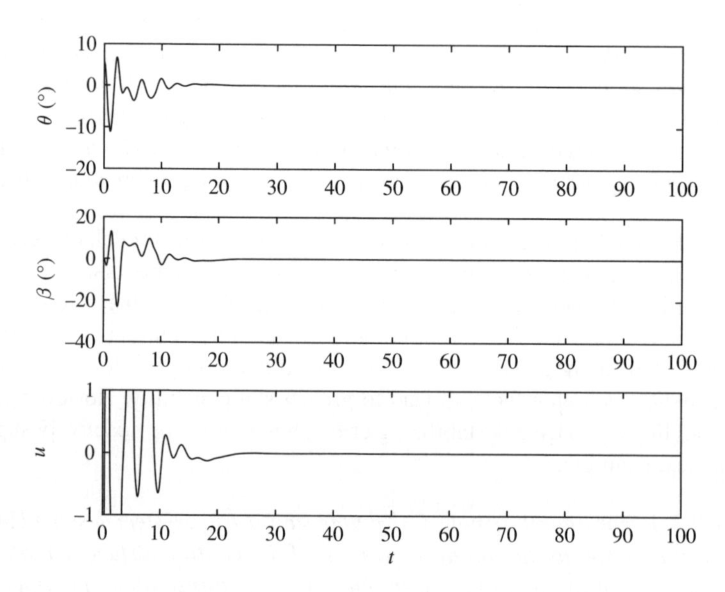

Figure 10.5 Closed-loop response of the reprogrammed integral adaptive controller with $\Gamma = 5I$, for an initial pitch perturbation at $U = 50$ m/s, standard sea-level condition, with a nonlinearly hardening pitch spring and a saturated control-surface actuator of unmodelled characteristics, $u_m = 1, \beta_m = 28.65°, a = 515, \bar{\theta} = 0.573°$

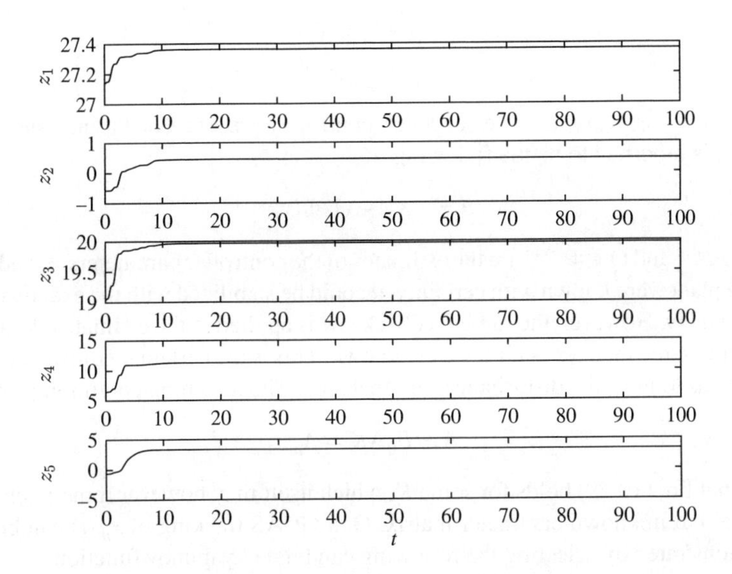

Figure 10.6 Variation of adaptation variables of the reprogrammed integral adaptive controller with $\Gamma = 5I$, for an initial pitch perturbation at $U = 50$ m/s, standard sea-level condition, with a nonlinearly hardening pitch spring and a saturated control-surface actuator of unmodelled characteristics, $u_m = 1, \beta_m = 28.65°, a = 515, \bar{\theta} = 0.573°$

where $a, \bar{\theta}$ are unknown parameters. If the adaptive integral controller designed in Example 10.2.2 were to be applied to this actual system, it would be unstable, as demonstrated by the simulation plotted in Fig. 10.4 for $U = 50$ m/ss, $\rho = 1.225$ kg/m^3, $a = 515, \bar{\theta} = 0.573°$. The integral adaptive controller designed in Example 10.2.2 is now reprogrammed by changing the adaptation gain matrix to $\Gamma = 5I$. The resulting closed-loop system is now asymptotically stable for the given initial condition, as seen by the plots of Figs. 10.5 and 10.6.

10.3 Model Reference Adaptation of Nonlinear Plant

When an uncertain, nonlinear perturbation, $f(x) : \mathbb{R}^n \rightarrow \mathbb{R}^m$, is present in the input channels of an otherwise linear (but uncertain) ASE plant of order n with m inputs, its state-space representation can be represented as follows (Lavretsky and Wise 2013):

$$\dot{x} = Ax + B_0\Lambda(u + f(x)), \tag{10.25}$$

where $A \in \mathbb{R}^{n\times n}$ is uncertain (but constant), $B_0 \in \mathbb{R}^{n\times m}$ is constant and known and $\Lambda \in \mathbb{R}^{m\times m}$ is an unknown, constant, diagonal matrix with positive elements, $\lambda_{ii} > 0, i = 1, \cdots, m$. Such a representation implies that the uncertainties are matched with the applied inputs. To begin the adaptation design process, it is assumed that $f(x)$ can be expressed as a linear combination of r locally Lipschitz, continuous and known functions, $\Phi(x) : \mathbb{R}^n \rightarrow \mathbb{R}^r$, given by

$$f(x) = \Theta^T\Phi(x), \tag{10.26}$$

where $\Theta \in \mathbb{R}^{r\times m}$ is a constant, but unknown coefficient matrix. The task of model reference adaptation is to globally, uniformly and asymptotically track the state of a reference model, given by

$$\dot{x}_m = A_m x_m, \tag{10.27}$$

where $A_m \in \mathbb{R}^{n\times n}$ is Hurwitz. The feedback control in presence of the nonlinear regressor vector, $\Phi(x)$, is modified to be the following:

$$u = -\hat{K}x - \hat{\Theta}^T\Phi(x); \tag{10.28}$$

where $\hat{K} \in \mathbb{R}^{m\times n}$ and $\hat{\Theta} \in \mathbb{R}^{r\times m}$ are the estimates of the controller parameters, K and Θ, respectively. If the plant were known with certainty, it could be stabilized with the exactly determined values of K and Θ. However, the same feedback law is applied to the estimates, $\hat{K}$ and $\hat{\Theta}$, treating them in the same manner as if they were known. This is the certainty equivalence principle, which works as long as the disturbances are matched. The matching condition is given by

$$A - B_0\Lambda K = A_m. \tag{10.29}$$

Assuming that Eq. (10.29) holds for some K, which itself may however remain unknown, and with an ideal but unknown parameter matrix, Θ, a GUAS tracking of $x_m(t)$ can be achieved. This is demonstrated by selecting the following candidate Lyapunov function:

$$V = e^T Pe + \mathrm{tr}\left(\delta K R_x^{-1}\delta K^T\Lambda + \delta\Theta^T R_\Theta^{-1}\delta\Theta\Lambda\right), \tag{10.30}$$

with $e(t) = x(t) - x_m(t)$, P, R_x and R_Θ being symmetric and positive-definite cost matrices, and

$$\begin{aligned}\hat{K} &= K + \delta K\\ \hat{\Theta} &= \Theta + \delta\Theta.\end{aligned} \tag{10.31}$$

The function V is radially unbounded in $(e, \delta K, \delta\Theta)$ and has the following time derivative:

$$\begin{aligned}\dot{V} &= \dot{e}^T Pe + e^T P\dot{e} + 2\mathrm{tr}\left(\delta K R_x^{-1}\dot{\hat{K}}^T\Lambda + \delta\Theta^T R_\Theta^{-1}\dot{\hat{\Theta}}\Lambda\right)\\ &= e^T\left(A_m^T P + PA_m\right)e + 2e^T PB_0\Lambda\left[\delta K - \delta\Theta^T\Phi(x)\right]\\ &\quad + 2\mathrm{tr}\left(\delta K R_x^{-1}\dot{\hat{K}}^T\Lambda + \delta\Theta^T R_\Theta^{-1}\dot{\hat{\Theta}}\Lambda\right).\end{aligned} \tag{10.32}$$

Since A_m is Hurwitz, the matrix P satisfies the following Lyapunov identity (Chapter 6):

$$A_m^T P + PA_m = -Q, \tag{10.33}$$

where Q is a positive-definite matrix. Substitution of Eq. (10.33) into Eq. (10.32), and by the following use of the trace property,

$$\begin{aligned}e^T PB_0\Lambda\delta Kx &= \mathrm{tr}(\delta Kxe^T PB_0\Lambda)\\ e^T PB_0\Lambda\delta\Theta^T\Phi(x) &= \mathrm{tr}(\delta\Theta^T\Phi(x)e^T PB_0\Lambda),\end{aligned} \tag{10.34}$$

results in the following

$$\begin{aligned}\dot{V} &= -e^T Qe + 2\mathrm{tr}\{\delta K(R_x^{-1}\dot{\hat{K}}^T - xe^T PB_0)\Lambda\}\\ &\quad + 2\mathrm{tr}\{\delta\Theta^T(R_\Theta^{-1}\dot{\hat{\Theta}}^T - \Phi(x)e^T PB_0)\Lambda\}.\end{aligned} \tag{10.35}$$

In order to make $\dot{V}$ negative definite for all $x(t)$, the following adaptation laws are selected:

$$\begin{aligned}\dot{\hat{K}}^T &= R_x x e^T P B_0 \\ \dot{\hat{\Theta}}^T &= R_\Theta \Phi(x) e^T P B_0.\end{aligned} \tag{10.36}$$

This choice results in the following:

$$\dot{V} = -e^T Q e \leq 0 \tag{10.37}$$

and

$$\ddot{V} = -2e^T Q \dot{e}, \tag{10.38}$$

which by Barbalat's lemma guarantees that the error, $e(t)$, uniformly converges to 0 in the limit, $t \to \infty$. Thus, global uniform asymptotic stability is guaranteed, without the knowledge of the uncertain plant parameters, (A, Λ, Θ).

10.4 Robust Model Reference Adaptation

The MRAS adaptive control procedure can be applied to robust control in the presence of bounded random disturbance inputs. In such a case, the uncertainty in the dynamics matrix, A, is replaced by the process noise signal, $p(t)$, leading to the following state-space representation of the plant:

$$\begin{aligned}\dot{x} &= A_m x + B_0 \Lambda (u + f(x)) + p(t) \\ y &= C_m x,\end{aligned} \tag{10.39}$$

where A_m, C_m (being the reference model's coefficients) and $B_0 \in \mathbb{R}^{n \times m}$ are constants and known, $f(x)$ is a matched nonlinear perturbation of the structure given by Eq. (10.26), and $\Lambda \in \mathbb{R}^{m \times m}$ is an unknown, constant, diagonal matrix with positive elements, $\lambda_{ii} > 0, i = 1, \cdots, m$. The unforced reference model is represented as follows:

$$\begin{aligned}\dot{x}_m &= A_m x_m \\ y_m &= C_m x_m.\end{aligned} \tag{10.40}$$

The process noise vector is assumed to be bounded by an known upper bound, p_m, as follows:

$$| p(t) | \leq p_m. \tag{10.41}$$

It is further assumed that A_m is Hurwitz and the pair $(A_m, B_0 \Lambda)$ is controllable. The objective is to find a stabilizing feedback controller given by

$$u = -\hat{\Theta}^T \Phi(x), \tag{10.42}$$

where $\hat{\Theta} \in \mathbb{R}^{r \times m}$ is the estimate of the perfect parameter gain matrix, Θ, which would stabilize the plant if $p(t) = 0$ and Λ was known with certainty. The closed-loop system is then given by

$$\begin{aligned}\dot{x} &= A_m x - B_0 \Lambda \delta\Theta^T \Phi(x) + p(t) \\ y &= C_m x,\end{aligned} \tag{10.43}$$

where $\delta\Theta = \hat{\Theta} - \Theta$, or as follows in terms of the tracking error, $e(t) = x(t) - x_m(t)$:

$$\dot{e} = A_m e - B_0 \Lambda \delta\Theta^T \Phi(x) + p(t). \tag{10.44}$$

In order to derive an adaptation law, the following radially unbounded, candidate Lyapunov function is defined:

$$V = e^T P e + \text{tr}(\delta\Theta^T R^{-1} \delta\Theta \Lambda), \tag{10.45}$$

where P, R are constant, symmetric, positive-definite weighting matrices, with R being invertible. Taking the time derivative of Lyapunov function along a trajectory, we have

$$\dot{V} = e^T (A_m^T P + P A_m) e - 2e^T P B_0 \Lambda \delta\Theta^T \Phi(x) + 2e^T P p(t) + 2\text{tr}(\delta\Theta^T R^{-1} \dot{\hat{\Theta}} \Lambda). \tag{10.46}$$

By substituting the Lyapunov identity, Eq. (10.33), into Eq. (10.46), along with the trace property, $a^T b = \text{tr}(b a^T)$, we have

$$\dot{V} = -e^T Q e + 2e^T P p(t) + 2\text{tr}\left\{\delta\Theta^T \left(R^{-1}\dot{\hat{\Theta}} - \Phi e^T P B\right) \Lambda\right\}. \tag{10.47}$$

If the standard MRAS adaptation law, $\dot{\Theta}^T = R\Phi(x) e^T P B$, is applied here, the error, $e(t)$, might remain bounded by Barbalat's lemma, but $\delta\Theta(t)$ may become unbounded because of the noise, $p(t)$, because $\delta\Theta(t)$ grows steadily in proportion to the time integral of $e(t)$, even for small values of $|\, e \,|$ for which $\dot{V} = 0$. This is known as parameter divergence (or drift) and must be avoided by introducing a dead zone in the adaptation law as follows:

$$\dot{\hat{\Theta}}^T = \begin{cases} R\Phi(x) e^T P B, & (|\, e \,| > e_m) \\ 0, & (|\, e \,| \leq e_m) \end{cases} \tag{10.48}$$

The error bound used for the dead zone can be expressed as the following function (Khalil 1996) of the known process noise bound, p_m:

$$e_m = \frac{2 p_m \rho(P)}{\rho(Q)}, \tag{10.49}$$

where $\rho(.)$ denotes the spectral radius (see Chapter 2). Since both P, Q are positive definite, their spectral radii are the same as their maximum eigenvalues.

An adaptation law based on the dead zone, Eq. (10.49), is likely to lead to chattering in the closed-loop response. A way to avoid the problem of chattering is to modify the adaptation law as follows (Lavretsky and Wise 2013):

$$\dot{\hat{\Theta}}^T = R\Phi(x) e^T P B \mu(|\, e \,|), \tag{10.50}$$

where

$$\mu(|\, e \,|) = \max\left\{0, \min\left(1, \frac{|\, e \,| - \delta e_m}{(1 - \delta) e_m}\right)\right\} \tag{10.51}$$

is a continuous modulation function with $0 < \delta < 1$ a constant. Thus a boundary layer is added to the dead zone in order to have a smooth behaviour of the closed-loop response near the dead zone. (Tewari 2013) arrived at a similar boundary-layer-type, dead-zone modification for a robust adaptive control for spacecraft de-orbiting problem using a gradient-type adaptation.

Ioannou and Kokotovic (Ioannou 1983) proposed a robust adaptive MRAS scheme for the case when the upper bound on the process noise, p_m, is unknown. Such a method, called the σ-modification, is more practical for an ASE problem than the one given above, because the unsteady aerodynamics model can have perturbations of unknown bound when flow-separation and shock-wave effects are involved. The σ-modification adaptation law is the following:

$$\dot{\hat{\Theta}}^T = R\left[\Phi(x)e^T PB - \sigma\hat{\Theta}^T\right], \tag{10.52}$$

where $\sigma > 0$ is a constant gain. By applying Lyapunov's direct method, the UUB (Chapter 8) of all signals of the closed-loop system can be shown for this method (Ioannou 1983). Thus robustness is guaranteed without knowing the bound on $p(t)$.

An improvement to σ-modification has been also proposed. The new scheme, called e-modification (Narendra and Annaswamy 1987), addresses a particular drawback of the σ-modification algorithm when $\mid e \mid$ is particularly small. In that case, Eq. (10.52) approximately becomes the following:

$$\dot{\hat{\Theta}}^T \simeq -R\sigma\hat{\Theta}^T, \tag{10.53}$$

which drives $\hat{\Theta}$ to small values, thereby reducing their effectiveness. This can be termed as 'unlearning' of the parameter gain matrix. In such an event, were the disturbance $p(t)$ to be removed, and the reference model be excited by an external reference input, the parameter error matrix, $\delta\Theta$, may not converge. Therefore, e-modification multiplies σ by a term proportional to the tracking error as follows:

$$\dot{\hat{\Theta}}^T = R\left[\Phi(x)e^T PB - \sigma \mid e^T PB \mid \hat{\Theta}^T\right]. \tag{10.54}$$

This makes both the terms on the right-hand side contribute to the adaptation law, even when $\mid e \mid$ is small.

Example 10.4.1 *Let us reconsider the MRAS typical-section flutter-suppression design of Chapter 8 in the presence of a normally distributed process noise, $p(t)$, in all the state variables, with a standard deviation of 10^{-5}. This could be considered a case of time-dependent variation in model properties due to effects such as atmospheric gusts, structural nonlinearity and aerodynamic flow separation/reattachment. A direct simulation of an MRAS designed with $P = 10^7 I, R = I$ and using linear state feedback $\Phi(x) = x, \hat{\Theta} = \hat{k}$ without robustness considerations, is shown in Figs. 10.7–10.10 for an initial pitch angle perturbation at the supercritical condition of $U = 65$ m/s and standard sea level. The plant parameters A, Λ are assumed to be known in this simulation. The state variables, $h(t)/b, \theta(t), \beta(t)$, the normal acceleration output, $y(t)$, and the input, $u(t)$, are all seen to respond relatively smoothly in the presence of the process noise. While the system's response converges to a steady state for larger values of time, the response in the controller parameter estimates is especially prone to noise (Figs. 10.9 and 10.10), and some of them are seen to grow unboundedly ($\hat{K}_1, \hat{K}_7$), which indicates an overall unstable adaptation system. The use of high adaptation gain thus results in spurious oscillations in the integration scheme (Runge–Kutta, fourth-order) due to the random process noise. Although the system's tracking error is maintained at small values, the controller parameters are adapted by an unstable process. In such a case, rate-limited and saturated input torque will*

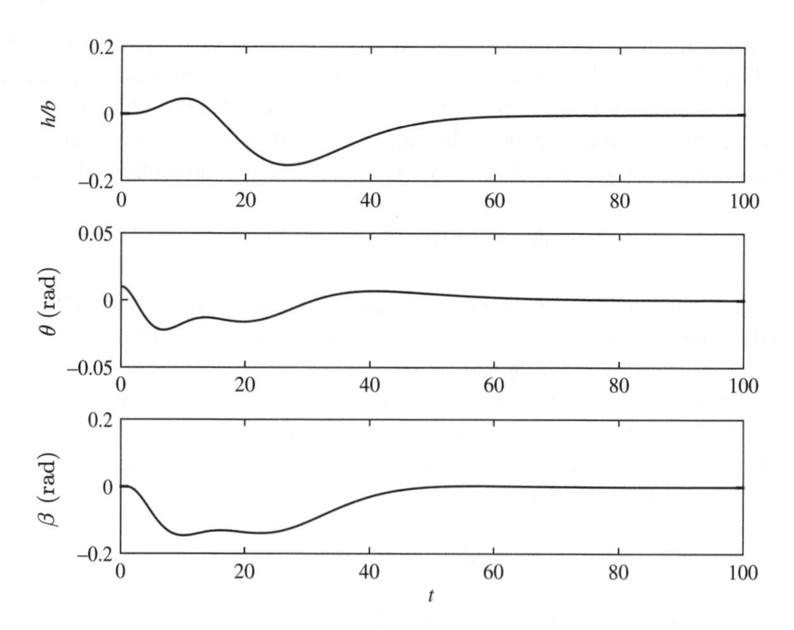

Figure 10.7 Simulated initial response of plunge displacement, h/b, pitch angle, θ, and control-surface deflection, β, of an MRAS ($P = 10^7 I, R = I$) flutter-suppression system for a typical section in the presence of random process noise at a supercritical speed and standard sea level

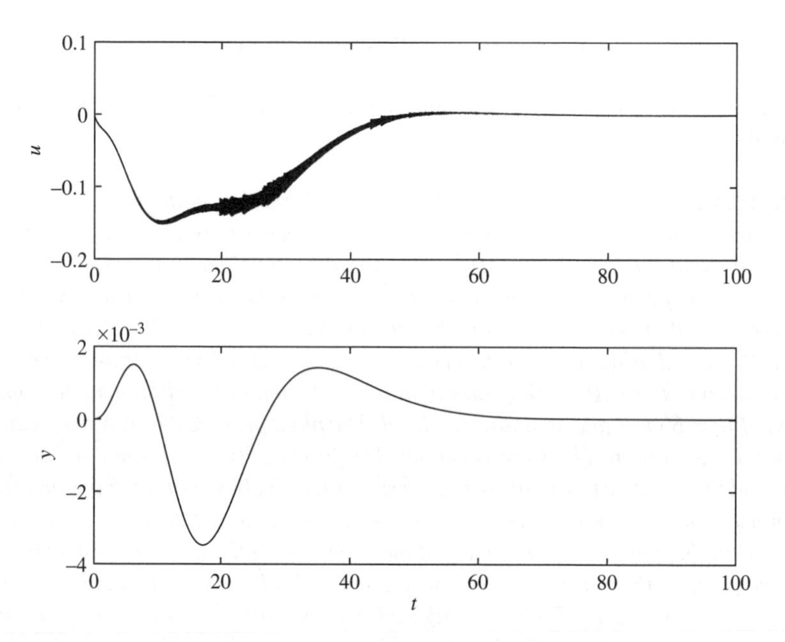

Figure 10.8 Simulated initial response of input torque, u, and normal acceleration output, y, of an MRAS ($P = 10^7 I, R = I$) flutter-suppression system for a typical section in the presence of random process noise at a supercritical speed and standard sea level

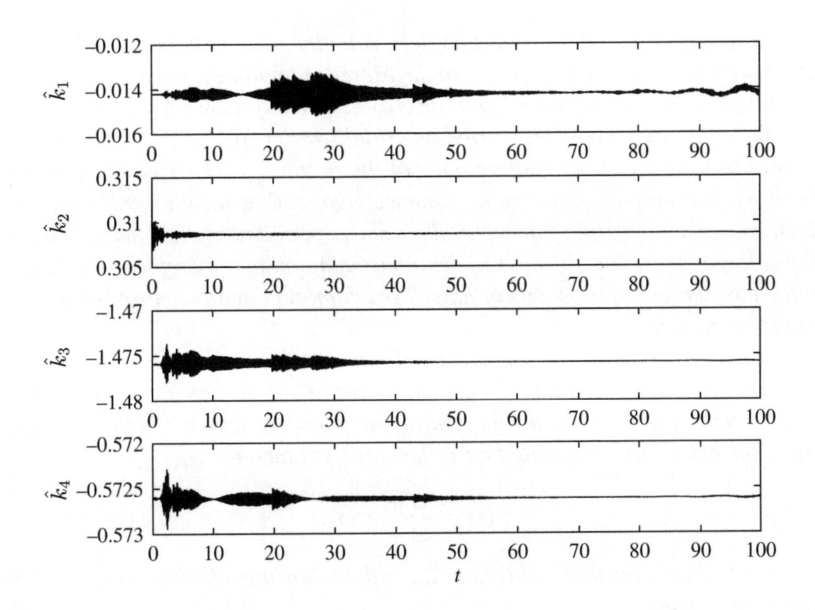

Figure 10.9 Simulated variation of state-feedback adaptation gains, $\hat{k}_1 - \hat{k}_4$, of an MRAS ($P = 10^7 I$, $R = I$) flutter-suppression system for a typical section in the presence of random process noise at a supercritical speed and standard sea level

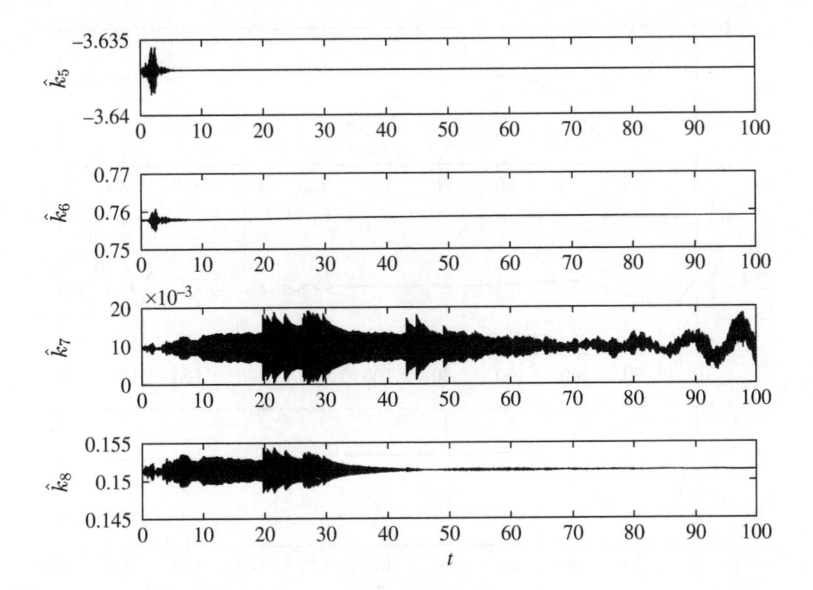

Figure 10.10 Simulated variation of state-feedback adaptation gains, $\hat{k}_5 - \hat{k}_8$, of an MRAS ($P = 10^7 I$, $R = I$) flutter-suppression system for a typical section in the presence of random process noise at a supercritical speed and standard sea level

lead to an instability of the adaptive ASE system. Adaptive flutter suppression by this method would therefore be deemed to have failed in the presence of process noise.

A possible way of avoiding noisy adaptation is by decreasing the weightage of the estimation error, P, in the adaptation law. This is tried by using $P = P = 10^4 I, R = I$, *and the simulation is carried out for* $t = 200$. *The resulting plots of the response (Figs. 10.11–10.14) show that while the noisy adaptation is now avoided, the weightage, P, is insufficiently large in magnitude, which causes a slow drift of the control parameters, and ultimately leads to the system's instability at large times. The objective of the robust adaptation is therefore to strike a balance between the conflicting requirements of noiseless adaptation and the closed-loop stability of the controller parameters.*

The simplest way to suppress the rapid control actuation demand due to noise feedback is by putting a low-pass filter in the control loop, usually just before the actuator (Tewari 2011). This is effectively carried out by replacing the control input u by a variable z in the state equations, and relating the two by the following first-order transfer function:

$$u(s) = \frac{as+1}{bs+1} z(s), \tag{10.55}$$

where $a > 0, b > 0$ *are constants. This introduces the following additional state equation into the closed-loop system:*

$$\dot{\xi} = -\frac{\xi}{a} + \left(\frac{1-b}{a}\right)\frac{u}{a}, \tag{10.56}$$

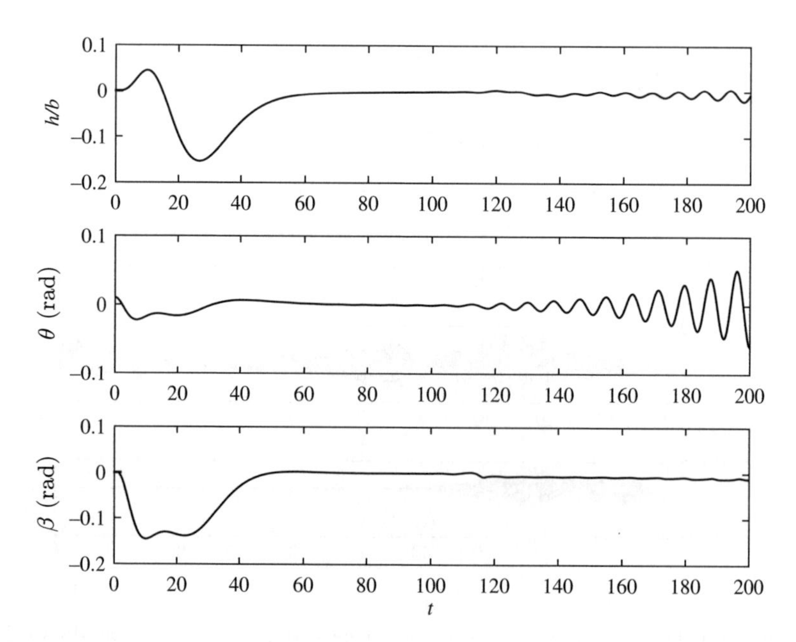

Figure 10.11 Simulated initial response of plunge displacement, h/b, pitch angle, θ, and control-surface deflection, β, of an MRAS ($P = 10^4 I, R = I$) flutter-suppression system for a typical section in the presence of random process noise at a supercritical speed and standard sea level

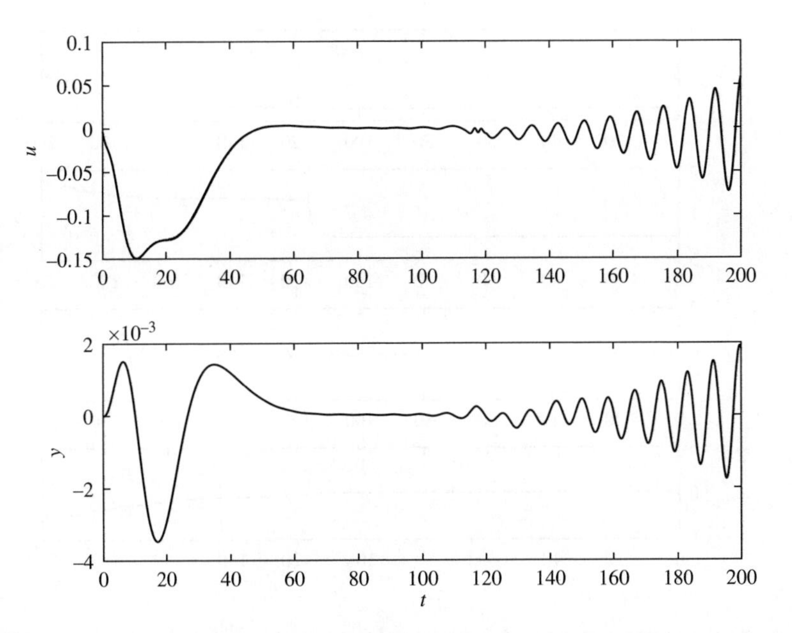

Figure 10.12 Simulated initial response of input torque, u, and normal acceleration output, y, of an MRAS ($P = 10^4 I, R = I$) flutter-suppression system for a typical section in the presence of random process noise at a supercritical speed and standard sea level

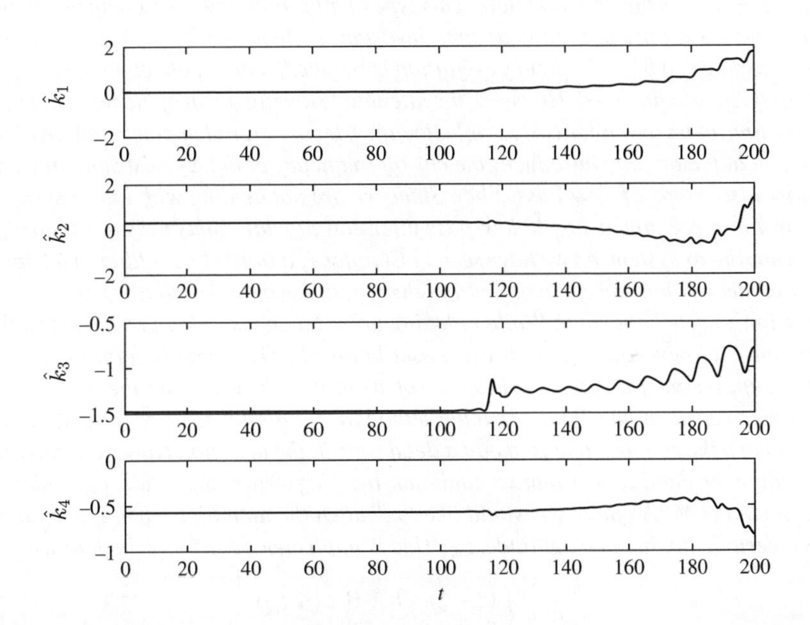

Figure 10.13 Simulated variation of state-feedback adaptation gains, $\hat{k}_1 - \hat{k}_4$, of an MRAS ($P = 10^4 I, R = I$) flutter-suppression system for a typical section in the presence of random process noise at a supercritical speed and standard sea level

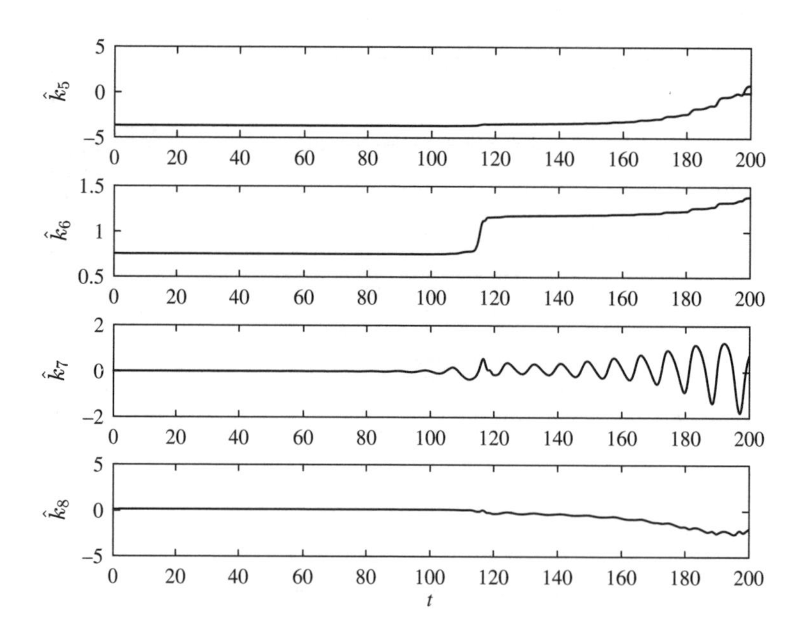

Figure 10.14 Simulated variation of state-feedback adaptation gains, $\hat{k}_5 - \hat{k}_8$, of an MRAS ($P = 10^4 I, R = I$) flutter-suppression system for a typical section in the presence of random process noise at a supercritical speed and standard sea level

where $\xi = z - bu/a$ is the actuator state. This type of filter is called a lag compensator (Tewari 2002), and moves a pole of the linear time-invariant system from $s = -1/a$ to $s = -1/b$. If $a < b$, the response to high-frequency excitation is reduced, without affecting the DC gain (i.e. the zero-frequency behaviour). However, the attendant slowing down of the response increases the settling time of an asymptotically stable linear system. A digital implementation of the analogue system automatically introduces the cut-off frequency of the Nyquist sampling rate, and is therefore also a type of 'low-pass' filter. Since we are not dealing with linear systems here, it is important to see how a simple low-pass filter can provide robustness to random process noise in a nonlinear system. As we have seen in Chapter 7, a stable linear filter in feedback with a static nonlinearity has the effect of limiting the amplitude of the oscillatory response. This is called the limit-cycle behaviour. While reducing noise feedback at high frequencies, the filter can thus cause a low-frequency, limit-cycle oscillation (LCO), which is certainly undesirable in a flutter-suppression problem. The best way of avoiding LCO is to make the control actuation less sensitive to noise inputs. While various methods are available to do so (as discussed above in this chapter), the simplest one is to add a dead zone in the actuator. However, instead of formally building the dead zone around a minimum-tracking-error magnitude, e_m, which is commonly done in the MRAS field, we would like to deaden the actuation when the input demand exceeds a certain maximum magnitude, z_m. This is implemented here by the following logic:

$$u = \begin{cases} (z - \xi)a/b, & (|z| \leq z_m) \\ 0, & (|z| > z_m) \end{cases} \tag{10.57}$$

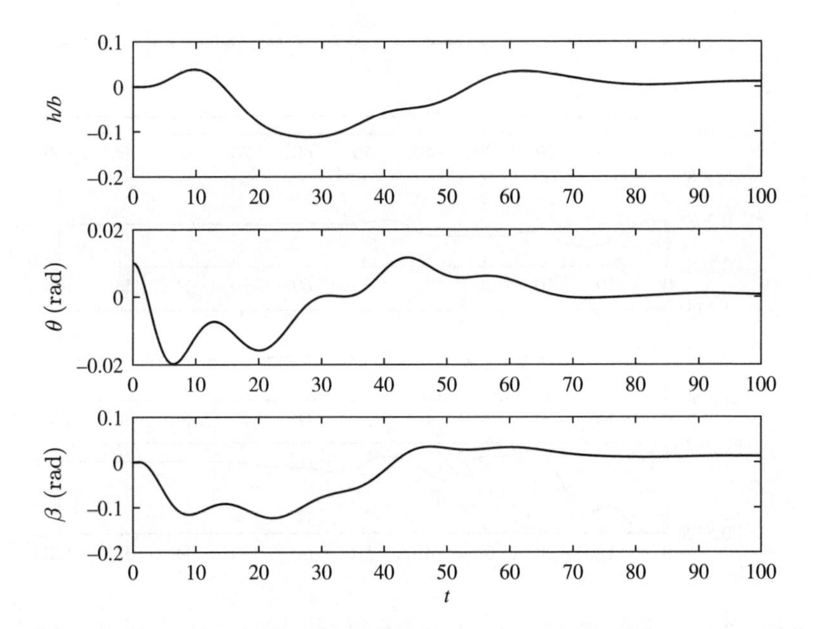

Figure 10.15 Simulated initial response of plunge displacement, h/b, pitch angle, θ, and control-surface deflection, β, of a robust MRAS flutter-suppression system with a filtered actuation dead zone, for a typical section in the presence of random process noise at a supercritical speed and standard sea level

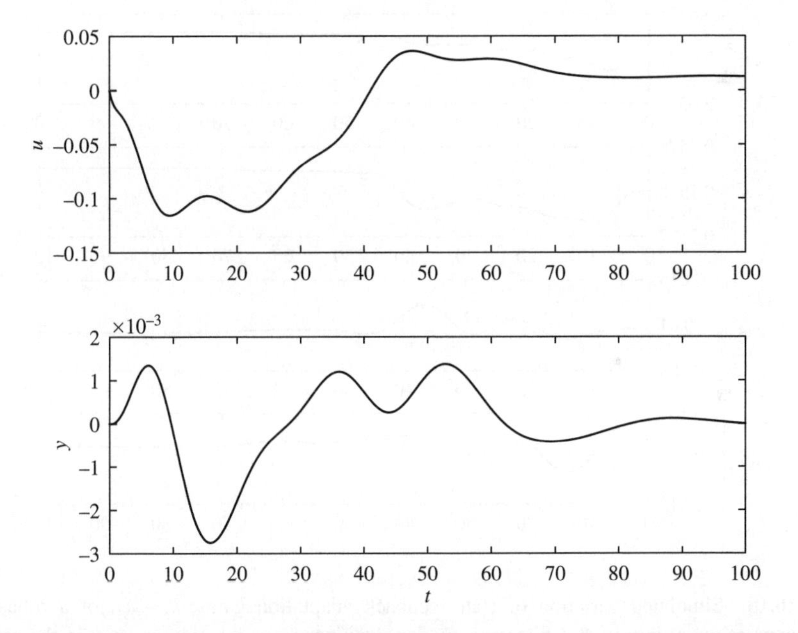

Figure 10.16 Simulated initial response of input torque, u, and normal acceleration output, y, of a robust MRAS flutter-suppression system with a filtered actuation dead zone, for a typical section in the presence of random process noise at a supercritical speed and standard sea level

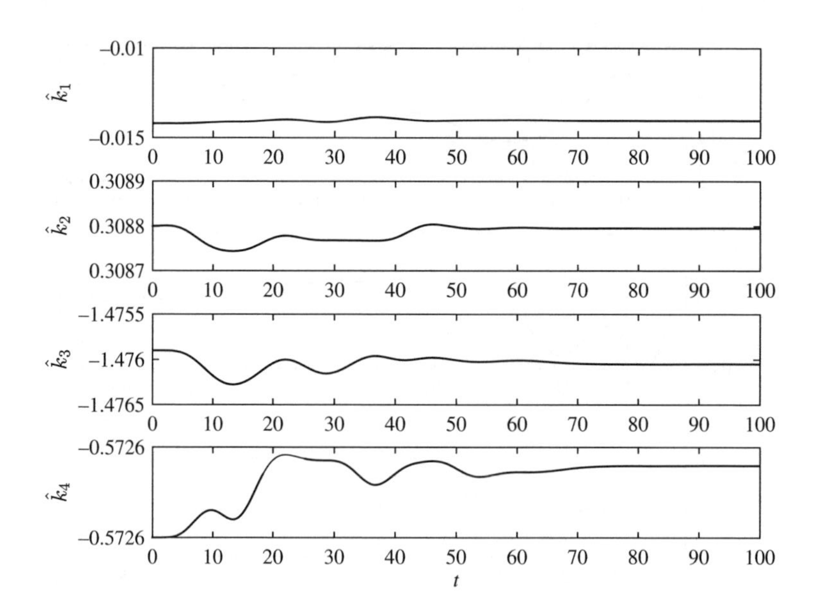

Figure 10.17 Simulated variation of state-feedback adaptation gains, $\hat{k}_1 - \hat{k}_4$, of a robust MRAS flutter-suppression system with a filtered actuation dead zone, for a typical section in the presence of random process noise at a supercritical speed and standard sea level

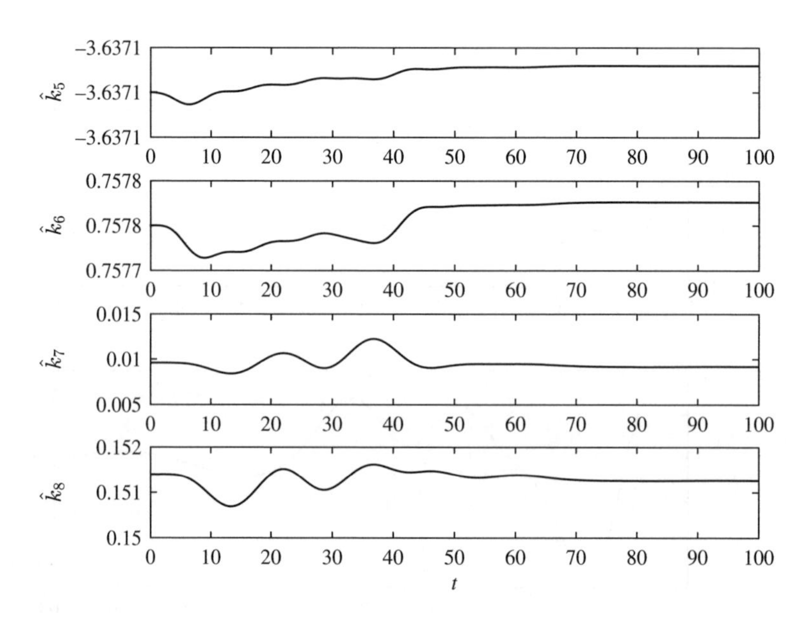

Figure 10.18 Simulated variation of state-feedback adaptation gains, $\hat{k}_5 - \hat{k}_8$, of a robust MRAS flutter-suppression system with a filtered actuation dead zone, for a typical section in the presence of random process noise at a supercritical speed and standard sea level

The effectiveness of this simple method is seen by the simulation of the MRAS with a low-pass, actuation dead zone using $a = 0.01, b = 0.1, z_m = 0.2$ *for the supercritical condition given above, and for a normally distributed process noise of intensity* 10^{-5}. *Owing to the filtered dead zone, now there is no need of excessively weighting the state error with a large magnitude of P, in order to have a stable response. Hence,* $P = 0.1I, R = I$ *are used in the adaptation law. The results plotted in Figs. 10.15–10.18 show a stable response, free from either chatter or a large transient overshoot, and converging to a steady state in all state variables and controller parameters. There is however a slight non-zero steady-state error in the state variables due to the dead zone. Such a scheme therefore offers the best promise of a practical application for adaptive and robust flutter suppression, because of its simplicity. The design parameters,* a, b, z_m, P, R *can be fine-tuned to the best possible closed-loop response in the presence of process noise of a maximum expected intensity.*

10.4.1 Output-Feedback Design

When only the measurement signals, $y(t) \in \mathbb{R}^p$, are available for feedback, an observer-based output-feedback MRAS method is required. Consider the following transfer-matrix realization of the plant:

$$G(s) = C(sI - A)^{-1}B + D \in \mathbb{C}^{p\times m}, \tag{10.58}$$

with an output-feedback law given by

$$u(t) = -\Theta(t)y(t) \in \mathbb{R}^m, \tag{10.59}$$

where $\Theta \in \mathbb{R}^{m\times p}$ is a matrix of time-varying controller parameters consisting of the regulator and observer gains. The controller parameters, Θ, can be arranged in a single column vector, $\theta \in \mathbb{R}^{mp}$. The closed-loop system has the following transfer matrix representation:

$$y(s) = G_c(s, \theta)r(s), \tag{10.60}$$

where $r(s)$ is a reference input. An update law is required for the controller parameters such that the output error,

$$e(t) = y(t) - y_m(t), \tag{10.61}$$

is driven to zero in the steady state. The error is expressed as follows in the Laplace domain:

$$e(s) = [G_c(s, \theta) - G_m(s)]r(s), \tag{10.62}$$

where $G_m(s)$ is the transfer matrix of the reference aeroelastic plant. A gradient adaptation law for this problem is given by

$$\frac{d\theta}{dt} = \Gamma\left(\frac{\partial e}{\partial \theta}\right)^T e, \tag{10.63}$$

where $\Gamma \in \mathbb{R}^{mp\times mp}$ is a constant adaptation gain matrix. The selection of a suitable sensitivity matrix, $\partial e/\partial \theta \in \mathbb{R}^{mp\times p}$, is the crux of the gradient-based adaptation, and must be carried out as discussed above for the state-feedback problem. However, it requires a structure for the dependence of the closed-loop transfer matrix, $G_c(s, \theta)$, on the controller parameters, θ.

For a Lyapunov-based adaptation, consider the following candidate Lyapunov function:

$$V = e^T e + (\theta - \theta_0)^T (\theta - \theta_0), \tag{10.64}$$

with $P \in \mathbb{R}^{p\times p}$ being symmetric and positive definite and θ_0 satisfying the following matching condition:

$$G_c(s, \theta_0) = G_m(s). \tag{10.65}$$

The function V is radially unbounded in e and $(\theta - \theta_0)$ and has the following time-derivative:

$$\dot{V} = 2e^T \dot{e} + 2(\theta - \theta_0)^T \dot{\theta}. \tag{10.66}$$

Since $G_m(s)$ is strictly positive real, and the parametric uncertainty $(\theta - \theta_0)$ appears in a feedback loop with the matched system, Eq. (10.65), any adaptation law minimizing $(\theta - \theta_0)$ is dissipative in terms of the error, e, by the passivity approach (Chapter 6). However, a design of such an adaptation law requires assuming a structure for the closed-loop system, $G_c(s, \theta)$, by either an observer-based or a direct feedback control law. The latter is of the form, $u = Fy$, where F is a controller parameters matrix.

Direct output-feedback designs must be based on the passivity formalism for achieving input–output stability. Unfortunately, a passivity-based design requires a square plant (see Chapter 6), while a practical ASE plant may not be square. Although the number of inputs, m, is limited to the number of available control surfaces, any number of sensors (accelerometers, optical sensors) can be mounted on the wing in order to yield a redundant data for aeroelastic mode-shape determination. Such redundancy is necessary if a parameter estimation-based controller (such as the STR of Chapter 7) is to be designed for the ASE system. Thus, the number of outputs, p, is always greater than the number of inputs. It would then seem impossible to apply the passivity-based theorems of Chapter 5 for ASE plants. However, recent developments in control theory (Misra 1998) allow the 'squaring-up' of a non-square plant by adding fictitious inputs. Application of such an approach to MRAS design by a projection-based output-feedback method is presented in Chapter 14 of (Lavretsky and Wise 2013). Here we are concerned with an observer-based structure of the MRAS system. Both the methods can be considered equivalent, as the parameter adaptation in both is superimposed on a baseline linear quadratic regulator. The difference lies in the derivation of the adaptation mechanism.

An output feedback method can be applied to a non-square plant by rendering it virtually square through the addition of fictitious inputs. Consider a proper ($D = 0$) plant represented by

$$\begin{aligned}\dot{x} &= Ax + B_0 \Lambda u \\ y &= Cx \\ z &= C_2 x,\end{aligned} \tag{10.67}$$

where $x \in \mathbb{R}^n$ is the state, $u \in \mathbb{R}^m$ the input, $y \in \mathbb{R}^p$ the output, and $z \in \mathbb{R}^\ell$ the regulated output vectors. Here it is assumed that $A \in \mathbb{R}^{n\times n}$, $B_0 \in \mathbb{R}^{n\times m}$, $C \in \mathbb{R}^{p\times n}$ and $C_2 \in \mathbb{R}^{\ell\times n}$ are a known matrices, while $\Lambda \in \mathbb{R}^{m\times m}$ is an unknown, constant, diagonal matrix with positive elements, $\lambda_{ii} > 0, i = 1, \cdots, m$. Thus the uncertainty comes in with the applied inputs, while the unforced part is known with certainty. The following assumptions are applied to carry out the direct output-feedback design for a non-square plant:

(a) The plant is minimum phase (i.e. does not have any zeros in the right-half plane).
(b) The pair (A, B_0) is controllable and the pair (A, C) is observable.
(c) The rank of the matrix B_0 is equal to m, the number of inputs.
(d) The rank of the matrix C is equal to p, the number of outputs.
(e) The rank of the matrix CB_0 is equal to m, the number of inputs.
(f) The number of outputs is greater than the number of inputs ($p > m$).

For a further discussion of the direct (non-observer-based) output-feedback method, the reader is referred to Lavretsky and Wise.

For an observer-based design, consider a general (non-square) plant with the following state-space representation:

$$\dot{x} = Ax + B_0\Lambda u, \tag{10.68}$$

$$y = Cx + Du. \tag{10.69}$$

to be stabilized by the control law,

$$u = -K\hat{x}, \tag{10.70}$$

where the estimated state, $\hat{x}$, is generated by the following observer:

$$\dot{\hat{x}} = (A - LC)\hat{x} + (B - LD)u + Ly, \tag{10.71}$$

based on the measured output vector, y. Here we use the abbreviation, $B = B_0\Lambda$, with the understanding that B_0 is known, while the diagonal perturbation, Λ is unknown. The regulated system is thus represented by

$$\begin{Bmatrix} \dot{x} \\ \dot{\hat{x}} \end{Bmatrix} = \begin{pmatrix} A & -BK \\ LC & A - BK - LC \end{pmatrix} \begin{Bmatrix} x \\ \hat{x} \end{Bmatrix}. \tag{10.72}$$

An MRAS approach cannot be applied to simultaneously adapt both $\hat{K}$, the estimate of regulator gains, and $\hat{L}$, the estimate of observer gains, by a single adaptation law. This is because the compensator is designed by the separation principle, where the regulator and observer are designed independently of each other. Thus we must look for separate adaptation laws of the regulator and observer gains, which also fit in with the certainty equivalence approach where a part of the control system is designed assuming that the other part is known with certainty. Any variation of $\hat{L}$ during adaptation for $\hat{K}$ can be regarded as process noise, which is handled in the manner given above for robust MRAS. Similarly, any variation of $\hat{K}$ in adapting $\hat{L}$ is treated as a bounded process noise.

For the derivation of an adaptation law of regulator gains, $\hat{K}$, we express Eq. (10.72) as follows:

$$\dot{z} = F_1 z - G_1 \Lambda \Theta_1^T \Phi_1(z), \tag{10.73}$$

where

$$F_1 = \begin{pmatrix} A & 0 \\ LC & A - LC \end{pmatrix}, \quad G_1 = \begin{pmatrix} B_0 \\ B_0 \end{pmatrix}, \quad \Phi_1(z) = (0, \quad I)z, \tag{10.74}$$

$\Theta_1^T = \hat{K}$ and $z = \left(x^T, \hat{x}^T\right)^T$. The uncertainties in the plant matrices A, B cause uncertainties in the matrices F_1 and Λ. Note that as Eq. (10.73) is in the form of Eqs. (10.25) and (10.26), the following adaptation law will render the control system GUAS stable, provided the observer gains, L, are known with certainty:

$$\dot{\hat{K}} = G_1^T P_1 e \Phi_1^T(z) \Gamma_1, \tag{10.75}$$

where $e = z - x_m, x_m$ being the reference state, and P_1, Γ_1 are symmetric, positive-definite gain matrices, with Γ_1 also being non-singular.

Similarly, for the adaptation of the observer gains, $\hat{L}$, we write

$$\dot{z} = F_2 z + G_2 \Theta_2 \Phi_2(z), \tag{10.76}$$

where

$$F_2 = \begin{pmatrix} A & -BK \\ 0 & A - BK \end{pmatrix}, \quad G_2 = \begin{pmatrix} 0 \\ I \end{pmatrix}, \quad \Phi_2(z) = (C, -C)z, \tag{10.77}$$

and $\Theta_2 = \hat{L}$. The uncertainties in A, B are treated as that in the matrix F_2, whereas $G_2, \Phi_2(z)$ are known. The GUAS-stabilizing adaptation law for the observer gains is therefore the following:

$$\dot{\hat{L}} = -\Gamma_2 \Phi_2(z) e^T P_2 G_2, \tag{10.78}$$

with P_2, Γ_2 being symmetric, positive-definite gain matrices, and Γ_2 also being non-singular.

10.4.2 *Adaptive Flutter Suppression of a Three-Dimensional Wing*

Application of model-reference adaptation to a 3-D wing is much more cumbersome than that for a typical section, primarily due to the large number of degrees of freedom involved. Even when a few structural modes are retained by a balanced realization of the mass and stiffness matrices, the necessity of rational-function approximations (RFAs) for both wing and the control surfaces increases the size of the aeroelastic plant by an order of magnitude, when compared to a typical section. Furthermore, as an observer is invariably required in a 3-D application, there is a further increase (nearly doubling) of the order of the system. Both the regulator and observer parameters must be updated by the adaptation laws, which implies twice as many controller parameters (unless a reduced-order observer is used) as the state-feedback case considered above.

Consider the ASE system with the following state-space representation:

$$\dot{X} = \bar{A}X + \bar{B}u, \tag{10.79}$$

$$y = \bar{E}X + \bar{D}u, \tag{10.80}$$

with state vector,

$$X = (x^T, x_c^T)^T,$$

and the following coefficient matrices:

$$\bar{A} = \begin{pmatrix} A & BC_c \\ 0 & A_c \end{pmatrix}, \quad \bar{B} = \begin{pmatrix} BD_c \\ B_c \end{pmatrix}, \tag{10.81}$$

$$\bar{E} = \begin{pmatrix} E & DC_c \end{pmatrix}, \quad \bar{D} = DD_c, \tag{10.82}$$

with A, B, E, D being the coefficients of the aeroelastic plant, and A_c, B_c, C_c, D_c those of the control-surface actuators. The regulator control law,

$$u = -K\hat{X}, \tag{10.83}$$

employs the estimated state, $\hat{X}$, generated by the following linear observer:

$$\dot{\hat{X}} = (\bar{A} - L\bar{E})\hat{X} + (\bar{B} - L\bar{D})u + Ly, \tag{10.84}$$

based on the measured output vector, y. Alternatively, a reduced-order observer can be designed. The regulated ASE system is thus represented by

$$\begin{Bmatrix} \dot{X} \\ \dot{\hat{X}} \end{Bmatrix} = \begin{pmatrix} \bar{A} & -\bar{B}K \\ L\bar{E} & \bar{A} - \bar{B}K - L\bar{E} \end{pmatrix} \begin{Bmatrix} X \\ \hat{X} \end{Bmatrix}, \tag{10.85}$$

which can be expressed in the following form amenable to the adaptation of regulator gains, $\hat{K}$:

$$\dot{Z} = F_1 Z + G_1 \Theta_1^T \Phi_1(Z), \tag{10.86}$$

where

$$F_1 = \begin{pmatrix} \bar{A} & 0 \\ L\bar{E} & \bar{A} - L\bar{E} \end{pmatrix}, \quad G_1 = \begin{pmatrix} \bar{B} \\ \bar{B} \end{pmatrix}, \quad \Phi_1(Z) = (0, \quad -I)Z, \tag{10.87}$$

$\Theta_1^T = \hat{K}$ and $Z = \left(X^T, \hat{X}^T\right)^T$. This gives the following adaptation law:

$$\dot{\hat{K}} = G_1^T P_1 e \Phi_1^T(Z) R_1, \tag{10.88}$$

where $e = Z - x_m$, x_m being the state of a reference model, and P_1, R_1 are symmetric, positive-definite gain matrices, with R_1 also being non-singular.

For the adaptation of the observer gains, $\hat{L}$, we have

$$\dot{Z} = F_2 Z + G_2 \Theta_2 \Phi_2(Z), \tag{10.89}$$

where

$$F_2 = \begin{pmatrix} \bar{A} & -\bar{B}K \\ 0 & \bar{A} - \bar{B}K \end{pmatrix}, \quad G_2 = \begin{pmatrix} 0 \\ I \end{pmatrix}, \quad \Phi_2(Z) = (\bar{E}, \quad -\bar{E})Z, \tag{10.90}$$

and $\Theta_2 = \hat{L}$. This leads to the following adaptation law:

$$\dot{\hat{L}} = R_2 \Phi_2(Z) e^T P_2 G_2, \tag{10.91}$$

with P_2, R_2 being symmetric, positive-definite gain matrices, and R_2 also being non-singular.

The reference model characteristics can be chosen to be the ASE model at a reference flight condition, with the regulator and observer gains pre-selected to yield a constant, Hurwitz dynamics matrix, A_m. The observer and regulator gains are then adapted by an MRAS scheme such that random static variations in the plant parameters, $\bar{A}$ and $\bar{B}$, do not cause instability in the overall system. This is illustrated in the following example.

Example 10.4.2 *The modified DAST-ARW1 wing (Appendix C) used for deriving a 3-D active flutter-suppression system in Chapter 4 for a reference, supercritical flight condition (standard*

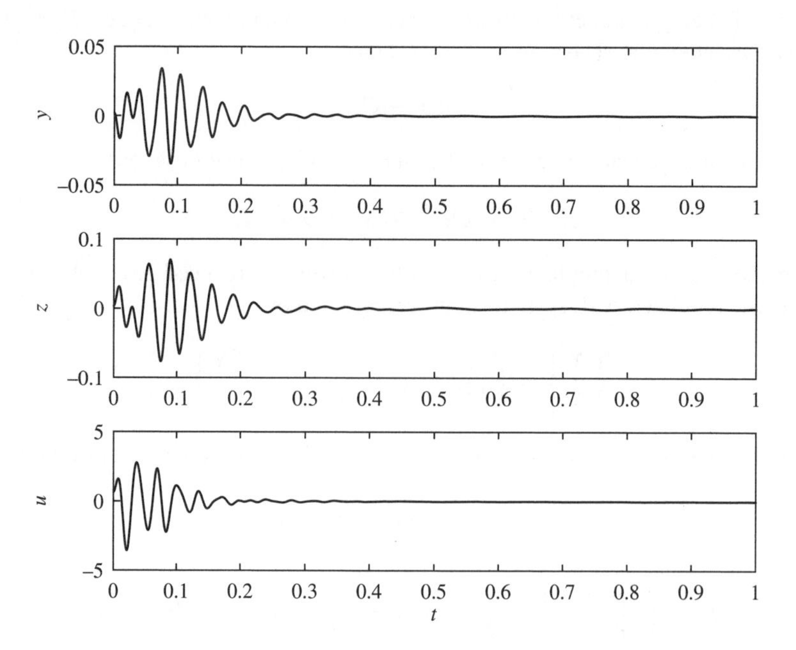

Figure 10.19 Simulated initial response of normal acceleration, y, tip displacement, z, and control torque input, u, of a model reference adaptive flutter-suppression system for the modified DAST-ARW1 wing in the presence of constant random perturbations in the lag parameters at a supercritical flight condition

altitude, 7.6 km, and flight Mach number, 0.95) with selected regulator and observer gains is now to be rendered adaptive with respect to random static variations in the two lag parameters, b_1, b_2, as well as a diagonal perturbation, Λ, in the matrix $\bar{B} = B_0\Lambda$, with $B_0 = B_m$. This is achieved by choosing $P_1 = 10^{-6}I, R_1 = I, P_2 = I, R_2 = I$ for the overall ASE system of order 112, comprising 28 aeroelastic states, $x(t)$, 28 observer states, $\hat{x}$, and a controller parameters vector, $\Theta = (K, L^T)^T$ of order 56. When the random perturbation in the lag parameters is assumed to be uniformly distributed with a standard deviation of 10^{-5} and an initial tip-displacement (the sixth state variable) $z(0) = 0.01$ m is applied, the system's response is plotted in Figs. 10.19 and 10.20. The dimensionless normal acceleration, $y(t)$, and the tip displacement, $z(t)$, show a settled closed-loop response, and the required control torque, $u(t)$, is within reasonable limits (Fig. 10.19). Owing to their large magnitudes, the regulator gains, K, are hardly changed by the adaptation law (e.g. Fig. 10.20). However, the observer gains, L, show a slight variation (Fig. 10.21) that converges to a steady state. The scheme can handle larger variations in the lag parameters, but would take a longer computational time to simulate and is not being considered here.

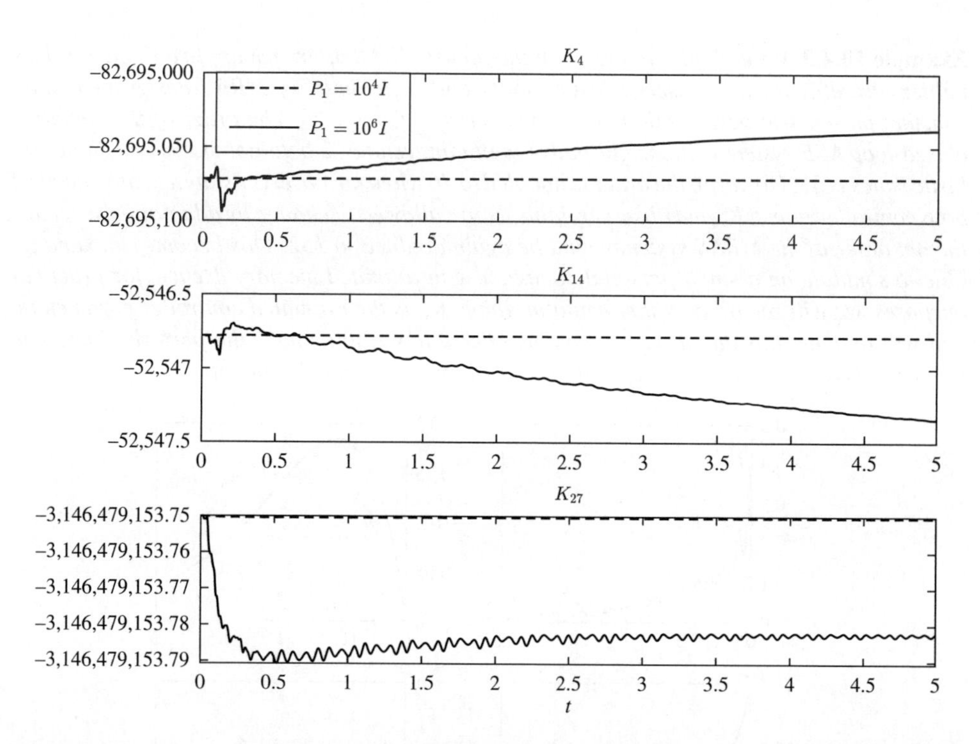

Figure 10.20 Simulated variation in the selected regulator gains, K, of a model reference adaptive flutter-suppression system for the modified DAST-ARW1 wing in the presence of constant random perturbations in the lag parameters at a supercritical flight condition

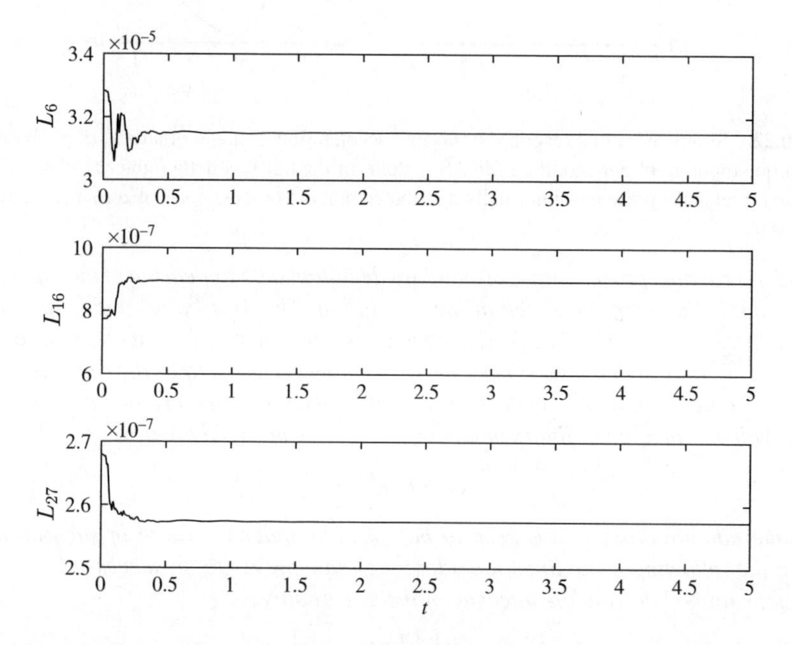

Figure 10.21 Simulated variation in the selected observer gains, L, of a model reference adaptive flutter-suppression system for the modified DAST-ARW1 wing in the presence of constant random perturbations in the lag parameters at a supercritical flight condition

Example 10.4.3 *We end this chapter with the robust MRAS approach applied to the tail-less fighter aircraft, whose aeroservoelastic model and a preliminary MRAS design to handle constant parametric perturbations was considered in Chapter 8. The order of the complete closed-loop ASE system with 2 rigid-body longitudinal states, 2 actuator states and 34 aeroelastic states is 38. However, the order is doubled to 76 when the observer states,* $\hat{x}$*, are added. If both controller gains, K, and observer gains, L, are allowed to change by adaptation schemes, the net order of the MRAS system would be again doubled to 152. This becomes unmanageable to simulate on a small personal computer with a limited memory. Hence, for practical purposes, we will regard only the regulator gains,* $\hat{K}$*, as the estimated controller parameters, and will assume that the observer remains adequately stable under the perturbations. The basis of this assumption is the separation (certainty equivalence) principle.*

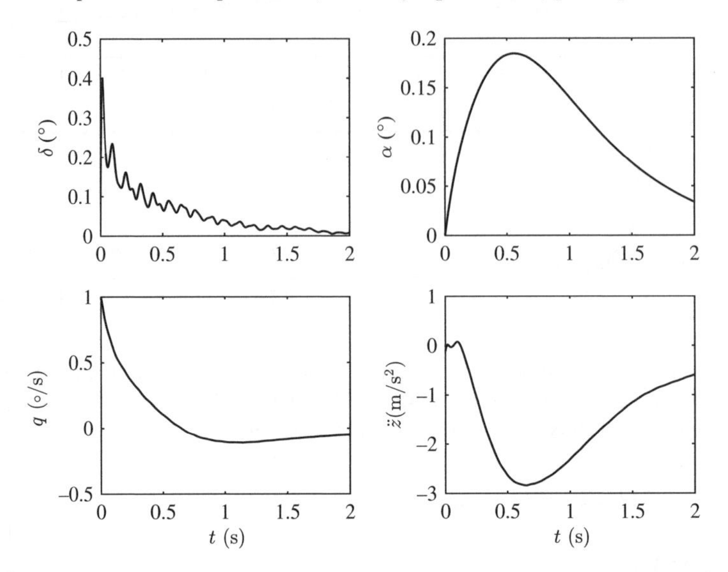

Figure 10.22 Simulated initial response of normal acceleration, $\ddot{z}$, angle of attack, α, pitch rate, q, and control torque input, u, of a σ-modified MRAS system for the tail-less delta fighter at $U = 306$ m/s and standard sea level in the presence of normally distributed random process noise of zero mean and standard deviation 0.05

Consider a random process noise of standard deviation 0.05 appearing in the input channel, which is updated at every time step of the simulation. This is a large random perturbation, which can arise because of such effects as severe turbulent gusts, periodically separated flows, and unsteady normal shock waves. The reference model is chosen to be the same as in Chapter 8 with fixed values of K and L. With $\Theta^T = \hat{K}$ *and L taken to be a constant vector at its model reference value. When the ordinary model reference scheme of Chapter 8,*

$$\dot{\hat{\Theta}} = Rxe^T PB, \tag{10.92}$$

is used, the adaptation system is seen to be unstable under the noise of such an intensity. However, the following σ*-modification adaptation law proves to be successful in stabilizing the system, without knowing the intensity of the disturbance:*

$$\dot{\hat{\Theta}} = R\left[xe^T PB - \sigma\hat{\Theta}\right], \tag{10.93}$$

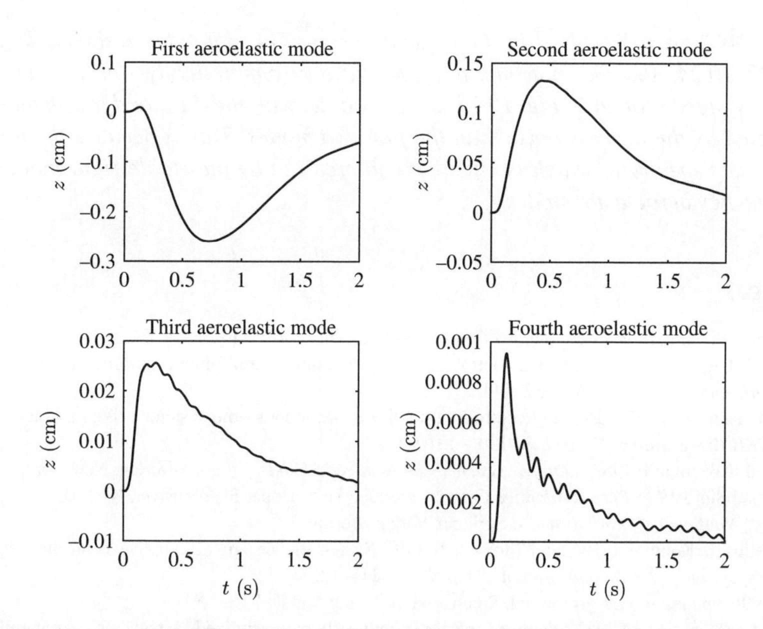

Figure 10.23 Simulated initial response of the vertical deflection, z, produced by the first four aeroelastic modes of a σ-modified MRAS system for the tail-less delta fighter at $U = 306$ m/s and standard sea level in the presence of normally distributed random process noise of zero mean and standard deviation 0.05

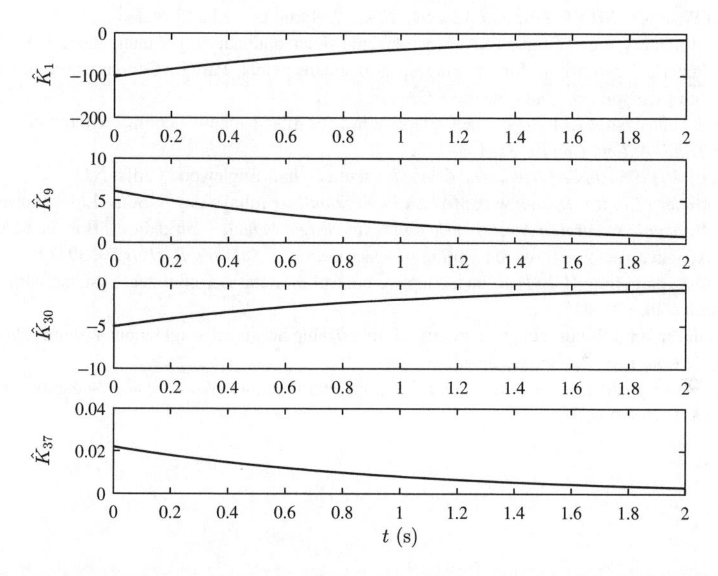

Figure 10.24 Simulated variation in the selected regulator gains, K, of a σ-modified MRAS system for the tail-less delta fighter at $U = 306$ m/s and standard sea level in the presence of normally distributed random process noise of zero mean and standard deviation 0.05

where $P = 10^{-9}I, \sigma = 1$ is used in the present design. The simulated response is plotted in Figs. 10.22–10.24, showing not only a UUB, but an asymptotically stable behaviour. The aeroelastic system, plotted in Fig. 10.23, is also stable, with the third and fourth modes being more affected by the process noise than the first two modes. This is because the noise has a high-frequency spectrum, which is effectively filtered out by the stable rigid modes and the lower-frequency aeroelastic modes.

References

Aström KJ and Wittenmark B 1995 *Adaptive Control*, 2nd ed. Addison-Wesley, New York.

Chang Y-C 2001 An adaptive H_∞ tracking control for a class of nonlinear multi-input-multi-output (MIMO) systems. *IEEE Trans. Autom. Control.* **46**, 1432–1437.

Chen F-C and Liu C-C 1994 Adaptively controlling nonlinear continuous-time systems using multilayer neural networks. *IEEE Trans. Autom. Control.* **39** 1306–1310.

Ioannou P and Kokotovic P 1983 *Adaptive Systems with Reduced Models*. Springer-Verlag, New York.

Ioannou PA and Sun J 1996 *Robust Adaptive Control*. Prentice-Hall, Upper Saddle River, NJ.

Isidori A 1989 *Nonlinear Control Systems*. Springer-Verlag, Berlin.

Kanellakopoulos I, Kokotovic PV, and Morse AS 1991 Systematic design of adaptive controllers for feedback linearizable systems. *IEEE Trans. Autom. Control* **36**, 1241–1253.

Khalil HK 1996 *Nonlinear Systems*, 3rd ed. Prentice-Hall, Upper Saddle River, NJ.

Kojic A and Annaswamy AM 2002 Adaptive control of nonlinearly parameterized systems with a triangular structure. *Automatica* **38**, 115–123.

Kokotovic PV 1991 *Foundations of Adaptive Control*. Springer-Verlag, Berlin.

Kokotovic PV and Arcak M 2001 Constructive nonlinear control: a historical perspective. *Automatica* **37**, 637–662.

Krstic M, Ioannis K, and Kokotovic P 1995 *Nonlinear and Adaptive Control Design*. John Wiley & Sons, Inc., New York.

Lavretsky E and Wise KA 2013 *Robust and Adaptive Control*. Springer-Verlag, London.

Lee BHK 1990 Oscillatory shock motion caused by transonic shock boundary-layer interaction. *AIAA J.* **28**, 942–944.

Misra P 1998 Numerical algorithms for squaring-up non-square plants, Part II: General case. *Proceedings of the American Control Conference*, San Fransisco, CA.

Narendra KS and Annaswamy AM 1987 A new adaptive law for robust adaptive control without persistency of excitation. *IEEE Trans. Autom. Control* **32**, 134–145.

Slotine JJE and Li W 1995 *Applied Nonlinear Control*. Prentice-Hall, Englewood Cliffs, NJ.

Tewari A 2002 *Modern Control Design with MATLAB and Simulink*. John Wiley & Sons, Ltd, Chichester.

Tewari A 2011 *Automatic Control of Atmospheric and Space Flight Vehicles*. Birkhäuser, Boston, MA.

Tewari A 2013 Adaptive vectored thrust deorbiting of space debris. *J. Spacecr. Rockets* **50**, 394–401.

Wang X-H, Su C-Y, and Hong H 2004 Robust adaptive control of a class of nonlinear systems with unknown dead zone. *Automatica* **40**, 407–413.

Xu H and Ioannou PA 2002 Robust adaptive control of linearizable nonlinear single input systems. *Proceedings 2002 XV IFAC World Congress*, Barcelona, Spain.

Yu S-H and Annaswamy AM 1997 Adaptive control of nonlinear dynamic systems using ?-adaptive neural networks. *Automatica* **33**, 1975–1995.

11

Adaptive Transonic Aeroservoelasticity

The transonic flight regime can be a hazardous flight condition. Many of the catastrophic aeroelastic instabilities occur at transonic speeds. These include a dip in the flutter dynamic pressure, control surface buzz, buffet and limit-cycle oscillations (LCOs), all caused by the presence of normal shock waves on the wing and the tails. High-speed aircraft spend a majority of their design life flying near the transonic regime. They cruise at speeds just below the sonic speed, either for an ideal combination of speed and fuel economy or for better manoeuvrability in the high-subsonic regime where the rate of turn is maximized. These airplanes are designed to fly just below the critical speed at which the flow first becomes sonic at any point on the aircraft. However, this design condition is often crossed inadvertently, and the shock waves are generally the result. An observant passenger with a window seat over the wing can sometimes see shock waves dancing on the upper surface in the cruise condition, either when flying through clouds or when excessive moisture is present in air. Why is a shock wave such an important factor in aircraft design? The answer to this question has two parts: (i) shock waves are pressure discontinuities that cause a large and sudden increase in drag (called wave drag), which can decrease the fuel economy by as much as a factor of 2. (ii) Shock waves are also indicators of a sudden onset of nonlinear and unsteady aerodynamic conditions prevailing in the transonic regime, which are capable of exciting large pressure fluctuations and creating an unstable aeroelastic coupling. For an ASE engineer, the latter factor is more important, because it represents a potential source of instability.

Controlling transonic aeroelasticity is a formidable task, not only because of the difficulty in modelling the nonlinear aerodynamic behaviour but also due to an inherent loss of controllability caused by the mixed subsonic–supersonic flow prevailing on the lifting surfaces. A normal shock wave is the boundary line between such characteristically different flows, which in the steady state is also the mathematical border between elliptic and hyperbolic types of partial differential equations (see Chapter 3). Since pressure disturbances cannot travel upstream of a normal shock wave, there is no possibility of a trailing-edge control surface affecting the supersonic flow ahead of the shock barrier. Hence, controllability of the flow is limited to the subsonic region extending from the control surface to the shock wave. This region vanishes completely when the shock wave moves downstream and sits astride the control-surface

Adaptive Aeroservoelastic Control, First Edition. Ashish Tewari.

hinge line. Consequently, the only hope of controlling the transonic ASE plant is through an aeroelastic coupling provided by the bending and twisting deformation produced by the movement of the control surface. In a typical-section model, this translates into controlling the pitch and plunge of the elastic axis by control-surface deflection. The transonic plant model has many peculiarities that are absent at both subsonic and supersonic speeds. These include an inherent coupling of the steady and unsteady aerodynamics to such an extent that a small change in the steady flow conditions can completely modify the unsteady behaviour. Consequently, a slight increase in the supercritical Mach number may decrease the control effectiveness by as much as an order of magnitude, thereby requiring a 10-fold increase of control deflection for maintaining a given equilibrium. When coupled with the flow separation caused by the shock waves, leading-edge vortices and large control movements, the highly uncertain and nonlinear transonic characteristics border on chaotic behaviour, which is nearly impossible to model accurately. Therefore, designing an adequate control system for transonic ASE applications has historically proved to be challenging. This is an important fact to keep in mind while reading the present chapter, whose main objective is to investigate how an adaptive control system can be designed to compensate for the lack of modelling accuracy in the transonic plant.

Advanced stability analysis by the methods presented in the later part of Chapter 6 becomes necessary when dealing with uncertain nonlinear systems. Passivity analysis and derivation of Lyapunov functions and Lyapunov-like functions by the Kalman–Yakubovich lemma is the basis of nonlinear adaptive control (Chapter 10) and is the thrust of transonic ASE techniques used in this chapter. Adaptive integral backstepping (Chapter 9) and describing functions for oscillatory nonlinear analysis (Chapter 7) are also applied to control transonic LCOs. As for a typical ASE plant, the transonic aeroelastic plant is also time invariant (autonomous), therefore the design procedure in all cases is restricted to closed-loop stabilization (set-point regulation). By guaranteeing nonlinear damping (or dissipation) due to adaptive control in various ways, asymptotic stability is ensured. In such an approach, it is not necessary to know the exact plant model, but only that the plant is controllable (or at least stabilizable).

11.1 Steady Transonic Flow Characteristics

Transonic flow is dominated by large pressure variations caused by mixed subsonic–supersonic local regions, usually with shock waves embedded in the flowfield. A shock wave is a non-isentropic, pressure discontinuity that appears when the local flow becomes supersonic. Downstream of a shock wave, the flow speed decreases and the static pressure rises abruptly. There are two basic types of shock waves: normal shock and oblique shock. A normal shock is accompanied by decrease of flow speed from supersonic to subsonic and a much larger pressure rise compared to an oblique shock of the same upstream Mach number. Which of the two waves is actually encountered depends on the local flow geometry, wherein the continuity of the flow demands an appropriate shock wave angle (either normal or inclined to the upstream flow). Usually thin and sharp-edged (or pointed) bodies flying at supersonic speeds encounter an oblique shock wave, whereas a blunt-nosed or thick object at the same speed would have a normal shock appearing in front of it. The angle presented by the shock wave to oncoming flow is inversely proportional to the flow Mach number for any given object, while its strength (pressure rise) increases with the Mach number. Since transonic flow on an aircraft wing contains a mixture of subsonic and supersonic regions on curved surfaces, the flow continuity requires that any shock waves appearing on a smoothly contoured airfoil must be essentially curved, with the local inclination varying with a changing flow deflection. Whenever flow acceleration is required in a locally supersonic region (such as over a convex surface), the associated phenomenon is not a discontinuity, but a smooth increase of flow speed via a series of isentropic Mach waves called an expansion fan. The reader may consult a textbook on gas dynamics (Shapiro 1958) for a detailed description of compressible flows, shock waves and expansion fans.

In order to understand the physical flow characteristics on a wing operating at the transonic speeds, consider an airfoil equipped with a trailing-edge flap at a high-subsonic, steady freestream Mach number, M_∞, and a constant angle of attack. In the subcritical case, the flow everywhere remains subsonic and the analysis can be carried out by a linear aerodynamic theory. However, as the Mach number is increased, the critical condition is reached when the local flow speed approaches the speed of sound at a particular point on the wing (usually on the upper surface). A further increase of M_∞ produces supercritical flow where local supersonic regions appear on the airfoil (Fig. 11.1). The supersonic bubbles are terminated by normal shock waves, which may have unsymmetrical locations on the upper and lower surfaces depending on the angle of attack and the flap deflection. For a positively cambered airfoil at a small angle of attack, the local flow acceleration is larger on the upper surface, resulting in a larger supersonic region there, along with a higher local Mach number upstream of the shock (thus a stronger shock wave), when compared to that on the lower surface (Fig. 11.1(a)). The supersonic bubbles are located near the maximum thickness point and can change in position and extent with changes in the angle of attack, flap deflection and freestream Mach number. When either the angle of attack is increased or the flap is deflected downwards, there is an increase of flow acceleration on the upper surface, resulting in the upper supersonic region becoming larger and moving further downstream (Fig. 11.1(b)). In the extreme case, the upper-surface supersonic region may extend to beyond the flap hinge line, thereby creating an expansion wave pattern on the convex corner, as indicated in Fig. 11.1(b). The stronger normal shock on the upper surface can lead to an appreciable thickening and even separation of the boundary layer on the flap, which causes a divergence of pressure distribution at the trailing edge. On the lower surface, the decreased flow speed results in a smaller, more upstream supersonic bubble terminated by a weaker shock wave. In an extreme case, the lower-surface supersonic bubble can entirely vanish. If the freestream Mach number is sufficiently large, the subsonic flow acceleration caused by the concave corner at the lower-surface hinge line can produce a very small supersonic region astride the hinge line followed by a shock wave.

An upward flap deflection (or a reduction in the angle of attack) causes the associated supersonic region to move upstream and an oblique shock wave on the flap hinge line (Fig. 11.1(c)). Of course, the oblique shock is combined with a normal shock terminating the supersonic bubble. Such a shock formation, termed the *lambda shock* due to its shape, could also cause localized flow separation near the trailing edge. The opposite tendency is seen on the lower surface, where the supersonic flow extends further downstream, and in an extreme case could cross the flap hinge line, resulting in an expansion fan on the convex corner, as well as a normal shock at the downstream edge of the supersonic bubble. Clearly, the flow conditions encountered on the flap depend on the freestream conditions (Mach number, angle of attack) as well as on the flap deflection, and are very important if the flap is to be used as a control device.

The corresponding pressure distributions typical for each of the supercritical flow conditions encountered in Fig. 11.1 are shown in Fig. 11.2. The abrupt increase in pressure coefficient

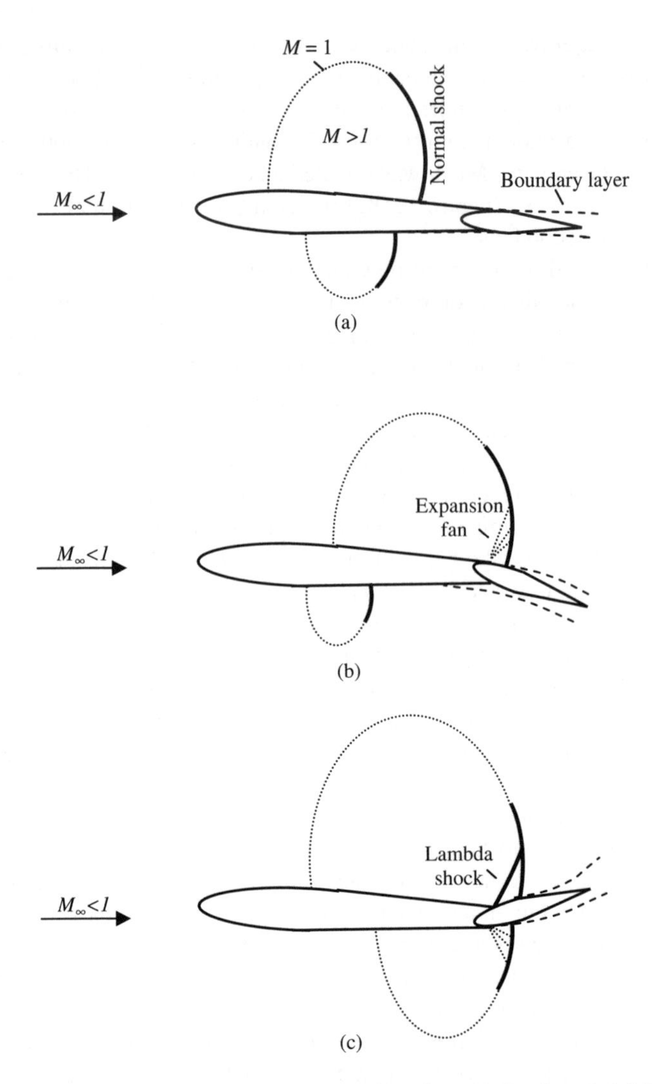

Figure 11.1 Supercritical flow patterns on an airfoil with a flap. (a) Cambered airfoil at small angle of attack. (b), (c) Cambered airfoil at moderate angle of attack with flap deflection

after a shock wave and the smooth decline in pressure in an expansion fan are quite clear. The hinge line of the flap also experiences a sharp change in the pressure in the subsonic flow condition (lower surface, case (b)). Shock-induced separation at the trailing edge is responsible for the flow not leaving the airfoil in the same direction on the two surfaces, which means the Kutta condition is not satisfied. While the pressure distribution is qualitatively changed for each flap deflection, the effect on the integrated pressure and the overall lift, pitching moment and hinge moment is expected to be only quantitative, provided the extent of flow separation (i.e. flap deflection) is small. This qualitative examination of steady flow characteristics in the transonic case offers an insight into what can be expected in the unsteady case, where both freestream conditions and flap deflection are time dependent.

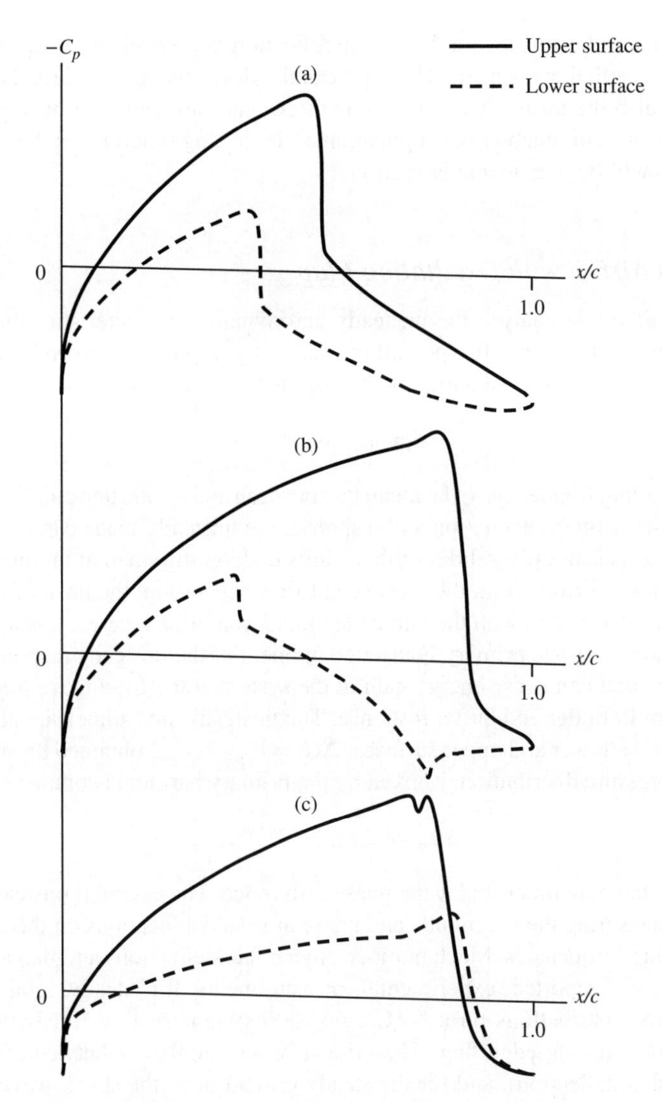

Figure 11.2 Supercritical pressure distributions corresponding to cases shown in Fig. 11.1. (a) Cambered airfoil at small angle of attack. (b), (c) Cambered airfoil at moderate angle of attack with flap deflection

11.2 Unsteady Transonic Flow Characteristics

When encountering unsteady flow conditions (such as due to wing vibration or a changing flap deflection), the flow on a wing becomes time dependent. In a transonic case, the unsteady effects can be nonlinear functions of the Mach number as well as the forcing frequency and amplitude. When large motion amplitudes are present, there may also be a significant dependence on Reynolds number due to a periodic separation and reattachment of the boundary layer. Consequently, unsteady flow analysis in the transonic case can quickly become a nightmare for an aeroelastician. Fortunately, a major simplification occurs when the approximation of small amplitude oscillations is made. Not only can the flow separation effects be neglected in such a case but also the unsteady aerodynamic forces and moments can be approximated to be simple harmonic. This enables the application of describing functions in the transonic ASE plant models, as will be seen in a later section.

11.2.1 Thin Airfoil with Oscillating Flap

In order to qualitatively analyse the unsteady aerodynamic characteristics, we begin with thin airfoils of the fighter aircraft type, either oscillating in pitch or forced by an oscillating trailing-edge flap. For the simple harmonic flap oscillation, we have

$$\delta = \delta_s + \delta_0 e^{i\omega t}, \tag{11.1}$$

where ω is the forcing frequency, δ_s the mean flap position and δ_0, the flap amplitude. If a linear relationship exists between excitation and response, the unsteady loads can also be assumed to be harmonic. This is usually valid for thin airfoils undergoing small amplitude oscillations. However, for thicker airfoils or in the presence of flow separation, the unsteady airloads will have a nonlinear relationship with the flap deflection. Even in such cases, it is a good approximation to consider only the primary Fourier component of the unsteady response, because it is the component that can cause energy gain in the system at the forcing frequency, which is the main concern in flutter and buffet response. The unsteady, non-dimensional pressure difference between the lower and upper surfaces, $\Delta C_p = C_{p\ell} - C_{pu}$, obtained by subtracting the steady (mean) pressure distribution, is given by the primary harmonic component as follows:

$$\Delta C_p = \mid \Delta C_p \mid e^{(i\omega t + \phi)}, \tag{11.2}$$

where $\mid \Delta C_p \mid$ is the magnitude and ϕ the phase difference. The essential unsteady analysis is carried out by measuring the magnitude and phase at selected locations on the airfoil surface for various forcing frequencies, Mach numbers, mean flap deflections and flap amplitudes.

Tijdeman (1977) reported experimental measurements for steady, quasi-steady and unsteady pressure distributions on a NACA 64A006 symmetrical airfoil section equipped with a 25% chord trailing-edge flap. The critical Mach number is between 0.82 and 0.85 (depending on flap deflection), and for the steady critical flow, the shock wave is located at 45% chord, which becomes stronger and moves downstream as the Mach number is increased. The shock wave reaches the hinge line at freestream Mach number of 0.92, slightly after which extensive flow separation takes place on the flap and a complicated flow pattern (alternating lambda-shock/expansion fan) is observed on the hinge line. Therefore, the flap's effectiveness as a control device is confined to Mach numbers less than 0.92. Figures 11.3–11.5 show unsteady pressure-difference plots for a fixed flap frequency of $\omega = 120$ Hz and flap deflection amplitude $\delta_0 = 1.0°$ about $\delta_s = 0$ and $\alpha = 0$ mean position, for a subcritical Mach number 0.8 and two supercritical cases ($M = 0.85, 0.90$). The experimental data for the plots is derived from Tijdeman's report (Tijdeman 1977). Since the mean flow is symmetrical, the upper

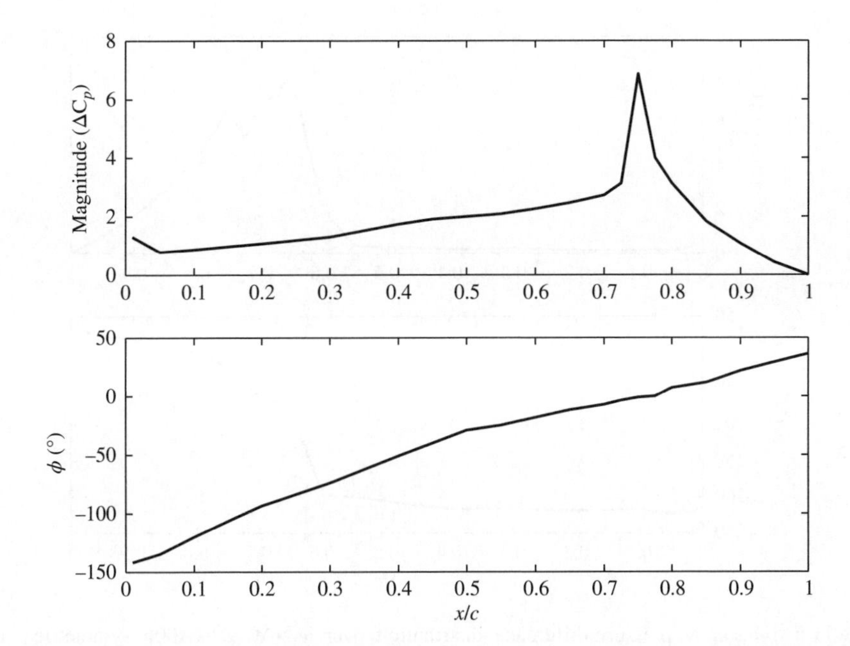

Figure 11.3 Unsteady pressure-difference distribution over a NACA 64A006 symmetrical airfoil with an oscillating flap, $\delta_0 = 1.0°$, $\omega = 120$ Hz, about the zero mean position for subcritical Mach number 0.8

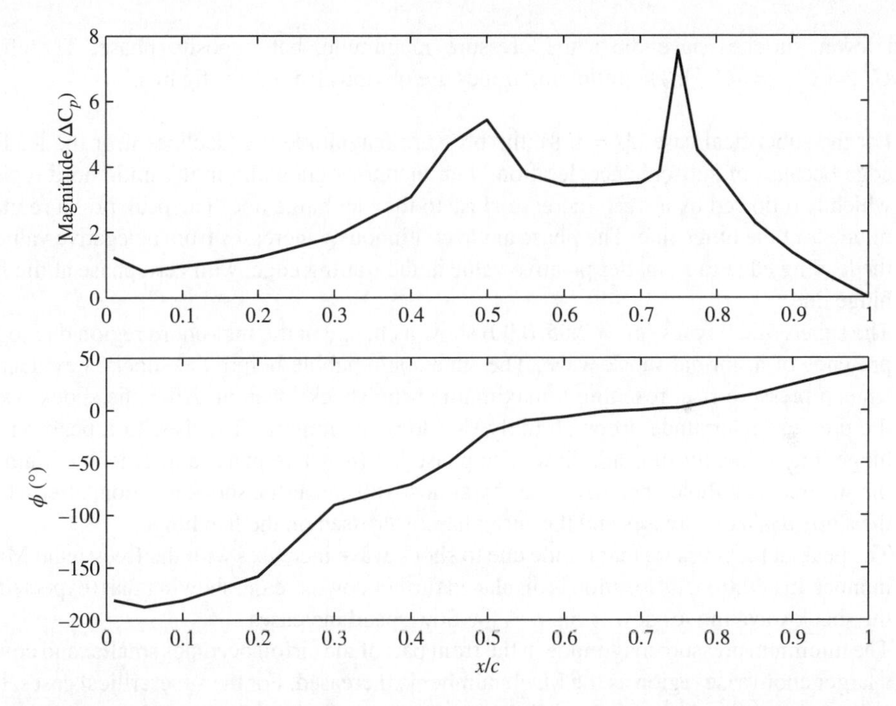

Figure 11.4 Unsteady pressure-difference distribution over a NACA 64A006 symmetrical airfoil with an oscillating flap, $\delta_0 = 1.0°$, $\omega = 120$ Hz, about the zero mean position for supercritical Mach number 0.85

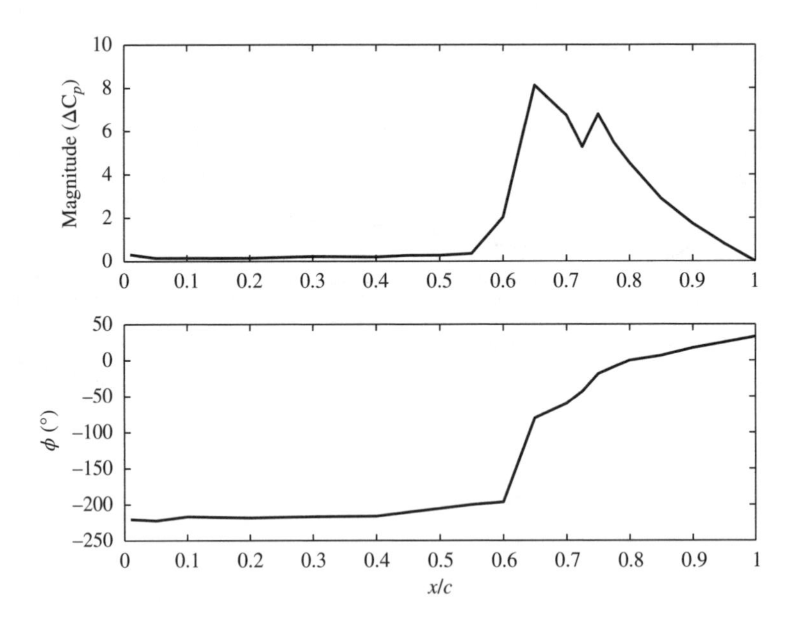

Figure 11.5 Unsteady pressure-difference distribution over a NACA 64A006 symmetrical airfoil with an oscillating flap, $\delta_0 = 1.0°$, $\omega = 120$ Hz, about the zero mean position for supercritical Mach number 0.9

and lower surfaces have the same pressure magnitude, but opposite phase. Therefore, $| \Delta C_p |=| C_{p\ell} |=| C_{pu} |$. The following trends are obvious from these figures:

1. For the subcritical case ($M = 0.8$), the pressure magnitude first declines near the leading edge because of subsonic acceleration, then increases gradually in the mid-chord region, which is followed by a steep increase close to the flap hinge line. The peak pressure magnitude is at the hinge line. The phase angle continuously increases from a negative value at the leading edge to a smaller positive value at the trailing edge, with zero phase at the flap hinge line.
2. The supercritical cases ($M = 0.85, 0.90$) show a change in the mid-chord region due to the presence of a normal shock wave. The supersonic bubble before the shock is evident as a steep pressure rise, reaching a maximum at the shock location. After the shock wave, the pressure magnitude drops abruptly to a local minimum, then rises to a peak on the hinge line in locally subsonic flow. The phase lag (negative phase angle) is maximum in the supersonic bubble, then decreases by almost 180°. near the shock location, after which the slope $\partial\phi/\partial x$ decreases and the zero phase is crossed on the flap hinge.
3. The peak in the pressure magnitude due to shock wave increases with the freestream Mach number. In addition, its location is displaced further downstream, showing that (expectedly) the shock wave moves downstream as the flow speed increases.
4. The minimum pressure magnitude in the front part of the airfoil becomes smaller and covers a larger chordwise region as the Mach number is increased. For the supercritical cases, this indicates an expanding supersonic bubble with the increase of flow speed.

5. The maximum phase lag before the shock persists for a larger chordwise extent as the Mach number is increased. This again indicates an expanding supersonic bubble with increasing flow speed.
6. The peak in pressure magnitude and the near 180°. change in the phase close to the shock location indicate that there may be a natural mode associated with the shock wave in the unsteady transfer function. However, this can be confirmed only by a frequency response Bode plot of $\Delta C_p/\delta$ (Chapter 1). Unfortunately, data is provided at only a few frequency points, which does not enable a systematic frequency response analysis for the pressure distributions, which could have clearly identified the unsteady aerodynamic natural modes due to shock wave motion.

Tijdeman (1977) compared the unsteady pressures at flap frequencies of 30, 90 and 120 Hz. An ever-increasing peak magnitude with increasing frequency for the slightly supercritical case of $M = 0.85$ is observed in Fig. 11.6. For the higher Mach number $M = 0.90$ (Fig. 11.7) when the shock wave reaches the hinge line, there is no effect of the flap on the pressure distribution upstream of the shock wave, and a flat pressure magnitude and phase are observed in the front part of the airfoil. However, the aerodynamic natural frequency due to unsteady shock wave is undetermined, which could have been explored by taking more frequency points in the spectrum between 90 and 120 Hz until a peak maxima could be obtained, indicating a natural mode. Such a frequency response analysis has not been reported even elsewhere in

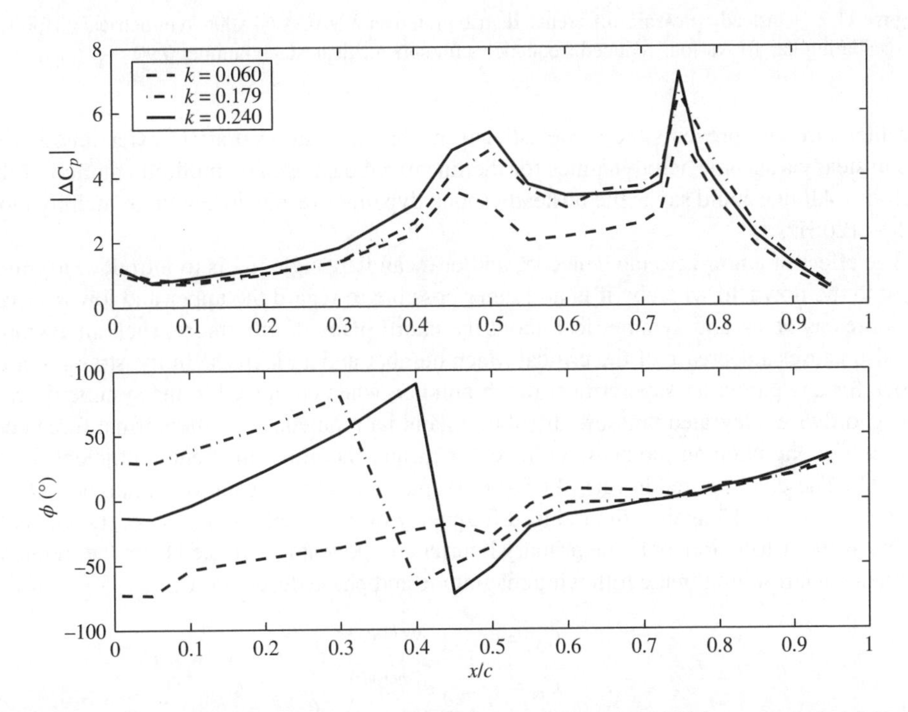

Figure 11.6 Unsteady pressure-difference distribution over a NACA 64A006 symmetrical airfoil with an oscillating flap for various reduced frequencies for supercritical Mach number 0.85

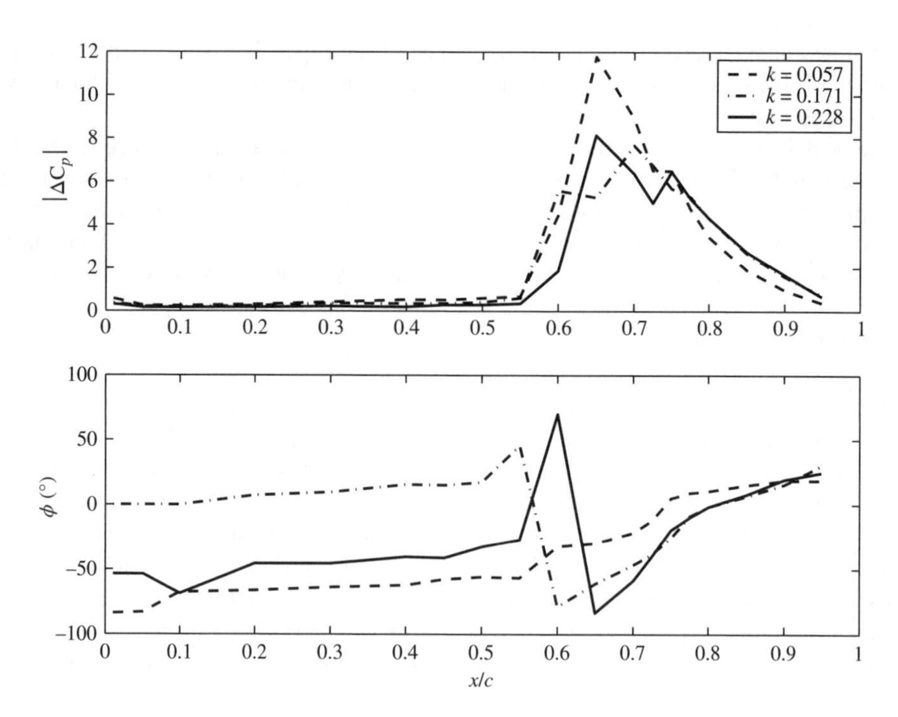

Figure 11.7 Unsteady pressure-difference distribution over a NACA 64A006 symmetrical airfoil with an oscillating flap for various reduced frequencies for supercritical Mach number 0.90

the literature and prevents the aeroelastician in obtaining an accurate transfer function for the unsteady transonic aerodynamics for the linearized case (small amplitude motion of thin airfoils). All one could say is the unsteady shock dynamics results in a natural pitching mode below 120 Hz.

The effect of a non-zero incidence, α, and/or mean flap angle, δ_s, is to introduce asymmetry into the flowfield whereby it is no longer possible to regard the upper and lower surface pressures as being anti-symmetrical about the chord plane. Furthermore, such an asymmetry also causes a decrease of the critical Mach number and an increase in the strength of the shock for any particular supercritical Mach number, when compared to the symmetric case. The chordwise integrated pressure distribution and its moments give the normal force coefficient, C_N, the pitching moment coefficient, C_m, and the hinge moment coefficient for the flap, C_h. These are plotted in Figs. 11.8–11.10 against the freestream Mach number, M, and in Figs. 11.11–11.13 against the reduced frequency of flap oscillation, $k = \omega c/(2U)$. As for the pressure distribution, only the primary Fourier component is extracted from the frequency domain data, resulting in the following magnitude and phase descriptions:

$$C_N = | C_N | e^{(i\omega t + \phi_N)}$$
$$C_m = | C_m | e^{(i\omega t + \phi_m)}$$
$$C_h = | C_h | e^{(i\omega t + \phi_h)}. \tag{11.3}$$

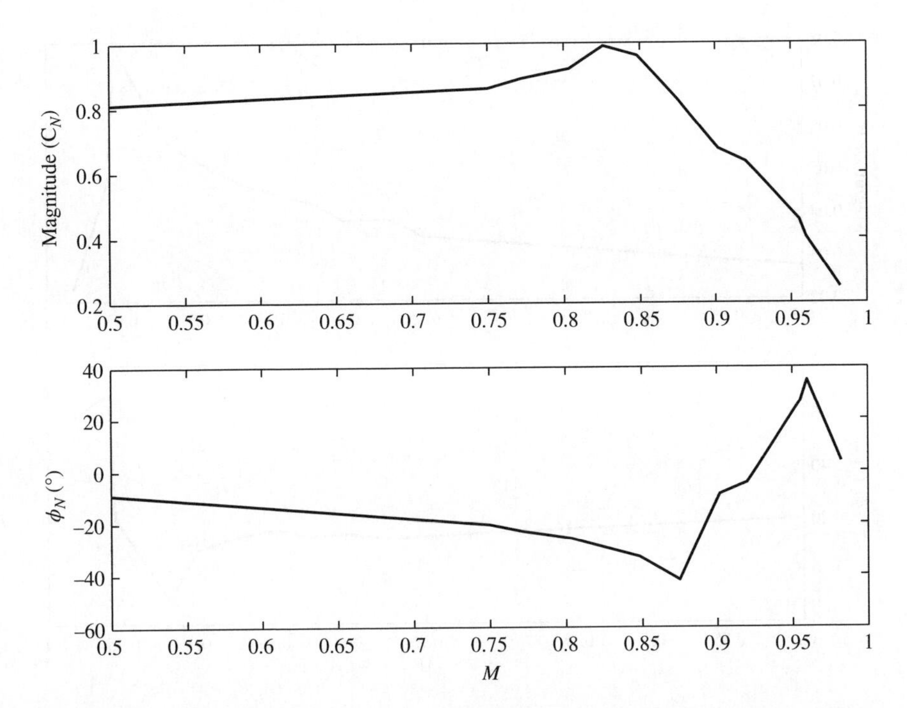

Figure 11.8 Unsteady normal force coefficient of a NACA 64A006 symmetrical airfoil with an oscillating flap, $\delta_0 = 1.0°$, $\omega = 120$ Hz, about zero mean position, as a function of freestream Mach number

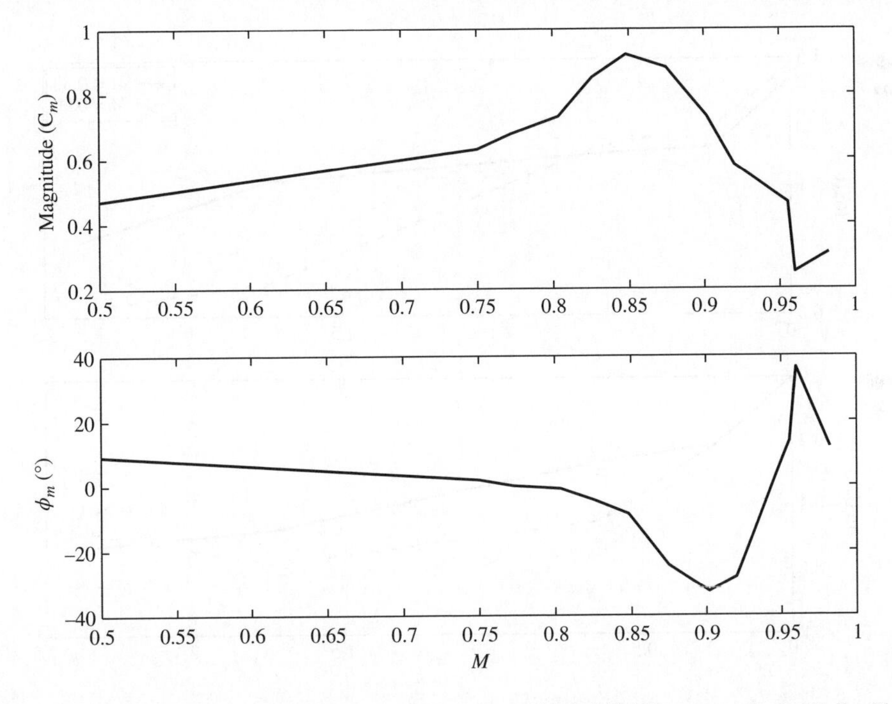

Figure 11.9 Unsteady pitching moment coefficient of a NACA 64A006 symmetrical airfoil with an oscillating flap, $\delta_0 = 1.0°$, $\omega = 120$ Hz, about zero mean position, as a function of freestream Mach number

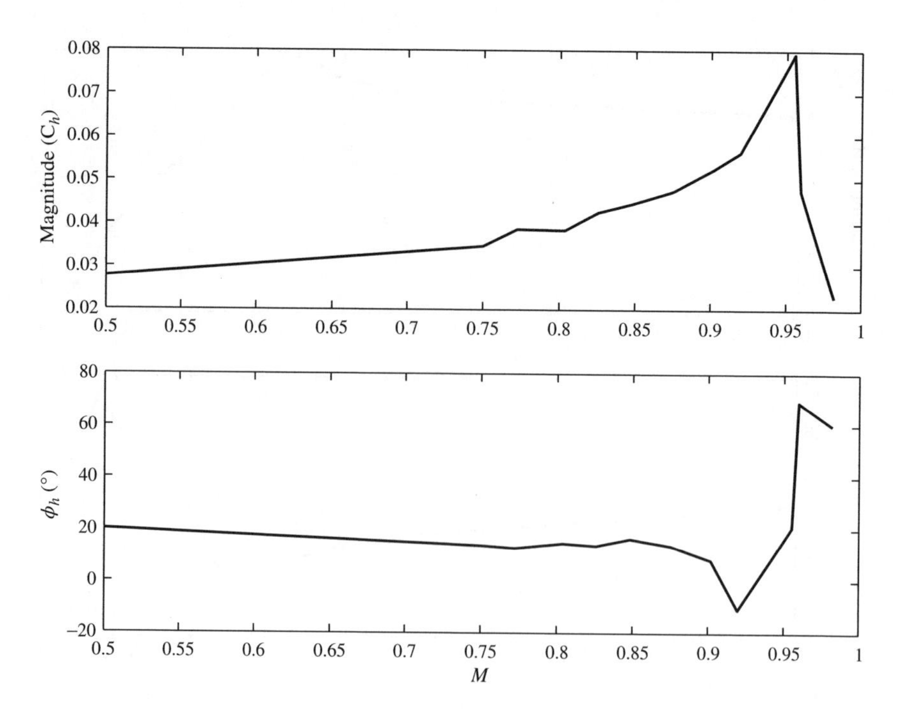

Figure 11.10 Unsteady hinge moment coefficient of a NACA 64A006 symmetrical airfoil with an oscillating flap, $\delta_0 = 1.0°$, $\omega = 120$ Hz, about zero mean position, as a function of freestream Mach number

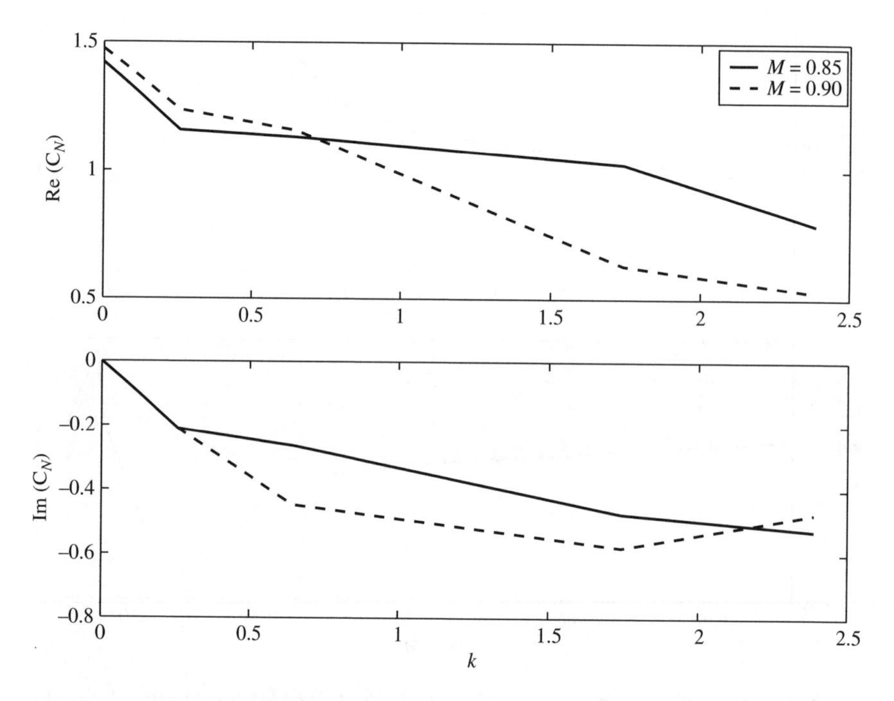

Figure 11.11 Unsteady normal force coefficient of a NACA 64A006 symmetrical airfoil with an oscillating flap, $\delta_0 = 1.0°$, $\omega = 120$ Hz, about zero mean position, as a function of reduced frequency

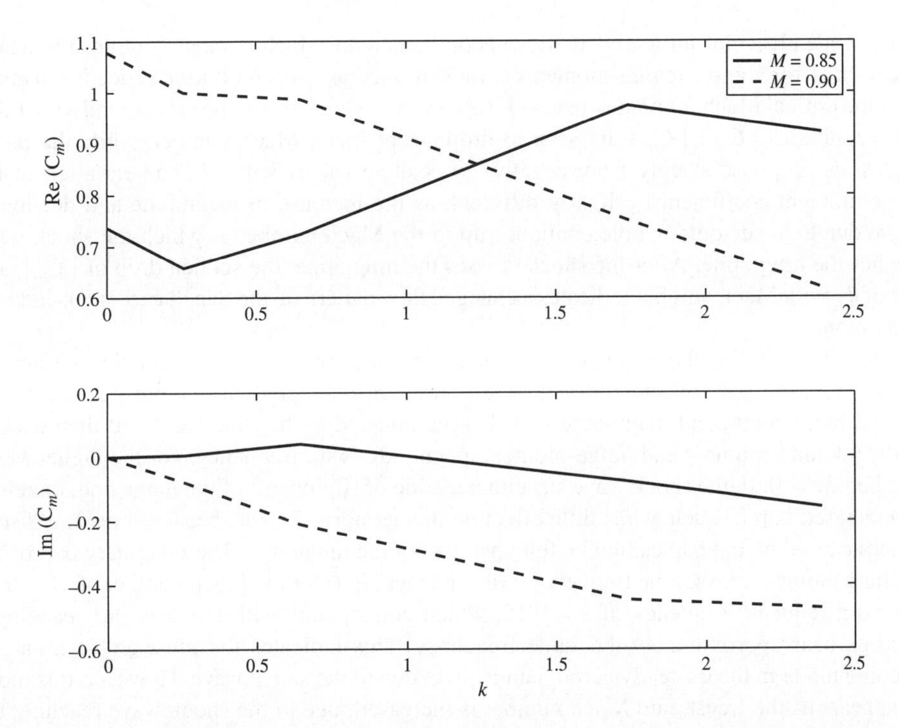

Figure 11.12 Unsteady pitching moment coefficient of a NACA 64A006 symmetrical airfoil with an oscillating flap, $\delta_0 = 1.0°$, $\omega = 120$ Hz, about zero mean position, as a function of reduced frequency

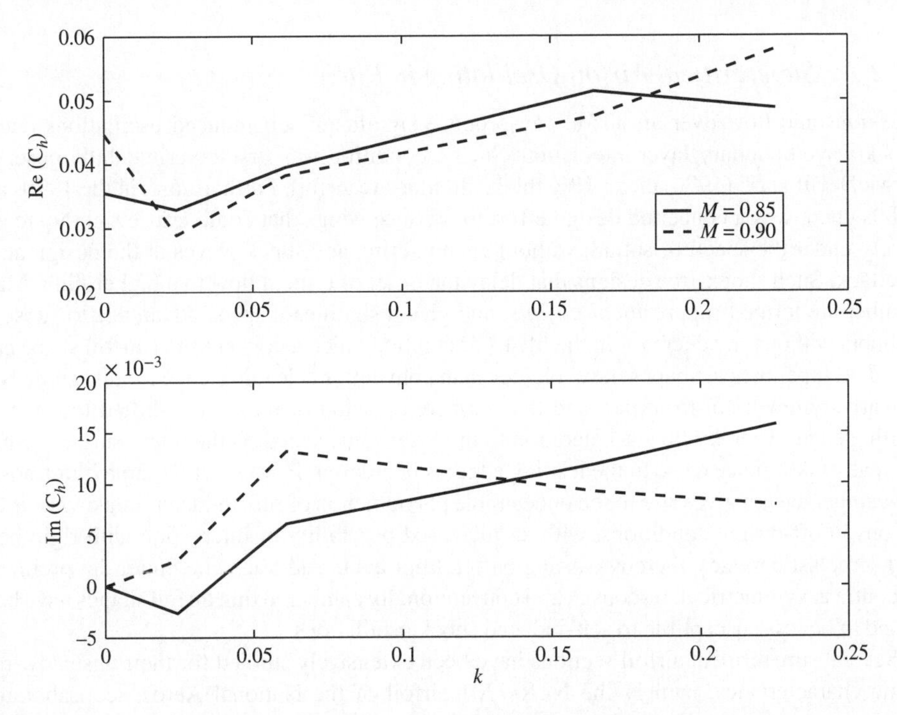

Figure 11.13 Unsteady hinge moment coefficient of a NACA 64A006 symmetrical airfoil with an oscillating flap, $\delta_0 = 1.0°$, $\omega = 120$ Hz, about zero mean position, as a function of reduced frequency

The dependence of unsteady coefficients on freestream Mach number is quite significant. The normal-force and pitching-moment coefficients increase in magnitude and decline in phase until the critical Mach number is reached. As soon as a shock wave appears around $M = 0.85$, the magnitudes $| C_N |, | C_m |$ are seen to drop steeply with Mach number, while the phase angles, ϕ_N, ϕ_m, rise steeply from negative to positive values with M. The variation of the hinge-moment coefficient is slightly different, as the increase in magnitude and decline in phase due to supersonic bubble continues up to the Mach number at which the shock wave reaches the hinge line. After the shock crosses the hinge line, the sudden drop of $| C_h |$, and rise of ϕ_h with Mach number indicates a changed flow pattern on the flap due to shock-induced separation.

The effect of the flap's reduced frequency on the unsteady aerodynamic coefficients (Figs. 11.11–11.13) for supercritical Mach numbers is evident in a rapid decline of the normal-force coefficient magnitude with k, accompanied by first increase, and then decline of the pitching-moment and hinge-moment magnitudes with frequency. For the higher Mach number ($M = 0.9$), the shock wave sits either astride or slightly aft of the hinge line, therefore an increased flap frequency has little effect on the net normal force, because pressure disturbances caused by the flap cannot be felt upstream of the hinge line. The imaginary part of the pitching moment crosses the frequency axis (changes sign) for the Mach number of $M = 0.85$ around a reduced frequency of $k = 0.15$, which corresponds with the real part reaching a maxima near approximately the same frequency. This indicates the presence of a natural pitching mode in the unsteady aerodynamic plant due to the shock wave. However, this mode disappears as the freestream Mach number is increased, due to the shock wave reaching the hinge line. Unfortunately, the exact natural frequency of this shock-pitch-coupled mode cannot be identified because of a paucity of data points at low frequencies (as remarked earlier).

11.2.2 Supercritical Airfoil Oscillating in Pitch

The transonic flow over an airfoil can produce significant self-induced oscillations due to shock-wave/boundary-layer interaction. Such a condition was first experimentally observed by McDevitt *et al.* (1976) on an 18% thick, circular-arc airfoil. Furthermore, in the 1970s and 1980s, there was a concerted design effort to produce wings that could cruise right up to and barely under the speed of sound, without encountering any shock waves at the design angle of attack. Such shock-free designs that delay the onset of critical flow to a higher flight Mach number are termed *supercritical airfoils*, and give a significant speed advantage to subsonic airliners without any decrease in the lift-to-drag ratio, which a conventional airfoil shape cannot. The supercritical shapes have a larger than conventional leading-edge radius, along with a nearly symmetrical front part and the cambered portion of the airfoil shifted towards the trailing edge, whereby flow acceleration to the local sonic speed on the upper surface (which normally takes place close to the leading edge) is decreased. However, this same blunt-nosed, aft-camber design gives rise to the undesirable phenomenon of self-induced shock wave oscillations in off-design conditions, with an increased possibility of interaction with rigid body and aeroelastic modes, thereby causing buffet, limit-cycle and transonic flutter dip problems. In contrast, symmetrical, bi-convex and conventionally cambered thin airfoil shapes have been found to be less susceptible to self-induced shock oscillations.

Several supercritical airfoil sections have been extensively studied for their unsteady transonic characteristics, namely the NLR-7301 airfoil of the National Aerospace Laboratory

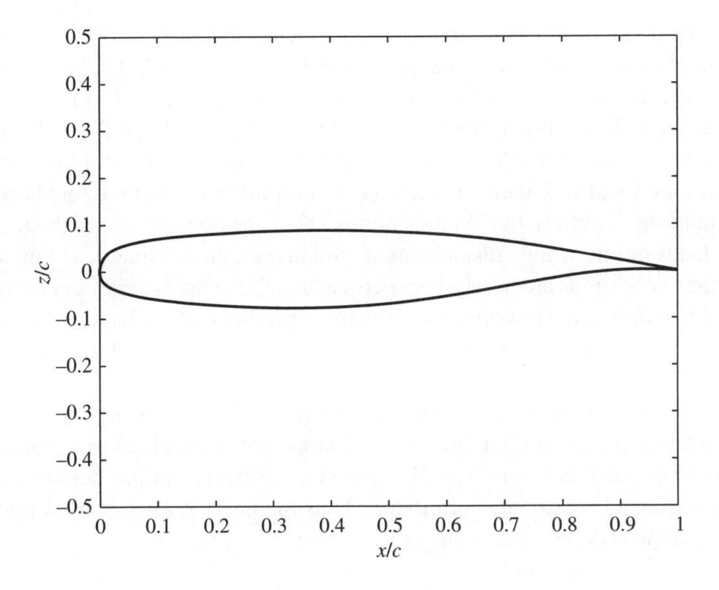

Figure 11.14 The NLR-7301 supercritical airfoil

(NLR), Amsterdam (Tijdeman 1977), the Bauer–Garabedian–Korn (BGK) No.1 airfoil of the National Aeronautical Establishment, (Ottawa Lee 1990) and the Royal Aeronautical Establishment (RAE) No. 2822 airfoil (Cook *et al.* 1979).

Consider a supercritical airfoil ((Fig. 11.14) with chord c at a steady freestream Mach number M_∞ with a mean angle of attack, α_s, mean lift coefficient, $C_{\ell s}$, and mean pitching moment about quarter chord point, C_{ms}. An oscillatory pitching motion with small amplitude, α_0, and frequency, ω, is now superimposed on the mean flow, such that the net angle of attack is given by

$$\alpha = \alpha_s + \alpha_0 \sin(\omega t). \tag{11.4}$$

The lift and pitching moment of an oscillating airfoil at transonic speeds depend on the oscillation frequency, ω, and amplitude, α_0, as well as the mean flow, α_s. The dependence of unsteady transonic flow on the mean flow conditions makes it very difficult to analyse, in contrast with moderate subsonic or supersonic flows where such dependence is absent, hence linear superposition of steady and unsteady flows is possible. Experimental investigations into transonic unsteady aerodynamics of oscillating airfoils have revealed the following basic characteristics (Tijdeman 1977):

1. The presence of nearly normal shock waves on the upper and lower surfaces of a wing causes a change in the magnitude and phase of the unsteady air loads from those that are predicted by linearized aerodynamic theories.
2. The shock wave effects on pressure distribution are nearly linear when the oscillatory motion is confined to small amplitude about a symmetrical equilibrium condition (e.g. symmetric airfoil with a zero angle of attack). Such a shock dynamics is termed Type A motion (Tijdeman 1977).

3. When the flight Mach number is reduced slightly below that of Type A motion, there is a significant nonlinear dependence of the shock strength on the shock-displacement amplitude. The nonlinear displacement of the normal shock wave can be small (Type B motion) for a moderate reduction in flight Mach number, but can become large (Type C motion) if the Mach number is reduced further (Tijdeman 1977). However, as the nonlinear effects of the shock wave are localized, their integrated effect on the aerodynamic forces and moments is often quite small even in the Type C motion. Thus the dependence of the unsteady aerodynamic loads on the wing's displacement remains essentially linear, as long as the mean (steady) flow conditions are nearly symmetrical and the wing's oscillatory motion is of relatively small amplitude. However, it is difficult to predict under what precise combinations of the mean flow conditions and motion amplitudes a particular airfoil can have nonlinear aerodynamics.
4. If the mean flow conditions are highly unsymmetrical (e.g. a wing operating at a large angle of attack), or the motion amplitude is large, the unsteady flow separation induced by stronger normal shock waves results in a nonlinear relationship between the air loads and the generalized coordinates of motion. Thus the unsteady behaviour in the nonlinear, transonic regime is dependent on the mean flow conditions.

Either strong shock waves or large displacements of weak normal shock waves can cause a nonlinear relationship between generalized aerodynamic forces and generalized coordinates. Since strong normal shocks are undesirable from drag considerations, the supercritical airfoils are designed to be shock-free (or with weak normal shocks) at steady Mach numbers in the conventionally transonic regime where most modern airliners typically cruise. Therefore, not only is the drag penalty associated with strong shocks avoided but also a linearized aerodynamic relationship can be employed via the unsteady aerodynamic influence coefficients (Chapter 3) for aeroelastic analysis. In a basic ASE model, one can thus use a linearized unsteady aerodynamic influence coefficient matrix, provided the mean flow conditions are essentially shock-free and the unsteady motion of the shock waves is of small amplitude.

11.3 Modelling for Transonic Unsteady Aerodynamics

Accurate modelling of transonic aerodynamics is possible only by a computational fluid dynamics (CFD) method. It was seen in Chapter 3 that the transonic small-disturbance (TSD) model is a convenient and accurate representation of unsteady transonic aerodynamics for ASE purposes. Not only can the TSD equation model the inherent nonlinearity associated with the mixed subsonic/supersonic characteristics but it can also capture the unsteady behaviour of weak, normal shock waves present at the transonic speeds. It is therefore natural to treat the TSD equation as a basic model for unsteady transonic aerodynamics of ASE systems, written as follows:

$$\left[1 - M_\infty^2 - \frac{(\gamma+1)M_\infty^2}{U_\infty}\phi_x\right]\phi_{xx} + \phi_{zz} = \frac{2M_\infty^2}{U_\infty}\phi_{xt} + \frac{M_\infty^2}{U_\infty^2}\phi_{tt}, \tag{11.5}$$

or, in terms of a frame convecting with the freestream velocity, as the following nonlinear wave equation:

$$\nabla^2\phi - \frac{(\gamma+1)M_\infty^2}{U_\infty}\phi_x\phi_{xx} = \frac{1}{a_\infty^2}\left.\frac{D^2\phi}{Dt^2}\right|_\infty . \tag{11.6}$$

Equations (11.5) and (11.6) are alternative representations of the same equation, but indicate two entirely different ways of attacking the TSD problem. The first (Eq. (11.5)) is the finite-difference (or finite-volume) approximation applied in a body-fixed reference frame using a body-fitted grid, while the second is the iterative solution of the nonlinear wave equation with a finite-element approximation in an appropriate domain. The boundary conditions in each case must be carefully applied in order to faithfully model the physical flow. These include the far-field conditions, the solid wall condition on the normal velocity component and additional conditions applied to obtain a unique and physically meaningful solution. The latter include the Kutta condition at the trailing edge and an artificial viscosity (or entropy) condition to introduce dissipation in an inviscid model. The grid geometry is crucial in both the time marching of the solution, as well as in the application of the boundary conditions. Finite-difference-type CFD solution procedures are based on structured grids, which either conform to a changing solid boundary (body-fitted grid), or remain fixed to the freestream flow (space-fixed grid). The application of time-marching and boundary conditions in each case is different. An unstructured grid which can adapt itself to the time-varying boundary is the most favoured of all, but it is very difficult to generate for a given problem. Finite-element type solutions generally employ unstructured grids, because they do not require a spatial marching inside the computational domain. For further details of CFD solution procedures and grid geometries, the reader is referred to the textbooks devoted to the topic (Hirsch 1990, Tannehill *et al.* 1997).

Solutions to the TSD equation can be used to derive interesting analytical results, such as an indicial (or Duhamel's) approximation for unsteady lift and moment coefficients, the Volterra–Wiener method and describing functions for shock-induced LCOs. These approximate functions can be employed in either a regressive parameter identification scheme for the self-tuning regulator (STR) (Chapter 5) or in a reference model for the model reference adaptation scheme (MRAS) (Chapter 8). Thus the TSD equation can be the kernel of future growth in adaptive transonic aeroservoelasticity.

11.3.1 Indicial Method

As the rational-function approximations (RFAs) are applied in the subsonic and supersonic small-perturbation flows to derive a linear, time-invariant (LTI) state-space representation of the aeroelastic system, an indicial approximation can be similarly applied to derive an equivalent approximation for the transonic regime. (Marzocca *et al.* 2005) suggest using the following Wagner-type functions for generalized indicial transonic aerodynamics of an airfoil:

$$\phi(\tau) = \sum_{i=1}^{n} A_i e^{-\beta_i \tau}, \tag{11.7}$$

whose coefficients $A_i, \beta_i, i = 1, \dots, n$ are identified by a CFD (Euler) scheme of the finite-volume type. The indicial functions are nonlinearly dependent on the induced downwash, $w(t)$. For example, the indicial lift-response function, ϕ_L, is the following Frèchet derivative of the downwash:

$$\begin{aligned}\phi_L[w(\zeta), \tau, \sigma] &= \lim_{\Delta w \to 0} \frac{C_L(\tau)}{\Delta w} \\ &= \lim_{\Delta w \to 0} \left\{ \frac{C_L[w(\zeta) + H(\zeta - \sigma)\Delta w] - C_L[w(\zeta)]}{\Delta w} \right\},\end{aligned} \tag{11.8}$$

where $C_L(t)$ is the unsteady lift coefficient, whose time evolution is given by the following nonlinear generalization of Duhamel's integral:

$$C_L(\tau) = C_L(\tau)\big|_{w=w(0)} + \int_0^\tau \frac{\mathrm{d}w}{\mathrm{d}\sigma}\phi_L(w(\zeta), \tau, \sigma)\mathrm{d}\sigma. \tag{11.9}$$

A similar indicial function, ϕ_M, can be defined for the pitching moment. For the incompressible flow, both ϕ_L and ϕ_M equal Wagner's function. The indicial approach is shown to compare well (Marzocca *et al.* 2005) with the direct numerical solutions obtained by an Euler solver (with two kinds of grid evolution) for symmetric NACA series airfoils undergoing step changes in the angle of attack near unity Mach number. However, as might be expected, discrepancies are observed at low reduced frequencies where the shock-induced nonlinear effects are predominant. Hence the indicial approach appears to be promising for ASE applications, except at very low reduced frequencies.

11.3.2 Volterra–Wiener Method

The Volterra method approximates the response of a nonlinear, autonomous (time-invariant) system as a series of convolution integrals of increasing order, with the first term being the LTI convolution integral. The kernel of each convolution integral (called Volterra kernel) can be identified by correlation functions derived from the Volterra–Wiener theory (Rugh 1981) for nonlinear electrical systems. Such kernel identification schemes have been applied in the past for separated flow aerodynamics computed by Navier–Stokes solvers, and offer a great promise for shock-dominated flows as well.

For an illustration of the method, consider a single-input, single-output, autonomous system, whose response to an arbitrary input, $u(t)$, which begins to act at a time, $t = 0$, is exactly given by the following infinite series:

$$\begin{aligned} y(t) = &\int_0^t h(t-\tau)u(\tau)\mathrm{d}\tau + \int_0^t \int_0^t h_2(t-\tau_1, t-\tau_2)u(\tau_1)u(\tau_2)\mathrm{d}\tau_1\mathrm{d}\tau_2 \\ &+ \int_0^t \int_0^t \cdots \int_0^t h_n(t-\tau_1, t-\tau_2, \cdots, t-\tau_n)u(\tau_1)u(\tau_2)\cdots u(\tau_n)\mathrm{d}\tau_1\mathrm{d}\tau_2\cdots\mathrm{d}\tau_n \\ &+ \cdots \end{aligned} \tag{11.10}$$

where $h(t)$ is the impulse response function and $h_n(.)$ denotes a symmetric kernel function of order n with the symmetry exemplified by

$$h_2(t-\tau_1, t-\tau_2) = h_2(t-\tau_2, t-\tau_1). \tag{11.11}$$

Since the system is causal, we have $h_n(t-\tau_1, t-\tau_2, \cdots, t-\tau_n) = 0$ if $\tau_i < 0$ for any $i = 1, \cdots, n$. The Volterra kernel, $h_n(.)$, denotes the response of the nonlinear system to n unit impulses applied at n different instants. The identification of the kernels is performed by comparing the response with a given finite-order Volterra series with an actually computed signal.

The Volterra–Wiener method was applied by Silva (1993) to a TSD solution with a pulse-rate change in the angle of attack for a two-dimensional airfoil, assuming only the first two terms of the Volterra series. This amounts to a bilinear (or weakly nonlinear) approximation for the transonic aeroelastic system. This application illustrated the feasibility of the approach, but more work needs to be done for identifying the nonlinear kernels for arbitrary inputs and to test the validity of the weak nonlinearity assumption for a practical system.

11.3.3 Describing Function Method

The lift and pitching moment coefficients for a pitching airfoil in the linearized case, such as that of Type A shock motion, can be expressed as follows:

$$C_\ell = C_{\ell s} + C_{\ell 0}\sin(\omega t + \phi_\ell), \tag{11.12}$$

$$C_m = C_{ms} + C_{m0}\sin(\omega t + \phi_m), \tag{11.13}$$

where $C_{\ell 0}, C_{m0}$ are the unsteady amplitudes and ϕ_ℓ, ϕ_m the phase lag angles of lift and pitching moment, respectively. In the unsteady case where strong shock waves are absent, the amplitude ratios $C_{\ell 0}/\alpha_0, C_{m0}/\alpha_0$ and the phase lag angles depend only on the frequency ω, and not on the mean flow condition α_s. Hence, the following frequency response relationship exists for the linearized, transonic, unsteady aerodynamics of an airfoil with an oscillating angle of attack:

$$\begin{Bmatrix} C_{\ell 0}e^{i(\omega t+\phi_\ell)} \\ C_{m0}e^{i(\omega t+\phi_m)} \end{Bmatrix} = \begin{pmatrix} G_\ell(\omega) \\ G_m(\omega) \end{pmatrix} \alpha_0 e^{i\omega t}, \tag{11.14}$$

where

$$\mid G_\ell(\omega) \mid = \frac{C_\ell(\omega)}{\alpha_0}$$

$$\mid G_m(\omega) \mid = \frac{C_m(\omega)}{\alpha_0}, \tag{11.15}$$

$$\phi_m(\omega) = \tan^{-1}\frac{\mathrm{Im}\{G_m(\omega)\}}{\mathrm{Re}\{G_m(\omega)\}}. \tag{11.16}$$

By carrying out a wind-tunnel test in which the amplitude ratios and phase angles are determined at a range of oscillation frequencies, the linear transfer functions for lift and pitching moment are derived by analytic continuation, which involves substituting the Laplace variable by the fundamental harmonic, $s = i\omega$, in Eqs. (11.15) and (11.16). Such investigations have been conducted on several airfoils, such as the NLR-7301 supercritical airfoil (Tijdeman 1977) shown in Fig. 11.14.

For nonlinear shock behaviour, such as the Type B and C motions, the assumption of linearity breaks down. However, in an ASE application, the aerodynamic behaviour of an oscillating shock wave is hardly to be seen in isolation, but must be studied in a closed loop with a structural dynamic system, as well as with a feedback control system. This fact is illustrated in Fig. 11.15 for an ASE regulation system, where the linear structural dynamics block,

$$\dot{x} = Ax + Bu + FQ, \tag{11.17}$$

with state vector, $x(t)$, aerodynamic loads vector, $Q(t)$, and coefficient matrices, A, B, F, is in a closed loop with a nonlinear feedback regulator,

$$u = g(x), \tag{11.18}$$

and a nonlinear, transonic aerodynamics block,

$$Q = f(x, u). \tag{11.19}$$

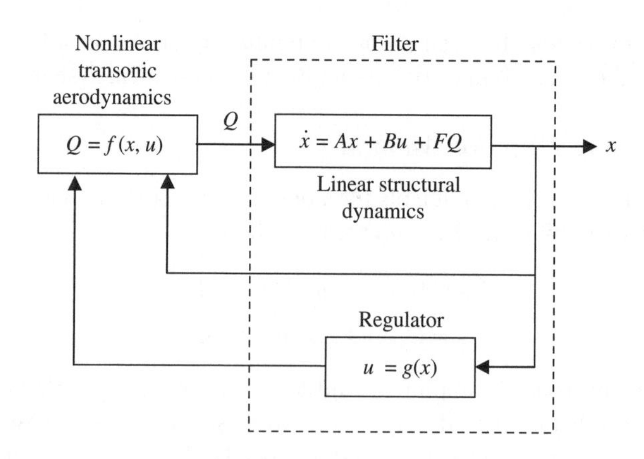

Figure 11.15 Nonlinear aeroservoelastic regulation system for the transonic regime

The system of Fig. 11.15 is in the classic configuration of a nonlinear block (comprising the aerodynamics and controller subsystems) in a feedback loop with a linear, structural dynamics subsystem. Its stability can therefore be analysed by the circle criteria (Chapter 7), which is the extended Nyquist stability criterion for nonlinear systems. While a control law, $u = g(x)$, can be designed to stabilize a known aerodynamic nonlinearity, even in the absence of the control inputs, $(u = 0)$, the linear, stable structural dynamics system is capable of filtering out higher harmonic signals from the aerodynamic spectrum, resulting in only a few fundamental peaks in the aeroelastic frequency response. Consider the following Fourier series representation of the aerodynamic loads vector:

$$Q(t) = Q_0 + \sum_{i=1}^{\infty} Q_{a_i} \cos(i\omega t) + Q_{b_i} \sin(i\omega t), \tag{11.20}$$

where the steady-state aerodynamic loads vector is given by

$$Q_0 = \frac{1}{\pi} \int_{-\pi}^{\pi} Q(t) \mathrm{d}(\omega t) \tag{11.21}$$

and the unsteady harmonic coefficients by

$$Q_{a_i} = \frac{1}{\pi} \int_{-\pi}^{\pi} Q(t) \cos(i\omega t) \mathrm{d}(\omega t), \tag{11.22}$$

$$Q_{b_i} = \frac{1}{\pi} \int_{-\pi}^{\pi} Q(t) \sin(i\omega t) \mathrm{d}(\omega t). \tag{11.23}$$

Since the ASE applications are mainly concerned with suppressing oscillations about a given steady state, it is reasonable to assume that the steady part of the nonlinearity, Q_0, does not contribute to the unsteady aerodynamic spectrum. Hence, Q_0 is dropped from Eq. (11.20), which implies that the unsteady aerodynamic nonlinearity is an odd function of the frequency. Furthermore, the structural dynamics subsystem is designed to be stable. Therefore, by the circle criterion (Chapter 7), it follows that the unforced aeroelastic system is stable and all the higher harmonics of the aerodynamics load spectrum except the first few are suppressed by the structural system function acting as a low-pass filter in a feedback loop. Consequently, the output signal, $x(t)$, of the unforced aeroelastic system can be approximated by only the first few (N) harmonics, given by

$$x(t) = \sum_{i=1}^{N} c_i \cos(i\omega t) + d_i \sin(i\omega t). \tag{11.24}$$

The low-frequency TSD equation provides an excellent computational model for identifying describing functions for transonic aerodynamics (Ueda and Dowell 1984). Consider the typical-section model for an airfoil oscillating in pitch, $\theta(t)$, and plunge, $h(t)$, with an aerodynamic transfer matrix, $G(s)$, governing the linear part of the unsteady lift and pitching moment coefficients, $C_L(t), C_m(t)$, as follows:

$$\begin{Bmatrix} C_L(s) \\ C_m(s) \end{Bmatrix} = G(s) \begin{Bmatrix} \theta(s) \\ h(s) \end{Bmatrix}, \tag{11.25}$$

where s is the non-dimensional Laplace operator. For the low-frequency TSD, model the following effective angle of attack can be defined for the airfoil:

$$\alpha(t) = \theta + \frac{\dot{h}}{U}, \tag{11.26}$$

which governs the lift and pitching moment by nonlinear functional relationships, $C_L(\alpha, \dot{\alpha})$ and $C_m(\alpha, \dot{\alpha})$, respectively. These nonlinear aerodynamic effects can be represented by an operator, $f(\alpha)$, in a feedback loop with the linear aerodynamic subsystem (Fig. 7.1). If the system is undergoing harmonic oscillations,

$$\theta(t) = \theta_0 e^{ikt}, \quad h(t) = h_0 e^{i(kt+\phi)}, \tag{11.27}$$

where k is the reduced frequency and ϕ the phase difference between pitch and plunge, the following Fourier series expansion can be employed for the aerodynamic lift and moment (Ueda and Dowell 1984):

$$\begin{Bmatrix} C_L \\ C_m \end{Bmatrix} = \begin{pmatrix} D_{L_R} & D_{m_I}/k \\ D_{m_R} & D_{L_I}/k \end{pmatrix} \begin{Bmatrix} \alpha \\ \dot{\alpha} \end{Bmatrix}, \tag{11.28}$$

with $\alpha_0 = \sqrt{\theta_0^2 + k^2 h_0^2}$, and

$$\begin{aligned} D_{L_R} &= \frac{1}{\alpha_0 \pi} \int_{-\pi}^{\pi} C_L(\alpha, \dot{\alpha}) \sin(k\tau) \mathrm{d}(k\tau) \\ D_{L_I} &= \frac{1}{\alpha_0 \pi} \int_{-\pi}^{\pi} C_L(\alpha, \dot{\alpha}) \cos(k\tau) \mathrm{d}(k\tau) \\ D_{m_R} &= \frac{1}{\alpha_0 \pi} \int_{-\pi}^{\pi} C_m(\alpha, \dot{\alpha}) \sin(k\tau) \mathrm{d}(k\tau) \\ D_{m_I} &= \frac{1}{\alpha_0 \pi} \int_{-\pi}^{\pi} C_m(\alpha, \dot{\alpha}) \cos(k\tau) \mathrm{d}(k\tau). \end{aligned} \tag{11.29}$$

The complex operator notation applied to Eqs. (11.28) and (11.29) yields the following:

$$C_L(\alpha, \dot{\alpha}) = D_L(\alpha_0)\alpha, \tag{11.30}$$

where

$$D_L(\alpha_0) = D_{L_R} + iD_{L_I}, \tag{11.31}$$

and so on. For a constant angle-of-attack amplitude, α_0, the form of these equations is identical with those for a linear system, hence a Nyquist-like stability analysis is possible. The describing function, $N(ik, \alpha_0)$, is then defined by

$$f(\alpha(t)) = N(ik, \alpha_0)\alpha(t) = N(ik, \alpha_0)\alpha_0 e^{ikt}, \tag{11.32}$$

where $f(\alpha)$ is the operator of nonlinear aerodynamic effects. Equation (11.32) can be represented in the non-dimensional Laplace domain as follows:

$$f(\alpha(s)) = N(s, \alpha_0)\frac{\alpha_0}{s - ik}, \tag{11.33}$$

which, for $k = 0$, gives the following indicial response function:

$$\phi(s) = N(s, \alpha_0)\frac{\alpha_0}{s}. \tag{11.34}$$

The indicial response can be derived from the low-frequency TSD solution by curve fitting, which directly yields the describing function, $N(s, \alpha_0)$, to be used in the extended Nyquist stability analysis (see Chapter 7). Ueda and Dowell (1984) successfully applied this method to determine the open-loop flutter condition of a transonic airfoil by the low-frequency TSD model. A similar application is possible for a closed-loop flutter-suppression system design.

11.4 Transonic Aeroelastic Plant

Consider the following aeroelastic equations of motion for the transonic regime:

$$M\ddot{q} + C\dot{q} + Kq = Q_a + Fu. \tag{11.35}$$

Here, $q(t)$ is the generalized coordinates vector, M, C, K the generalized structural mass, damping and stiffness matrices, respectively, $Q_a(t)$ is the generalized unsteady aerodynamic force vector. The generalized control input vector $u(t)$ is applied via a coefficient matrix, F. The aerodynamic model can be assumed to be a separated into a linear aerodynamic plant and a nonlinear shock dynamics plant given by

$$Q_a = M_\ell\ddot{q} + C_\ell\dot{q} + K_\ell q + N_a x_a + f(q, \dot{q}), \tag{11.36}$$

where $M_\ell, C_\ell, K_\ell, N_a$ are the matrix parameters of the linear aerodynamic behaviour (see the ideal aerodynamics plant above) with the associated aerodynamic state vector, $x_a(t)$, with aerodynamic state equation

$$\dot{x}_a = F_a x_a + \Gamma_a \begin{Bmatrix} q \\ \dot{q} \end{Bmatrix} \tag{11.37}$$

and $f(q,\dot{q})$ is the nonlinear aerodynamic generalized force vector associated with the unsteady shock motion. The overall state-space representation can then be given by the following:

$$\dot{x} = Ax + (0,f,0)^T + Bu, \tag{11.38}$$

where

$$A = \begin{pmatrix} 0 & I & 0 \\ -\bar{M}^{-1}\bar{K} & -\bar{M}^{-1}\bar{C} & -\bar{M}^{-1}N_a \\ & \Gamma_a & F_a \end{pmatrix} \tag{11.39}$$

and

$$B = (0, \bar{M}^{-1}F)^T, \tag{11.40}$$

where $\bar{M} = M - M_\ell, \bar{C} = C - C_\ell, \bar{K} = K - K_\ell$ are the aeroelastic mass, damping and stiffness matrices, respectively, and 0 and I are null and identity matrices, respectively, of appropriate dimensions.

The challenge is to design an adaptive feedback control law in the complete absence of any mathematical model for the nonlinear aerodynamic forcing term, $f(q,\dot{q})$. The methods presented above for approximating the nonlinear forcing term by an indicial function, a Volterra integral or a describing function can be applied for the derivation of the control law. However, even if such a functional approximation is unavailable, the nonlinear adaptation methods of Chapter 10 could be applied directly to yield a workable adaptive control system.

11.5 Adaptive Control of Control-Surface Nonlinearity

The adaptive control of aerodynamic nonlinear behaviour caused by the presence of normal shock waves near the hinge line of the control surfaces is an important aeroservoelastic application. Apart from causing a nonlinear increase in the control stiffness and damping, shock waves can excite a dynamic phenomenon called control-surface buzz. Before the vagaries of transonic flight were understood and accounted for, the nonlinear effects of shock waves on the control-surface hinge line often caused a 'freezing-up' of the elevator control in a dive, with disastrous consequences. The solution was to remove the elevator altogether and to replace it by an all-moving tail. This design is now in place in every high-speed aircraft designed after the 1950s. The actuator dynamics in the presence of shock waves is nonlinear, which can be approximated as follows:

$$\ddot{\delta} = -\frac{c_\delta}{I}\dot{\delta} - \frac{k_\delta}{I}\delta - k_2\delta^2 - k_3\delta^3 - c_2\dot{\delta}^2 + \delta_c, \tag{11.41}$$

where k_2, k_3, c_2 are positive, but unknown constants. A regressor form of the nonlinearity is the following:

$$f(x) = \Theta^T\Phi(x), \tag{11.42}$$

where $\Theta^T = (k_2, k_3, c_2)$, $\Phi^T(x) = (-x_{37}^2, -x_{37}^3, x_{38}^2)$ and $x_{37} = \delta, x_{38} = \dot{\delta}$ are the actuator states. When this model is implemented in the MRAS with σ-modification (see Chapter 10), we have the following adaptation law:

$$\dot{\hat{\Theta}} = R\left[\Phi(x)e^T PB - \sigma\hat{\Theta}\right], \tag{11.43}$$

where $\sigma > 1$. However, as the control-surface actuator response is usually governed by servos, this type of nonlinearity leads to instability of the overall ASE system due to an adaptation of the servo gains. The solution is to remove the servo gains from the nonlinear adaptation scheme. Let us consider an application of this method to a fighter-type aircraft equipped with elevons.

Example 11.5.1 *Consider the tail-less, delta-winged fighter aircraft (Chapter 8) equipped with elevons, a pitch-rate gyro and an accelerometer. The aircraft is statically unstable in the manoeuvre flight condition of Mach 0.9 and standard sea level, for which an adaptive stabilization system was designed in Chapters 8 and 10 by including the actuator and the aeroelastic modes in the feedback law. However, as the gains of the three coupled subsystems (rigid aircraft, actuator and aeroelastic aircraft) are evolved by a common scheme, there is a possibility of overall instability when any subsystem behaves 'abnormally'. This is true when the shock-driven dynamics of the actuator becomes inherently nonlinear, while the adaptation scheme is based on the actuator being a linear system. We first consider the response to an initial perturbation in the pitch rate of the MRAS system designed in Example 10.5 in the presence of actuator nonlinearity of the form given by Eq. (11.42) with* $k_1 = 0.001, k_2 = 0.002, c_2 = 0.0005$. *The results plotted in Figs. 11.16 and 11.17 show an unstable adaptive ASE system due to a divergent rigid-body closed-loop subsystem. The coupled aeroelastic response is also seen to be divergent (Fig. 11.17), which is particularly catastrophic. The MRAS system fails to stabilize the rigid-body dynamics due to the dependence of the regulator gains on the nonlinear system's tracking error, e, from the linear reference model, which grows without bound.*

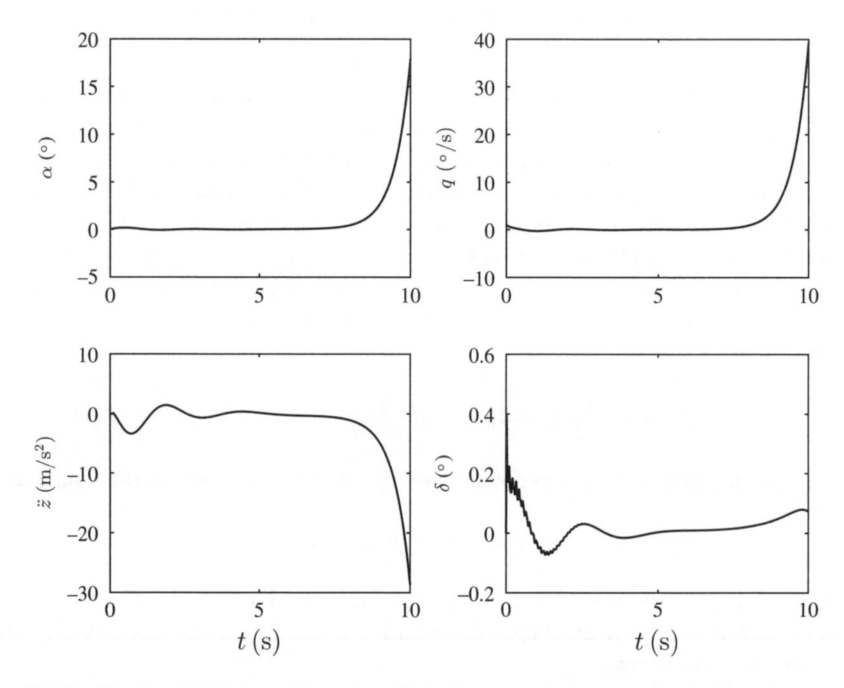

Figure 11.16 Simulated initial response $(\alpha, q, \ddot{z}, \delta)$ due to a pitch rate perturbation of a σ-modified MRAS system for the tail-less delta fighter at $U = 306$ m/s and standard sea level in the presence of shock-induced actuator nonlinearity

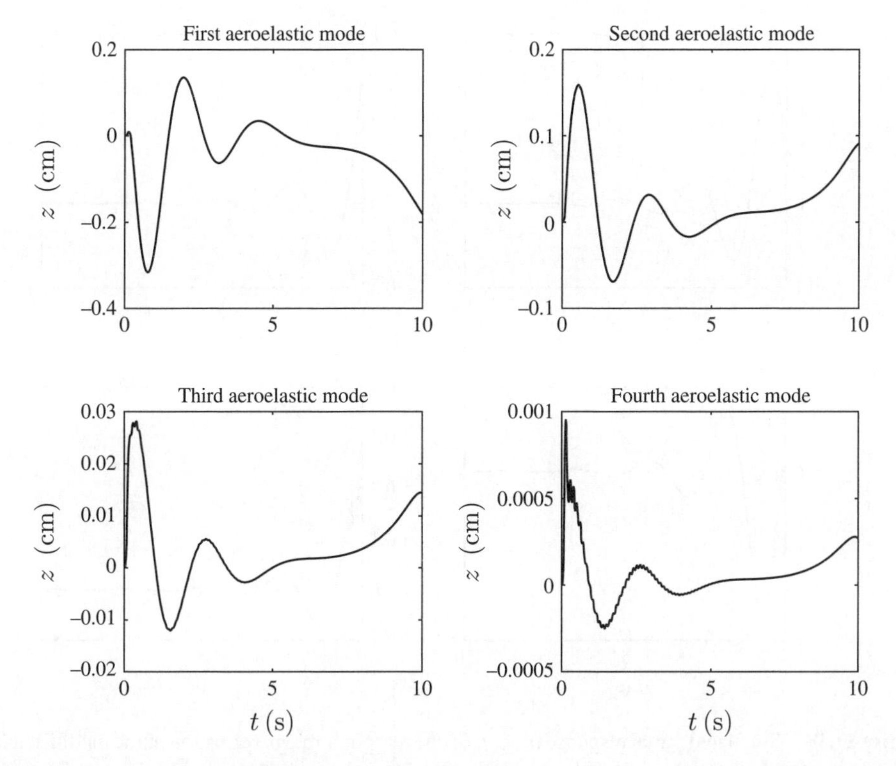

Figure 11.17 Simulated initial response of the aeroelastic modes due to a pitch-rate perturbation of a σ-modified MRAS system for the tail-less delta fighter at $U = 306$ m/s and standard sea level in the presence of shock-induced actuator nonlinearity

Since the divergent rigid-body response is due to the adaptive coupling with the nonlinear actuator dynamics, a simple solution is to change the adaptation law such that the rigid-body regulator gains ($\hat{K}_1, \hat{K}_2$) adapt only with the rigid states, $\alpha(t)$, $q(t)$, rather than with the tracking error. This is carried out as follows:

$$\left\{ \begin{matrix} \dot{\hat{K}}_1 \\ \dot{\hat{K}}_2 \end{matrix} \right\} = - \left\{ \begin{matrix} \alpha \\ q \end{matrix} \right\}. \tag{11.44}$$

The other controller parameters are kept constant at their nominal values. The resulting closed-loop response of the modified adaptation law, plotted in Figs. 11.18–11.20, seen to be asymptotically stable with all the state variables settling to zero, is less than 10 s. This example illustrates the advantage of simple feedback laws over more complex ones, especially when dealing with unknown nonlinear forcing terms.

11.5.1 Transonic Flutter Mechanism

The phenomenon of a sudden dip in the flutter dynamic pressure when traversing the sonic regime from either the subsonic or supersonic Mach numbers is associated with the nonlinear effects of unsteady shock waves. Since it is difficult to model the unsteady aerodynamic

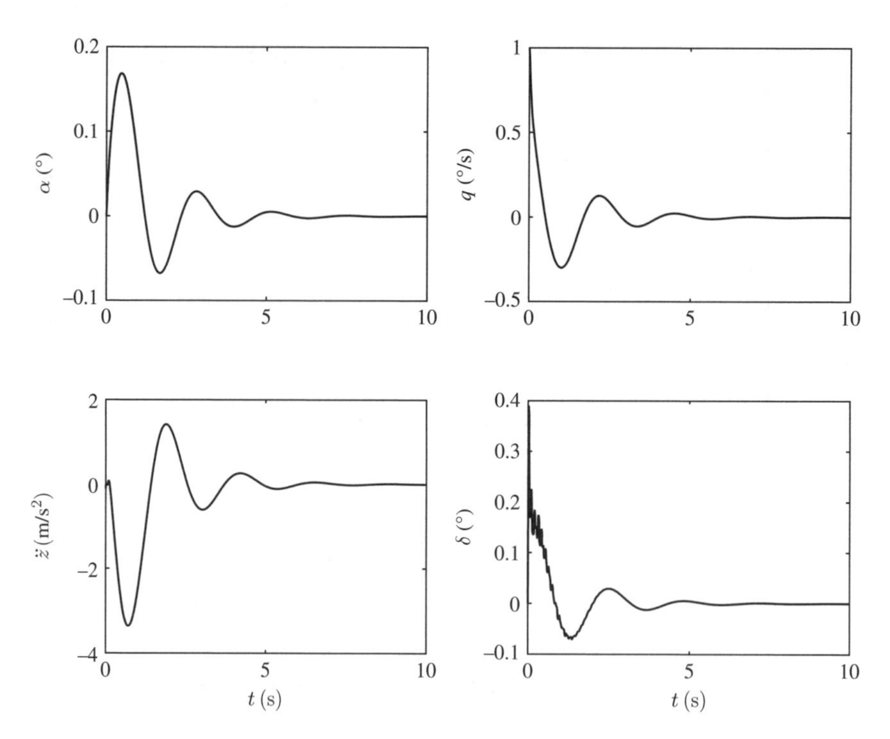

Figure 11.18 Simulated initial response ($\alpha, q, \ddot{z}, \delta$) due to a pitch-rate perturbation of the modified adaptive scheme for the tail-less delta fighter at $U = 306$ m/s and standard sea level in the presence of a shock-induced actuator nonlinearity

behaviour in the presence of strong shock waves, a self-adaptive, backstepping control is proposed here for transonic flutter suppression. The alternative method of smoothly interpolating the gain scheduler of the subsonic and supersonic speeds (see the previous section) in the transonic region is fraught with danger due to its highly uncertain nature. Even a small deviation from the expected interpolated behaviour can have catastrophic consequences. The mechanisms behind single degree-of-freedom pitch or control surface flutter and the bending-torsion flutter in the presence of unsteady shock waves have been a focus of intense research. While both classical analysis and experiments have predicted nonlinear flutter in transonic/low-supersonic regimes, the precise mechanism behind such an instability has remained an object of speculation. To quote Ashley (1980):

> It is the opinion of the author and others (cf. Sec. 4.2 of McGrew et al.) that a predominant factor in these anomalies is the presence of shock waves located part-way back along the chords of the upper, and sometimes also, the lower wing surfaces. These shocks may move periodically in harmony with the oscillatory angle-of-attack changes, and even at very low reduced frequencies, lag significantly in phase behind what would be estimated on a quasi-steady basis.

Hence, a phase lag in pitch caused by shock-wave dynamics is primarily considered to be responsible for transonic flutter. In a linearized subsonic flow, the flutter condition is caused

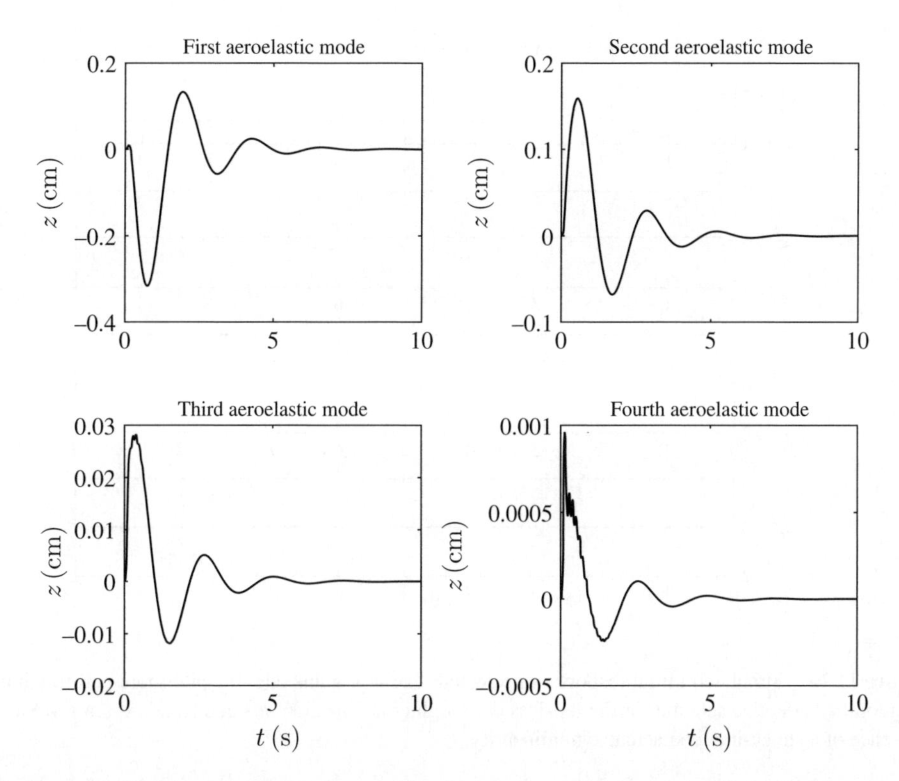

Figure 11.19 Simulated initial response of the aeroelastic modes due to a pitch-rate perturbation of the modified adaptive scheme for the tail-less delta fighter at $U = 306$ m/s and standard sea level in the presence of a shock-induced actuator nonlinearity

by a phase lag in pitch due to a circulatory wake (see Chapter 4). Therefore, it can be surmised that an oscillating shock wave interacting with a circulatory wake accelerates the flutter condition, thereby causing the transonic flutter dip. A decline in the aerodynamic damping due to the shock wave's interaction with the circulatory flow can be calculated from the time lag of what is termed 'Kutta waves' (Lee 1990) travelling back and forth between the shock and the trailing edge, giving rise to another suggested mechanism for transonic flutter (as well as control-surface buzz). (Ashley 1980) proposed a useful approximate model based on his insight for flutter corrections due to the shock-wave oscillation, which has been applied by several authors (Mabey 1980) in their own analyses of transonic flutter. He suggested a simple analytical correction in lift and pitching moment due to shock-wave oscillation, to be applied to classical solutions of the two-dimensional, oscillatory subsonic flow (cf. the low-frequency solution of (Kemp and Homicz 1976) to Possio's integral equation (Chapter 3)). Such a correction was derived from the low-frequency TSD solution. Alternative subsonic methods are available (see Appendix B) for use as the basis of such a correction. For the three-dimensional wing, Green's function method (Tewari 2015) appears to be especially suitable in transonic flutter analysis, as it utilizes a similar grid geometry on the wing as the classical doublet-lattice method.

Pioneering experimental work by Tijdeman (1977), Zwaan (1985), and others at NLR-Amsterdam, revealed that the effects of unsteady shock waves are such that a

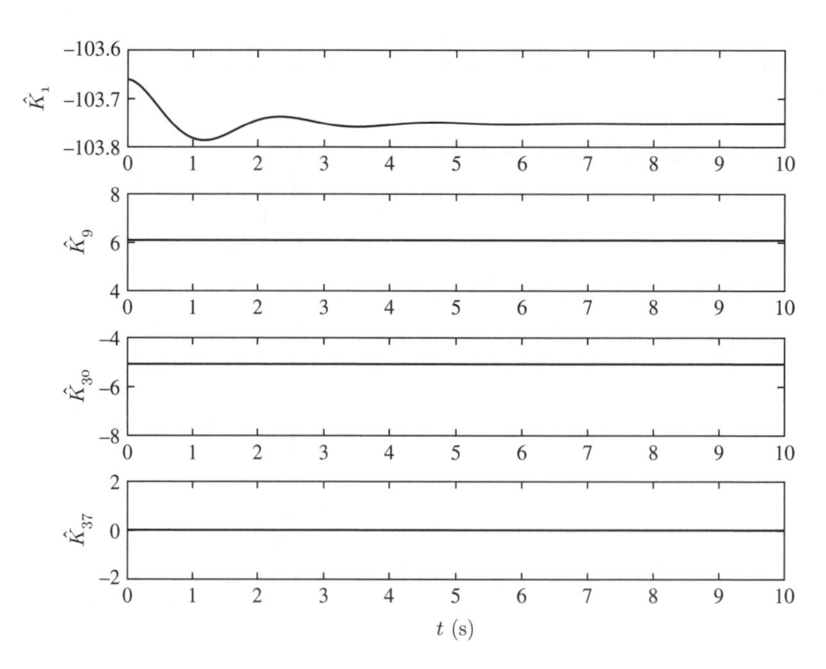

Figure 11.20 Simulated initial response of selected regulator gains due to a pitch-rate perturbation of the modified adaptive scheme for the tail-less delta fighter at $U = 306$ m/s and standard sea level in the presence of a shock-induced actuator nonlinearity

quasi-linear flutter model could be applicable even for thick supercritical wings. The flutter parameters are of course the functions of the Mach number, hence the flutter analysis should be carried out at several Mach numbers in order to obtain the decrease in flutter margin (flutter dip) encountered at transonic speeds. Furthermore, approximate models based on the low-frequency TSD solutions can be employed to represent the fundamental harmonic of shock-induced dynamics. Then a Nyquist-like analysis can reveal the LCO expected at the flutter condition.

11.6 Adaptive Control of Limit-Cycle Oscillation

An aircraft manoeuvring at a transonic Mach number is susceptible to LCO due to the presence of shock waves on the wing. This is due to a sustained oscillation in the primary wing torsion mode, with the twist angle, $\theta(t)$, feeding the shock-wave displacement in such a way as to maintain a nearly constant amplitude. The precise model of such nonlinear aeroelastic behaviour requires a sophisticated unsteady aerodynamic model based on the Navier–Stokes equations of viscous, turbulent, transonic flow. Since it is not always possible to derive such a model, an adaptive control system is necessary for identifying the concerned dynamics and applying a corrective control input in order to suppress it.

Cunningham (1989) proposed a nonlinear spring model of transonic LCO involving wing torsion based on the experimental observation of the shock-induced trailing-edge separation (SITES) phenomenon on a variable-sweep wing with a supercritical airfoil on the F-111 transonic active controls technology (TACT) aircraft. It was seen in the experiment that SITES leads to the development of the self-sustained pitching motion due to a hysteresis like, nonlinear dependence of the pitching moment on the sign of the pitch rate as a critical angle of attack (twist angle) is crossed. A positive pitch rate delays the shock-induced separation, thereby providing a negative pitching moment, which in turn results in a negative pitch rate, and hence a subsequent onset of positive pitching moment due to SITES. A coupling of such a nonlinear aerodynamic spring with the torsional wing mode is therefore the essential mechanism of the LCO at a constant torsional frequency.

Example 11.6.1 *We briefly discuss the basic mathematical model of LCO inspired by Cunningham's work (Cunningham 1989). Let a torsional wing mode of natural frequency, ω, and damping ratio, $\zeta > 0$, be responsible for the transonic LCO. The unsteady aerodynamic moment feeding and sustaining the LCO is due to a periodic shock-induced flow separation and reattachment near the trailing edge (SITES) of the wing. This nonlinear aerodynamic pitching moment can be represented by an angular acceleration, $\alpha(\theta, \dot{\theta})$, where $\theta(t)$ is the twist angle, which is equal to the change in the angle of attack of the wing from the steady-state equilibrium. A basic model of such an aerodynamic nonlinearity is the following:*

$$\alpha(\theta, q) = \begin{cases} 0, & \theta < \bar{\theta} \\ a\dot{\theta}, & \theta \geq \bar{\theta} \end{cases} \tag{11.45}$$

where $a, \bar{\theta}$ (thus α) are unknown aerodynamic parameters.

The equation of motion of the torsional LCO plant is the following:

$$\ddot{\theta} + 2\omega\zeta\dot{\theta} + \omega^2\theta = \alpha(\theta, \dot{\theta}) + u, \tag{11.46}$$

where $u(t)$ is the angular acceleration control input provided by a trailing-edge control surface. An adaptive control law based on pitch rate $q = \dot{\theta}$ measured by a rate gyro (Fig. 11.21) is to be designed such that the LCO is suppressed. This control law must necessarily be nonlinear, such as that given by

$$u = g(q, x), \tag{11.47}$$

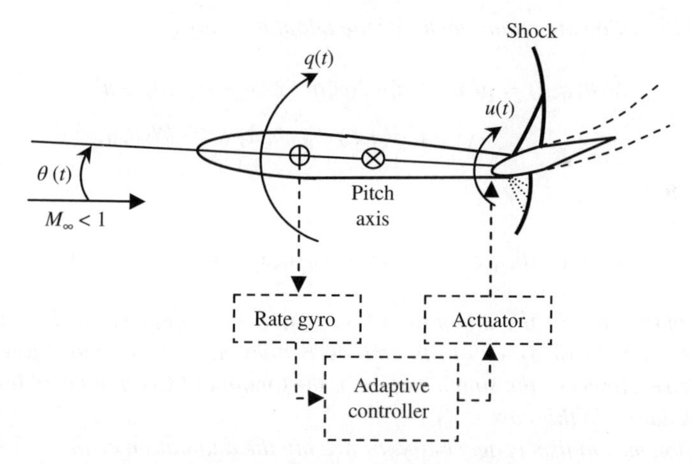

Figure 11.21 Adaptive aeroservoelastic system for automatically suppressing the transonic limit-cycle oscillation (LCO) via a pitch-rate gyro and a trailing-edge control surface input

where $g(.)$ is based on identification of the unknown aeroelastic plant behaviour in terms of x, which is an estimate of the unknown nonlinear parameter α. Thus, x is the additional variable necessary for feedback identification of the unsteady transonic aeroelastic behaviour represented by $\alpha(\theta, q)$, which may also vary with the flight Mach number in an unpredictable manner.

The plant state equations are expressed as follows:

$$\dot{\theta} = q$$
$$\dot{q} = -2\omega\zeta q - \omega^2\theta + \alpha(\theta, q) + u,$$

along with the adaptive, backstepping adaptation law:

$$\dot{x} = f(q, x), \tag{11.48}$$

where $f(q, x)$ must be determined such that the closed-loop system is asymptotically stable.

For global asymptotic stability of the closed-loop system, consider the following control Lyapunov function:

$$V(\theta, q, x) = \frac{1}{2}\omega^2\theta^2 + \frac{1}{2}q^2 + \frac{1}{2}[x - \alpha(\theta, q)]^2. \tag{11.49}$$

The time derivative of V is given by

$$\dot{V}(\theta, q, x) = \omega^2\theta\dot{\theta} + q\dot{q} + (x - \alpha)f(q, x) - (x - \alpha)\dot{\alpha}. \tag{11.50}$$

A sufficient condition for uniform, global asymptotic stability is given by the LaSalle–Yoshizawa theorem (Chapter 8) as the following inequality with a positive-definite function of the state variables, $W(\theta, q, x) > 0$,

$$\dot{V}(\theta, q, x) \leq -W(\theta, q, x). \tag{11.51}$$

Thus we have the following requirement for the adaptation law:

$$\begin{aligned}\dot{V}(\theta, q, x) &= \omega^2\theta q + q[-2\omega\zeta q - \omega^2\theta + \alpha(\theta, q) + u] \\ &\quad + (x - \alpha)f(q, x) - (x - \alpha)\dot{\alpha} \leq -W(\theta, q, x)\end{aligned} \tag{11.52}$$

or

$$\dot{V}(\theta, q, x) = -2\omega\zeta q^2 + \alpha(\theta, q)q + uq + (x - \alpha)f(q, x) - (x - \alpha)\dot{\alpha} \leq -W(\theta, q, x). \tag{11.53}$$

Neither the control, u, nor the adaptation law, $f(q, x)$, must depend on the unknown aerodynamic nonlinearity, $\alpha(\theta, q)$, or on its time derivative, $\dot{\alpha}$. Such a control law is generally difficult to derive. However, the simple nature of the torsional LCO plant enables us to obtain a stabilizing solution in this case.

An adaptation mechanism is necessary for driving the adaptation error,

$$e = x - \alpha,$$

to zero, irrespective of the magnitude and sign of the nonlinear disturbance, α. This implies that the desired adaptation error rate, $\dot{e}$, must have the opposite sign to that of the error, e, such that the error dynamics is given by

$$\dot{e} = -ce,$$

where $c > 0$. This allows us to write the unknown term in Eq. (11.53) as follows:

$$-(x - \alpha)\dot{\alpha} = -ce^2 - \dot{x}e.$$

On substituting this expression into the inequality Eq. (11.53), we have

$$\dot{V} = -2\omega\zeta q^2 + \alpha q + uq + (x - \alpha)f(q, x) - ce^2 - \dot{x}e \leq -W. \tag{11.54}$$

Now it is a simple matter to choose an adaptation law for satisfying the sufficient stability condition given by Eq. (11.54). An example is the law $\dot{x} = f(q, x) = q$, which yields the following:

$$\dot{V} = -2\omega\zeta q^2 + \alpha q + uq + (x - \alpha)f(q, x) - ce^2 - qe \leq -W < 0. \tag{11.55}$$

A simple choice of the stabilizing, feedback control law is the following:

$$u = g(q, x) = -kq - x, \tag{11.56}$$

which automatically satisfies Eq. (11.55) for all $k > 1/(4c)$ and $\zeta > 0$, with the following identity:

$$\dot{V} \leq -W = -(q, e)\begin{pmatrix} k & 1/2 \\ 1/2 & c \end{pmatrix}\begin{Bmatrix} q \\ e \end{Bmatrix}. \tag{11.57}$$

The feedback gain, k, is to be selected by trial based on the desired rate of error decay, as c is not known a priori and could depend on the initial condition exciting the limit cycle.

However, we immediately realize that the adaptation gain, x, can be easily replaced by the twist angle, θ, and the control law of Eq. (11.56) is none other than a linear state feedback law (or the proportional integral control if q is the sole measured output). The adaptation mechanism in such a case can thus be regarded as adding integral action to the linear feedback of pitch rate, q.

The resulting closed-loop system is of second order, and is given by

$$\begin{aligned} \dot{\theta} &= q \\ \dot{q} &= -\omega^2\theta - (2\omega\zeta + k)q - \theta + \alpha(\theta, q). \end{aligned} \tag{11.58}$$

Since the closed-loop system is guaranteed to be uniformly, globally asymptotically stable for $k > 1/(4c)$ by the Lyapunov stability theorem, all initial perturbations decay to zero as $t \to \infty$. This fact is illustrated by the simulated response of the closed-loop system with $\omega = 10$ rad/s, $\zeta = 0.01$, $a = 0.7$, $\bar{\theta} = 2°$ and $k = 1$ to the initial condition, $\theta = 5.73°$, $q = 0$, and is plotted in Figs. 11.22 and 11.23. The open-loop and closed-loop responses of the nonlinear SITES aerodynamic function, $\alpha(\theta, q)$, are plotted in Fig. 11.24. Clearly, the aerodynamic torque responsible for feeding the LCO is broken by the feedback controller in one and a half cycles, by driving the twist angle below $\bar{\theta}$. Note that this simple self-adaptation mechanism regulates the ASE system without any knowledge of the system parameters $\omega, \zeta, a, \bar{\theta}$. The smallest value

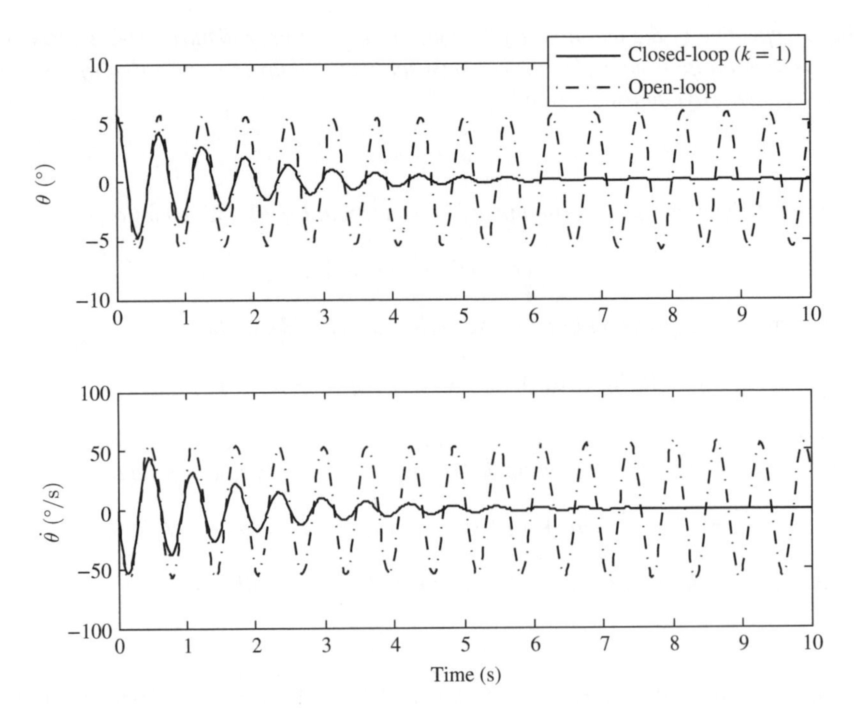

Figure 11.22 Simulated pitch response of the simple linear adaptive aeroservoelastic system for transonic LCO suppression

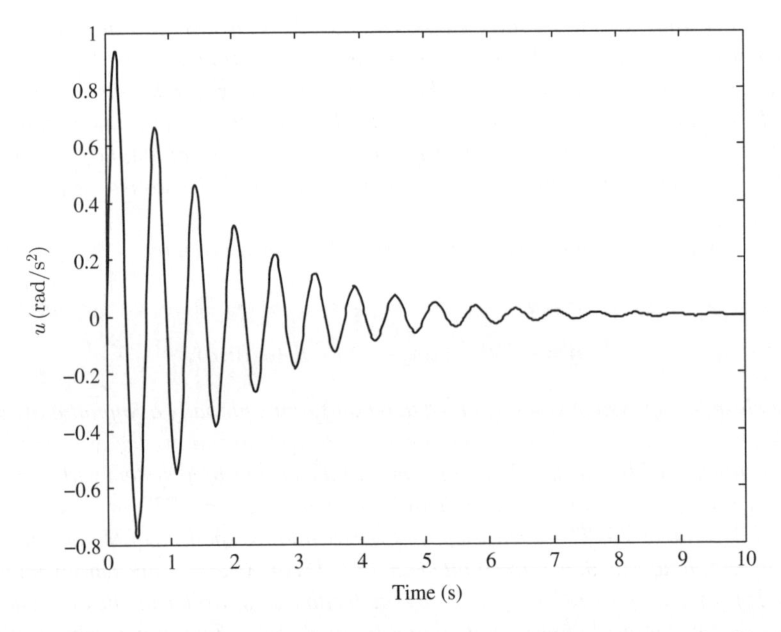

Figure 11.23 Simulated control input of the simple linear adaptive aeroservoelastic system for transonic LCO suppression

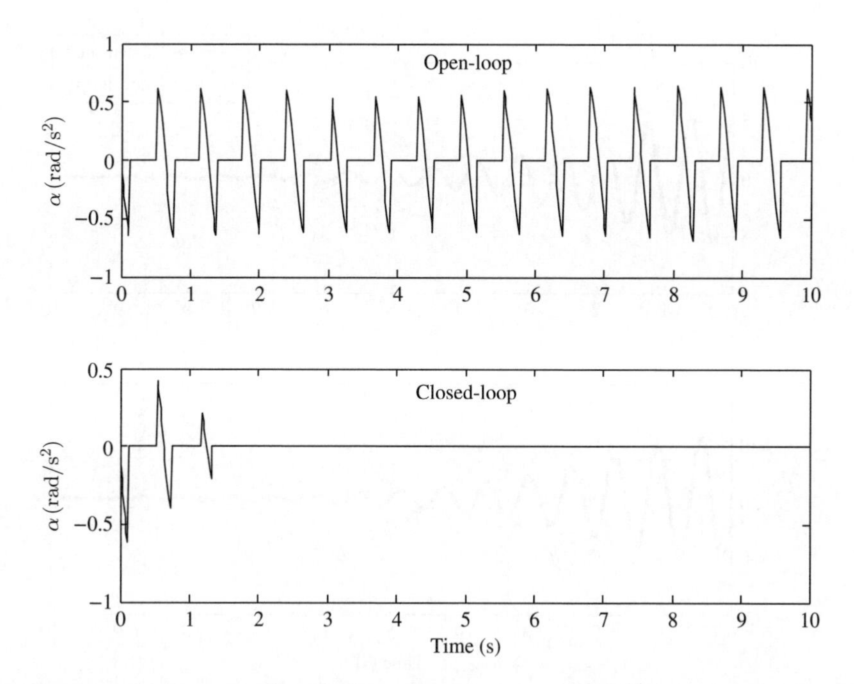

Figure 11.24 Nonlinear, aerodynamic forcing acceleration due to shock-induced trailing-edge separation (SITES) responsible for transonic LCO

of k that stabilizes the given limit-cycle is 0.036, which results in the error decaying to zero in about 100 s.

Now consider a more sophisticated adaptation mechanism, which utilizes the structure of aerodynamic nonlinearity given by Eq. (11.45), where the new control Lyapunov function is now the following:

$$V = \frac{1}{2}\omega^2\theta^2 + \frac{1}{2}q^2 + \frac{1}{2}(x-a)^2. \tag{11.59}$$

Here the constant, a, as well as the limiting pitch angle, $\bar{\theta}$, are unknown aerodynamic parameters, with a to be estimated by the adaptation gain x. The time derivative of V is thus given by

$$\dot{V} = \omega^2\theta\dot{\theta} + q\dot{q} + (x-a)f(q,x). \tag{11.60}$$

For satisfying the LaSalle–Yoshizawa theorem we choose as before, $W(\theta, q) = kq^2$, where $k > 0$, resulting in

$$\dot{V} \leq -kq^2 \tag{11.61}$$

or

$$\dot{V} = q[-2\omega\zeta q + \alpha(\theta, q) + u] + (x-a)f(q,x) \leq -kq^2. \tag{11.62}$$

Since the aerodynamic forcing term, $\alpha = aq$, comes into play only for $\theta > \bar{\theta}$, we attempt to cancel it by adaptive feedback as follows:

$$\dot{V} = aq^2 + uq + (x-aq)f(q,x) \leq -(k-2\omega\zeta)q^2. \tag{11.63}$$

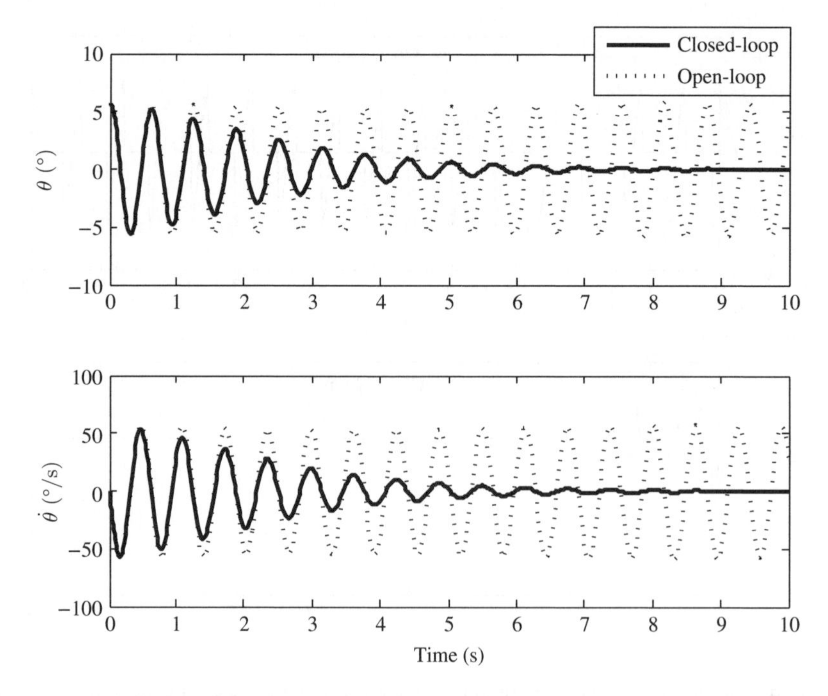

Figure 11.25 Simulated pitch response of nonlinear adaptive aeroservoelastic system for transonic LCO suppression

Since neither the control, u, nor the adaptation, law $f(q,x)$, must depend on the unknown aerodynamic parameter, a, we select an adaptation law $\dot{x} = f(q,x) = q^2$, which yields the following:

$$\dot{V} = uq + xq^2 \leq -(k - 2\omega\zeta)q^2. \tag{11.64}$$

A likely choice of the stabilizing, nonlinear feedback control law is then the following:

$$u = g(q,x) = -(k + x)q, \tag{11.65}$$

which automatically satisfies Eq. (11.64) for all $k > 0$ and $\zeta > 0$. The resulting nonlinear closed-loop system is now of third order, given by

$$\begin{aligned}
\dot{\theta} &= q \\
\dot{q} &= -\omega^2\theta - (2\omega\zeta + k)q - \theta + \alpha(\theta, q) \\
\dot{x} &= q^2.
\end{aligned} \tag{11.66}$$

The asymptotically stable simulated response of the nonlinear closed-loop system with the same parameters and initial conditions as considered above ($\omega = 10$ rad/s, $\zeta = 0.01$, $a = 0.7$, $\bar{\theta} = 2°$, $k = 0.1$, $\theta = 5.73°$, $q = 0$) is plotted in Figs. 11.25–11.27. Note that the required input magnitude is now only about half of that required by the linear adaptation and feedback control system (Fig. 11.23). However, the nonlinear adaptation loop requires an integration of the square of the measured pitch rate, q^2, for obtaining the adaptation gain, x.

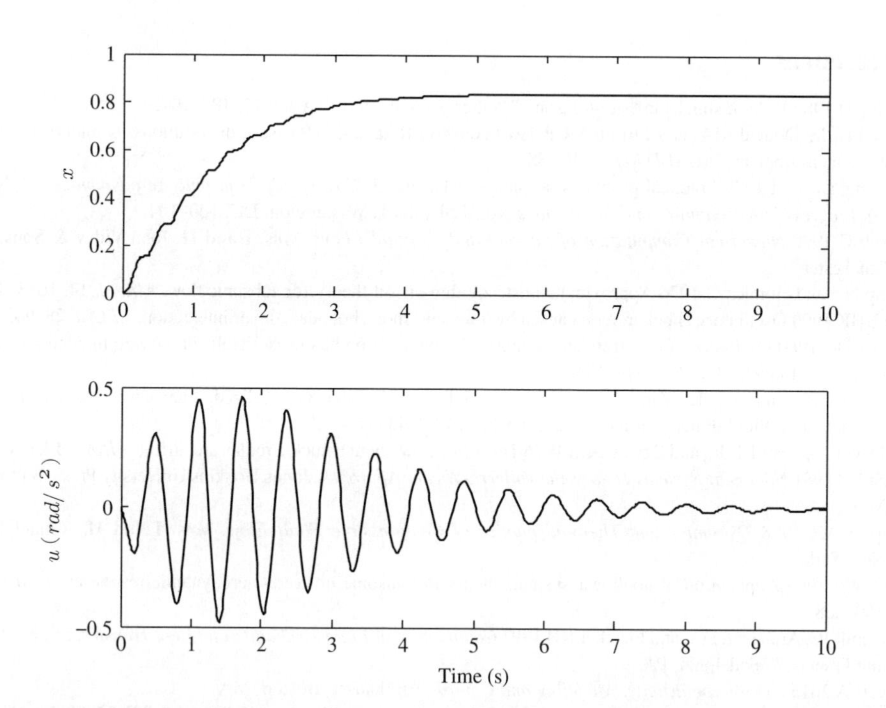

Figure 11.26 Simulated adaptation gain and control input of nonlinear adaptive aeroservoelastic system for transonic LCO suppression

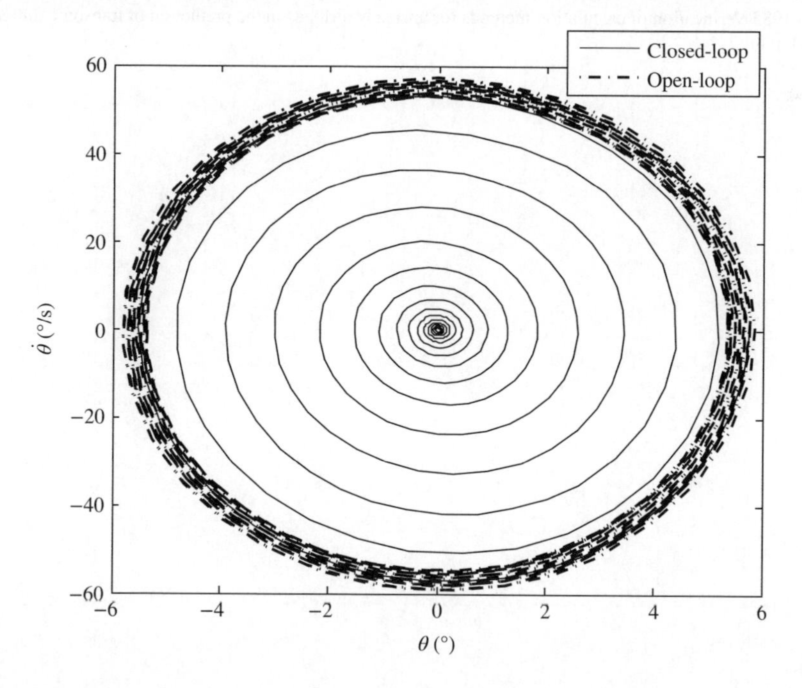

Figure 11.27 Phase portrait of the nonlinear adaptive aeroservoelastic system for transonic LCO suppression

References

Ashley H 1980 Role of shocks in "sub-transonic" flutter phenomenon. *J. Aircr.* **17**, 187–197.

Cook PH, McDonald MA, and Firmin MCP 1979 Aerofoil RAE 2822–Pressure distributions, boundary layer and wake measurements. *AGARD Rept.* **AR-138**.

Cunningham AM 1989 Practical problems: airplanes (Chapater 3 of *Unsteady Transonic Aerodynamics*, ed. Nixon D). *Progress in Astronautics and Aeronautics*, vol. **120**, AIAA, Washington, DC, 100–104.

Hirsch C 1990 *Numerical Computation of Internal and External Flows*. Vols. **I and II**, John Wiley & Sons, Ltd, Chichester.

Kemp NH and Homicz G 1976 Approximate unsteady thin-airfoil theory for subsonic flow. *AIAA J.* **14**, 1083–1089.

Lee BHK 1990 Oscillatory shock motion caused by transonic shock boundary-layer interaction. *AIAA J.* **28**, 942–944.

Mabey DG 1980 Oscillatory flows from shock-induced separation on biconvex airfoils of varying thickness in ventilated wind-tunnels. *AGARD Rept.* **CP-296**.

Marzocca P, Librescu L, Kim D, Lee I, and Schober S 2005 Generalized transonic aerodynamics via computational-fluid-dynamics/ indicial approach. *AIAA J.* **43**, 915–917.

McDevitt JB, Levy LL Jr., and Deiwert GS 1976 Transonic flow about a thick circular-arc airfoil. *AIAA J.* **14**, 606–613.

Rugh WJ 1981 *Nonlinear Systems Theory, the Volterra-Wiener Approach*. Johns Hopkins University Press, Baltimore, MD.

Shapiro AH 1958 *Dynamics and Thermodynamics of Compressible Fluid Flow*. Vols. **I and II**, Ronald Press, New York.

Silva WA 1993 Application of nonlinear systems theory to transonic unsteady aerodynamic responses. *J. Aircr.* **30**, 660–668.

Tannehill JC, Anderson DA, and Pletcher RH 1997 *Computational Fluid Mechanics and Heat Transfer*, 2nd ed. Taylor and Francis, Philadelphia, PA.

Tewari A 2015 *Aeroservoelasticity: Modeling and Control*. Birkhäuser, Boston, MA.

Tijdeman H 1977 Investigations of the transonic flow around oscillating airfoils. *NLR* **TR-77090 U**, National Aerospace Laboratories, The Netherlands.

Ueda T and Dowell EH 1984 Flutter analysis using nonlinear aerodynamic forces. *J. Aircr.* **21**, 101–109.

Zwaan RJ 1985 Verification of calculation methods for unsteady airloads in the prediction of transonic flutter. *J. Aircr.* **22**, 833–839.

Appendix A

Analytical Solution for Ideal Unsteady Aerodynamics

Low-speed, unsteady flows are modelled by elementary solutions to the incompressible, irrotational (ideal) flow problem, which is governed by the Laplace equation for the perturbation velocity potential, $\phi(x, z, t)$, given by

$$\nabla^2 \phi = 0. \tag{A.1}$$

The perturbation is superimposed over the steady, uniform freestream flow of speed U far away from the perturbing object. Unsteady, two-dimensional flow past thin airfoils is effectively approximated by flow over a flat chord plane. Let x be a coordinate measured from the mid-chord point along the flat-plate airfoil of chord $2b$ in the freestream direction, and rendered non-dimensional such that the leading edge is at $x = -1$, and the trailing edge at $x = 1$. The direction normal to the chord plane is measured by the non-dimensional coordinate, z, and a complex Joukowski conformal map from the physical plane, (x, z), to the transformed plane, (ξ, η), is defined by

$$Z = \frac{1}{2}\left(\zeta + \frac{1}{\zeta}\right), \tag{A.2}$$

where $Z = x + iz$ and $\zeta = \xi + i\eta$. The conformal mapping transforms the line segment $-1 \leq x \leq 1$, $y = 0$ (represented as a slit in the Z-plane) into a unit circle centred at $\xi = 0, \eta = 0$, while preserving the velocity potential, $\phi(x, z, t)$, at the mapped points. The entire Z-plane maps into the region outside the circle, the upper surface of the chord plane (slit) into the upper half arc of the circle and the lower surface of the slit into its lower half arc. The origin of the ζ-plane is the transform of all the points at infinity in the Z-plane. The slit being the representation of a solid airfoil, the flow is not allowed to pass across the slit, but can only go around it. In the process of going around the slit, a vortical flow pattern (or circulation) is created. The time-dependent nature of the unsteady flow causes the circulation to evolve with time. Since the flow is ideal and hence does not have a dissipation mechanism, Kelvin's theorem states that the circulation formed by a closed curve of fluid particles is conserved. As the fluid particles convect downstream, they produce a wake whose net circulation must

Adaptive Aeroservoelastic Control, First Edition. Ashish Tewari.

be equal and opposite to the circulation around the airfoil at a given time instant. Thus the wake influences the circulation around the airfoil. If the airfoil has unsteady motion normal to the chord plane, it causes a non-circulatory flow (normal flow or upwash), which must be cancelled by the upwash induced by the circulation due to the unsteady wake at all points on the chord plane.

Since the governing Laplace equation is linear, its solution can be expressed as a linear superposition of non-circulatory (wake-independent) and circulatory (wake-induced) parts, as follows:

$$\phi = \phi_{\rm nc} + \phi_c. \tag{A.3}$$

A source and a sink are placed at diametrically opposite ends of the circle in the ζ-plane in order to simulate the non-circulatory (n.c.) flow pattern caused by a normal velocity disturbance, $w(a,t)$, applied on an element, Δa, located at the point, $x = a$, on the chord plane. Note that although the source and sink are vertically separated by a zero distance in the physical plane at $x = a$, their effects do not cancel each other, because the flow cannot pass normal to the slit. Hence they effectively represent a doublet placed at $x = a$ with its axis parallel to the z-axis. The change in the n.c. velocity potential at a point x on the upper surface is derived from the doublet solution to be the following:

$$\Delta\phi_{\rm nc}(x,t) = \frac{1}{2\pi} w(a,t) b \Delta a \ell(x,a), \tag{A.4}$$

where

$$\ell(x,a) = 2\log \frac{1 - xa - \sqrt{1-x^2}\sqrt{1-a^2}}{x-a}. \tag{A.5}$$

The resulting n.c. pressure distribution is then given by the unsteady Bernoulli equation (Chapter 3) to be the following:

$$\begin{aligned}\Delta p_{\rm nc}(x,t) &= -2\rho\left(\frac{\partial}{\partial t} + \frac{U}{b}\frac{\partial}{\partial x}\right)\Delta\phi_{\rm nc} \\ &= -\frac{2}{\pi}\rho\left(\frac{U}{b}\frac{\sqrt{1-a^2}}{(x-a)\sqrt{1-x^2}}\right) bw\Delta a \\ &\quad -\frac{\rho}{\pi} b\Delta\xi\ell(x,a)\frac{\partial w}{\partial t}.\end{aligned} \tag{A.6}$$

This result can also be obtained by the application of Green's integral theorem, which is an alternative to the conformal mapping method.

The circulatory part of the velocity potential, $\Delta\phi_c$, is required to cancel the non-circulatory upwash, $w(a,t)$, by an equal and opposite induced upwash, such that there is no pressure singularity, either at the arbitrary point, $x = a$, or at the trailing edge, $x = 1$. The circulation on the airfoil, $\Delta\Gamma$, required for inducing the necessary upwash is provided by an element, Δx_w, of the wake behind the trailing edge, located at $x = x_w > 1$, with a vorticity distribution, $\gamma_w(x_w,t)$. The wake is a force-free surface, hence cannot sustain a pressure difference across itself. By either conformal mapping or Green's integral solution with doublet (or vortex) distribution, the circulatory perturbation potential integrated over the wake is derived to be the following:

$$\Delta\phi_c(x,t) = \frac{b}{2\pi}\int_1^s \gamma_w(x_w,t)\tan^{-1}\frac{\sqrt{1-x^2}\sqrt{x_w^2-1}}{1-xx_w}{\rm d}x_w. \tag{A.7}$$

The integration for the wake is carried out from $x_w = 1$ to $x_w = s$ and is consistent with the flow started at a previous time, $t = (s-1)b/U$. The vorticity generated at the trailing edge convects downstream with the flow, and reaches a position $x = x_w$ after a time $t = (x_w - 1)b/U$, without any change in its strength (Kelvin's theorem). This implies that

$$\gamma_w\left(x_w, t\right) = \gamma_w\left[1, \left(x_w - 1\right) b/U\right],$$

which simplifies the solution of the integral in Eq. (A.7).

The circulation around the airfoil (and the vorticity of the wake) should be such that the flow leaves smoothly at $x = 1$ (Kutta condition). This condition translates into the requirement of a finite tangential velocity component at the trailing edge,

$$\frac{\partial}{\partial x}\left(\Delta\phi_{\text{nc}} + \Delta\phi_c\right)\big|_{x=1}. \tag{A.8}$$

This results in the following integral equation to be solved for $\gamma_w(x_w, t)$, subject to the upwash boundary condition $w(a, t)$ on the solid airfoil surface:

$$\frac{1}{\pi} w(a,t)\Delta a\sqrt{\frac{1+a}{1-a}} = -\frac{1}{2\pi}\int_1^s \gamma_w(x_w, t)\sqrt{\frac{x_w+1}{x_w-1}}\mathrm{d}x_w. \tag{A.9}$$

The circulatory pressure difference on the airfoil is derived from the assumption that the unsteadiness in the velocity potential is caused only by the motion of the wake convecting downstream with the freestream speed, U,

$$\frac{\partial}{\partial t}(.) = \left(\frac{U}{b}\right)\frac{\partial}{\partial x_w}(.),$$

which results in the following:

$$\begin{aligned}\Delta p_c(x,t) &= -2\rho\frac{U}{b}\left(\frac{\partial}{\partial x} + \frac{\partial}{\partial x_w}\right)\Delta\phi_c \\ &= -\frac{\rho U}{\pi\sqrt{1-x^2}}\int_1^s \gamma_w\left(x_w, t\right)\frac{x + x_w}{\sqrt{x_w^2 - 1}}\mathrm{d}x_w\end{aligned} \tag{A.10}$$

or

$$\Delta p_c(x,t) = \frac{2\rho U}{\sqrt{1-x^2}}\frac{\int_1^s \gamma_w\left(x_w,t\right)\frac{x+x_w}{\sqrt{x_w^2-1}}\mathrm{d}x_w}{\int_1^s \gamma_w\left(x_w,t\right)\frac{\sqrt{1+x_w}}{\sqrt{x_w-1}}\mathrm{d}x_w}\Delta Q(a,t), \tag{A.11}$$

where

$$\Delta Q(a,t) = \frac{1}{\pi}w(a,t)\Delta a\sqrt{\frac{1+a}{1-a}}, \tag{A.12}$$

is the forcing term in the integral equation due to the non-circulatory upwash prescribed at a point $x = a$ on the wing. The pressure distribution is to be derived for a given type of airfoil motion, for which the upwash distribution on the chord plane is specified.

The discussion given here closely parallels the development by Wagner (1925) for a unit step change in the upwash, $w(a, t) = u_s(t)$, due to an airfoil impulsively started from rest as well as of the development by Theodorsen (1934) for the simple harmonic (oscillatory) upwash, $w(a,t) = \bar{w}e^{i\omega t}$. While Wagner's analysis of indicial airfoil motion is relevant in the transient aerodynamics modelling, the remainder of the discussion here is confined to the simple harmonic motion, which is necessary for deriving the unsteady aerodynamic transfer function. In the harmonic limit, we have

$$w(x,t) = \bar{w}(x)e^{i\omega t}, \tag{A.13}$$

where ω is the frequency of oscillation and $\bar{w}(x)$ the (complex) upwash amplitude at a given point (prescribed by rigid and elastic motion). Furthermore, the wake response is also harmonic, and is given by the vorticity at a point $x = x_w$:

$$\gamma_w(x_w,t) = \bar{\gamma}e^{i\omega[t-(x_w-1)b/U]} = \bar{\gamma}e^{i[\omega t - k(x_w-1)]}, \tag{A.14}$$

where $\bar{\gamma}$ is the complex wake vorticity amplitude and

$$k = \frac{\omega b}{U} \tag{A.15}$$

is the reduced frequency representing the number of waves in a wake length of 2π semi-chords, b. Hence, k is the governing parameter of circulatory incompressible, irrotational flow. In the harmonic case, the wake is assumed to have developed to its full extent ($s \to \infty$) before small amplitude perturbation, $\bar{\gamma}$, is applied. This is analogous to an infinite sheet of vortices fixed to the wing and oscillating at the excitation frequency. A change in the vorticity of the wake element at $x = x_w$ affects the circulation around the wing only after time $t = (x_w - 1)b/U$, therefore a phase lag is inherent in the circulatory pressure distribution. However, in the limit $s \to \infty$, the exponential term on the right-hand side of Eq. (A.14) vanishesand Δp_c can be expressed as follows:

$$\Delta p_c(x,t) = \Delta\bar{p}_c(x)e^{i\omega t}, \tag{A.16}$$

where

$$\Delta\bar{p}_c(x) = \frac{2\rho U}{\sqrt{1-x^2}} \frac{\int_1^s \frac{x+x_w}{\sqrt{x_w^2-1}} e^{-ikx_w}\mathrm{d}x_w}{\int_1^s \frac{1+x_w}{\sqrt{x_w^2-1}} e^{-ikx_w}\mathrm{d}x_w} \Delta\bar{Q}(a), \tag{A.17}$$

where

$$\Delta\bar{Q}(a) = \frac{1}{\pi}\bar{w}(a)\Delta a\sqrt{\frac{1+a}{1-a}}. \tag{A.18}$$

The main difficulty with the formulation of Eq. (A.17) is the evaluation of the improper integrals in the limit $s \to \infty$ due to the oscillatory nature of the integrands. The difficulty is partly resolved by writing

$$\frac{\int_1^s \frac{x_w}{\sqrt{x_w^2-1}} e^{-ikx_w}\mathrm{d}x_w}{\int_1^s \frac{1+x_w}{\sqrt{x_w^2-1}} e^{-ikx_w}\mathrm{d}x_w} = 1 - \frac{\int_1^s \frac{e^{-ikx_w}}{\sqrt{x_w^2-1}}\mathrm{d}x_w}{\int_1^s \frac{e^{-ikx_w}}{\sqrt{x_w^2-1}}\mathrm{d}x_w + \int_1^s \frac{x_w e^{-ikx_w}}{\sqrt{x_w^2-1}}\mathrm{d}x_w}, \tag{A.19}$$

which leaves the only the following problematic integral:

$$I(k) = \int_1^s \frac{xe^{-ikx}}{\sqrt{x^2-1}}\mathrm{d}x. \tag{A.20}$$

In the limit $s \to \infty$, this improper integral can be evaluated by considering the reduced frequency to be a complex number, such that $ik = a + ib, a > 0$, for which the integrand

$$\frac{xe^{-ikx}}{\sqrt{x^2 - 1}} = \frac{xe^{bx}e^{-iax}}{\sqrt{x^2 - 1}}$$

vanishes identically in the limit $x \to \infty$. Thus we write

$$I(k) = \frac{\mathrm{d}}{\mathrm{d}k}\int_1^{\infty} \frac{ie^{-ikx}}{\sqrt{x^2 - 1}}\mathrm{d}x = \frac{\mathrm{d}}{\mathrm{d}k}\frac{\pi}{2}H_0^{(2)}(k) = -\frac{\pi}{2}H_1^{(2)}(k), \tag{A.21}$$

where $H_n^{(2)}(.)$ is the Hankel function of second kind and order n. Such a method of evaluating an improper, harmonic integral by converting the frequency to a complex number is called analytic continuation, and is equivalent to extending the simple harmonic motion to a more general one of either a growing ($ik = \sigma + i\kappa, \sigma > 0$) or decaying oscillation ($ik = \sigma + i\kappa, \sigma < 0$) at the given frequency k. Alternatively, a quasi-steady term can be added and subtracted from the integrand, which corresponds to the assumption of the unsteady upwash being equal to its steady-state value. The additional quasi-steady term vanishes in the limit $s \to \infty$, resulting in a convergent integral.

The integral $I(k)$ substituted into Eq. (A.19) yields the following:

$$\begin{aligned} C(k) &= \frac{\int_1^s \frac{x_w}{\sqrt{x_w^2-1}}e^{-ikx_w}\mathrm{d}x_w}{\int_1^s \frac{1+x_w}{\sqrt{x_w^2-1}}e^{-ikx_w}\mathrm{d}x_w} \\ &= 1 - \frac{H_0^{(2)}(k)}{H_0^{(2)}(k) - iH_1^{(2)}(k)} \\ &= \frac{H_1^{(2)}(k)}{H_1^{(2)}(k) + iH_0^{(2)}(k)}, \end{aligned} \tag{A.22}$$

where $C(k)$ is called Theodorsen's function. This completes the derivation of analytical lifting pressure distribution due to a prescribed upwash distribution, $\bar{w}(a)$, in the following closed-form expression:

$$\begin{aligned} \Delta\bar{p}_c(x) &= \frac{2\rho U}{\sqrt{1 - x^2}}\{C(k) + x[1 - C(k)]\}\Delta\bar{Q}(a) \\ &= \frac{2\rho U}{\pi}\bar{w}(a)\Delta a\sqrt{\frac{1+a}{1-a}}\sqrt{\frac{1-x}{1+x}}\left[C(k) - 1 + \frac{1}{1-x}\right]. \end{aligned} \tag{A.23}$$

A result equivalent to Eq. (A.23) was first derived by Theodorsen (1934).

A.1 Pure Heaving Oscillation

For the special case of an airfoil in a pure heaving oscillation, $z = z_0e^{i\omega t}$, Biot (Fung 1955) derived the equivalent result through acceleration potential formulation. Here, the non-dimensional upwash amplitude is constant at all points, $x = a$, given by

$$\bar{w}(a) = -ikUz_0. \tag{A.24}$$

The total lift amplitude per unit span is

$$\bar{L} = \int_{-1}^{1} (\Delta p_{nc}(x) + \Delta p_c(x))\mathrm{d}x = \rho\pi U^2 y_0 k^2 \left[1 - i\frac{2}{k}C(k)\right] \tag{A.25}$$

and the net pitching moment amplitude per unit span about the mid-chord is derived to be

$$\bar{M} = -\int_{-1}^{1} (\Delta p_{nc}(x) + \Delta p_c(x))x\mathrm{d}x = -i\rho\pi U^2 y_0 kC(k). \tag{A.26}$$

These expressions can be verified by substituting Eq. (A.24) into Eqs. (A.6) and (A.23) and performing the chordwise integrations. From Eqs. (A.25) and (A.26), it is evident that the circulatory part of the pressure distribution (the part containing the factor $C(k)$) has its centroid at the quarter-chord point, while the non-circulatory pressure is uniformly distributed about the mid-chord location. The non-circulatory lift is seen to be independent of the flight speed, U.

A.2 Küssner–Schwarz Solution for General Oscillation

For a general oscillation of a point, x, on an airfoil chord measured from the mid-chord given by

$$z(x,t) = f(x)e^{i\omega t}, \tag{A.27}$$

the upwash is expressed as follows:

$$w(x,t) = \frac{\partial z}{\partial t} + U\frac{\partial z}{\partial x} = U\left(ikz + \frac{\partial z}{\partial x}\right), \tag{A.28}$$

with the understanding that the coordinates are non-dimensionalized by the semi-chord, b. If a Fourier cosine series expansion is employed for the upwash, given by

$$w(\theta,t) = Ue^{i\omega t}\left(P_0 + 2\sum_{n=1}^{\infty} P_n \cos n\theta\right), \tag{A.29}$$

where $x = \cos\theta$ and

$$\begin{aligned} P_0 &= \frac{1}{U\pi}e^{-i\omega t}\int_0^{\pi} w(\theta,t)\mathrm{d}\theta \\ P_n &= \frac{1}{U\pi}e^{-i\omega t}\int_0^{\pi} w(\theta,t)\cos n\theta\mathrm{d}\theta, \end{aligned} \tag{A.30}$$

then the resulting chordwise lift distribution per unit span is expressed as follows:

$$\ell(\theta,t) = \rho U^2 e^{i\omega t}\left(2a_0 \tan\frac{\theta}{2} + 4\sum_{n=1}^{\infty} a_n \sin n\theta\right). \tag{A.31}$$

Note that the Kutta condition at the trailing edge is satisfied by the lift distribution. The coefficients of the lift distribution are derived by Küssner and Schwarz (1940) to be the following:

$$a_0 = C(k)(P_0 + P_1) - P_1$$
$$a_n = P_n - \frac{ik}{2n}\left(P_{n+1} - P_{n-1}\right). \tag{A.32}$$

Hence, there is an equivalence between the oscillatory solution derived by Theodorsen and that of Küssner and Schwarz. When Eq. (A.32) is substituted into Eq. (A.31), the following general lift distribution is obtained (Küssner and Schwarz 1940):

$$\ell(\theta, t) = \frac{2}{\pi}\rho U \int_0^{\pi} w(\sigma, t)\left\{[C(k)(1+\cos\sigma) - \cos\sigma]\tan\frac{\theta}{2} + ikS(\cos\theta, \cos\sigma)\sin\sigma + \frac{\sin\theta}{\cos\sigma - \cos\theta}\right\} d\sigma, \tag{A.33}$$

where

$$S(\cos\theta, \cos\sigma) = \log\frac{\sin\frac{\theta+\sigma}{2}}{\sin\frac{\theta-\sigma}{2}}. \tag{A.34}$$

The integral in Eq. (A.34) is singular, hence its Cauchy principal value is taken. The final expressions for the lift and pitching-moment amplitudes about the mid-chord point are derived to be the following:

$$\bar{L} = 2\pi\rho U^2\left[\left(P_0 + P_1\right)C(k) + \frac{ik}{2}\left(P_0 - P_2\right)\right]$$
$$\bar{M} = \pi\rho U^2\left\{P_0C(k) - P_1[1 - C(k)] - P_2 - \frac{ik}{4}\left(P_1 - P_3\right)\right\}. \tag{A.35}$$

References

Fung YC 1955 *An Introduction to the Theory of Aeroelasticity*. John Wiley & Sons, Inc., New York.

Küssner HG and Schwarz L 1940 Der schwingende flügel mit aerodynamisch ausgeglichenem ruder. *Luftfahrtforschung* **17**, 377–384.

Theodorsen T 1934 General theory of aerodynamic instability and the mechanism of flutter. *NACA Rept.*, **496**.

Wagner H 1925 Über die entstehung des dynamischen auftriebs von tragflügeln. *Z. Angew. Math. Mech.* **5**, 17–35.

Appendix B

Solution to Possio's Integral Equation for Subsonic, Unsteady Aerodynamics

Subsonic, unsteady aerodynamic modelling is crucial in aeroservoelastic design, because it provides a linearized model while capturing the essential features of the flowfield. Flutter estimates of most aircraft require a subsonic model to which the nonlinear effects of shock waves and flow separation can be added. In an ASE design, these same effects are identified and controlled by an adaptive feedback scheme. Hence, it is necessary to consider simplified solution procedures for subsonic flow, which can be utilized in a baseline model. For illustration, only the two-dimensional flow will be considered here, as the three-dimensional subsonic unsteady formulation is well explored by doublet lattice, doublet point and kernel function collocation methods (Tewari 2015).

The linearized governing equation for the unsteady subsonic flow of freestream Mach number, $M < 1$, past a thin airfoil is expressed in terms of the disturbance acceleration potential as follows (see Chapter 3):

$$(1 - M^2)\psi_{xx} + \psi_{zz} = \frac{2M}{a}\psi_{xt} + \frac{1}{a^2}\psi_{tt}, \tag{B.1}$$

where a is the speed of sound approximated to be constant over the flowfield. The governing equation can be transformed into the following wave equation:

$$\nabla^2\psi = \frac{1}{a^2}\left.\frac{\mathrm{D}^2\psi}{\mathrm{D}t^2}\right|_\infty, \tag{B.2}$$

where

$$\left.\frac{\mathrm{D}}{\mathrm{D}t}\right|_\infty (.) = \frac{\partial}{\partial t}(.) + U\frac{\partial}{\partial x}(.) \tag{B.3}$$

is the Eulerian derivative representing the time derivative seen by an observer moving with the freestream velocity, $(U, 0)^T$. Hence, in the linearized case, the wave propagation speed is a, the constant speed of sound.

Adaptive Aeroservoelastic Control, First Edition. Ashish Tewari.

The integral equation relating the upwash amplitude to that of the pressure difference across the mean surface of an airfoil oscillating in a subsonic flow is derived by Possio (1938) to be the following:

$$\bar{w}(x) = -\frac{k}{\rho U}\int_{-1}^{1} K[M, k(x-\xi)]\Delta\bar{p}(\xi)\mathrm{d}\xi, \tag{B.4}$$

where k is the reduced frequency. In terms of the non-dimensional bound-vorticity (or lift-coefficient) amplitude, $\bar{\gamma}(\xi)$, and the angle-of-attack amplitude, $\bar{\alpha} = \bar{w}/U$, the integral equation is expressed as follows:

$$\bar{\alpha}(x) = -k\int_{-1}^{1} K[M, k(x-\xi)]\bar{\gamma}(\xi)\mathrm{d}\xi. \tag{B.5}$$

Thus the kernel function of the integral equation represents the important aerodynamic influence coefficient between non-dimensional lift and angle of attack. The integral equation must be solved for $\bar{\gamma}(\xi)$, given the angle-of-attack distribution, $\bar{\alpha}(x)$, on the airfoil's chord plane. The kernel function is expressed as follows in terms of the non-dimensional variables (Garrick 1957):

$$\begin{aligned} K[M, k(x-\xi)] = {} & -\frac{1}{4\beta}e^{i\mu(x-\xi)}\left\{ H_0^{(2)}(\kappa \mid x-\xi \mid) - iM\frac{(x-\xi)}{\mid x-\xi \mid}H_1^{(2)}(\kappa \mid x-\xi \mid) \right. \\ & - i\beta^2 e^{-ik(x-\xi)/\beta^2}\left(\frac{2}{\pi\beta}\log\frac{1+\beta}{M} \right. \\ & \left.\left. + \int_0^{k(x-\xi)/\beta^2} e^{i\lambda}H_0^{(2)}(M \mid \lambda \mid)\,\mathrm{d}\lambda \right)\right\}, \end{aligned} \tag{B.6}$$

where $\beta = \sqrt{1-M^2}$, $\kappa = kM/\beta^2$, $\mu = M\kappa$ and $H_n^{(2)}(.)$ are Hankel functions of the second kind and order n. The integral equation must be solved for the complex bound vorticity amplitude, $\bar{\gamma}(\xi)$, given a prescribed angle-of-attack distribution, $\bar{\alpha}(x)$, on the wing. The Kutta condition at the trailing edge requires that $\bar{\gamma}(1) = 0$, which is generally satisfied by selecting the bound vorticity distribution as the following infinite series:

$$\bar{\gamma}(x) = U\left(2a_0\cot\frac{\theta}{2} + 4\sum_1^{\infty} a_n\frac{\sin n\theta}{n} \right), \tag{B.7}$$

where $x = -\cos\theta$.

B.1 Dietze's Iterative Solution

Dietze (1947) proposed that there is only a small difference between the subsonic kernel function, $K(M, kr)$, and that in the incompressible limit, $K(0, kr)$, where $r = x - \xi$, for the practical range of reduced frequencies, $0 \le k \le 1$, and for Mach numbers not in the vicinity of 1.0. Therefore, a first-order expansion of the kernel function about the incompressible case is reasonable. The incompressible kernel function is obtained from Eq. (B.6) to be the following:

$$K(0, kr) = \frac{1}{2\pi}\left\{ \frac{1}{r} - ike^{-ikr}\left[C_i(kr) + i\left(S_i(kr) + \frac{\pi}{2} \right) \right] \right\}, \tag{B.8}$$

where C_i and S_i are the following cosine and sine integral functions:

$$C_i(y) = -\int_{-y}^{\infty} \frac{\cos u}{u} \mathrm{d}u \tag{B.9}$$

$$S_i(y) = \int_0^y \frac{\sin u}{u} \mathrm{d}u. \tag{B.10}$$

In the steady-flow limit, $k \to 0$, the incompressible kernel is seen to have the following value:

$$K[0, k(x-\xi)] = \frac{\beta}{2\pi(x-\xi)}. \tag{B.11}$$

Hence, the subsonic kernel is a well-behaved function of reduced frequency (for $x \neq \xi$) and can be linearly expanded as follows (Dietze 1947):

$$K[M, k(x-\xi)] = K[0, k(x-\xi)] + \Delta K. \tag{B.12}$$

The numerical scheme of Dietze begins with the computation of analytical vorticity distribution, $\overline{\gamma}_0$, for the incompressible case (discussed below) and the given angle-of-attack distribution, $\overline{\alpha}(x)$. This amounts to replacing the kernel function in Eq. (B.4) by $K[(0, k(x-\xi)]$. Next, the upwash distribution is corrected by writing

$$\overline{\gamma}_1 = \overline{\gamma}_0 + \Delta\overline{\gamma}_0, \tag{B.13}$$

where

$$\Delta\overline{\alpha}_1(x) = \int_{-1}^{1} K[M, k(x-\xi)]\Delta\overline{\gamma}_0(\xi)\mathrm{d}\xi = \int_{-1}^{1} \Delta K\overline{\gamma}_0(\xi)\mathrm{d}\xi. \tag{B.14}$$

In Eq. (B.14), the difference in the kernel function, ΔK, is computed by Eq. (B.12) and used to evaluate the correction in the incompressible angle of attack, $\Delta\overline{\alpha}_1$, which is then prescribed to analytically update the incompressible vorticity distribution, $\overline{\gamma}_1$, for the next iteration. This process is repeated until convergence is obtained in the vorticity distribution. Each successive step should produce a smaller change in the analytical, incompressible solution. Unfortunately, this is not always guaranteed, especially if a singularity exists in the kernel function, such as that at the leading edge of a control surface, $x = \xi$. Furthermore, as M approaches unity, increased number of steps are necessary for a converged solution.

B.2 Analytical Solution by Fettis

In order to avoid the problems of singularity and convergence in Dietze's method, Fettis (1952) modified the kernel evaluation procedure by replacing the singular part of the kernel function by an approximate algebraic expression. Thus analytical, incompressible expressions are utilized in each step without the need for any iteration. The kernel of Possio's integral equation, Eq. (B.5), is expressed as follows:

$$K[M, k(x-\xi)] = \frac{1}{\sqrt{1-M^2}}K[0, k(x-\xi)] + \frac{M^2}{2\pi k(x-\xi)\sqrt{1-M^2}} + \hat{K}[M, k(x-\xi)], \tag{B.15}$$

which (as in Dietze's method) utilizes the availability of the incompressible kernel, $K[0, k(x-\xi)]$. The non-singular part of the kernel, $\hat{K}[M, k(x-\xi)]$, is approximated by the following polynomial series:

$$\hat{K}[M, k(x-\xi)] = a_0 + a_1 k(x-\xi) + a_2 k^2 (x-\xi)^2 + \cdots + a_n k^n (x-\xi)^n, \tag{B.16}$$

whose coefficients, $a_0, \cdots, a_n$, are determined from the upwash boundary conditions. Hence, a series expansion of the lift distribution, which is utilized in other methods, is unnecessary. Fettis provided an alternative expansion for the non-singular kernel as follows:

$$\hat{K}[M, k(x-\xi)] = u_0(x) + u_1(x)\xi + u_2(x)\xi^2 + \cdots + u_n(x)\xi^n, \tag{B.17}$$

which is more amenable to integration with respect to ξ. For the incompressible case, use is made of the following solution (Küssner 1940), to which the Kutta condition is applied:

$$f(x) = -k \int_{-1}^{1} K[0, k(x-\xi)] \bar{\gamma}_0(\xi) \mathrm{d}\xi, \tag{B.18}$$

resulting in the following incompressible lift:

$$\bar{\gamma}_0(x) = \int_{-1}^{1} G(x, \zeta) f(\zeta) \mathrm{d}\zeta, \tag{B.19}$$

where

$$G(x, \zeta) = \frac{2}{\pi} \left[ik\Lambda(x, \zeta) + \sqrt{\frac{1-x}{1+x}} \sqrt{\frac{1+\zeta}{1-\zeta}} \left\{ C(k) + \frac{1}{\zeta - x} \right\} \right], \tag{B.20}$$

with $C(k)$ being Theodorsen's function and

$$\Lambda(x, \zeta) = \frac{1}{2} \log \frac{1 - x\zeta + \sqrt{1-x^2}\sqrt{1-\zeta^2}}{1 - x\zeta - \sqrt{1-x^2}\sqrt{1-\zeta^2}}. \tag{B.21}$$

A rearrangement of the integral equation after the substitution of Eqs. (B.17) and (B.19) produces the following set of linear algebraic equations to be solved for the unknown aerodynamic load distribution from the known incompressible solution on the right-hand side:

$$\begin{aligned}
A_{00}X_0 + A_{01}X_1 + \cdots + A_{0n}X_n &= \frac{B_0}{\sqrt{1-M^2}} \\
A_{10}X_0 + A_{11}X_1 + \cdots + A_{1n}X_n &= \frac{B_1}{\sqrt{1-M^2}} \\
&\vdots \\
A_{n0}X_0 + A_{n1}X_1 + \cdots + A_{nn}X_n &= \frac{B_n}{\sqrt{1-M^2}},
\end{aligned} \tag{B.22}$$

where

$$X_n = \int_{-1}^{1} \bar{\gamma}(\xi) \xi^n d\xi, \tag{B.23}$$

$$A_{00} = e^{i\mu} - \frac{M^2C(k)}{\pi(1-M^2)}\int_{-1}^{1} e^{-i\mu\xi}\sqrt{\frac{1-\xi}{1+\xi}}\mathrm{d}\xi - \frac{i\mu}{\pi}\int_{-1}^{1} e^{-i\mu\xi}\cos^{-1}\xi\mathrm{d}\xi + \frac{k}{\sqrt{1-M^2}}\int_{-1}^{1} e^{-i\mu\xi}\phi_0(\xi)\mathrm{d}\xi \tag{B.24}$$

$$A_{0n} = \frac{k}{\sqrt{1-M^2}}\int_{-1}^{1} e^{-i\mu\xi}\phi_n(\xi)\mathrm{d}\xi, \quad (n > 0) \tag{B.25}$$

$$B_0 = \int_{-1}^{1} e^{-i\mu\xi}\overline{\gamma}_0(\xi)\mathrm{d}\xi \tag{B.26}$$

$$B_n = \int_{-1}^{1} \xi^{n-1}\overline{\gamma}_0(\xi)\mathrm{d}\xi, \quad (n > 0) \tag{B.27}$$

$$A_{10} = 1 + i\mu - \frac{M^2C(k)}{\pi(1-M^2)}\int_{-1}^{1}\sqrt{\frac{1-\xi}{1+\xi}}\mathrm{d}\xi - \frac{i\mu}{\pi}\int_{-1}^{1}\cos^{-1}\xi\mathrm{d}\xi + \frac{k}{\sqrt{1-M^2}}\int_{-1}^{1}\phi_0(\xi)\mathrm{d}\xi \tag{B.28}$$

$$A_{11} = i\mu + \frac{k}{\sqrt{1-M^2}}\int_{-1}^{1}\phi_1(\xi)\mathrm{d}\xi \tag{B.29}$$

$$A_{1n} = \frac{k}{\sqrt{1-M^2}}\int_{-1}^{1}\phi_n(\xi)\mathrm{d}\xi, \quad (n > 1) \tag{B.30}$$

$$A_{20} = -i\frac{\mu}{2} - \frac{M^2C(k)}{\pi(1-M^2)}\int_{-1}^{1}\xi\sqrt{\frac{1-\xi}{1+\xi}}\mathrm{d}\xi - \frac{i\mu}{\pi}\int_{-1}^{1}\xi\cos^{-1}\xi\mathrm{d}\xi + \frac{k}{\sqrt{1-M^2}}\int_{-1}^{1}\xi\phi_0(\xi)\mathrm{d}\xi \tag{B.31}$$

$$A_{21} = 1 + \frac{k}{\sqrt{1-M^2}}\int_{-1}^{1}\xi\phi_1(\xi)\mathrm{d}\xi \tag{B.32}$$

$$A_{22} = i\frac{\mu}{2} + \frac{k}{\sqrt{1-M^2}}\int_{-1}^{1}\xi\phi_2(\xi)\mathrm{d}\xi \tag{B.33}$$

$$A_{2n} = \frac{k}{\sqrt{1-M^2}}\int_{-1}^{1}\xi\phi_n(\xi)\mathrm{d}\xi, \quad (n > 2) \tag{B.34}$$

$$A_{30} = i\frac{\mu}{3} - \frac{M^2C(k)}{\pi(1-M^2)}\int_{-1}^{1}\xi^2\sqrt{\frac{1-\xi}{1+\xi}}\mathrm{d}\xi - \frac{i\mu}{\pi}\int_{-1}^{1}\xi^2\cos^{-1}\xi\mathrm{d}\xi + \frac{k}{\sqrt{1-M^2}}\int_{-1}^{1}\xi^2\phi_0(\xi)\mathrm{d}\xi \tag{B.35}$$

$$A_{31} = \frac{k}{\sqrt{1-M^2}} \int_{-1}^{1} \xi^2 \phi_1(\xi) \mathrm{d}\xi \tag{B.36}$$

$$A_{32} = 1 + \frac{k}{\sqrt{1-M^2}} \int_{-1}^{1} \xi^2 \phi_2(\xi) \mathrm{d}\xi \tag{B.37}$$

$$A_{33} = i\frac{\mu}{3} + \frac{k}{\sqrt{1-M^2}} \int_{-1}^{1} \xi^2 \phi_3(\xi) \mathrm{d}\xi \tag{B.38}$$

$$A_{3n} = \frac{k}{\sqrt{1-M^2}} \int_{-1}^{1} \xi^2 \phi_n(\xi) \mathrm{d}\xi, \qquad (n > 3), \tag{B.39}$$

and so on, with

$$\mu = \frac{kM^2}{1-M^2} \tag{B.40}$$

and

$$\phi_m(x) = \int_{-1}^{1} G(x,\zeta) u_m(\zeta) d\zeta, \quad (m = 0, 1, 2, \cdots, n). \tag{B.41}$$

The solution vector, $\{X\} = (X_0, X_1, \cdots, X_n)^T$, is obtained by Gaussian elimination. Note that X_0 gives the lift and X_1 the pitching moment about the mid-chord. The next two higher-order solution terms, X_2, X_3, relate to the lift and pitching-moment contributions of the control-surface located between $\xi = x$ to $\xi = 1$. Terms of even higher order give the contributions of a tab on the control surface, and so on. The main advantage of Fettis' derivation is its non-iterative application to produce the state-space coefficient matrices, $[A]$, $[B]$ (see Chapter 3) for ASE applications in a baseline, subsonic condition.

B.3 Closed-Form Solution

Another non-iterative solution of Possio's integral equation is derived by Balakrishnan (1999), Lin and Illif (2000), which can be considered as an alternative to the formulations by Dietze and Fettis given above. A Laplace transform of Possio's integral equation is taken as follows:

$$\alpha(x,s) = -s \int_{-1}^{1} G(M, x-\xi, s) \gamma(\xi, s) \mathrm{d}\xi, \tag{B.42}$$

where s is the non-dimensional Laplace variable and

$$G(M, x-\xi, s) = \frac{1}{2\pi\beta} \left\{ e^{\mu(x-\xi)} \left[K_0^{(2)}(\kappa \mid x-\xi \mid) - iM \frac{(x-\xi)}{\mid x-\xi \mid} K_1^{(2)}(\kappa \mid x-\xi \mid) \right] - s \int_{-\infty}^{(x-\xi)} e^{-s(x-\xi)+s\lambda/\beta^2} K_0^{(2)} \left(\frac{sM \mid \lambda \mid}{\beta^2} \right) \mathrm{d}\lambda \right\}, \tag{B.43}$$

where $\beta = \sqrt{1-M^2}$, $\kappa = sM/\beta^2$, $\mu = M\kappa$ and $K_n^{(2)}(.)$ are modified Bessel functions of the second kind and order n. The generalized kernel, $G(M, x-\xi, s)$, is analytic in the entire s-plane, except along the negative real axis, where the modified Bessel functions have singularities. The special case of oscillatory motion, $s = ik$, results in Possio's integral equation.

A closed-form solution to Eq. (B.42) is expressed (Balakrishnan 1999) as follows:

$$\gamma(x,s) = r(x,s) + \frac{s}{\beta^2}\int_{-1}^{x} \cosh\kappa(x-\sigma) r(\sigma,s)\mathrm{d}\sigma + \kappa\int_{-1}^{x} \sinh\kappa(x-\sigma) r(\sigma,s)\mathrm{d}\sigma, \tag{B.44}$$

where

$$\begin{aligned} r(x,s) &= g(x,s) - \kappa K(\kappa,x)\int_{-1}^{1} \sinh\kappa(1-\sigma) r(\sigma,s)\mathrm{d}\sigma \\ &+ \left[\frac{k\lambda}{\omega}p(s,x) - M\kappa K(\kappa,x)\right]\int_{-1}^{1} \cosh\kappa(1-\sigma) r(\sigma,s)\mathrm{d}\sigma \end{aligned} \tag{B.45}$$

and

$$f(x,s) = e^{-M\kappa x}\gamma(x,s) \tag{B.46}$$

$$g(x,s) = \frac{2}{\pi\beta}\sqrt{\frac{1-x}{1+x}}\int_{-1}^{1}\sqrt{\frac{1-\sigma}{1+\sigma}}e^{-M\kappa\sigma}\frac{\alpha(x,s)}{\sigma-x}\mathrm{d}\sigma \tag{B.47}$$

$$K(\kappa,x) = \frac{1}{\pi^2}\sqrt{\frac{1-x}{1+x}}\int_{-1}^{1}\sqrt{\frac{1-\sigma}{1+\sigma}}\frac{K_0^{(2)}[\kappa(1-\sigma)]}{\sigma-x}\mathrm{d}\sigma \tag{B.48}$$

$$p(s,x) = \frac{1}{\pi}\sqrt{\frac{1-x}{1+x}}\int_{0}^{\infty} e^{-\frac{s}{\beta^2}\sigma}\frac{1}{x-\sigma-1}\sqrt{\frac{\sigma+2}{\sigma}}\mathrm{d}\sigma. \tag{B.49}$$

Lift and pitching moment computed by this method are in agreement (Lin and Illif 2000) with those calculated by the previous methods of (Dietze 1947, Fettis 1952), as well as by the Mathieu function methods (Reissner 1951, Timman *et al.* 1951), which require a complex contour integration.

References

Balakrishnan AV 1999 Unsteady aerodynamics – subsonic compressible inviscid case. *NASA Contractor Report* **CR-1999-206583**.

Dietze F 1947 I. The air forces of the harmonically vibrating wing in a compressible medium at subsonic velocity. II. Numerical tables and curves. *Air Materiel Command, U.S. Air Force* **F-TS-506-RE and F-TS-948-RE**.

Fettis HE 1952 An approximate method for the calculation of nonstationary air forces at subsonic speeds. *Wright Air Develop. Center, U.S. Air Force* **Tech. Rept. 52-56**.

Garrick IE 1957 Nonsteady wing characteristics. In *Section F, Vol. VII of High Speed Aerodynamics and Jet Propulsion*, Princeton University Press, Princeton, NJ.

Küssner HG 1940 Allegemeine tragflächentheorie. *Luftfahrtforschung* **17**, 370–378. (Also Trans. *NACA Tech. Memo.* **979**, 1941.)

Lin J and Iliff KW 2000 Aerodynamic lift and moment calculations using a closed-form solution of the Possio equation. *NASA Technical Memorandum* **TM-2000-209019**.

Possio C 1938 Aerodynamic forces on an oscillating profile in a compressible fluid at subsonic speeds. *Aerotecnica* **18**, 441–458.

Reissner E 1951 On the application of Mathieu functions in the theory of subsonic compressible flow past oscillating airfoils. *NACA* **Tech. Note 2363**.

Tewari A 2015 *Aeroservoelasticity: Modeling and Control*. Birkhäuser, Boston, MA.

Timman R, van de Vooren AI, and Greidanus, JH, 1951 Aerodynamic coefficients of an oscillating airfoil in two-dimensional subsonic flow. *J. Aeronaut. Sci.* **18**, 717–802. (Also 1954 Tables of aerodynamic coefficients for an oscillating wing-flap system in subsonic compressible flow. *Natl. Aeronaut. Research Inst., Amsterdam* **NLL Rept. F151**.)

Appendix C

Flutter Analysis of Modified DAST-ARW1 Wing

NASA-Langley's *Drone for Aeroelastic Testing* (DAST-ARW1) wing is represented here by a trapezoidal plan form which does not include the leading-edge extension of the original wing, and a single trailing-edge control surface (flap), shown in Fig. C.1. The original DAST-ARW1 wing has two trailing-edge control surfaces of approximately 20% chord each, one inboard, and the other outboard. These are replaced by a single control surface with a 40% chord, spanning the outboard 76% and 98% semi-span locations. The moment of inertia of the flap about its hinge is $I_\delta = 0.1$ kg m^2, while its rotational stiffness about the hinge line is $k_\delta = 100$ Nm/rad, which implies a natural in vacuo flap mode of 31.623 rad/s. The in vacuo structural natural frequencies and damping ratios for the first six wing modes reported by Cox and Gilyard (1986) are listed in Table C.1.

The vertical deflection mode shapes of the six structural modes (Table C.1) at six selected spanwise locations are listed in the following matrix:

$$z = \begin{pmatrix} -0.0029 & -0.1437 & -0.0030 & -0.0793 & 0.0691 & -0.0418 \\ 0.0442 & -0.1775 & 0.2304 & 0.1076 & -0.3412 & -0.3048 \\ 0.1595 & -0.1179 & 0.4772 & 0.2880 & -0.5993 & -0.5098 \\ 0.2879 & 0.0763 & 0.6381 & 0.5552 & -0.5854 & -0.6427 \\ 0.4864 & 0.3901 & 0.5500 & 0.6535 & -0.3532 & -0.4567 \\ 0.8081 & 0.8809 & 0.0980 & 0.4048 & 0.2284 & 0.1542 \end{pmatrix}$$

while the corresponding mode shapes for the deflection slopes are the following:

$$dz/dx = \begin{pmatrix} 0.0199 & 0.0064 & -0.0211 & -0.0265 & 0.0004 & -0.0821 \\ 0.1017 & 0.0676 & -0.0162 & -0.0934 & -0.2300 & -0.2363 \\ 0.2032 & 0.1949 & -0.0832 & -0.1066 & -0.3250 & -0.3518 \\ 0.4406 & 0.4023 & -0.2004 & -0.0414 & -0.0047 & -0.3758 \\ 0.7306 & 0.6987 & -0.4231 & -0.0319 & 0.5398 & -0.4374 \\ 0.4692 & 0.5544 & -0.8793 & -0.9882 & -0.7417 & -0.6936 \end{pmatrix}$$

Adaptive Aeroservoelastic Control, First Edition. Ashish Tewari.

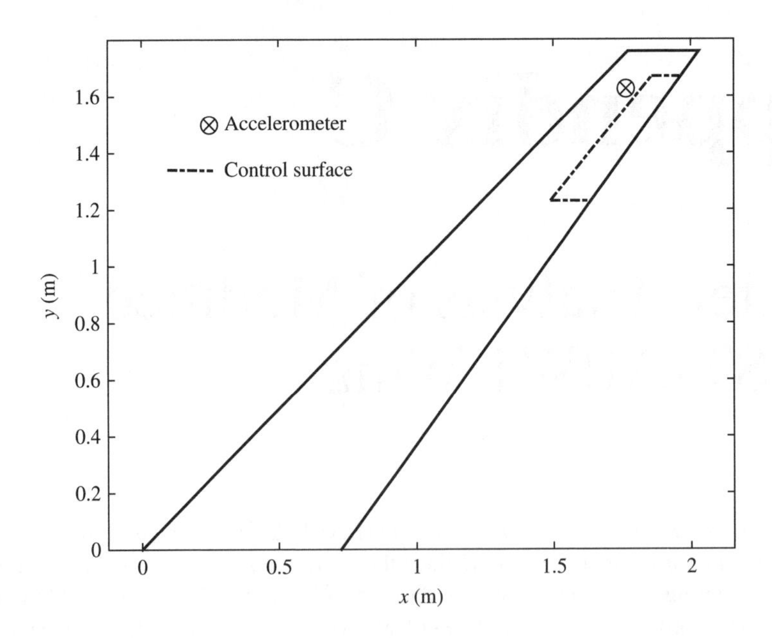

Figure C.1 Modified DAST-ARW1 wing plan form geometry

Table C.1 Structural vibration modes of the DAST-ARW1 wing

Natural frequency (Hz)	Damping ratio	Mode type
9.3	0.00588	Symmetric bending
13.56	0.00882	Anti-symmetric bending
30.30	0.00937	Symmetric bending/torsion
32.72	0.01943	Anti-symmetric bending/torsion
38.96	0.01447	Symmetric torsion
48.91	0.02010	Anti-symmetric torsion

These are combined to yield the following modal matrix for aerodynamic influence coefficients:

$$\Phi = ikz + \mathrm{d}z/\mathrm{d}x,$$

where $k = \omega\bar{c}/(2U)$ is the reduced frequency and the deflections are rendered non-dimensional by dividing by the characteristic length (mean semi-chord), $\bar{c}/(2U)$.

The modal masses, stiffnesses and damping coefficients are derived as follows (Tewari 2015), in order to fit the flight flutter test data for this wing reported by Bennett and

Abel (1982):

$$M = \text{diag.}(2.0967, 1.9164, 1.8036, 1.7360, 1.6007, 1.4429)\ \text{kg}$$

$$K = 10^5 \begin{pmatrix} 1.7861 & 1.9831 & 0.2042 & 1.1739 & 0.5544 & 0.4213 \\ 1.4578 & 2.3976 & -2.9320 & -2.4571 & 3.5898 & 3.5093 \\ -1.3326 & -4.6836 & 5.8085 & 2.6710 & -8.3447 & -6.9302 \\ -1.4480 & -1.4594 & 1.9777 & 2.2929 & -1.8597 & -2.4083 \\ -0.0841 & -0.1959 & 0.4750 & 0.3269 & -0.4011 & -0.5714 \\ -1.6214 & -3.8540 & 4.8391 & 3.1149 & -6.5868 & -5.7113 \end{pmatrix} \text{N/m}$$

$$C = \begin{pmatrix} 3.9877 & 18.0085 & -35.1898 & -19.0859 & 40.9523 & 42.3896 \\ 75.3328 & 50.9966 & 72.4875 & 66.8918 & -54.2124 & -63.2161 \\ -56.3060 & -60.5175 & -13.7980 & -35.6402 & -14.8539 & 2.2176 \\ -60.9295 & -38.6040 & -51.3028 & -42.4412 & 41.6701 & 42.9690 \\ 4.3764 & -0.8387 & 17.5343 & 14.1483 & -15.3260 & -18.6483 \\ -89.2669 & -66.3578 & -78.0079 & -80.2630 & 48.9853 & 66.8707 \end{pmatrix} \text{Ns/m}$$

For doublet-lattice calculations, a MATLAB code developed by Tewari (2015) is employed. Figure C.2 is a sample plot of the unsteady pressure distribution caused by a linear combination of the three symmetric structural modes, computed by using 30 spanwise and 10 chordwise grid divisions. The normalization of the deflections and slopes is carried out by dividing by the

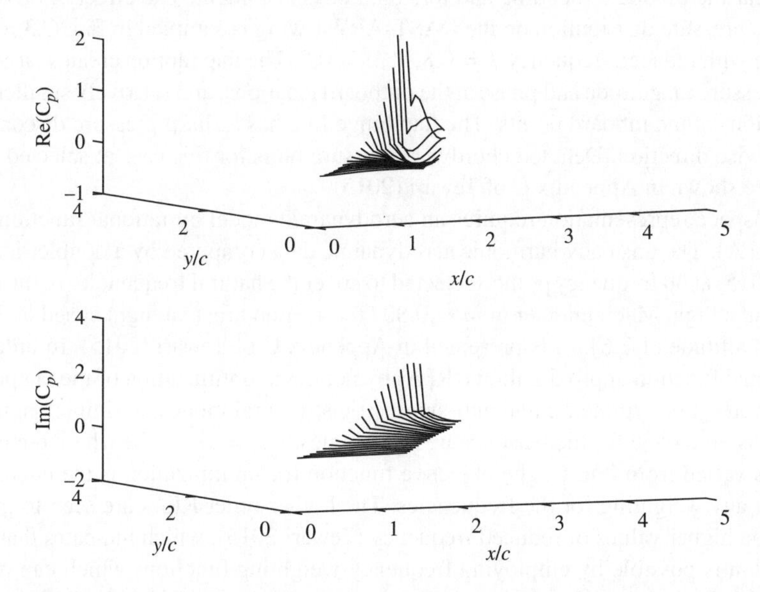

Figure C.2 Unsteady pressure distribution on the DAST-ARW1 wing due to three bending/torsion modes at $M = 0.7$ and $k = 0.6$

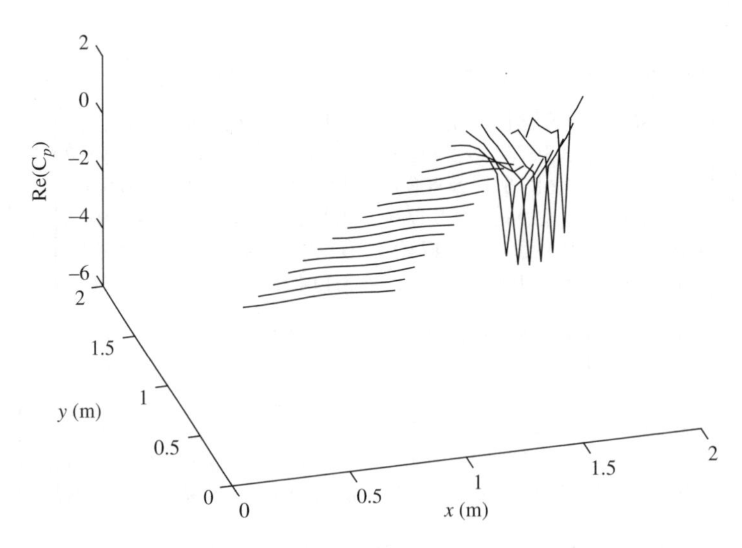

Figure C.3 Unsteady pressure distribution (real part) on the modified DAST-ARW1 wing due to the flap mode at $M = 0.807$ and $k = 0.8$

magnitudes $| z |$ and $| dz/dx |$, respectively. Here, the pressure waviness and increased magnitudes near the tip due to bending and torsion modes is evident. The effect of an oscillating flap on the pressure distribution on the DAST-ARW1 wing is sampled in Fig. C.3 for the flap oscillating with reduced frequency, $k = 0.8$, at $M = 0.8$. The flap motion creates large changes in both pressure magnitude and phase at the outboard locations, and relatively smaller pressure modifications at the inboard points. The flap hinge line has a sharp pressure discontinuity in the chordwise direction. Detailed chordwise pressure plots for this case at selected spanwise stations are shown in Appendix C of Tewari (2015).

A state-space representation requires an aerodynamic model by rational function approximation (RFA). The unsteady harmonic aerodynamic data computed by a doublet-lattice code (Tewari 2015) at 30 frequency points (selected to cover the natural frequencies of the structural modes) and a flight Mach number of $M = 0.807$ (corresponding to a flight speed of 250 m/s at a standard altitude of 7.6 km) is presented in Appendix C of Tewari (2015). In order to generate rational function approximations (RFA) by nonlinear optimization of the lag parameters for the given set of symmetric and anti-symmetric structural modes, a Simplex non-gradient optimizer is employed for the least-squares RFA poles, $b_j, j = 1, \cdots, N$, where the number of poles N is varied from 2 to 6. The objective function for minimization is the curve-fit error, ϵ, without any weighting for the frequencies. The higher-order RFAs are seen to give a better fit at the higher values of reduced frequency (Tewari 2015), which indicates that a further optimization is possible by employing frequency-weighting functions which can reduce the number of poles required for a given accuracy. A sample curve fit for the RFA is shown in Fig. C.4, while the optimized values of the lag parameters are listed in Table C.2.

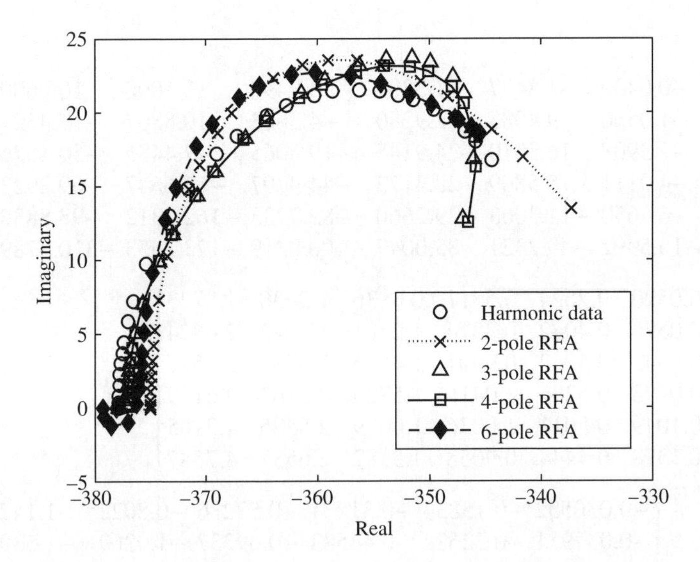

Figure C.4 Curve-fitting for element (6,6) of generalized aerodynamic transfer matrix of the modified DAST-ARW1 wing with six symmetric and anti-symmetric modes at $M = 0.807$

Table C.2 Optimized lag parameters for modified DAST-ARW1 wing at $U = 250$ m/s

N	b_j	ϵ
2	0.020077, 414503.75	0.00159
3	6.9187×10^{-5}, 3403.8425 5202.256	0.00093
4	0.001969, 0.230118 7869.2315, 7782.0475	0.0008935
6	0.238246, 0.05157 2.03284, 813.363 8570.242, 7980.97	0.00074

For brevity, the RFA coefficient matrices are presented here for the reference condition with only two lag parameters, whose values are the following:

$$b_1 = 10^{-6}(b/U), \quad b_2 = 1.85(b/U).$$

Such a choice allows for a scheduling of the lag parameters with the flight speed U – which is necessary for conducting a flutter analysis – without having to perform a nonlinear optimization at every speed. The aerodynamic coefficient matrices for the given flight condition are as follows:

$$A_0 = \begin{pmatrix} -0.0439 & -1.8487 & -1.3809 & 3.9490 & 13.1090 & 10.1600 \\ -1.0560 & -9.4088 & -12.9280 & -4.3663 & 10.8807 & -17.4327 \\ -2.3908 & -16.3010 & -24.9145 & -19.3065 & -7.4486 & -50.4676 \\ -4.0111 & -15.5809 & -28.4172 & -44.9707 & -66.6647 & -69.2923 \\ -6.5650 & -11.3063 & -29.2660 & -82.7733 & -162.4312 & -98.8830 \\ -10.5992 & -49.2321 & -85.0047 & -100.0319 & -123.2653 & -310.7789 \end{pmatrix} \text{N/m}$$

$$A_1 = \begin{pmatrix} 0.0520 & 0.2624 & 0.5311 & 0.8836 & 1.3630 & 1.8723 \\ 0.0576 & 0.4082 & 0.8074 & 1.2481 & 1.8042 & 2.4824 \\ 0.0643 & 0.5062 & 0.9901 & 1.4725 & 2.0781 & 3.0577 \\ 0.0793 & 0.5296 & 1.0416 & 1.5725 & 2.2962 & 3.6159 \\ 0.1049 & 0.5011 & 1.0036 & 1.6079 & 2.5485 & 4.2165 \\ 0.1318 & 0.4444 & 0.9058 & 1.5382 & 2.6537 & 4.7587 \end{pmatrix} \text{Ns/m}$$

$$A_2 = 10^{-3} \begin{pmatrix} -0.030532 & -0.15834 & -0.31831 & -0.52236 & -0.80225 & -1.142 \\ -0.037911 & -0.22624 & -0.44583 & -0.69237 & -1.0269 & -1.5394 \\ -0.046541 & -0.2666 & -0.52024 & -0.78915 & -1.1747 & -1.914 \\ -0.059263 & -0.27506 & -0.53894 & -0.83841 & -1.3151 & -2.2701 \\ -0.077774 & -0.27169 & -0.53823 & -0.88757 & -1.5016 & -2.6505 \\ -0.095264 & -0.28659 & -0.56147 & -0.93688 & -1.6284 & -2.9442 \end{pmatrix} \text{kg}$$

$$A_3 = \begin{pmatrix} -0.93425 & -4.3053 & -8.5849 & -13.636 & -21.377 & -37.71 \\ -1.0965 & -5.0894 & -10.107 & -16.025 & -25.031 & -43.459 \\ -1.2479 & -5.8489 & -11.583 & -18.324 & -28.489 & -48.958 \\ -1.4115 & -6.7693 & -13.283 & -20.635 & -31.657 & -55.203 \\ -1.5679 & -7.6391 & -14.872 & -22.774 & -34.585 & -60.877 \\ -1.6774 & -8.1089 & -15.734 & -24.037 & -36.443 & -63.821 \end{pmatrix}$$

$$A_4 = \begin{pmatrix} -97.801 & -508.18 & -1032.1 & -1717.8 & -2646.2 & -3603.9 \\ -105.13 & -773.73 & -1549.3 & -2437.5 & -3563 & -4743 \\ -115.29 & -949.42 & -1888.5 & -2880.3 & -4142.5 & -5880.7 \\ -145.1 & -980.48 & -1966.5 & -3042.1 & -4539.5 & -7127.2 \\ -198.92 & -905.88 & -1853.1 & -3030.2 & -4923.9 & -8493.1 \\ -245.95 & -854.23 & -1735.1 & -2903.1 & -5038.3 & -9282.9 \end{pmatrix}$$

For the determination of the controls coefficient matrix, B, the generalized aerodynamic forces (GAFs) and the hinge moment due to control surface are necessary. The GAF contribution of the trailing-edge flap is computed by the revised doublet-lattice grid and reported in Appendix C of Tewari (2015). Owing to the pressure discontinuity at the hinge line of the oscillating flap, the curve fit for the control-surface GAF with a given number of lag parameters is degraded (see Fig. C.5) when compared to that of the wing without the control surface. The RFA coefficients for the control surface GAF with two lag parameters are derived from the GAF data to be the following Tewari (2015):

$$B_0^T = (-1.1880,\ -1.5713,\ -2.9118,\ -10.4382,\ -101.6408,\ -120.6015)$$

$$B_1^T = (1.0286,\ 0.8900,\ 0.4865,\ 0.2060,\ 0.1702,\ 0.2863)$$

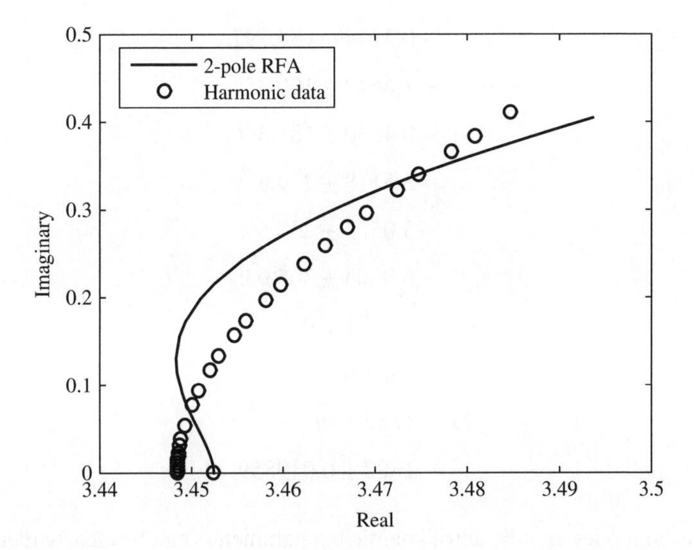

Figure C.5 Least-squares curve fit of unsteady hinge moment for the modified DAST-ARW1 wing at $M = 0.807$ with two lag parameters

$$B_2^T = 10^{-3}(-0.5646,\ -0.5204,\ -0.3229,\ -0.1787,\ -0.1288,\ -0.1616)$$
$$B_3^T = (0.6953,\ 0.4578,\ 0.3557,\ 0.3885,\ 0.4454,\ 0.3734)$$
$$B_4^T = 10^3(-1.9612,\ -1.7064,\ -0.9540,\ -0.4534,\ -0.4078,\ -0.5711).$$

The hinge-moment RFA is computed next, with the curve fit shown in Fig. C.5 using the same two lag parameters as those for the generalized aerodynamic forces. The average fit error per frequency point is only 0.027%. The numerator coefficients of hinge-moment RFA for $U = 250$ m/s are listed as follows:

$$a_0 = 3.4524$$
$$a_1 = -0.0036$$
$$a_2 = 8.6236 \times 10^{-7}$$
$$a_3 = -0.0040$$
$$a_4 = 4.4747$$

When the aeroelastic data given above is substituted into the characteristic equation,

$$|\ sI - A\ | = 0$$

the open-loop poles of the aeroelastic system at $M = 0.807$ and altitude 7.6 km are the following:

$$-0.00097699 + 0i$$
$$5\times -0.0010173 + 0i$$

$$
\begin{aligned}
&-0.31449 \pm 57.797i\\
&-1.6542 \pm 100.35i\\
&-0.4830 \pm 181.35i\\
&-5.5118 \pm 219.83i\\
&-3.0523 \pm 237.59i\\
&-6.0303 \pm 305.03i\\
&-1875.1 + 0i\\
&-1881.6 + 0i\\
2\times &-1882 + 0i\\
&-1882 \pm 0.019859i
\end{aligned}
$$

The aeroelastic modes and the aerodynamic lag parameters can be clearly identified in this list. All the stable, real poles result from the aerodynamic lag states, while the complex conjugate pairs (except the last one) are the aeroelastic modes. When the flight speed is increased, flutter is experienced by one of the aeroelastic modes becoming unstable. This is shown in Fig. C.6, which shows the variation of the natural frequency, ω_n, and damping ratio, ζ, of the second symmetric bending/torsion mode (of in vacuo natural frequency 30.3 Hz) in the speed range 250–295 m/s at 7.6 km standard altitude. This mode is seen to become unstable at a flutter speed of 284.7 m/s, which corresponds to a Mach number of 0.9192 at the given altitude. The flutter frequency for this mode corresponds to 28.691 Hz (181.272 rad/s) and the flutter Mach number is 0.92, both of which are matched with the flight flutter test (Bennett and Abel 1982).

When the flight speed is increased beyond the flutter velocity to $U = 295$ m/s, the nature of the RFA undergoes a transformation to a spiralling shape, as shown in Fig. C.7. This indicates the need for including more intermediate frequency points for a better curve fit. The hinge-moment RFA coefficients at $U = 295$ m/s (listed below) for aerodynamic damping and aerodynamic inertia change in sign, while the lag numerator coefficients are seen to change in sign and increase in magnitude, all of which indicate a stabilizing aerodynamic influence on the flap rotation at supercritical speed (above the open-loop flutter speed):

$$
\begin{aligned}
a_0 &= 4.4327\\
a_1 &= 0.0047\\
a_2 &= -4.3595 \times 10^{-6}\\
a_3 &= 0.0048\\
a_4 &= -11.1086
\end{aligned}
$$

Finally, we derive the coefficient matrices for the output equation. The sensor location for the modified DAST-ARW1 wing is shown in Fig. C.1. The outboard selection of accelerometer gives the best resolution of individual contributions from all the relevant modes to the acceleration output. The output coefficients for the given flight condition ($U = 250$ m/s, $M = 0.807$,

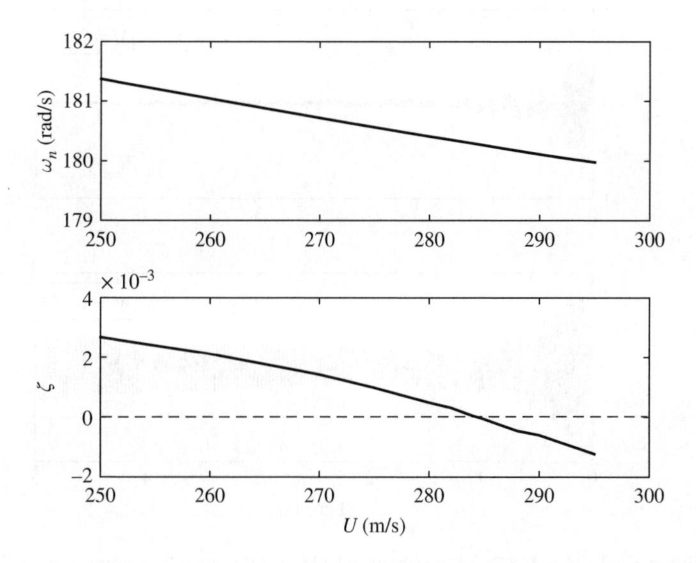

Figure C.6 Open-loop flutter analysis of the modified DAST-ARW1 wing at 7.6 km standard altitude (second symmetric bending/torsion mode of in vacuo natural frequency 30.3 Hz)

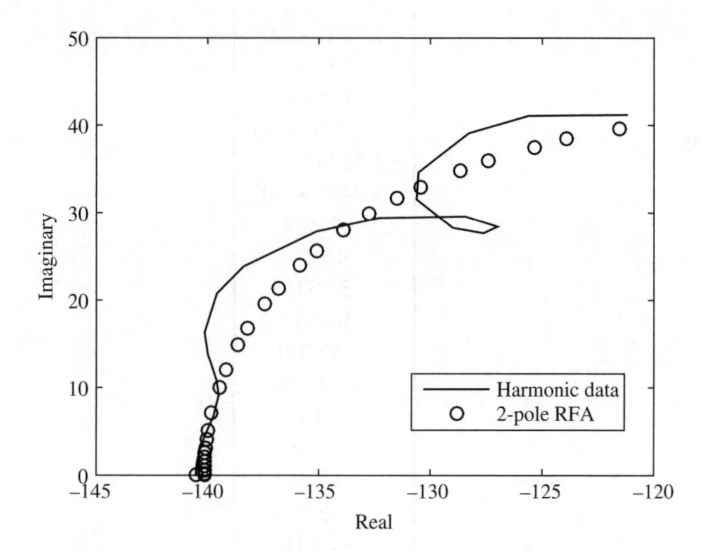

Figure C.7 Least-squares curve fit for element (1 × 5) of generalized aerodynamics forces due to flap mode for the modified DAST-ARW1 wing at $M = 0.95$ with two lag parameters

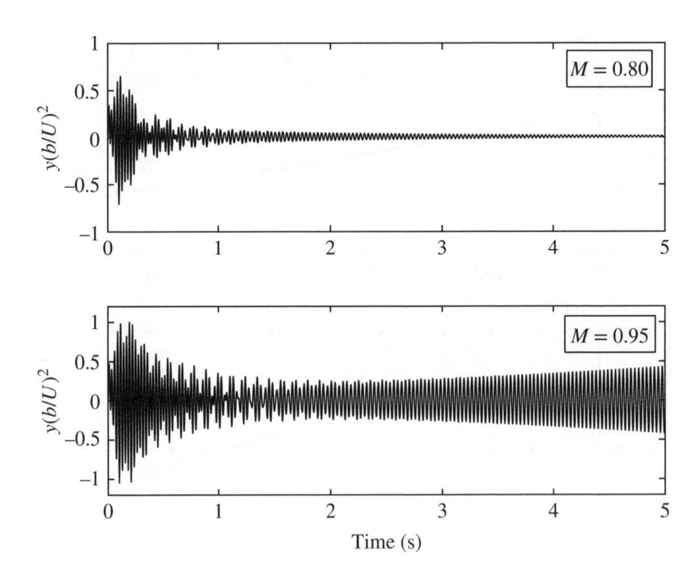

Figure C.8 Subcritical ($M = 0.8$) and supercritical ($M = 0.95$) accelerometer outputs for initial unit tip deflection of the modified DAST-ARW1 wing

altitude 7.6 km) are the following:

$$C^T = \left\{ \begin{array}{c} 1.1207 \times 10^5 \\ 2.6638 \times 10^5 \\ -3.3454 \times 10^5 \\ -2.1536 \times 10^5 \\ 4.5524 \times 10^5 \\ 3.9455 \times 10^5 \\ 61.807 \\ 46.18 \\ 54.575 \\ 56.567 \\ -32.051 \\ -42.966 \\ -1.1581 \\ -5.5986 \\ -10.863 \\ -16.595 \\ -25.16 \\ -44.061 \\ -169.87 \\ -589.51 \\ -1197.4 \\ -2003.7 \\ -3478.1 \\ -6409.9 \end{array} \right\}$$

$$D = (-3.138 \times 10^{-5},\ -0.00010323,\ -0.00021489,\ -0.00037257,\ -0.00070238,\ 0.69163)$$

The simulated normal acceleration response to an initial unit tip deflection at a subcritical (below flutter speed) Mach number of 0.8 and a supercritical Mach number 0.95 at 7.6 km altitude are compared in Fig. C.8 for the first 5 s. The stable (or decaying) subcritical response and an exponentially growing (unstable) supercritical response are evident.

References

Bennett RM and Abel I 1982 Flight flutter test and data analysis techniques applied to a drone aircraft. *J. Aircr.* **19**, 589–595.

Cox TH and Gilyard GB 1986 Ground vibration test results for drones for aerodynamic and structural testing (DAST)/aeroelastic research wing (ARW-1R) aircraft. *NASA Tech. Memo.*, **85906**.

Tewari A 2015 *Aeroservoelasticity - Modeling and Control*. Birkhäuser, Boston, MA.

Index

Adaptive Aeroservoelastic Control, First Edition. Ashish Tewari.